Functional Genomics and Proteomics

Functional Genomics and Proteomics

Edited by **Charles Malkoff**

SYRAWOOD
PUBLISHING HOUSE

New York

Published by Syrawood Publishing House,
750 Third Avenue, 9th Floor,
New York, NY 10017, USA
www.syrawoodpublishinghouse.com

Functional Genomics and Proteomics
Edited by Charles Malkoff

International Standard Book Number: 978-1-68286-122-6 (Hardback)

Printed in the United States of America.

Contents

Preface

Functional genomics and proteomics play a crucial role in analysing available genetic data and gathering key information for further use. The book emphasizes on the dynamic aspects of genomics and proteomics such as regulation of genes, transcription, translation and protein-protein interactions, large scale protein structures, etc. Researches and case-studies included in this book attempt to provide methods, models and techniques to analyze and gather information from large pool of available genomic data of various organisms. This book provides a detailed explanation on structure determination and structural genomics. Students and researchers will find this book beneficial.

The information shared in this book is based on empirical researches made by veterans in this field of study. The elaborative information provided in this book will help the readers further their scope of knowledge leading to advancements in this field.

Finally, I would like to thank my fellow researchers who gave constructive feedback and my family members who supported me at every step of my research.

Editor

Cellulose-hemicellulose networks as target for *in planta* modification of the properties of natural fibres

Olawole O. Obembe*, Evert Jacobsen, Richard G.F. Visser, and Jean-Paul Vincken

Graduate School Experimental Plant Sciences, Laboratory of Plant Breeding, Wageningen University P O Box 386, 6700 AJ Wageningen, The Netherlands

Plant cell wall polysaccharides are predominant components of fibres. Natural fibres have a wide range of industrial applications, such as in paper and textile industries. Furthermore, their demand for use as bio-composites in building and automotive applications is also increasing. For the various applications, a gain of control over fibre characteristics is important. Inherent fibre characteristics are largely determined by the ratio and interactions of cellulose and hemicelluloses. Two main strategies for bioengineering fibre properties are reviewed: (i) modifying the cellulose/hemicellulose ratio (by biosynthesis or biodegradation of specific polysaccharides), and (ii) interference with cellulose-hemicellulose interactions using carbohydrate-binding modules. These *in planta* approaches may have the potential of complementing the currently used surface modification approaches for modifying fibre characteristics.

Key words: natural fibre, cellulose, hemicellulose, interactions, cell wall modification, carbohydrate binding module, cellulose synthase

Table of contents

1. Introduction

The plant cell wall is a dynamic, highly organised extracytoplasmic matrix consisting of various polysaccharides, structural proteins, and aromatic substances, which are constantly remodelled and restructured during growth and development. These complex matrices define the shape and size of individual cells within the plant body and ultimately plant

---

*Correspondence author. E-mail: odun_wole@yahoo.co.uk;

morphology (O'Neill and York, 2003). Other divergent and vital functions provided by the cell wall include tensile strength, defending the plant against pathogens and mediating the communication with symbionts and neighbouring cells (Carpita and Gibeaut, 1993; John et al., 1997; Dhugga, 2001).

Plant cell wall polysaccharides have been receiving a lot of attention in recent years, because they constitute an important part of food and feed, and consequently they can be important in food processing. Cell wall polysaccharides are used commercially as gums, gelling agents and stabilizers (Morris and Wilde, 1997). Digestibility of feed products depends to a large extent on cell wall composition. Forage maize (*Zea mays*) is excellent roughage for ruminants because of its high energy content. The maize forage digestibility is, however, lowered by the presence of a considerable level of lignin and hemicelluloses, which create a physical and chemical barrier to digestion (Fontaine et al., 2003). Moreover, plant cell wall polysaccharides are the predominant components of fibres. Natural fibre is a thread-like material from plants, which can be used for making products such as cloth, paper, and rope. Fibre crops include cotton, hemp, flax, agave and tree species. Natural fibres have a wide range of industrial applications. Wood fibres have found application in pulp and paper industries whereas cotton, flax, hemp and agave fibres are used in the textile industries. Environmental considerations are giving room for the use of natural fibres as an alternative for synthetic polymers in industrial composites (Gustavsson et al., 2005). For the various applications, it is important to gain control over fibre characteristics, which in turn are determined by cell wall composition and interactions of wall components.

2. Polymer networks in plant cell walls

Plant cell walls can be divided into three distinct zones, the middle lamella, primary wall, and secondary wall. The primary cell wall is deposited between the plasma membrane and the middle lamella during cell growth and elongation. The secondary cell wall is added at the inner face of the primary wall after cessation of growth (Raven et al., 1992). The secondary cell wall is more important than primary wall with respect to fibre properties in that natural fibres from most fibre crops such as cotton (textile), flax (textile), and forest trees (paper) are derived from the secondary cell walls of specialized cells and tissues (Aspeborg et al. 2005; Gorshkova and Morvan, 2006, Kim and Triplett; 2001).

A number of models have been proposed for the structure and architecture of the primary cell wall (Carpita and Gibeaut, 1993; McCann and Roberts, 1994; Ha et al., 1997). These models emphasise three independent but interacting networks based on cellulose-hemicellulose

(>50% dry weight), pectin (25-40% dry weight) and structural glycoproteins (1-10% dry weight). Cellulose is a linear polymer of □-1,4-linked glucose (Figure 1), with each glucose residue oriented 180° to its neighbour such that the polymeric repeating unit is cellobiose (Brown et al., 1996). This allows the glucan chain to adopt a flat, ribbon-like structure. Hemicelluloses are wall polysaccharides that are not solubilized by hot water, but are solubilized by aqueous alkali (O'Neill and York, 2003). They are usually branched polysaccharides, which are structurally homologous to cellulose, in that they have a backbone composed of □-1,4-linked pyranosyl residues, such as glucose, mannose, and xylose (Figure. 1). This structural similarity facilitates a strong, non-covalent association of the hemicellulose with cellulose microfibrils. Xyloglucan is the most abundant hemicellulosic polysaccharide in the primary cell walls of non-graminaceous plants (about 20% dry weight). The cellulose-xyloglucan network is thought to be the principal load-bearing element in the primary cell wall.

Xyloglucan has a 'cellulosic' backbone of □-1,4-linked glucosyl (Glc) residues. Unlike the linear cellulose, however, the backbone residues bear α-linked xylopyranose branches attached to the O6 position of glucose, which may be further substituted by galactopyranosyl residues and other monosaccharides (Figure 1). Cross-linking pectic polysaccharides predominate the middle lamella, which is located between two contiguous cells (Raven et al., 1992) and constitute the embedding matrix for the cellulose-hemicellulose network of the primary cell wall (Carpita and Gibeaut, 1993). Extensin, the main structural glycoprotein in the primary cell wall, adds rigidity and strength to the wall by cross-linking with one another (Brady et al., 1996) or with pectins (Brady et al., 1996; MacDougall et al., 2001). The pectic polysaccharides and the extensin will not be discussed further, as they are not so important with respect to fibre characteristics. Xylans are also present in some primary walls. They are the cross-linking polysaccharides in monocots (Carpita and Gibeaut, 1993); their structure will be discussed in the next section.

The secondary wall synthesis is characterised by a marked increase in the synthesis of cellulose, non-cellulosic cross-linking glycans, and lignin (Awano et al., 2002; Raven et al., 1992). It should be noted, however, that unlike the primary walls, there are no good models of the secondary walls as yet. Xylans including arabinoxylans, glucuronoxylans, and glucuronoarabinoxylans are the major hemicellulosic polysaccharide in the secondary cell wall (Ebringerova and Heinze, 2000). They are composed of a β-1,4-D-xylosyl backbone (Figure 1), which is substituted to varying extent at the O2 or O3 position of the xylosyl residues with glucuronyl (GlcA), acetyl and arabinosyl groups (O'Neill and York, 2003). Mannose-containing hemicelluloses, including (galacto)mannans and

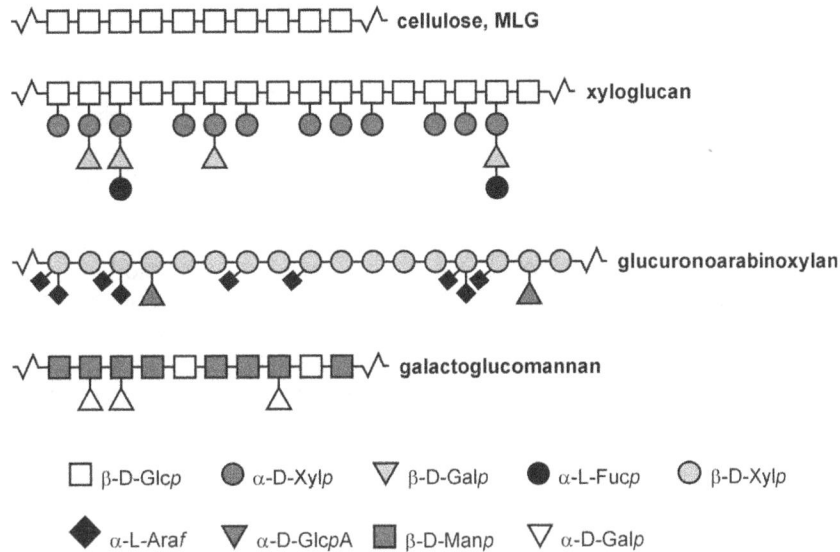

Figure 1. Schematic structures of cellulose and hemicelluloses. Like cellulose, most hemicelluloses have backbones, which are composed of β-1,4-linked-D-pyranosyl residues, such as glucose, mannose and xylose. The mixed-linkage glucan backbone is however, an exception, in that it is composed of β-D-Glc residues of which 30% are 1,3-linked and the remaining 70% are 1,4-linked.

(galacto)glucomannans, are found in considerable amounts in a variety of plant species as carbohydrate reserves. (Galacto)mannans have a β-1,4-linked D-mannosyl (Man) backbone that is substituted at the O6 position of certain Man residues (Stephen, 1982). Glucomannans are abundant in secondary cell walls of woody species; they have a backbone that contains both 1,4-linked Man and 1,4-linked Glc residues. Galactoglucomannans are particularly abundant in the primary cell walls of Solanaceous species (O'Neill and York, 2003). They have a similar backbone as the glucomannans, but some of the backbone Man residues bear single-unit galactosyl side chains at the O6 position (Stephen, 1982) (Figure 1). It should be noted that xyloglucans were also found in the thickened secondary walls of tamarind, where they occur as a reserve polysaccharides (Gidley et al., 1991).

3. Factors that can influence cellulose-hemicellulose networks

Two main approaches have been identified for the modification of the cellulose-hemicellulose networks in the plant: (i) by interfering with the biosynthesis or degradation of polysaccharide, and (ii) by directly interfering with the cellulose-hemicellulose interactions. In order to use the first approach, an understanding of the (hemi)cellulose biosynthesis is imperative. An overview of literature on (hemi)cellulose biosynthesis and its potential for fibre modification is given below, whereas the enzymic in planta degradation of cellulose-hemicellulose networks is only briefly discussed. Similarly, an overview of the potential of carbohydrate-binding modules (CBMs) with

respect to influencing the interactions between cellulose and hemicellulose is given.

3.1. Modification of the cellulose/hemicellulose ratio

3.1.1. Biosynthesis of wall polysaccharides

3.1.1.1. Cellulose biosynthesis.

The crystalline cellulose microfibril is formed by the spontaneous association of about 36 □-D-glucan chains, which are simultaneously synthesised by a large membrane-localised complex that has been visualised by microscopy (Tsekos and Reiss, 1992). The association of the membrane complex with cellulose microfibrils as revealed by freeze-fracture electron microscopy suggested that the complexes are the sites of cellulose synthesis (Kimura et al., 1999). In vascular plants, these complexes are rosette structures with six-fold symmetry and a diameter of 24-30 nm (Mueller and Brown, 1980). The rosettes have been proposed to consist of six subunits, each of which has six catalytic subunits molded in the rosette structure (Delmer and Amor, 1995; Brown and Saxena, 2000). In addition to freeze fracture evidence, mutation studies also showed positive correlation between cellulose content and numbers of rosettes in the mutants (Kokubo et al., 1991; Arioli et al., 1998), confirming that they are the biosynthetic machinery of cellulose. However, the underlying mechanisms of rosette assembly, the precise nature of cellulose biosynthesis, as well as the full identity of the components of the cellulose synthase complex are still not well understood.

Cellulose synthase (CesA). CesA proteins are thought to catalyse the polymerisation of glucose into glucan chains, using UDP-glucose as the donor substrate. The first two plant genes for cellulose biosynthesis, GhCesA-1 and GhCesA-2, were identified by random sequencing of cDNA libraries from developing cotton fibres (Pear et al., 1996). Antibody labelling of the membrane-bound rosette with an antibody raised against the cotton CesAs demonstrated that CesA proteins are indeed members of the multiple-enzyme cellulose synthesising complex (Kimura et al., 1999; Itoh and Kimura, 2001).

The GhCesA proteins had conserved regions surrounding the D, DxD, D and QxxRW motifs, (x stands for any amino acid), previously identified in *Acetobacter xylinum* CesA (Saxena et al., 1995). This motif is presumably involved in substrate binding and catalysis, and is characteristic of processive glycosyltransferases (GTs).

From hydropathy analysis plots, it was predicted that the deduced GhCesA proteins have two trans-membrane helices in the N-terminal region and six in the C-terminal region (Delmer, 1999). The central region of the protein, comprising the D, DxD, D and QxxRW motifs plus two plant-specific insertions, is predicted to be within the cytoplasm. This is consistent with the notion that these motifs are involved in binding the substrate UDP-glc and carry out catalysis on the cytoplasmic face of the enzyme. A zinc-binding domain at the N-terminal of GhCesA was found as another plant CesA-specific motif. Mutant studies have since demonstrated the involvement of CesA proteins in cellulose biosynthesis. Earliest evidence came from the studies on two cellulose-deficient *Arabidopsis* mutants, *rsw1* (Arioli et al., 1998), and *irx3* (Turner and Somerville, 1997; Taylor et al., 1999). There is a body of evidence that indicates that at least three CesA proteins are involved in cellulose synthesis within the same cell at the same developmental stage. Three *Arabidopsis* CesA isoforms (AtCesA4, 7 and 8) have been shown to interact and constitute the subunits of the same complex that synthesises cellulose during secondary wall deposition in xylem cells (Gardiner et al., 2003; Taylor et al., 2003). Similarly, three CesA isoforms (AtCesA1, 3 and 6) have been reported to be required for cellulose synthesis in primary-walled cells of *Arabidopsis* (Fagard et al., 2000; Scheible et al., 2001; Desprez et al., 2002; Robert et al., 2004).

CesA genes are members of a large multi-gene family. In *Arabidopsis* there are up to 40 genes that bear similarity to the original CesA genes. This super-family has been divided into one family of ten 'true' CesA genes (CesA) and six families of about 30 cellulose synthase-like genes (Csl) that are less closely related (Richmond and Somerville, 2001). A web site maintained by Richmond and Somerville (http://cellwall.stanford.edu) documents sequence data for cellulose synthase and cellulose synthase-like genes in several different plant species.

Other proteins involved in cellulose biosynthesis

Apart from the CesA proteins, other proteins whose functions have been linked to cellulose biosynthesis have been identified. For instance, the role of a membrane-anchored endo-1,4-□-glucanase (KORRIGAN) in deposition of cellulose in the cell walls has been reported (Nicol et al., 1998; Lane et al., 2001). The gene of this glucanase was isolated from the *kor* mutant that causes a decrease in the content of crystalline cellulose and abnormal cell wall formation. It was demonstrated that the catalytic domain of the enzyme is located on the outside of the plasma membrane (cell wall side) (Molhoj et al., 2001).

In addition to *korrigan*, a novel membrane-associated form of sucrose synthase was proposed to be involved in cellulose synthesis, by channelling UDP-glucose to cellulose synthase in a closely coupled reaction (Amor et al., 1995). Immunolocalisation of sucrose synthase at the site of cellulose deposition in tracheary elements was demonstrated by electron microscopy (Salnikov et al., 2001). Furthermore, the involvement of an enzyme, UDP-Glc:sterol glucosyltransferase (SGT), in cellulose biosynthesis was suggested recently (Peng et al., 2002). The authors observed that digestion of noncrystalline cellulose with cellulase released not only CesA proteins, but also small amounts of a sitosterol linked to glucose (Peng et al., 2001; Peng et al., 2002). Further metabolic studies led the authors to propose a biosynthetic pathway for cellulose that starts with a SGT-mediated transfer of a glucosyl residue from the soluble cytoplasmic substrate UDP-glucose onto sitosterol to form sitosterol-□-glucoside (SG) on the inner surface of the plasma membrane. The idea is that SG, being a hydrophobic glucoside, may serve as a primer for glucan chain elongation. A link between sterol biosynthesis and cellulose synthesis has indeed been established recently, through the analysis of sterol biosynthesis mutants (Schrick, 2004).

There were reports, which indicated that modulation of cellulose content, through altered expression of the CesA proteins, could alter the cellulose-hemicellulose networks. Analyses of the various primary and secondary cell wall mutants of *Arabidopsis* revealed that some of them had severe reduction in cellulose synthesis, with consequential abnormal cell wall assembly (Turner and Somerville, 1997; Arioli et al., 1998; Fagard et al., 2000; Taylor et al., 2004). Probably a lower cellulose/hemicellulose ratio yields more amorphous cellulose fibres, whereas a higher cellulose/hemicellulose ratio might favour cellulose crystallinity, and consequently fibre properties might be altered. The impact of modifications should be further established with biophysical, biochemical and X-ray crystallographic analyses.

3.1.1.2. Hemicellulose biosynthesis.

Unlike cellulose, which is synthesized at the plasma membrane by CesA proteins, the non-cellulosic matrix polysaccharides (hemicelluloses) are produced within the Golgi by glycan synthases and glycosyltransferase (Keegstra and Raikhel, 2001). Research on hemicellulose biosynthesis is trailing behind that on cellulose. Efforts to identify enzymes mediating the biosynthesis by using biochemical purification strategies were successful for Arabidopsis xyloglucan fucosyltranferase (Perrin et al., 1999), galactomannan galactosyltransferase from fenugreek (Edwards et al., 1999) and xyloglucan xylosyl transferases from Arabidopsis (Faik et al., 2002). Mutant screens and reverse genetics strategies have also led to the identification of Arabidopsis xyloglucan galactosyltransferase (Madson et al., 2003). The sequence similarity between the CesA genes and the Csl genes, especially with respect to the conserved D, DxD, D and QXXRW motifs, originally suggested that they also encode processive glycosyltransferases (Cutler and Somerville, 1997). Based on this similarity, it was suggested that the backbone of hemicelluloses might be biosynthesized by Golgi-resident Csl proteins (Richmond and Somerville, 2001; Hazen et al., 2002). This hypothesis was later supported by a biochemical evidence that a CslA gene from guar encodes β-mannan synthase, which is involved in the formation of the β-1,4-mannan backbone of galactomannan (Dhugga et al., 2004). This biological function of the CslA gene was confirmed by heterologous expression of the Arabidopsis CslA in Drosophila (Liepman et al., 2005). Burton et al. (2006) recently provided additional evidence for the involvement of Csl genes in hemicellulose biosynthesis. They used comparative genomics to link a major quantitative trait locus for (1,3;1,4)-β-D-glucan content in barley grain to a cluster of cellulose synthase-like CslF genes in rice. They then expressed the rice CslF genes in Arabidopsis and detected β-glucan in the transgenic Arabidopsis, thus confirming that the rice CslF proteins are involved in (1,3;1,4)-β-D-glucan biosynthesis.

Interference with hemicellulose biosynthesis might have implications on cellulose-hemicellulose interactions. Evidence for this was provided by the following observations, made with different cellulose-hemicellulose composites produced in the Acetobacter model system (Whitney et al., 1995; Whitney et al., 1998; Whitney et al., 2000). It was shown with the Acetobacter model system that the interactions of hemicelluloses with cellulose microfibrils during cellulose biosynthesis cause cellulose to lose its crystallinity (Whitney et al., 1995; Whitney et al., 1999). Varying levels of reduction in cellulose crystallinity of the native bacterial cellulose were observed in the composites. For instance in the cellulose-xyloglucan composite, a 29% reduction was observed, whereas in cellulose-glucomannan composite, a 57% reduction was observed (Whitney et al., 1998; Whitney et al., 1999). It was also demonstrated that the composites have lower stiffness, leading to dramatic reduction in mechanical strength, for example 80% reduction in composites with xyloglucan (Whitney et al., 1999; Whitney et al., 2000). From the foregoing, a gain of control over the hemicellulose deposition has great potential to alter cellulose-hemicellulose networks. With the recent breakthroughs of identifying the genes responsible for the biosyntheses of the backbones of two hemicelluloses (mannan and mixed-linkage glucan) and the unflinching efforts to identify the genes encoding other hemicellulose synthases, research activities, which would aim at modulating hemicellulose content, should commence, in earnest.

In addition to modulating hemicellulose content, there were reports, which indicated that fibre properties could be modified by interfering with the side chain substitution of the cross-linking polysaccharide. Studies on Arabidopsis mutants with mutations of the MUR2 and MUR3 genes, which encode xyloglucan-specific fucosyl and galactosyl transferases, respectively, revealed that tensile strength of the fibre was enhanced by increased galactosylation of the xyloglucan (Ryden et al., 2003; Pena et al., 2004).

3.1.2. Degradation of wall polysaccharides

Another approach that can be employed to modify the cellulose-hemicellulose network is in planta polysaccharide degradation. Cell wall disassembly is a common feature of many developmental processes such as, fruit softening, organ abscission and dehiscence, and seed germination. These processes are characterised by marked irreversible changes in wall structure and wall strength (Rose et al., 2003). Owing to the heterogeneous nature of the plant cell wall, many endogenous wall-degrading proteins act in synergy, where one class of protein enhances the activity of the other. For instance, exo-acting glycoside hydrolases removing polysaccharide side chains might expose the polymer backbone and enhance its rapid depolymerization by endo-acting glycoside hydrolases (Rose et al., 2003). Similarly, the non-hydrolytic disruption of non-covalent polysaccharide interactions by proteins such as expansins, may also facilitate easy accessibility of a range of substrates to their enzymes (Rose et al., 2003). With a repertoire of characterized bacterial glycoside hydrolases (see web site for carbohydrate active enzymes (CAZY), http://afmb.cnrsmrs.fr/~cazy/CAZY/index.html), it might be possible to modify cellulose-hemicellulose ratio by targeting the expression of the genes encoding these enzyme activities to the cell wall. Oomen et al. (2002) reported fragmentation of the backbone of a branched pectic polysaccharide, rhamnogalacturonan (RG) I by expressing rhamnogalacturonan lyase from Aspergillus aculeatus in potato cell wall. Moreover, Sørensen et al.

Table 1. Summary of carbohydrate-binding modules with relevance for binding cellulosic- and hemicellulosic polysaccharides. Data on specificity are approximate, and were taken from http://afmb.cnrsmrs.fr/~cazy/CAZY/index.html, and Najmudin et al. (2006).

Module	Approx. size (aa)	Occurrence	Specificity					
			CC	AC	X	XG	M	MLG
CBM1	~40	fungi	X	X				
CBM2a	~100	bacteria	X					
CBM2b	~100	bacteria		X	X			
CBM3	~150	bacteria	X					
CBM4	~150	bacteria		X	X	X		
CBM5	~60	bacteria	X	X				
CBM6	~120	bacteria		X	X			
CBM9	~170	bacteria	X	X	X	X		
CBM10	~50	bacteria	X	X				
CBM11	180-200	bacteria		X		X		
CBM13	~150	plants			X			
CBM15	~150	bacteria			X			
CBM17	~200	bacteria		X		X		
CBM22	~160	plants			X			X
CBM27	~122	bacteria					X	
CBM28	~178	bacteria		X		X		X
CBM29	~124	fungi	X	X	X	X	X[a]	
CBM30	~174	bacteria	X					X
CBM31	~124	bacteria			X			
CBM35	~130	bacteria			X		X	
CBM36	120-130	bacteria			X			
CBM37	~100	bacteria	X	X	X			
CBM43	90-100	plants			X			X
CBM44	~150	bacteria	X	X		X	X[b]	X

Legend: CC = crystalline cellulose; AC = amorphous cellulose; X = xylan; M = mannan; XG = xyloglucan; MLG = mixed linkage □-glucan,
[a]Binds both glucomannan and galactomannan. [b]Binds only glucomannan, not galactomannan.

(2000) had earlier reported the removal of part of galactan side chains that were attached to the RG I, by expressing an endo-galactanase from *Aspergillus aculeatus* in potato.

3.2. Interference with cellulose-hemicellulose interactions

Polysaccharide-binding proteins might also modify cellulose-hemicellulose networks. Of particular interest are proteins or parts thereof, which specifically bind polysaccharides without exerting an activity towards them. In nature, numerous organisms express a range of glycoside hydrolases, esterases, and polysaccharide lyases. Cell wall polysaccharide hydrolases from aerobic micro-organisms are generally modular in structure comprising a catalytic module appended to one or more non-catalytic carbohydrate-binding modules (CBMs) (Boraston et al., 2004). However, in anaerobic bacteria the plant cell wall degradative enzymes assemble into large multi-protein complexes that bind tightly to cellulose (Bayer et al., 1998). The main function of CBMs is to attach the enzyme to the polymeric surface and thereby increase the local concentration of the enzyme, leading to more effective degradation (Bolam et al., 1998; Gill et al., 1999).

In addition to binding, some CBMs may also display functions such as substrate disruption or the sequestering and feeding of single glycan chains into the active site of the adjacent catalytic module (Din et al., 1994; Southall et al., 1999).

3.2.1. Family classification and ligand specificity of CBMs.

CBMs are divided into 45 families based on amino acid sequence similarities, details of which can be found in the regularly updated web site for carbohydrate active enzymes (CAZY), http://afmb.cnrsmrs.fr/~cazy/CAZY/index.html. The classification has predictive value for binding specificity and structure. The ligand specificity CBMs with relevance to cell wall polysaccharide binding is presented in Table 1. The CBMs exist in different sizes, ranging from 40-60 to 200 amino acids. Families 1, 5 and 10 are examples of the small CBMs while families 11 and 17 represent the large CBMs. CBMs can accommodate the heterogeneity of the plant cell wall polysaccharides (Boraston et al., 2004). For example, most CBMs that recognise cellulose, bind to both crystalline and amorphous cellulose but with differing binding affinities. It has been shown that CBMs that bind single cellulose chains can also accommodate xyloglucan side chains, such as, for instance CBM44, which bind with equal affinity to cellulose and xyloglucan (Najmudin et al. 2006). In addition to that, some of them can accommodate backbone heterogeneity through selective flexibility, as exhibited by those of the family 29, which recognises the β-1,4-linked backbone of mannose and glucose and to a lesser extent, those of xylan and xyloglucan (Table 1) (Freelove et al., 2001; Charnock et al., 2002).

3.2.2. Relationship between structure and function of the CBMs.

NMR and X-ray crystal structures have revealed that the CBMs that bind soluble polysaccharides are grooved and that the depth of the clefts varies from very shallow to being able to accommodate the entire width of a pyranose ring (Boraston et al., 2004). Examples of these CBMs include family 29 (Freelove et al., 2001), family 2b (Bolam et al., 2001) and family 22 (Charnock et al., 2000). The 3-D structures of many of these CBMs show a characteristic groove (Figure. 2A-D). Aromatic residues (Trp, Tyr) play a pivotal role in ligand binding and the orientations of these amino acids are key determinants of specificity of these CBMs (Simpson et al., 2000). Alternatively, CBMs binding insoluble crystalline cellulose have a flat surface, which enables them to attach to cellulose. CBM3 is a typical flat-surface-binding CBM (Figure. 2E). Its 3-D structure shows that the residues, which are involved in binding, are oriented in a geometry that is complementary to the flat surface of cellulose. Generally, at least two aromats are required for interaction with the target ligand, the binding of which is often reinforced by hydrogen bonding interactions between the CBM and the carbohydrate. The number of glycosyl residues bound by the CBM can be different, i.e.

six for CBM2b (Simpson et al., 2000) and CBM29 (Charnock et al., 2002), and four for CBM22 (Charnock et al., 2000). It has been suggested that short and shallow grooves might better accommodate polysaccharides with side chains (Boraston et al., 2004). This is important with respect to hemicelluloses because they are often heavily branched.

3.2.3. Binding affinity of CBMs.

Some polysaccharide-degrading enzymes may possess more than one CBM, in order to facilitate increased affinity of the enzymes for the polysaccharide. This idea is supported by many binding affinity studies involving one versus two CBMs. For example, an artificial protein construct, consisting of two covalently linked family 1 CBMs, had 6-to-10-fold higher affinity for insoluble cellulose, as compared to the individual modules (Linder et al., 1996). Also, two CBMs of the family 29, CBM29-1 and CBM29-2, exist naturally in tandem as component of the Piromyces equi cellulase-hemicellulase complex (Freelove et al., 2001). The tandem CBM29-1-2 was shown to possess much higher binding affinity than the single CBM29-1 and CBM29-2 modules, indicating a synergy between the two single modules. Another possible role for the multiple CBMs is that they increase the diversity of polysaccharides that the parent enzyme can interact with (Gill et al., 1999). Prime example of such enzymes is the Cellulomonas fimi xylanase 11A, which contains two family 2b CBMs, CBM2b-1 and CBM2b-2. CBM2b-1 specifically binds to xylan while CBM2b-2 additionally binds to cellulose (Bolam et al., 2001). As for the tandem CBM29-1-2, the two family 2b CBMs were also shown to have higher affinity when incorporated into a single protein species, than when expressed as discrete entities. Another interesting CBM, CBM22 from Clostridium thermocellum (Charnock et al., 2000), which has affinity for xylan also exist in Arabidopsis xylanases in multiple copies (Henrissat et al., 2001; Suzuki et al., 2002).

3.2.4. CBMs can modify properties of composites.

In Acetobacter xylinum, which has long been regarded as the model system for cellulose biosynthesis, polymerisation and crystallization of cellulose are coupled processes. It was observed that interference with crystallization in the model system results in acceleration of polymerisation (Benziman et al., 1980). A number of cellulose-binding, organic substances like carboxymethyl cellulose (CMC) and fluorescent brightening agents (FBAs, e.g. calcofluor white) prevent microfibril crystallization in the Acetobacter model system, thereby enhancing polymerization (Haigler, 1991). These molecules bind to the polysaccharide chains immediately

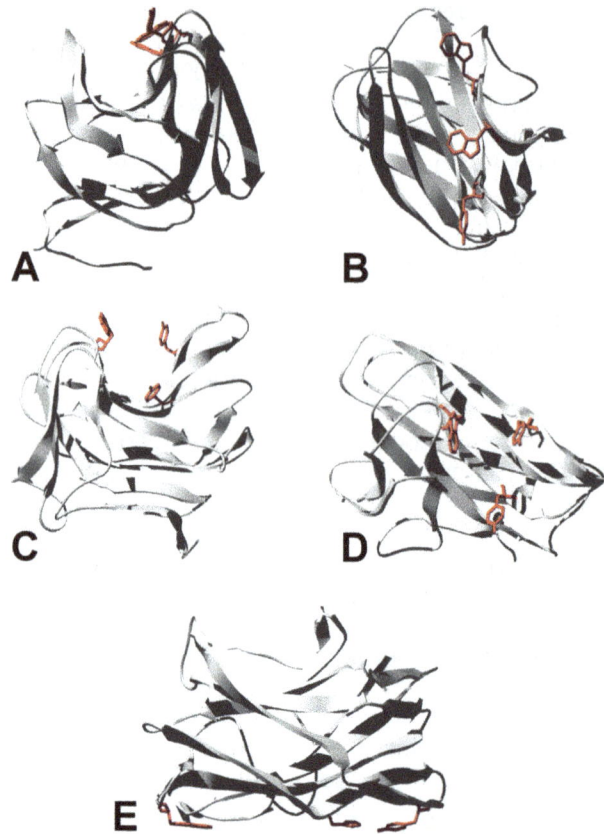

Figure 2. 3-D structures of selected CBMs that bind hemicelluloses (A-D) and crystalline cellulose (E), showing topology and residues involved in binding. Aromatic amino acids residues, which have been implicated in binding glycosyl residues, are indicated in red. The CBM29-2 (1GWK) with aromats Trp24, Trp26 and Tyr46 has a shallow groove for polysaccharide binding (A: front view; B: structure has been rotated 90° with the front view facing up) than CBM22-2. CBM22-2 (1DYO) with aromats Trp53, Tyr103 and Tyr134, has a less shallow polysaccharide-binding groove (C: front view; D: structure has been rotated 90° with the front view facing up) than CBM29-2. It can be seen that the aromats in CBM29-2 are positioned differently in the groove as compared to those in CBM22. CBM3 (1NBC) with residues His57, Tyr67 and Trp118 (E), reveal the coplanar orientation of the residues involved in binding, which is characteristic of CBMs that bind the flat surface of cellulose. 1DYO, 1GWK, and 1NBC represent the codes for the files containing the atomic coordinates for building the structural models.

after their extrusion from the cell surface, preventing normal assembly of microfibrils and cell walls (Haigler, 1991). It was also demonstrated that microbial CBMs could modulate cellulose biosynthesis, by achieving an up to 5-fold increase in the rate of biosynthesis as compared with the controls (Shpigel et al., 1998). A hypothetical model of the physico-mechanical mechanism of action has been proposed, whereby a flat-surface, cellulose-recognizing CBM slides between cellulose fibers and separates them in a wedge-like action (Levy and Shoseyov, 2002). The authors speculated that when the interaction occurs during the initial stages of crystallization, the result is increased rate of synthesis and splayed fibrils. Post-synthesis interaction results in non-hydolytic fiber disruption (Levy and Shoseyov, 2002).

Furthermore, it was indicated that CBMs could interfere with the attachment of hemicellulose to cellulose, as reported for CBM3, which competed with xyloglucan for binding sites when it was added first to *Acetobacter* cellulose (Shpigel et al., 1998). The competition decreased the amount of the xyloglucan that bound to the cellulose in the absence of CBM by about 12%. This indicates that the CBMs can be used to prevent cellulose-hemicellulose interactions in plants, leading to the production of cellulose fibres with higher crystallinity, as discussed earlier under hemicellulose biosynthesis.

In plant systems, it was shown that the elongation growth of *Arabidopsis* seedlings and peach (*Prunus persica* L.) pollen tubes could be affected *in vitro* by exogenous supply of a recombinant bacterial cellulose-recognising CBM3 (Levy et al., 2002). Furthermore, it has been shown that CBMs can modulate cell wall structure and growth of transgenic plants (Kilburn et al., 2000; Quentin, 2003); Safra-Dassa et al., 2006; Shoseyov, 2001. Introduction of the CBM3 gene from *Clostridium cellulovorans* into potato plants (Safra-Dassa et al., 2006) and poplar tree plants (Shoseyov, 2001) was reported to enhance the growth rate of the transgenic plants. A similar enhancement in plant growth was reported for transgenic plants expressing a mannan-recognising CBM27 (Kilburn et al., 2000). Most of these research activities relating to the potential use of CBMs for plant cell wall modification have been on the cellulose-specific CBMs. Using immunohistochemistry to investigate the specificity of CBMs that recognise xylan polysaccharides in cell walls, McCartney et al. (2006) showed that these CBMs display significant variation in specificity for xylans in both primary and secondary cell walls. This, together with the summary of ligand specificity in Table 1, reveals that, if one were to use CBMs for *in planta* wall modification, the choice of CBM for influencing the interaction between cellulose and hemicellulose is likely to be critical. With the detailed characterization of CBMs known to date, it is now time to investigate them more intensively in plants for modification purposes. The results so far seem very promising. By introducing a grooved CBM, which exhibits promiscuous recognition for different hemicelluloses and for cellulose, more especially in the secondary wall, evidence was obtained that grooved CBMs also can yield transgenic plants with altered wall structures and plant growth (Obembe et al., unpublished results). The ultimate goal is to provide valuable information for modification of the interactions of cellulose with non-cellulosic polysaccharides for various industrial applications, such as for instance the production of cellulose fibre with high tensile strength for textile manufacturing and the production of fibres with less attachment of lignin for paper manufacturing.

4. Concluding Remarks

A number of in-roads have been made into *in planta* modification of cell wall for enhancing cellulose fibre properties. However, at the moment the tools to do this in a deliberate manner are not in our hands as yet, but might be in the future, especially with the increasing understanding of the plant cell wall biosynthesis. It is envisaged that the strategies, when developed, can be adapted for modifying fibre properties in fibre crops like flax and hemp, with a view to achieving higher quality fibre for use in the various applications. Besides chemical derivatization, *in vitro* enzyme-mediated modification of fibres is the trend nowadays for tailoring cellulose fibres with enhanced properties for specific industrial applications (Gustavsson et al., 2004; Gustavsson et al., 2005). This approach has the advantage that plant development is not compromised. However, it has its own limitation as well, in that it cannot modify inherent fibre properties. This is only possible during cell wall polysaccharide biosynthesis, and as such, with an *in planta* approach that ensures minimal effect on plant development. Hence, both approaches might complement each other in future. The in-roads that have been made in the 2 lines of investigations (modification of cellulose/hemicellulose ratio and interference with cellulose-hemicellulose interactions) could be combined to create a sort of synergy to achieve a more specific *in planta* modification of the fibre properties.

5. Acknowledgements

The plant cell wall modification project was supported by the Netherlands Foundation for the Advancement of Tropical Research (WOTRO), the Netherlands.

6. References

Amor Y, Haigler CH, Johnson S, Wainscott M, Delmer DP (1995). A membrane-associated form of sucrose synthase and its potential role in synthesis of cellulose and callose in plants. Proc. Natl Acad. Sci. USA 92: 9353-9357.

Arioli T, Peng LC, Betzner AS, Burn J, Wittke W, Herth W, Camilleri C, Hofte H, Plazinski J, Birch R, Cork A, Glover J, Redmond J, Williamson RE (1998). Molecular analysis of cellulose biosynthesis in *Arabidopsis*. Science 279: 717-720.

Aspeborg H, Schrader J, Coutinho PM, Stam M, Kallas Å, Djerbi S, Nilsson P, Denman S, Amini B, Sterky F, Master E, Sandberg G, Mellerowicz E, Sundberg B, Henrissat B, Teeri TT (2005). Carbohydrate-active enzymes involved in the secondary cell wall biogenesis in hybrid aspen. Plant Physiol. 137: 983-997.

Awano T, Takabe K, Fujita M (2002). Xylan deposition on secondary wall of *Fagus crenata* fibre. Protoplasma 219: 106-115.

Bayer EA, Chanzy H, Lamed R, Shoham Y (1998). Cellulose, cellulases and cellulosomes. Curr. Opin. Struct. Biol. 8: 548-557.

Benziman M, Haigler Candace H, Brown RM, R. WA, Cooper KM (1980). Polymerization and crystallization are coupled processes in *Acetobacter xylinum*. Proc. Natl Acad. Sci. USA 77, 6678-82.

Bolam DN, Ciruela A, McQueen-Mason S, Simpson P, Williamson MP, Rixon JE, Boraston A, Hazlewood GP, Gilbert HJ (1998). Pseudomonas cellulose-binding domains mediate their effects by increasing enzyme substrate proximity. Biochem. J. 331: 775-781.

Bolam DN, Xie HF, White P, Simpson PJ, Hancock SM, Williamson MP, Gilbert HJ (2001). Evidence for synergy between family 2b carbohydrate binding modules in *Cellulomonas fimi* xylanase 11A. Biochemistry 40: 2468-2477.

Boraston AB, Bolam DN, Gilbert HJ, Davies GJ (2004). Carbohydrate-binding modules: fine-tuning polysaccharide recognition. Biochem. J. 382: 769-781.

Brady JD, Sadler IH, Fry SC (1996). Di-isodityrosine, a novel tetrameric derivative of tyrosine in plant cell wall proteins: A new potential cross-link. Biochem. J. 315: 323-327.

Brown RM, Saxena IM (2000). Cellulose biosynthesis: A model for understanding the assembly of biopolymers. Plant Physiol. Biochem. 38: 57-67.

Brown RM, Saxena IM, Kudlicka K (1996). Cellulose biosynthesis in higher plants. Trends Plant Sci. 1:149-156.

Burton RA, Wilson SM, Hrmova M, Harvey AJ, Shirley NJ, Medhurst A, Stone BA, Newbigin EJ, Bacic A and Fincher GB (2006). Cellulose synthase-like CslF genes mediate the synthesis of cell wall (1,3;1,4)-β-D-glucans. Science 311: 1940-1942.

Carpita NC, Gibeaut DM (1993). Structural models of primary-cell walls in flowering plants - consistency of molecular-structure with the physical-properties of the walls during growth. Plant J. 3: 1-30.

Charnock SJ, Bolam DN, Nurizzo D, Szabo L, McKie VA, Gilbert HJ, Davies GJ (2002). Promiscuity in ligand-binding: The three-dimensional structure of a Piromyces carbohydrate-binding module, CBM29-2, in complex with cello-and mannohexaose. Proc. Natl Acad. Sci. USA 99: 14077-14082.

Charnock SJ, Bolam DN, Turkenburg JP, Gilbert HJ, Ferreira LMA, Davies GJ, Fontes C (2000). The X6 "thermostabilizing" domains of xylanases are carbohydrate-binding modules: Structure and biochemistry of the Clostridium thermocellum X6b domain. Biochemistry 39: 5013-5021.

Cutler S, Somerville C (1997). Cellulose synthesis: Cloning in silico. Curr. Biol. 7: R108-R111.

Delmer DP (1999). Cellulose biosynthesis: Exciting times for a difficult filed of study. Annu. Rev. Plant Physiol. Plant Mol. Biol. 50: 245-276.

Delmer DP, Amor Y (1995). Cellulose Biosynthesis. Plant Cell 7: 987-1000.

Desprez T, Vernhettes S, Fagard M, Refregier G, Desnos T, Aletti E, Py N, Pelletier S, Hofte H (2002). Resistance against herbicide isoxaben and cellulose deficiency caused by distinct mutations in same cellulose synthase isoform CESA6. Plant Physiol. 128: 482-490.

Din N, Damude HG, Gilkes NR, Miller RC, Warren RAJ, Kilburn DG (1994). C-1-C-X revisited - intramolecular synergism in a cellulase. Proc. Natl Acad. Sci. USA 91: 11383-11387.

Dhugga KS, Barreiro R, Whitten B, Stecca K, Hazebroek J, Randhawa GS, Dolan M, Kinney AJ, Tomes D, Nichols S, Anderson P (2004). Guar seed beta-mannan synthase is a member of the cellulose synthase super gene family. Science 303: 363-366.

Ebringerova A, Heinze T (2000). Xylan and xylan derivatives - biopolymers with valuable properties, 1 - Naturally occurring xylans structures, procedures and properties. Macromol. Rapid Comm. 21: 542-556.

Edwards ME, Dickson CA, Chengappa S, Sidebottom C, Gidley MJ, Reid JSG (1999). Molecular characterisation of a membrane-bound galactosyltransferase of plant cell wall matrix polysaccharide biosynthesis. Plant J. 19: 691-697.

Fagard M, Desnos T, Desprez T, Goubet F, Refregier G, Mouille G, McCann M, Rayon C, Vernhettes S, Hofte H (2000). PROCUSTE1 encodes a cellulose synthase required for normal cell elongation specifically in roots and dark-grown hypocotyls of Arabidopsis. Plant Cell 12: 2409-2423.

Faik A, Price NJ, Raikhel NV, Keegstra K (2002). An Arabidopsis gene encoding an alpha xylosyltransferase involved in xyloglucan biosynthesis. Proc. Natl Acad. Sci. USA 99: 7797-7802.

Fontaine AS, Bout S, Barriere Y, Vermerris W (2003). Variation in cell wall composition among forage maize (Zea mays L.) inbred lines and its impact on digestibility: Analysis of neutral detergent fibre composition by pyrolysis-gas chromatography-mass spectrometry. J. Agric. Food Chem. 51: 8080-8087.

Freelove ACJ, Bolam DN, White P, Hazlewood GP, Gilbert HJ (2001). A novel carbohydrate binding protein is a component of the plant cell wall-degrading complex of Piromyces equi. J. Biol. Chem. 276: 43010-43017.

Gidley MJ, Lillford PJ, Rowlands DW, Lang P, Dentini M, Crescenzi V, Edwards M, Fanutti C, Reid JSG (1991) Structure and solution properties of tamarind seed polysaccharide. Carbohydr Res 214: 299-314

Gill J, Rixon JE, Bolam DN, McQueen-Mason S, Simpson PJ, Williamson MP, Hazlewood GP, Gilbert HJ (1999). The type II and X cellulose-binding domains of Pseudomonas xylanase A potentiate catalytic activity against complex substrates by a common mechanism. Biochem. J. 342: 473-480.

Gorshkova T, Morvan C (2006). Secondary cell-wall assembly in flax phloem fibres: role of galactans. Planta 223:149-58.

Gustavsson MT, Persson PV, Iversen T, Hult K, Martinelle M (2004). Polyester coating of cellulose fibre surfaces catalyzed by a cellulose-binding module-Candida antarctica lipase B fusion protein. Biomacromolecules 5: 106-112.

Gustavsson MT, Persson PV, Iversen T, Martinelle M, Hult K, Teeri TT, Brumer H (2005). Modification of cellulose fibre surfaces by use of a lipase and a xyloglucan endotransglycosylase. Biomacromolecules 6: 196-203.

Ha MA, Apperley DC, Jarvis MC (1997). Molecular rigidity in dry and hydrated onion cell walls. Plant Physiol. 115: 593-598.

Haigler CH (1991). Relationship between polymerization and crystallization in microfibril biogenesis, in Biosynthesis and biodegradation of cellulose (eds. Haigler, C. H., Weimer, P.J.). Marcel Dekker, New York.

Hazen SP, Scott-Craig JS, Walton JD (2002). Cellulose synthase-like genes of rice. Plant Physiol. 128: 336-340

Henrissat B, Coutinho PM, Davies GJ (2001). A census of carbohydrate-active enzymes in the genome of Arabidopsis thaliana. Plant Mol. Biol. 47: 55-72.

Itoh T, Kimura S (2001). Immunogold labeling of terminal cellulose-synthesizing complexes. J. Plant Res. 114: 483-489.

John M, Rohrig H, Schmidt J, Walden R, Schell J (1997). Cell signalling by oligosaccharides. Trends Plant Sci. 2: 111-115.

Keegstra K, Raikhel N (2001). Plant glycosyltransferases. Curr. Opin. Plant Biol. 4: 219-224

Kilburn DG, Warren AJ, Stoll D, Gilkes NR, Shoseyov O, Shani Z (2000). Expression of a mannan binding domain to alter plant morphology. WO 2000/068391.

Kim HJ and Triplett BA (2001). Cotton Fiber Growth in Planta and in Vitro. Models for Plant Cell Elongation and Cell Wall Biogenesis. Plant Physiol. 127:1361-1366.

Kimura S, Laosinchai W, Itoh T, Cui XJ, Linder CR, Brown RM (1999). Immunogold labeling of rosette terminal cellulose-synthesizing complexes in the vascular plant Vigna angularis. Plant Cell 11: 2075-2085.

Kokubo A, Sakurai N, Kuraishi S, Takeda K (1991). Culm brittleness of barley (Hordeum-Vulgare L) mutants is caused by smaller number of cellulose molecules in cell-wall. Plant Physiol. 97: 509-514.

Lane DR, Wiedemeier A, Peng LC, Hofte H, Vernhettes S, Desprez T, Hocart CH, Birch RJ, Baskin TI, Burn JE, Arioli T, Betzner AS, Williamson RE (2001). Temperature-sensitive alleles of RSW2 link the KORRIGAN endo-1,4-beta-glucanase to cellulose synthesis and cytokinesis in Arabidopsis. Plant Physiol. 126: 278-288.

Levy I, Shani Z, Shoseyov O (2002). Modification of polysaccharides and plant cell wall by endo- 1,4-beta-glucanase and cellulose-binding domains. Biomol. Eng. 19: 17-30.

Levy I, Shoseyov O (2002). Cellulose-binding domains biotechnological applications. Biotechnol. Adv. 20: 191-213.

Liepman AH, Wilkerson CG, Keegstra K (2005). Expression of cellulose synthase-like (Csl) genes in insect cells reveals that CslA family members encode mannan synthases Proc. Natl Acad. Sci. USA 102: 2221-2226.

Linder M, Salovuori I, Ruohonen L, Teeri TT (1996). Characterization of a double cellulose-binding domain - Synergistic high affinity binding to crystalline cellulose. J. Biol. Chem. 271: 21268-21272.

MacDougall AJ, Brett GM, Morris VJ, Rigby NM, Ridout MJ, Ring SG (2001). The effect of peptide behaviour-pectin interactions on the gelation of a plant cell wall pectin. Carb. Res. 335: 115-126.

Madson M, Dunand C, Li XM, Verma R, Vanzin GF, Calplan J, Shoue DA, Carpita NC, Reiter WD (2003). The MUR3 gene of Arabidopsis encodes a xyloglucan galactosyltransferase that is evolutionarily related to animal exostosins. Plant Cell 15: 1662-1670.

McCann MC, Roberts K (1994). Changes In Cell-Wall Architecture During Cell Elongation. J. Exp. Bot. 45: 1683-1691.

McCartney, L, Blake AW, Flint J, Bolam DN, Boraston AB, Gilbert HJ, Pnox JP (2006). Differential recognition of plants cell walls by microbial xylan-specific carbohydrate-binding modules. Proc. Natl Acad. Sci. USA 103: 4765-4770.

Molhoj M, Ulvskov P, Dal Degan F (2001). Characterization of a functional soluble form of a Brassica napus membrane-anchored endo-1,4-beta-glucanase heterologously expressed in Pichia pastoris. Plant Physiol. 127: 674-684.

Morris VJ, Wilde PJ (1997). Interactions of food biopolymers. Curr. Opin. Colloid Interface Sci. 2: 567-572.

Mueller SC, Brown RM (1980). Evidence for an intramembrane component associated with a cellulose microfibril-synthesizing complex in higher plants. J. Cell Biol. 84: 315-326.

Najmudin S, Guerreiro CI, Carvalho AL, Prates JA, Correia MA, Alves VD, Ferreira LM, Romao MJ, Gilbert HJ, Fontes CM, Bolam DN (2006). Xylodlucan is recognised by carbohydrate-binding modules that interact with beta-glucan chains. J. Biol. Chem. 281: 8815-8828

Nicol F, His I, Jauneau A, Vernhettes S, Canut H, Hofte H (1998). A plasma membrane-bound putative endo-1,4-beta-D-glucanase is required for normal wall assembly and cell elongation in Arabidopsis. Embo J. 17: 5563-5576.

O'Neill MA, York WS (2003). The composition and structure of plant primary cell wall. In The Plant Cell Wall. (ed. J.K.C. Rose). Blackwell Publishing, Oxford, UK.

Oomen RJFJ, Doeswijk-Voragen CHL, Bush MS, Vincken JP, Borkhardt B, Van de Broek LAM, Corsar J, Ulvskov P, Voragen AGJ, McCann MC, Visser RGF (2002). In muro fragmentation of the rhamnogalacturonan I backbone in potato (Solanum tuberosum L.) results in a reduction and altered location of the galactan and arabinan side-chains and abnormal periderm development Plant J. 30: 403-413.

Pear JR, Kawagoe Y, Schreckengost WE, Delmer DP, Stalker DM (1996) Higher plants contain homologs of the bacterial celA genes encoding the catalytic subunit of cellulose synthase. Proc. Natl Acad. Sci. USA 93: 12637-12642.

Pena MJ, Ryden P, Madson M, Smith AC, Carpita NC (2004). The galactose residues of xyloglucan are essential to maintain mechanical strength of the primary cell walls in Arabidopsis during growth. Plant Physiol. 134: 443-451.

Peng LC, Kawagoe Y, Hogan P, Delmer D (2002). Sitosterol-beta-glucoside as primer for cellulose synthesis in plants. Science 295: 147-150.

Peng LC, Xiang F, Roberts E, Kawagoe Y, Greve LC, Kreuz K, Delmer DP (2001). The experimental herbicide CGA 325'615 inhibits synthesis of crystalline cellulose and causes accumulation of noncrystalline beta-1,4-glucan associated with CesA protein. Plant Physiol. 126: 981-992.

Perrin RM, DeRocher AE, Bar-Peled M, Zeng WQ, Norambuena L, Orellana A, Raikhel NV, Keegstra K (1999). Xyloglucan fucosyltransferase, an enzyme involved in plant cell wall biosynthesis. Science 284: 1976-1979.

Quentin MGE (2003). Cloning, characterisation and assessment of the wall modifying ability of a fungal cellulose-binding domain. (PhD Thesis). Wageningen University, Wageningen.

Raven PH, Evert RF, Eichhorn SE (1992). Biology of plants. Fifth edition. Worth Publishers, New York, USA.

Richmond TA, Somerville CR (2001). Integrative approaches to determining Csl function. Plant Mol. Biol. 47: 131-143.

Robert S, Mouille G, Hofte H (2004). The mechanism and regulation of cellulose synthesis in primary walls: lessons from cellulose-deficient Arabidopsis mutants. Cellulose 11: 351-364.

Rose JKC, Catalá C, Gonzalez-Carranza ZH, Roberts JA (2003). Cell wall disassembly. In The Plant Cell Wall. (ed. J.K.C. Rose). Blackwell Publishing, Oxford, UK.

Ryden P, Sugimoto-Shirasu K, Smith AC, Findlay K, Reiter WD, McCann MC (2003). Tensile properties of Arabidopsis cell walls depend on both a xyloglucan cross-linked microfibrillar network and rhamnogalacturonan II-borate complexes. Plant Physiol. 132: 1033-1040.

Safra-Dassa L, Shani Z, Danin A, Roiz L, Shoseyov O, Wolf S (2006). Growth modulation of transgenic potato plants by heterologous expression of bacterial carbohydrate-binding module. Mol. Breeding DOI. 1007/s11032-006-9007-4.

Salnikov VV, Grimson MJ, Delmer DP, Haigler CH (2001). Sucrose synthase localizes to cellulose synthesis sites in tracheary elements. Phytochemistry 57: 823-833.

Saxena IM, Brown RM, Fevre M, Geremia RA, Henrissat B (1995). Multidomain architecture of beta-glycosyl transferases - implications for mechanism of action. J. Bacteriol. 177: 1419-1424.

Scheible WR, Eshed R, Richmond T, Delmer D, Somerville C (2001). Modifications of cellulose synthase confer resistance to isoxaben and thiazolidinone herbicides in Arabidopsis Ixr1 mutants. Proc. Natl Acad. Sci. USA 98: 10079-10084.

Schrick K (2004). A link between sterol biosynthesis, the cell-wall, and cellulose in Arabidopsis Plant J. 38: 562-562.

Shoseyov OY, K; Shani, Z; Rehovoth; Shpigel, E; Medigo, K (2001). Transgenic Plants of Altered Morphology. US patent 6,184,440, 2001.

Shpigel E, Roiz L, Goren R, Shoseyov O (1998). Bacterial cellulose-binding domain modulates in vitro elongation of different plant cells. Plant Physiol. 117: 1185-1194.

Simpson PJ, Xie HF, Bolam DN, Gilbert HJ, Williamson MP (2000). The structural basis for the ligand specificity of family 2 carbohydrate-binding modules. J. Biol. Chem. 275: 41137-41142.

Sørensen SO, Pauly M, Bush MS, Skjot M, McCann MC, Borkhardt B, Ulvskov P (2000). Pectin engineering: Modification of potato pectin by in vivo expression of an endo-1,4-β-D-galactanase. Proc. Natl Acad. Sci. USA 97:7639-7644

Southall SM, Simpson PJ, Gilbert HJ, Williamson G, Williamson MP (1999). The starch-binding domain from glucoamylase disrupts the structure of starch. Febs Lett. 447: 58-60.

Stephen AM (1982). Other plant polysaccharides. In the polysaccharides, Vol 2 (ed. G.O. Aspinall). Academic Press, New York, USA.

Suzuki M, Kato A, Nagata N, Komeda Y (2002). A xylanase, AtXyn1, is predominantly expressed in vascular bundles, and four putative xylanase genes were identified in the Arabidopsis thaliana genome. Plant Cell Physiol. 43: 759-767.

Taylor NG, Gardiner JC, Whiteman R, Turner SR (2004). Cellulose synthesis in the Arabidopsis secondary cell wall. Cellulose 11: 329-338.

Taylor NG, Scheible WR, Cutler S, Somerville CR, Turner SR (1999). The irregular xylem3 locus of Arabidopsis encodes a cellulose synthase required for secondary cell wall synthesis. Plant Cell 11: 769-780.

Tsekos I, Reiss HD (1992). Occurrence of the putative microfibril-synthesizing complexes (linear terminal complexes) in the plasma-membrane of the epiphytic marine red alga Erythrocladia-Subintegra Rosenv. Protoplasma 169: 57-67.

Turner SR, Somerville CR (1997). Collapsed xylem phenotype of Arabidopsis identifies mutants deficient in cellulose deposition in the secondary cell wall. Plant Cell 9: 689-701.

Whitney SEC, Brigham JE, Darke AH, Reid JSG, Gidley MJ (1995). In-Vitro Assembly of Cellulose/Xyloglucan Networks - Ultrastructural And Molecular Aspects. Plant J. 8: 491-504.

Whitney SEC, Brigham JE, Darke AH, Reid JSG, Gidley MJ (1998). Structural aspects of the interaction of mannan-based polysaccharides with bacterial cellulose. Carb. Res. 307: 299-309.

Whitney SEC, Gidley MJ, McQueen-Mason SJ (2000). Probing expansin action using cellulose/hemicellulose composites. Plant J. 22: 327-334.

Whitney SEC, Gothard MGE, Mitchell JT, Gidley MJ (1999). Roles of cellulose and xyloglucan in determining the mechanical properties of primary plant cell walls. Plant Physiol. 121: 657-663.

2

Gene pyramiding-A broad spectrum technique for developing durable stress resistance in crops

Raj Kumar Joshi and Sanghamitra Nayak

Centre of Biotechnology, Siksha O Anusandhan University, Bhubaneswar, India.

The development of molecular genetics and associated technology like MAS has led to the emergence of a new field in plant breeding-Gene pyramiding. Pyramiding entails stacking multiple genes leading to the simultaneous expression of more than one gene in a variety to develop durable resistance expression. Gene pyramiding is gaining considerable importance as it would improve the efficiency of plant breeding leading to the development of genetic stocks and precise development of broad spectrum resistance capabilities. The success of gene pyramiding depends upon several critical factors, including the number of genes to be transferred, the distance between the target genes and flanking markers, the number of genotype selected in each breeding generation, the nature of germplasm etc. Innovative tools such as DNA chips, micro arrays, SNPs are making rapid strides, aiming towards assessing the gene functions through genome wide experimental approaches. The power and efficiency of genotyping are expected to improve in the coming decades. The present review discusses the design parameters in a gene pyramiding scheme, potential application of gene pyramiding in crop plant improvement, and the prospect and challenges in integrating MAS based gene pyramiding with conventional plant breeding programmes.

Key words: Gene pyramiding, marker-assisted selection, durable resistance.

TABLE OF CONTENT

INTRODUCTION

Since the beginning of agriculture, humans have sought to improve crops by selecting for desired traits. Genetics have played an important part in this field. With the advent of genetic engineering and biotechnology, plant breeding has got a new dimension to produce crop verities with more desirable characters. Marker assisted selection (MAS) which involves indirect selection of traits

---

*Corresponding author: E-mail: sanghamitran@yahoo.com.

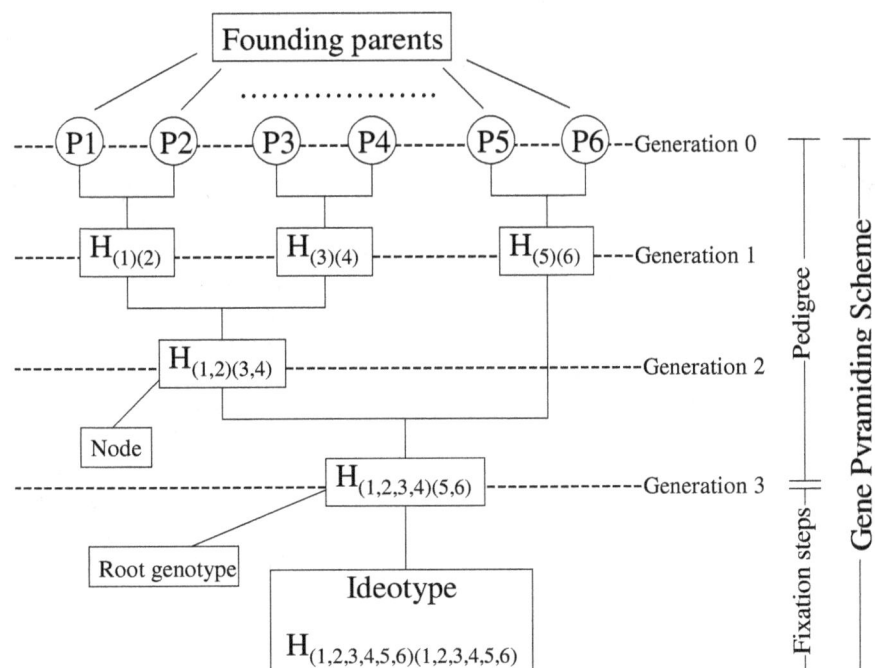

Figure 1. A distinct gene pyramiding scheme cumulating six target genes. (Hospital et al., 2004)

by selecting the marker linked to the gene of interest has become a reality with development and availability of an array of molecular markers and dense molecular genetic maps in crop plants. Molecular markers are especially advantageous for agronomic traits that are otherwise difficult to tag such as resistance to pathogens, insects and nematodes, tolerance to abiotic stresses, quality parameters and quantitative traits. Molecular markers studies using near isogenic lines (NILs), bulk segregant analysis or recombinant inbred lines (RILs) have accelerated the mapping of many genes in different plant species. Sequence tagged sites have been developed using RAPD and RFLP markers in tomato and rice and microsatellite markers in rice, wheat and cereals().In other words, there is now a large amount of research that addresses marker-aided selection in some form. However, although the process is now more efficient and sophisticated, it still mostly based on field selection and data analysis. Moreover, many MAS based improved traits have broken up in the past few years due to lack of durable resistance effect.

The challenge now is to develop new efficient marker assisted selection strategies aimed at plant improvement. Gene pyramiding holds greater prospects to attain durable resistance against biotic and abiotic stresses in crops. Different resistance genes often confer resistance to different isolates, races or biotypes. Combining their resistance broadens the number of races or isolates that a more than one character in a variety at the same time. In general, the development of pyramid lines is a long and

costly affair in addition to the epistatic effect. However MAS based gene pyramiding could facilitate in pyramiding of genes effectively into a single genetic background. When hybrids crops are the goal, additional options for pyramiding different resistance gene combinations into different parents also exist.

A DISTINCT GENE PYRAMIDING SCHEME

In a gene pyramiding scheme, strategy is to cumulate into a single genotype, genes that have been identified in multiple parents. The use of DNA markers, which permits complete gene identification of the progeny at each generation, increases the speed of pyramiding process. In general, the gene pyramiding aims at the derivation of an ideal genotype that is homozygous for the favorable alleles at all n loci. The gene pyramiding scheme can be distinguished into two parts (Figure 1). The first part is called a *pedigree*, which aims at cumulating of all target genes in a single genotype called the root genotype. The second part is called the *fixation step* which aims at fixing the target genes into a homozygous state i.e. to derive the ideal genotype from the one single genotype. Each node of the tree is called an intermediate genotype and has two parents. Each of this intermediate genotype variety can resist. Moreover, pyramiding can also improve becomes a parent in the next cross. The intermediate genotypes are not just an arbitrary offspring of a given cross but it is a particular genotype selected from among the offspring in which all parental target genes are present. Although the pedigree step may be common, several different procedures can be used to undergo fixation in gene pyramiding.

Generation of a population of doubled haploids from the root genotype is a possible procedure for the fixation steps. Here, a population of gametes is obtained from the genotypes and their genetic material is doubled. This leads to a population of fully homozygous individuals, among which the ideotype can be found. Using this process, the ideal genotype can be obtained in just one additional generation after the root genotype is obtained. However, producing large population of doubled haploid is difficult and cumbersome in certain plant species.

A possible alternative to this method is to self the root genotype directly to obtain the ideal genotype. However, selfing the root genotype will result in the breakage of linkage between the desired alleles and it will be difficult to derive this breaks as the linkage phase is rarely visible in selfed populations. As a result, it may span too many generations thereby stretching the gene pyramiding scheme.

Another alternative to all this methods would be to obtain a genotype carrying all favorable alleles in coupling by crossing the root genotype with a parent containing none of the favorable alleles. This confirms that the linkage phase of the offspring is known and the genotype can be derived without any mixing. The ideal genotype will be reached within two generations after the root genotype. However, instead of crossing with a blank parent, a more simplified method would be to cross the root genotype with one of the founding parents. In such programs, the linkage will still be known, and the selection will be for genotypes that are homozygous for the target gene brought by the founding parent but heterozygous for other regions. The desired genes need not be fixed subsequently, thereby increasing the probability of getting the ideal genotype. This is called as marker assisted backcross gene pyramiding. By far this is the most accepted and efficient method to do the gene pyramiding.

MARKER-ASSISTED BACKCROSSING

Breeders transfer a target allele from a donor variety to a popular cultivar by a repetitive process called backcrossing; which, unfortunately, is slow and uncertain. Breeding a plant that has the desired donor allele but otherwise looks just like the popular cultivar usually takes four years or longer. Worse, the augmented variety may look just like the popular cultivar, but it inevitably retains stray chromosome segments from the donor. Consequently, to a greater or lesser extent, it will fail to perform exactly like the popular cultivar, thus limiting its appeal to farmers.

Marker-assisted breeding tackles both problems by

allowing breeders to identify young plants with the desired trait and by facilitating the removal of stray donor genes from intermediate backcrosses. The result, in about two years, is an improved variety exactly like the popular cultivar except that it possesses the transferred advantageous gene. In principle, this technique can be applied to the breeding of any crop or farm animal. So far, however, breeders of trees and rice have dominated the field. Because markers allow breeders to select immature plants, the time saved in breeding slow-growing trees is immense. In the case of rice, the crop's relatively advanced state of genetic mapping has facilitated the application of molecular marker techniques.

Markers are effective aids to selection in backcrossing in three ways. First, markers can aid selection on target alleles whose effects are difficult to observe phenotypically. Examples include recessive genes, multiple disease resistance gene pyramids combined in one genotype (where they can epistatically mask each other's effects), alleles that are not expressed in the selection environments (e.g., genes conferring resistance to a disease that is not regularly present in environments), etc. Second, markers can be used to select for rare progeny in which recombination near the target gene have produced chromosomes that contain the target allele and as little possible surrounding DNA from the donor parent. Third, markers can be used to select rare progeny that are the result of recombination near the target gene, thus minimizing the effects of linkage drag.

In general, the marker assisted backcross based gene pyramiding can be performed in three strategies (Figure 2). In the first method, the recurrent parent (RP1) is crossed with donor parent (DP1) to produce the F1 hybrid and backcrossed up to third backcross generation (BC3) to produce the improved recurrent parent (IRP1). This improved recurrent parent is then crossed with other donor parent (DP2) to pyramid multiple genes. This strategy is less acceptable as it is time taking but pyramiding is very precise as it involve one gene at one time. In the second strategy, the recurrent parent (RP1) is crossed with donor parents (DP1, DP2, etc.) to get the F1 hybrids which are then intercrossed to produce improved F1 (IF1). This improved F1 is then backcrossed with the recurrent parent to get the improved recurrent parent (IRP). As such, the pyramiding is done in the pedigree step itself. However, when the donor parents are different, this method is less likely to be used because there is chance that the pyramided gene may be lost in the process. The third strategy is an amalgamation of the first two which involve simultaneous crossing of recurrent parent (RP1) with many donor parents and then backcrossing them up to the BC3 generation. The backcross populations with the individual gene are then intercrossed with each other to get the pyramided lines. This is the most acceptable way as in this method not only time is reduced but fixation of genes is fully assured.

Marker assisted backcrossing to be effective, depends

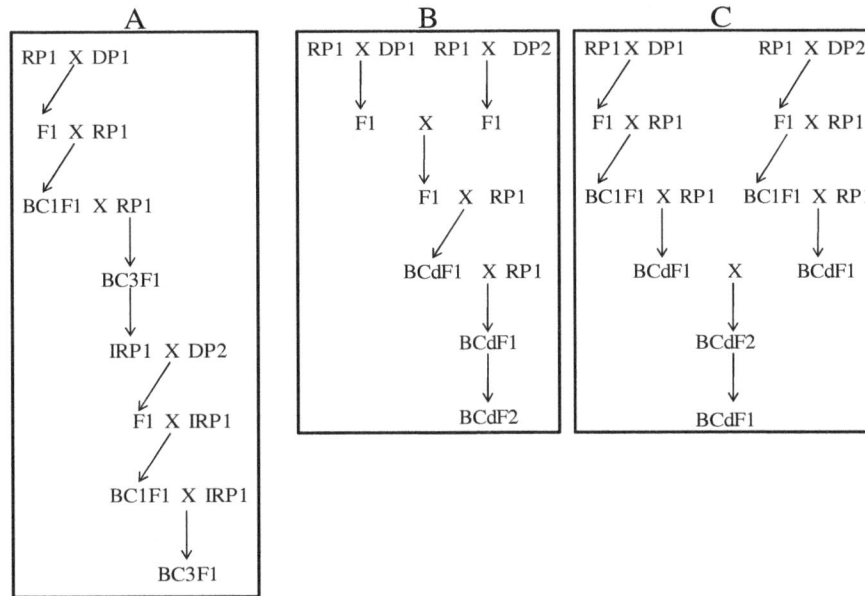

Figure 2. Different schemes of backcrossing for gene pyramiding. RP- Recurrent parent; DP- Donor parent; BC- Backcross; IRP- Improved recurrent parent. A. Stepwise transfer; B. Simultaneous transfer; C. Simultaneous and stepwise transfer.

upon several factors, including the distance between the closest markers and the target gene, the number of target genes to be transferred, the genetic base of the trait, the number of individuals that can be analyzed and the genetic background in which the target gene has to be transferred, the type of molecular marker(s) used, and available technical facilities (Weeden et al., 1992; Francia et al., 2005). When these entire selection criterions are maintained properly, only then a well acceptable MAB based gene pyramiding scheme can lead to durable crop improvement.

EFFICIENCY OF GENE PYRAMIDING

Computer simulations and theoretical calculations have provided powerful tools for analyzing the efficiency of gene pyramiding programmes. Three different gene pyramiding schemes, one based on a cascading pedigree, and two based on the order of crosses of the founding parents were evaluated to check the transmission probabilities of the target genes and the cumulated population size needed in each scheme (Ribaut and Hiosington, 1998; Ribaut et al., 2001; Hospital et. al 2004). The simulation was based on identical recombination fractions between adjacent loci and spaced about 20 cM. The major conclusions from this experiment are as follows:

A MAS based gene pyramiding scheme based on a cascading pedigree is less expensive as it spans five generations in general and requires the smallest cumulated

population size of all the schemes. The average transmission probability is 0.9975.

Gene pyramiding scheme based on the crosses of founding parents spans four generations but the population size is somewhat higher. The average transmission probability is 0.9967.

When gene pyramiding is carried out involving a larger number of target genes, each trait starts as a founding parent resulting in intermediate genotypes by subsequent crossing. It is based on a cascading pedigree and span one or two less generation in general.

QUALITATIVE IMPROVEMENT THROUGH GENE PYRAMIDING- SOME CASE STUDIES

Marker aided pyramiding of rice genes for BLB and blast disease

The successful effort on gene pyramiding in rice includes resistance to blight, blast, gall midge etc. Bacterial blight (BB) caused by *Xanthomonas oryzae* pv. *Oryzae* (Xoo) is one of the most destructive diseases of rice throughout the world and in some areas of Asia it is responsible for yield loss of more than 60%. The most efficient approach to overcome bacterial blight caused by *X. oryzae* is to produce resistant varieties; more than 25 BLB resistant genes have been identified and subsequently transferred into modern rice cultivars by cross breeding. However, the recent breakdowns of many resistant genes to BLB pathogens have significantly affected the rice production in many countries. One way to delay such a breakdown is

Table 1. Selected examples of MAS based gene pyramiding for important traits in major crops.

Crop	Trait	Pyramided genes	Reference
Rice	Blight resistance	*Xa4,xa5,xa13,Xa21*	Huang et al., 1997, Singh et al., 2001, Narayanan et al., 2002
	Blast resistance	*Pi(2)t,Piz5,Pi(t)a*	Hittalmani et al., 2000
	Gallmidge resistance	*Gm1,Gm4*	Kumaravadivel et al., 2006
Wheat	Leaf rust resistance	*Lr41, Lr42, Lr43*	Cox et al., 1994
	Powdery mildew resistance	*Pm-1, Pm-2*	Liu et al., 2000
Cotton	Insect pest resistance	*Cry 1Ac, Cry 2Ac*	Jackson et al., 2003, Gahan et al., 2005
Pea	Nodulation ability	*Sym9, Sym10*	Schneider et al., 2002
Barley	Yellow mosaic virus resistance	*rym4, rym5, rym9, rym11*	Werner et al., 2005
Soybean	Soybean mosaic virus resistance	*Rsv1, Rsv3, Rsv4*	Zhu et al., 2006

to pyramid multiple resistance genes in to rice varieties. It is practically difficult to transfer genes through conventionally gene transfer process due to vertifolia effect. International Rice Research Institute (IRRI) have successfully used the MAS based gene pyramiding to transfer four genes *Xa21,xa5,xa4* and *xa13* in elite rice cultivars (Huang et al., 1997). (Table 1) The pyramided lines showed a wider spectrum and a higher level of resistance than lines with only a single gene. Similarly, Sanchez, et al. 2000 successfully transferred three bacterial blight resistance genes into three susceptible rice lines possessing desirable agronomic characteristics via a marker-aided backcrossing procedure. In India, at Punjab Agricultural University (PAU), three BB resistance genes xa5, xa13 and Xa 21 were pyramided in PR106 (Singh et al., 2001) and Pusa 44 background and two of the PR1106 have been included in all India Coordinated testing during 2002. A similar work has also been successfully carried out in Central Rice Research Institute to pyramid three genes *xa5, xa13 and Xa21* in to elite rice cultivars Lalat and Tapaswini. All combinations of the three resistance genes were pyramided using STS markers.

Narayanan et al 2002 improved an elite indica rice line IR50 by pyramiding blast resistance gene Piz5 and bacterial blight resistance gene Xa21 through marker-assisted selection and genetic transformation. Ramalingam et al 2002 made four cross combinations of IRBB21 and successfully obtained improved lines pyramided with Xa21 and Wx (waxy) gene showing durable resistance to bacterial leaf blight and high amylose content.

Rice blast caused by the fungal pathogen *Magnaporthe grisea* is another devastating disease that provides constant challenge to rice production. The most effective way to reduce the crop yield is to breed for resistance to disease (Zeigler et al., 1994). Recently, many genes for qualitative blast resistance have been mapped using molecular markers and some of them have also been tried in MAS for blast resistance. Hittalmani et al. (2002) have successfully pyramided three genes, Pi1, Piz5 and Pita in a susceptible rice variety, Co39 using RFLP and PCR based markers for durable blast resistance.

Asian rice Gall Midge, *Orseolia oryzae* (Wood-mason) is a serious dipteran pest in major rice growing areas, causing an annual yield loss of US $550 million in Asia (Herdt, 1991). In India, the pest is widely distributed and is considered a major constraint to rice production (Bentur et al., 2003). Since no effective chemical control measure for gall midge is available, growing resistance varieties is a viable strategy which is not only economical, but is also ecologically, a friendly approach. On a similar note with BLB and Blast, extensive research has been undergone which have resulted in many mapped genes resistant to Gall Midge (Kumar et al., 2005). Katiyar et al. (2001) successfully did the genetic analysis and pyramiding of two gall midge resistance genes Gm2 and Gm6t in rice. Kumaravadivel et al. (2006) are in a process of pyramiding two dominant resistant genes *Gm1* and *Gm4* into the locally popular varieties of Tamil nadu. A similar work is also in progress Central Rice Research institute, Cuttack to pyramid *Gm1* and *Gm4* gene into popular cultivars like Swarna and Tapaswini. Recently, the Govt of India through Indian Council of Agricultural Research have started an extensive network project on gene pyramiding to produce multiple biotic stress resistance rice cultivars.

Molecular marker-facilitated gene pyramiding for powdery mildew resistance in wheat

The fungal pathogen *Blumeria graminis f. sp. tritici* is the causal agent of the powdery mildew disease in wheat (*Triticum aestivum* L.). Resistance to this pathogen is mediated by the Pm genes (Chen et al 2005). Since race-specific resistance is restricted to pathogens that carry the matching avirulence (avr)-gene, this type of resistance can be overcome in the field. For breeders, it is therefore desirable to create plants with more broad-spectrum and long-lasting resistance features. One strategy to achieve this goal is to combine different resistance genes by classical breeding. However, this is a time-consuming approach. MAS based gene pyramiding provides a more rapid tool to introduce new disease resistance specificities into crop plants. Liu et al. (2000), have underwent a gene pyramiding approach in which three powdery mildew resistance gene combinations, *Pm2 + Pm4a*, *Pm2 + Pm21*, *Pm4a + Pm21* were successfully integrated into an elite wheat cultivar 'Yang158'. Double homozygotes were selected from a small F_2 population with the help of molecular markers. As the parents were near-isogenic lines (NILs) of 'Yang158', the progenies showed good uniformity in morphological and other non-resistance agronomic traits. The present work illustrates the bright prospects for the utilization of molecular markers in breeding for host resistance.

Gene pyramiding of rust-resistance genes Lr41, Lr42 and Lr43 in common wheat

Leaf rust is one of the most important diseases of wheat worldwide, particularly in the Great Plains region of the USA. Gene has been advocated as a long-term strategy for the control of this disease in the recent times. Cox et al. (1993) has successfully pyramided three leaf rust-resistance genes *Lr41*, *Lr42* and *Lr43* into the common wheat (*Triticum aestivum* L.). Here, In order to diversify the genetic base of resistance in hard red winter wheat (*T. aestivum* L.) to leaf rust (caused by *Puccinia recondita* Rob. ex Desm.), five genes for resistance were transferred from the diploid goatgrass *Triticum. tauschii* (Coss.) Schmal. to hexaploid wheat lines. One of the derived lines, KS90WGRC10, had a very low infection type when inoculated with 23 cultures of *P. recondita*. The others, KS91WGRC11, KS92WGRC16, U1865, and U1866, had low to intermediate infection types with three cultures. Their infection types varied similarly to those of lines carrying previously transferred alleles of *Lr21*. WGRC10 carries a completely dominant gene, *Lr41*, on chromosome ID that segregates independently of any other *T. tauschii*-derived leaf rust-resistance genes. WGRC11 carries the partially dominant gene, *Lr42*, also on ID, which is linked to *Lr21* with a recombination value of 0.286 +- 0.023. WGRC16 carries a partially dominant gene, *Lr43* that segregates independently of all known genes for seedling resistance from T. tauschii; its chromosome location is not known. The genes carried by

U1865 and U1866 are allelic to *Lr21*. WGRC10, WGRC11, and WGRC16 have been released as germplasm by the Wheat Genetics Resource Center.

Stripe rust is another of the most devastating diseases of wheat worldwide. Santra et al. (2006) have successfully pyramided two single, dominant genes *Yr5* and *Yr15*, which independently confer complete resistance to all stripe rust races found in North America. The cereal cyst nematode (CCN) *Heterodera avenae* is a significant pathogen of wheat. The wild grass *Aegilops variabilis* Accession No.1 has been found to be resistant to pathotypes of CCN; at least two genes transferred to wheat, designated as *CreX* and *CreY*, are involved in the resistance response. Barloy et al. (2006) pyramided the two CCN resistance genes in a wheat background through marker-assisted selection. The completely linked RAPD marker of *Rkn-mn1* (*CreY*), OpY16-$_{1065,}$ previously obtained, was converted into a SCAR. All these dominant markers were used to incorporate in the same genotype the two *Ae. variabilis* chromosome segments carrying the two genes for resistance. CCN bioassays with the Ha12 pathotype showed that the level of resistance of the pyramided line was significantly higher than that of *CreX* and *CreY* single introgression lines, but lower than that of *A. variabilis*.

Gene pyramiding as a Bt resistance management strategy in cotton

Reports on the emergence of insect resistance to *Bacillus thuringiensis* delta endotoxins have raised doubts on the sustainability of Bt-toxin based pest management technologies. Corporate industry has responded to this challenge with innovations that include gene pyramiding among others. Pyramiding entails stacking multiple genes leading to the simultaneous expression of more than one toxin in a transgenic variety. Recently gene pyramiding has been hailed as a lasting Bt resistance management strategy (Jackson et al., 2003, Shelton et al., 2002). The strategy of Bt gene pyramiding rests on three core assumptions (Gahan et al., 2005). The first assumption is that insects resistant to only one toxin can be effectively controlled by a second toxin produced in the same plant.

This assumption forms the basis for the Bollgard® II cotton variety which has two toxins namely, Cry 1Ac and Cry 2Ac. The Cry 1Ac toxin controls tobacco budworm and pink bollworm while the Cry 2Ac toxin controls corn earworm (Jackson et al., 2003; Ferry et al., 2004; Purcell et al., 2004). The second assumption is that strains resistant to two toxins with independent actions can not emerge through selection pressure with one toxin alone. The third assumption underlying the strategy of Bt gene pyramiding is that a single gene will not confer resistance to two toxins that are immunologically distinct and that have different binding targets (Gahan et al., 2005).

Second generation pyramided dual- Bt gene cottons

Bollgard II® (Cry 1Ac + Cry 2Ab) and WideStrike™ (Cry1Ac + Cry 1F) express two Bt endotoxins and were introduced successfully by Monsanto in USA and India in order to raise the level of control for *H. zea*, which was not satisfactorily controlled by the Cry 1Ac toxin alone (Jackson et al., 2003; Ferry et al., 2004; Bates et al., 2005; Gahan et al., 2005). The Cry 1Ac and 2Ab toxins have different binding sites in the larval midgut and are considered to be a good combination to deploy in delaying resistance evolution. This is due to the fact that a species cannot easily evolve resistance to both toxins because that would require two simultaneous, independent mutations in genes encoding the receptors (Jackson et al., 2003). Future pest management practices will have to rely on the introduction of transgenic cottons that express other insecticidal toxins in addition to the Cry toxins (Ferry et al., 2004; Wu and Guo, 2005). Biological pest control using parasitoids and predators, cultural practices and other pest management tactics are all essential tactics in preserving the efficacy of Bt based products. But gene pyramiding approaches have definitely proven as effective method in broadening the scope and mode of action of toxins thereby providing growers with more options in their overall resistance management efforts (Manyangarirwa et al 2006).

Pyramiding resistance genes against the barley yellow mosaic virus complex (BaMMV, BaYMV, BaYMV-2)

Barley Yellow Mosaic Virus disease caused by different strains of BaYMV and BaMMV is a major threat to winter barley cultivation in Europe. Pyramiding of resistance genes has been effectively used as a promising strategy to avoid the selection of new virus strains and to create more durable resistances by Werner et al. (2005). For pyramiding of resistance genes rym4, rym5, rym9 and rym11, located on chromosomes 3H and 4H of barley, two different strategies have been developed. These strategies are based on doubled haploid lines (DHs) and marker assisted selection procedures. On the one hand F1 derived DH-plants of single crosses were screened by molecular markers for genotypes being homozygous recessive for both resistance genes. These genotypes were crossed to lines carrying one resistance gene in common and an additional third gene, leading to a DH-population of which 25% carry three resistance genes, 50% have two resistance genes and 25% possess a single resistance gene homozygous recessively. Alternatively, F1 plants having one resistance gene in common were directly inter-crossed [e.g. (rym4 · rym9) · (rym4 · rym11)] and about 100 seeds were produced per combination. Within these complex cross progenies plants were identified by markers being homozygous at the common resistance locus and heterozygous at the others. From such plants, theoretically present at a frequency of 6.25%, DH-lines were produced, which were screened for the presence of genotypes carrying three or two recessive resistance genes in a homozygous state.

Gene pyramiding for soybean mosaic virus resistance using microsatellite markers

Gene pyramiding has been used as an effective approach to achieve multiple and durable resistance to various strains of *Soybean Mosaic Virus* (SMV) in soybean [*Glycine max* (L.) Merr.]. Zhu et al. (2006) have successfully pyramided three genes *Rsv1*, *Rsv3*, and *Rsv4* for SMV resistance with the aid of microsatellite markers in order to develop new soybean lines containing multiple resistance genes. A population of 84 lines derived from J05 (*Rsv1*, *Rsv3*) x V94-5152 (*Rsv4*) were developed, and six specific SSR markers were identified for SMV resistance genes. Two SSR markers Sat154 and Satt510 were used for selecting lines having the *Rsv1* gene, Satt560 and Satt726 for *Rsv3*, and Sat_254 and Satt542 for *Rsv4*. These SSR markers allowed for identification and selection of specific lines and individual plants containing different genes and for distinction of the homozygous and heterozygous lines or individual plants for all three resistance loci. Individual plants with homozygous alleles at three genetic loci (*Rsv1 Rsv1*, *Rsv3Rsv3* and *Rsv4Rsv4*) have been identified and new soybean germplasm is expected to be released with three genes combined for SMV resistance.

POLYGENIC TRAIT IMPROVEMENT BY GENE PYRAMIDING- A STEP FORWARD

Many economically important traits such as yield, quality and tolerance to abiotic stresses are of a quantitative nature. Genetic variations affecting such traits are controlled by a relatively large number of loci each of which can make a small positive or negative contribution to the final phenotypic value of the traits. These loci are termed QTLs. Molecular markers provide the opportunity to manipulate QTLs as Mendelian entities. Several QTLs for traits of economic importance like rice blast resistance (Wang et al., 1994) black mold resistance in tomato (Robert et al., 2001), flour colour in wheat (Parker et al., 2000), have been tagged with molecular markers. There have been some successful uses of MAS for polygenic traits in plants (Lande et al 1990; Johnson and Mumm 1996; Schneider et al. 1997; Stuber et al. 1998; Tanksley et al. 1996; Yousef and Juvik 2001). However, the improvement of polygenic traits through MAS raises more questions. Infact, no experiment has clearly demonstrated whether using DNA markers for quantitative trait improvement is superior to conventional breeding selection (Beavis 1998). This is because of the complexity of the process as several genes are involved in the expression of polygenic traits and generally have smaller individual effects on the plant phenotype. This implies that several regions (QTL) must be manipulated at the same time in order to have a significant impact, and that the effect of individual regions is not easily identified. This warrants repetitions of field test to characterize accurately the effects of QTLs and to evaluate their stability across environments. Even in presence of these constraints we still believe that solutions exist for it. Recently, progressive work have been carried out for quantitative trait improvement through MAS and subsequently gene pyramiding. MAS for polygenic traits has been integrated with varying levels of success into various breeding methods such as recurrent selection (Yousef and Juvik 2001), selection for a target genotype (Stuber et al., 1998), and introgression of exotic germplasm into elite lines using advanced backcrossed inbred selection (Tanksley et al., 1996; Tanksley and McCouch, 1997). However, polygenic traits like yield present additional complexity because, unlike oligogenic disease resistance, selection for yield is usually conducted exclusively in crosses between elite lines from a restricted germplasm pool. So QTLs mapped in one population will have little relevance to those mapped in other populations. So, it is better that marker-QTL linkage estimates will have to be updated regularly to account for recombination occurring between many linked QTLs as well as between QTL and markers (Holland, 2004).

Genetic enhancement, through AB-QTL strategy have been undergone by pyramiding various traits of agronomic importance, including fruit quality and black mould resistance in tomato were accomplished using wild relatives (Robert et. al 2001). A broad spectrum project is under progress at CIMMYT to pyramid major QTLs for durable physiological expression in maize.

Conclusion

With MAS based gene pyramiding, it is now possible for the breeder to conduct many rounds of selections in a year. Gene pyramiding with marker technology can integrate into existing plant breeding programmes all over the world to allow researchers to access, transfer and combine genes at a rate and with a precision not previously possible. However, lot of problems still persists in this field. Some of the difficulties encountered have to do with the need to have better scoring methods, larger population sizes, multiple replications and environments, appropriate quantitative genetic analysis, various genetic backgrounds and independent verification through advanced generations (Young et al., 1999). However taking into account the number of ongoing experiments and the explosion of new molecular technology, it is not surprising that new or improved selection schemes are being developed and applied as in case of maize (Ribaut et al., 2001). This will help breeders get around problems

related to larger breeding populations, replications in diverse environments, and speed up the development of advanced lines. Furthermore improved scoring methods and screening techniques can be developed and implemented, and much better choices about target traits can be made.

New technological developments such as automation, allele-specific diagnostics and diversity array technology (Jaccoud et al., 2001) will make MAS based gene pyramiding more powerful and effective. The main problem in front us to find out the most suitable way to use the genome information for biological intrigues including MAS based gene pyramiding. Any development in plant breeding is measured in terms of the contribution made to improvement in food production. Therefore plant breeders must be convinced on the advantages of MAS based gene pyramiding to implement it successfully in breeding programs. Recent success in MAS based gene pyramiding indicates that success was met because a good choice of target traits was made, information on the mode of inheritance was available, protocols to integrate MAS based gene pyramiding technology into breeding programs were developed with a multidisciplinary effort.

We have no doubt that MAS based gene pyramiding has the potential to increase the rate of genetic gain when used in conjunction with traditional breeding and the adoption of MAS by cereal breeders in Australia and the subsequent commercialization of pyramided lines of cultivars bred is testimony to this. The feasibility of gene pyramiding has been demonstrated, especially for pyramiding disease resistance genes, not only at one place but at several institutes in India like Punjab Agricultural University (PAU), Central Rice Research Institute (CRRI), University of Agricultural Sciences (Bangalore), Indira Gandhi Agricultural University (IGAU), and Tamil Nadu Agricultural University (TNAU). This was achieved more or less independent of plant breeders and mostly in well adapted varieties. Plant breeders simultaneously came up with new varieties that may be higher yielding, and hence the pyramided lines did not find their way to the farmers` fields even though they yield at par with the recurrent parents. The big question lies ahead is how to make MAS based gene pyramiding operational in the developing world to get maximum benefit from it. Some possible options are;

Since MAS is expensive and breeding programmes are mostly funded by the local governments, the national governments can start some MAS based gene pyramiding projects with committed funding. In India the Indian Council of Agricultural Research (ICAR) has already taken the initiative and MAS based gene pyramiding projects are successfully undergoing in rice, maize, wheat etc. This has been an integral part of the breeding programme and not just any other backcross programme.

Breeders are not much excited about gene pyramiding

for simply inherited traits, and not many QTL (especially the productivity related ones) with tightly linked markers are available. This will take some more time, especially the productivity related QTL from the wild species germplasm, to become available to breeders. However, with development and access to reliable PCR based markers like SSPs and SNPs in several crop plants, efficiency of pyramiding large populations or breeding materials has significantly increased. QTL pyramiding requires using better scoring methods, appropriate quantitative genetic analysis, and independent verifications through parallel populations. Appropriate DNA markers should be used at a definite stage to maximize the efficiency of MAS.

ACKNOWLEDGEMENTS

We gratefully acknowledge the financial support from Siksha O Anusandhan University. We thank Prof. M.R. Nayak, President, Siksha O Anusandhan University for his able guidance and support.

REFERENCES

Barloy D, Lemoine J, Paulette A, Tanguy M, Roger R, Joseph J (2006). Marker-assisted pyramiding of two cereal cyst nematode resistance genes from *Aegilops variabilis* in wheat. Mol. Breed. 20: 31-40.

Bates SL, Zhao J, Roush RT, Shelton AM (2005). Insect resistance management in GM crops: past, present and future. Nat. Biotechnol. 23: 57-62.

Beavis WD (1998). In: Molecular Dissection of Complex Traits. (Ed. Paterson AH). (CRC Press, Boca Raton, FL). pp 145-162.

Bentur JS, Pasalu IC, Sarma NP, Prasada RU, Mishra B (2003). DRR paper series 01/2003. Directorate of Rice Research, Hyderabad, p 20.

Chen XM, Luo YH, Xia HC, Xia LQ, Chen X, Ren ZL, He ZH and Jia JZ (2005). Chromosomal location of powdery mildew resistance gene Pm16 in wheat using SSR marker analysis. Plant Breeding 124:(3) 225-228.

Cox TS, Raupp WJ, Gill BS (1993). Leaf rust-resistance genes, Lr41, Lr42 and Lr43 transferred from Triticum tauschii to common wheat. Crop Sci. 34: 339-343.

Ferry N, Edwards MG, Mulligan EA, Emami K, Petrova AS, Frantescu M, Davison GM, Gatehouse AMR (2004). Engineering resistance to insect pests. In: Christou P, Klee H (eds). Handbook of Plant Biotechnology. Vol. 1. John Wiley and Sons, Chichester, pp. 373-394.

Francia E, Tacconi G, Crosatti C, Barabaschi D, Bulgarelli D, Dall'Aglio E, Vale G (2005). Marker assisted selection in crop plants. Plant. Cell. Tissue. Organ. Cult. 82: 317-342.

Gahan LJ, Ma YT, Cobble MLM, Gould F, Moar WJ, Heckel DG (2005). Genetic basis of resistance to Cry 1Ac and Cry 2Aa in *Heliothis virescens* (Lepidoptera: Noctuidae). J. Econ. Entomol. 98: 1357-1368.

Herdt RW. Research Priorities for rice Biotechnology. In: Khush, G.S (eds) Rice Biotechnology. International Rice Research Institute, Manila pp. 19-54

Hittalmani S, Foolad M, Mew T, Rodriguiz R, Huang N (1994).Identification of blast resistance gene Pi-2 (t) in rice plants by flanking DNA markers. Rice. Genet. News. 11: 144-146.

Hittalmani, S, Parco A, Mew TV, Zeigler RS and Huang N (2000). Fine mapping and DNA marker-assisted pyramiding of the three major genes for blast resistance in rice. Theor. Appl. Genet. 100: 1121-1128.

Holland JB (2004). Implementation of molecular markers for quantitative

traits in breeding programs - challenges and opportunities. In: Proceeding of the 4th International Crop Science Congress. Brisbane, Australia.

Huang N, Angeles ER, Domingo J, Magpantay G, Singh S, Zhang Q, Kumaravadivel N, Bennett J and Khush GS (1997). Pyramiding of bacterial blight resistance genes in rice: marker-assisted selection using RFLP and PCR. Theor. Appl. Genet. 95: 313-320.

Jaccoud D, Peng K, Feinstein D Kilian A (2001). Diversity arrays: a solid state technology for sequence information independent genotyping. Nucleic. Acid. Res. 29(4):1-7.

Jackson RE, Bradley JR and Van Duyn JW (2003). Field performance of transgenic cottons expressing one or two Bacillus thuringiensis endotoxins against bollworm, Helicoverpa zea (Boddie). J. Cotton Sci. 7: 57-64.

Johnson GR, Mumm RH (1996). Marker assisted maize breeding. Proceedings of the 51st Ann. Corn & Sorghum Res. Conf., Chicago, IL (Amer. Seed Trade Assoc.).

Katiyar S, Verulkar S, Chandel G, Zhang Y, Huang B and Bennet J (2001). Genetic analysis and pyramiding of two gall midge resistance genes (Gm 2 and Gm 6t) in rice (Oryzae sativa L.). Euphytica. 122: 327-334.

Kumar A, Jain A, Sahu RK, Srivastava MN, Nair S, Mohan M (2005). Genetic analysis of resistance genes for the rice Gall Midge in two Rice Genotypes. Crop Sci. 5:1631-1635.

Kumaravadivel N, Uma MD, Saravanan PA, Suresh H (2006). Molecular marker-assisted selection and pyramiding genes for gall midge resistance in rice suitable for Tamil Nadu Region. In: ABSTRACTS-2nd International Rice Congress 2006. pp 257.

Lande R, Thompson R (1990). Efficiency of MAS in the improvement of quantitative traits.Genetics.124: 743-756.

Liu J, Liu D, Tao W, Li W, Wang S, Chen P, Cheng S and Gao D (2000). Molecular marker-facilitated pyramiding of different genes for powdery mildew resistance in wheat. Plant Breeding 119 (1): 21–24.

Liu J, Liu D, Tao W, Li W, Wang S, Chen P, Cheng S Gao D (2000). Molecular marker-facilitated pyramiding of different genes for powdery mildew resistance in wheat. Plant Breeding. 119: (1)16-22.

Manyangarirwa W, Turnbull M, McCutcheon GS Smith JP (2006). Gene pyramiding as a Bt resistance management strategy: How sustainable is this strategy. African. J. Biotech. 5(10): 81-785.

Narayanan NN, Baisakh N, Vera Cruz CM, Gnanamanickam SS, Datta K, Datta SK (2002). Molecular breeding for the development of Blast and Bacterial Blight resistance in Rice cv.IR50.Crop Sci. 42: 2072-2079.

Parker GD, Langridge P (2000). Development of a STS marker linked to a major locus controlling flour colour in wheat (Triticum aestivum L.).Mol. breed.6:169-174.

Purcell JP, Oppenhuizen M, Wofford T, Reed AJ, Perlak FJ (2004). The story of Bollgard. In: Christou P, Klee H (eds). Handbook of Plant Biotechnology. Vol. 2. John Wiley and Sons, Chichester, pp. 1147-1163.

Ribaut JM, Jiang C and Hoisington D (2001). Simulation experiments on efficiencies of gene introgression by backcrossing. Crop Sci.42: 557-565.

Robert VJM (2001). Marker assisted introgression of black mold resistance QTL alleles from wild Lycopersicon cheesmanii to cultivated tomato (L. esculentum) and evaluation of QTL phenotypic effects. Mol. Breed. (8): 217-223.

Sanchez AC, Brar DS, Huang N, Li Z, Khush GS (2000). Sequence Tagged Site Marker-Assisted Selection for Three Bacterial Blight Resistance Genes in Rice. Crop Sci. 40: 792-797.

Santra D, DeMacon VK, Garland-Campbell K, Kidwell K (2006). Marker-assisted backcross breeding for simultaneous introgression of stripe rust resistance genes yr5 and yr15 into spring wheat (triticum aestivum l.). In 2006 international meeting of ASA-CSSA-SSSA. pp74-75.

Schneider A (2002). Mapping of a nodulation loci sym9 and sym10 of pea. Theor. Appl. Genet. (104): 1312-1316.

Schneider KA, Brothers ME, Kelly JD (1997). Marker-assisted selection to improve drought resistance in common bean. Crop Sci. 37: 51-60.

Shelton AM, Zhao J, Roush RT (2002). Economic, ecological, food safety and social consequences of the deployment of Bt transgenic plants. Ann. Rev. Entomol. 47: 845-881.

Singh S, Sidhu JS, Huang N, Vikal Y, Li Z, Brar DS, Dhaliwal HS, Khush GS (2001). Pyramiding three bacterial blight resistance genes (xa5, xa13 and Xa21) using marker-assisted selection into indica rice cultivar PR106. Theor. Appl. Genet. 102: 1011-1015.

Stuber CW (1998). In: Molecular Dissection of Complex Traits (Ed. Paterson AH). (CRC Press, Boca Raton, FL). pp 197-206

Tanksley SD, Grandillo S, Fulton TM, Zamir D, Eshed Y, Petiard V, Lopez J (1996). Advanced backross QTL analysis in a cross between an elite processing line of tomato and its wild relative L. pimpinellifolium. Theor. Appl Genet. 92: 213-224.

Wang GL, Mackill DJ, Bonman JM, McCouch SR, Champoux MC and Nelson RJ (1994). RFLP mapping of genes conferring complete and partial resistance to blast in a durably resistance rice cultivar. Genet. 136: 1421-1434.

Weeden NR, Muehlbauer FJ, Ladizinsky G (1992). Extensive conservation of linkage relationships between pea and lentil genetic maps. J. Hered. 83: 123-129.

Werner K, Friedt W, Ordon F (2005). Strategies for pyramiding resistance genes against the barley yellow mosaic virus complex (BaMMV, BaYMV, BaYMV-2). Mol. Breed.16: 45-55.

Wu KM, Guo YY (2005). The evolution of cotton pest management practices in China. Ann. Rev. Entomol. 50: 31-52.

Young ND (1999). A cautiously optimistic vision for marker-assisted breeding. Molecular Breeding. 5: 505-510.

Yousef GG, Juvik JA (2001). Comparison of phenotypic and marker-assisted selection for quantitative traits in sweet corn. Crop Sci. 41: 645-655.

Zeigler RS, Tohme J, Nelson J, Levy M Correa F (1994). Linking blast population analysis to resistance breeding: a proposed strategy for durable resistance. In: Zeigler R S, Leong S, Teng P S (eds) Rice blast disease. CAB Int, Wallingford, UK. pp 267-292.

3

Application of genomic technologies to the improvement of meat quality in farm animals

Hamed Kharrati Koopaei* and Ali Esmailizadeh Koshkoiyeh

Department of Animal Science, Shahid Bahonar University of Kerman, Iran.

Meat quality is one of the most important economic traits in farm animals. The goal of genomics technologies is to provide genetic map and other resources to identify loci responsible for genetic variation in quantitative traits such as meat quality. Candidate gene and genome scanning are two main techniques for this purpose. In the past decade, advances in molecular genetics led to identify these genes and markers linked to them. Sequencing of animal genome is important to distinguish gene function and molecular basis of meat quality determinants. Candidate gene considers relationship between the traits of interest and known genes, while genome scanning studies relationship between traits and pre-mapped markers. So far, several genes and sequences were detected which affect meat quality, for example quantitative trait loci (QTL) on chromosome 18 in sheep which causes muscle hypertrophy. The aim of this review is introduce and applications of genomic technologies to the improvement of meat quality.

Key words: Genomics technologies, meat quality, candidate gene, genome scanning.

INTRODUCTION

Meat quality is one of the most important economic traits in farm animals. Meat quality trait has a multifactorial background and is controlled by an unknown number of quantitative trait loci (QTL). Genome research in farm animals progressed rapidly in recent years, moving from linkage maps to genome sequence. The work on farm animal genome sequencing began in the early 1990s, and assists in the understanding of genomics function in various organisms (Fadiel et al., 2005). Genomic technologies are combination of different branches of genetics molecular genetics, quantitative genetics, Mendelian genetics and bioinformatics science. This approach can be useful for improvement of meat quality in animal breeding (Kharrati et al., 2009). In the past

decade, advances of genomic technologies using linkage mapping and DNA sequencing are manifold. The information of the meat quality trait loci can be applied in breeding programs by using marker-assisted selection (MAS) (Gao et al., 2007).

The goal of genomic technologies is the characterization and mapping of the loci that affected these traits. The main outcome of genomics technologies is the determination of physical effect gene on phenotype, physiology and biochemistry of them.

Generally, meat quality depended on color, water keeping, tenderness and resistance against oxidation. It is influenced by several factors, such as breed, genotype, feeding, fasting, pre slaughter handling, stunning, slaughter methods, chilling and storage conditions (Rosenvold and Andersen, 2003).Improvements of meat quality traits with traditional breeding programs are very difficult, because heritability of them is very low. A lot of work has been carried out in this field to find potential

---

*Corresponding author. E-mail: hmd_kh_ko@yahoo.com.

genes or chromosome regions associated with the meat quality trait in different farm animals, such as pig, cattle, sheep and chicken (Gao et al., 2007).

GENOMICS TECHNOLOGIES

Candidate gene approach

The candidate gene approach studies the relationship between the traits and known genes that may be associated with the physiological pathways underlying the trait (Liu et al., 2008). In other words, this approach assumes that a gene involved in the physiology of the trait could harbor a mutation causing variation in the trait. The gene or part of gene, are sequenced in a number of different animals, and any variation found in the DNA sequences, is tested for association with variation in the phenotypic trait. This approach has had some success. For example a mutation was discovered in the estrogen receptor locus (ESR) which results increased litter size in pig. There are 2 problems with the candidate gene approach.

Firstly, there are usually a large number of candidate genes affecting the trait, so many genes must be sequenced in several animals and many association studies carried out in a large sample of animals.

Secondly, the causative mutation may lie in a gene that would not have been regarded prior as an obvious candidate for this particular trait. Candidate gene approach is performed in 5 steps: 1) collection of resource population. 2) Phenotyping of the traits. 3) Selection of gene or functional polymorphism that potentially could affect the traits. 4) Genotyping of the resource population for genes or functional polymorphism.

Lastly, one is statistical analysis of phenotypic and genotypic data (Da, 2003). This is an effective way to find the genes associated with the trait. So far a number of genes have been investigated. Candidate gene approach has been ubiquitously applied for gene disease research, genetic association studies, biomarker and drug target selection in many organisms from animals (Tabor et al., 2002).

The traditional candidate gene approach is largely limited by its reliance on existing knowledge about the known or presumed biology of the phenotype under investigation, unfortunately the detailed molecular anatomy of most biological traits remain unknown (Zhu and Zhao, 2007).

Recently, a new developing method on candidate gene approach [digital candidate gene (DigiCGA) has emerged and is primarily applied to identify potential candidate gene in some studies. DigiCGA, which also named in silico candidate gene approach or computer facilitated candidate gene approach, is a novel web resource based candidate gene identification approach. DigiCGA approach is included to ontology–based identification approach, computation-based identification approach and integrated identification approach (including literature-based meta-analysis).

The ontology-based identification approach is mainly involved in the bioinformatics analyses for in silico identification of candidate gene for specific interest in case of the semi structured, structured and controlled vocabularies for systematic annotation of gene function information from biological ontology source available through internet. A typical example of this approach is the prioritization of positional candidate gene by using gene ontology (Harhay and Keele, 2003).

The computation-based identification includes those computational candidate gene identification methods that describe computational framework to prioritize the most likely candidate gene through a variety of web resource-based data sets. There are many statistical algorithms or computational methods, and of which some included data mining analysis (Perez et al., 2002), hidden Markov analysis (Pellegrini-Calace et al., 2006), cluster analysis (Freudenberg and Propping, 2002) and kernel-based data fusion analysis (De Bie et al., 2007). There have been reported many candidate gene prioritized by the integrated identification approach such as pathway and gene ontology combined analysis (Tiffin et al., 2005).

Genome scan approach

The genome scan approach studies the relationship between a trait and markers selected across the genome to identify chromosomal locations associated with the trait (Andersson, 2001). The genome scan will find out the map location of a trait locus with major effect, It involves the following steps: 1) design and construction of resource population, 2) phenotyping traits of resource population, 3) selection of genetic markers, 4) genotyping of the population for selected markers, 5) construction of linkage maps, and 6) statistical analysis of the phenotypic and genotypic data derived from the resource population (Da. 2003).

Using the genome scan, a large amount of QTL can be obtained in farm animals that can provide a useful bridge to link genome information with phenotype. There is an animal QTL data base, which contains all publicly available QTL data on farm animal species for the past decade (Hu et al., 2007). Several groups have worked on the identification of QTL controlling meat quality and most of them are about meat pork quality. QTL are located on almost every porcine chromosome, for instance in pig, there are 12 types of meat quality and total 1405 QTL for meat quality, such as 595 QTL for anatomy, 25 QTL for conductivity, 64 QTL for fat composition, 18 QTL for chemical, 1 QTL for enzyme activity, 439 QTL for fatness, 5 QTL for odor, 79 QTL for meat color, 26 QTL for flavor, 3 QTL for stiffening, 66 QTL for PH and 84 QTL for texture (http.www.animal genome.org/QTLdb/pig html).

In other studies, quantitative trait locus genome scans was performed for porcine muscle fiber traits. In this research, it was reported that a complete QTL scan of muscle fiber trait in 160 animal from a F_2 cross between Iberian and Landrace pigs using 139 markers and identified 20 genomes regions distributed along 15 porcine chromosome (1, 2, 3, 4, 6, 7, 8, 9, 10, 11, 12, 13, 14, 15 and X) with direct and epistatic effect. Epistatic was frequent and some interactions were highly significant. Chromosome 10 and 11 seemed to behave as hubs; they harbored 2 individual QTL, but also 6 epistatic regions. Numerous individual QTL effects had cryptic alleles, with positive effects to phenotypic pure breed difference (Estelle et al., 2002).

Genome scans for QTL affecting carcass trait is another application for this approach for improvement of meat quality in farm animals. For instance, genome wide scans were performed for QTL affecting carcass traits in Hereford x composite double backcross populations. Genome wide scan for chromosomal regions influencing carcass traits was conducted spanning 2.413 morgans on 29 bovine autosomes using 229 microsatellite markers.

Phenotypes measured at harvest were: carcass weight, fat depth, marbling, percentage kidney, pelvic and heart fat and rib eye area. The result indicated promising location for QTL affecting live weight on BTA 17 and marbling on BTA 2 that segregate in *Bos taursus* (MacNeil and Grosz, 2002). Much research was performed to map QTL for carcass composition and meat quality in sheep. For instance, a partial genome scan to map QTL was performed for carcass composition by X-ray computer tomography, and meat quality traits in Scottish Blackface sheep. The population studied was a double backcross between lines of sheep divergently selected for carcass lean content (LEAN and FAT lines), comprising nine half-sib families.

Carcass composition (600 lambs) was assessed nondestructively using computerized tomography (CT) scanning, while meat quality measurements (initial and final pH of semi-membraneous, color, shear force value, carcass weight, lamb flavor, juiciness, tenderness and overall liking) were taken on 300 male lambs. Lambs and their sires were genotyped across candidate regions on chromosomes 1, 2, 3, 5, 14, 18, 20 and 21. QTL analyses were performed using regression interval mapping techniques. In total, nine genome-wise was significant and 11 chromosome-wise and suggestive QTL were

detected in seven out of eight chromosomes. Genome-wise significant QTL were mapped for lamb flavor on sheep chromosome 1 (OAR1); for muscle densities (OAR 2 and OAR3); for color (redness) (OAR3); for bone density (OAR 1); for slaughter live weight (OAR 1 and OAR 2) and for the weights of cold and hot carcass (OAR 5). The QTL with the strongest statistical evidence affected the lamb flavor of meat and was on OAR 1, in a region homologous with a porcine chromosome 13 (SSC13) QTL identified for pork flavor. This QTL was segregated in four of the nine families.

This study provides new information on QTL affecting meat quality and carcass composition traits in sheep, which may lead to novel opportunities for genetically improving these traits (Karamichou et al., 2006). Although, a genome scan can give full genome coverage for a trait, it will fail to detect trait loci with smaller effects if they do not reach the stringent significance of the threshold (Gao et al., 2007).

Fine mapping

Fine mapping involves the identification of markers that are very tightly linked to a targeted gene. A genetic fine map of a specific locus will usually give us its goal by the identification and location of marker that flank the targeted gene within one or fewer cent Morgan (cM) with a fine map, and assisted selection precise. Several problems are associated with the generation of fine map around any animal gene.

First, mapping marker to a resolution of one cM requires the investigation of a few hundred sexual progeny. Second, marker polymorphism can be low in mapping population and third, recombination rate per chromosome arm in sexual generation. The ultimate goal of genome scan approach is to identify the genes that underlie polygenic trait and gain better understanding of their physiological and biochemical functions. In fact a region of QTL often spans 5 to 30 cM and it is too large to find the target genes, so fine mapping needs to be done. It is a step towards restricting the region of interest and the number of potential candidate genes. The goal of fine mapping is mapping QTL to a narrow chromosome region so that the physical QTL affecting the phenotype can be identified and cloned (Gao et al., 2007).

Finally fine mapping a gene is usually an essential step in map-based gene isolation. For example, fine mapping quantitative trait loci for body weight and abdominal fat trait in chicken was performed. Highly significant QTL for body weight and abdominal fat trait on chicken chromosome 1 were reported. This research genotyped 9 more microsatellite markers, including 6 novel ones. Linkage analyses were performed. The result of the linkage analyses showed that the confidence intervals for body weight and abdominal fat percentage were narrowed sharply to a small interval spanning 5.5 and 3.7 Mb, respectively (Harhay and Keele, 2003).

IMPORTANT GENES AFFECTING MEAT QUALITY IN FARM ANIMALS

Cattle

DGAT1 gene

The possible association between the activity of diacylglycerol acyltransferase (DGAT) and muscle fat content was examined by Sorensen et al. (2006) in samples of longissimus dorsi (LD) and semitendinosus (ST) muscles from Holstein and Charolais bulls. The Holstein bulls exhibited higher fat content in both muscles and higher marbling score. In Holstein, DGAT activity was enhanced in the LD muscle, and there was tentative positive relationship between DGAT activity and the fat content in ST muscle. When muscle DGAT activity was examined as a function of DGAT genotype for all animals, regardless of breed, the DGAT activity of LD muscle of the K/K genotype was about five-fold greater than either the K/A or A/A genotypes (Sorensen et al., 2006).

Calpastatin gene

Corva et al. (2007) showed that the activity of the calpastatin proteolytic system is closely related to the postmortem tenderization of meat. The association between beef tenderness and single nucleotide polymorphism (SNP) markers on the CAST gene 3' untranslated region (SNP2870, alleles A/G) was investigated. Samples were provided from nine slaughter groups comprising 313 which had been reared in beef production systems in Argentina between 2002 and 2004 from crosses between Angus, Hereford and Limousine cattle. The results indicated no detectable effects were demonstrated for meat shear in the CAST marker. Body weight, carcass weight and rib eye area were not affected by any of the markers (Corva et al., 2007).

Calpain gene

Single nucleotide polymorphism (SNP) in exon 9 of calpain (CAPN1) is related to tenderness of longissimus dorsi (LD) and semitendinousus (ST) muscles in Bos taursus. The SNP causes amino acid substitution of alanine for glycine in the μ - calpain enzyme. Results

demonstrated that the exon 9 SNP is significantly associated with the tenderness of both ST and LD muscles (Esmailizadeh et al., 2005).

Myostatin gene

Effects of the myostatin F94L substitution was studied by Esmailizadeh et al. (2008a) on beef traits. The experiment used crosses between the Jersey and Limousine, with the design being a backcross using first cross bulls of Jersey$_x$Limousine or Limousine$_x$Jersey breeding, mated to Jersey and Limousine cows. The progeny was genotyped for the myostatin SNP and phenotyped. The SNP is a cytosine to adenine transversion in exon 1, causing an amino acid substitution of leucine for phenylalanine (F94L). They reported that the F94L allele in Limousine backcross calves was associated with an increase in meat weight and reduction in fat depth on live calves (600 days) and carcasses. Meat tenderness, pH and cooking loss of the longissimus dorsi (LD) were not affected by the F94L variant.

The results provided strong evidence that this myostatin F94L variant provides an intermediate and more useful phenotype than the severe double-muscling phenotype caused by knockout mutation in the myostatin gene (Esmailizadeh et al., 2008a).

Other genes

In a research by Esmailizadeh et al. (2005), QTL were detected for meat color and pH in Bos Taurus cattle. An experimental cattle backcross between the Jersey and Limousine breeds was per- formed in Australia and New Zealand to QTL for diverse production traits. Six crossbreed sires and their progeny were genotyped for 253 informative microsatellite markers covering the 29 bovine autosomes. Results of the genome scan using regression interval mapping revealed evidence for QTL on BTA 10,18,19 and 27 for meat color and BTA2,3,5,6,11,12,13,16,24 and 27 for meat pH. A number of detected QTL were mapped to genomic regions likely to contain the RN or RYR1 genes, which are know to affect meat quality traits in pigs (Esmailizadeh et al., 2005).

In other research by Esmailizadeh, et al. (2008b), 2 QTL were detected affecting beef tenderness, one of them located on the Bovine chromosome 2 and other on Bovine chromosome 29 close to the map position of the growth differentiation factor 8(GDF8) and calpain genes, respectively. They showed molecular dissection of these

QTL and indicated significant association between meat tenderness and SNP 316 in the CAPN1 gene and SNP 433 in the GDF 8 gene.

In this study, three parental half-sib families comprising 357 animals were genotyped for 189 microsatellite DNA marker. Meat tenderness was measured as Warner-Bratzler shear force (Esmailizadeh et al., 2008).

Chicken

In chicken, many investigations focused on fat deposition, such as the percentage of hypodermal fat, abdominal fat, and intramuscular fat in breast and legs (Gao et al., 2007). The intramuscular- fat (IMF) was in positive correlation with meat flavor and succulence, especially tenderness of meat. Increasing IMF and controlling fatty deposition is an increased interest in improving - meat quality.

Extracellular fatty acid binding protein (EX-FABP) gene

Fattiness is an important parameter to estimate meat quality, which has high heritability. Wang et al. (2001) showed, EX-FABP gene could be a candidate locus or linked to a major gene to significantly affect abdominal fat traits in chicken. In this experiment, F2 chickens derived from Broilers crossing to Silky were used to study the effect of EX-FABP gene on abdominal fat accumulation. Then, single nucleotide polymorphisms (SNPs) were detected by the technique of single strand conformation polymorphism (SSCP) and confirmed by sequencing. The results of least square analysis suggested that the birds with BB genotype have a higher abdominal fat weight and abdominal fat percentage than the birds with the other genotypes (AA and AB) (Wang et al., 2001).

Liver fatty acid-binding protein (L-FABP) gene

Wang et al. (2006) studied association between L-FABP gene and abdominal fat weight and percentage of abdominal fat. Fatty acid-binding proteins belong to a super family of lipid-binding proteins that exhibit a high affinity for long-chain fatty acids and appear to function in metabolism and intracellular transportation of lipids. In this research, study was designed to investigate expression characterization and association with growth and composition traits of the L-FABP gene in the chicken.

The results indicated that the L-FABP gene polymorphisms were associated with abdominal fat weight and percentage of abdominal fat, and the L-FABP gene could be a candidate locus or linked to a major gene(s) that affects fatness traits in the chicken. The results of the current study provided basic molecular information for studying the role of the L-FABP gene in the regulation of lipid metabolism in avian species (Wang et al., 2006).

Sheep

Callipyge (CLPG1) gene

Frenking et al. (2002) reported that a small genetic region near the telomere of ovine chromosome 18 was previously shown to carry the mutation causing the callipyge muscle hypertrophy phenotype in sheep. Expression of this phenotype is the only known case in mammals of paternal polar overdominance gene action. A region surrounding two positional candidate genes was sequenced in animals of known genotype. Mutation detection focused on an inbred ram of callipyge phenotype postulated to have inherited chromosome segments identical-by-descent with exception of the mutated position. Initial functional analysis indicated sequence encompassing the mutation is part of a novel transcript expressed in sheep fetal muscle (Frenking et al., 2002).

Pig

Calpastatin (CAST) gene

Suggestive QTL affecting raw firmness scores and average tenderness, juiciness, and chewiness on cooked meat were mapped to pig chromosome 2 using a three-generation intercross between Berkshire and Yorkshire pigs. Based on its function and location, the calpastatin (CAST) gene was considered to be a good candidate for the observed effects. Several misses and silent mutations identified in CAST and haplotypes covering most of the coding region were constructed and used for association analyses with meat quality traits. Results demonstrated that one CAST haplotype was significantly associated with lower Instron force and cooking loss and higher juiciness and therefore, this haplotype is associated with higher eating quality (Ciobanu et al., 2004).

Heart fatty acid-binding protein (H-FABP) gene

Gerbens et al. (1999) studied the relationship between variation in the heart fatty acid-binding protein (H-FABP) gene and (IMF) content. To estimate the effect of H-FABP, pigs from two Duroc populations were selectively mated in such a way that at least two genotypes were present in each litter. In total, data from 983 pigs and pedigree information from three preceding generations were analyzed. Offspring were tested for IMF content as well as back-fat thickness (BFT), body weight (BW) and drip loss of the meat (DRIP). All pigs were assigned to H-FABP Restriction fragment length polymorphism (RFLP) genotype classes either by the assessed genotype (75%) or based on a probability score determined according to genotypic information of their relatives (25%). Contrasts were detected between homozygous H-FABP RFLP genotype classes for IMF content (0.4%, $P < 0.05$), BFT (0.6 mm, $P < 0.01$), and BW (2.4 kg, $P < 0.01$). No significant contrasts were detected for DRIP.

Results for IMF content, BFT, and BW were confirmed when only genotyped animals were analyzed. H-FABP RFLP can be used as markers to select for increased IMF content and growth in breeding programs (Gerben et al., 1999).

Adipocyte fatty acid-binding protein: A-FABP gene (FABP4)

In the first intron of the porcine A-FABP gene, a microsatellite sequence was detected by Gerbens et al. (1998) that was polymorphic for all six pig breeds tested. This genetic variation within the A-FABP gene was associated with differences in IMF content and possibly growth in a Duroc population, whereas no effect on back-fat thickness and drip loss of the meat was detected. A considerable and significant contrast of approximately 1% IMF was observed between certain genotype classes. It was concluded that the A-FABP locus is involved in the regulation of intramuscular fat accretion in Duroc pigs (Garbsen et al., 1998).

REFERENCES

Andersson L (2001). Genetic dissection of phenotypic diversity in farm animals. Nat. Genet., 2: 130–138.

Corva P, Soria L, Schor A, Villarreal E, Cenci MP, Motter M, Mezzadra C, Depetris LM, Miquel C, Paván EG, Santini F, Naon JG (2007). Association of CAPN1 and CAST gene polymorphisms with meat tenderness in Bos Taurus beef cattle from Argentina. Genet. Mol. Biol., 30: 1064-1069.

Ciobanu DC, Bastiaansen JWM, Lonergan SM, Thomsen H, Dekkers JCM, Plastow GS, Rothschild MF (2004). New alleles in calpastatin gene are associated with meat quality traits in pigs. J. Anim. Sci., 82: 2829-2839.

Da Y (2003). Statistical analysis and experimental design for mapping genes of complex traits in domestic animals. Bioinformatics, 30(12): 1183–1192.

De Bie T, Tranchevent LC, Oeffelen LM (2007). Kernel-based data fusion for gene prioritization. Bioinformatics, 23: 125-132.

Esmailizadeh KA, Bottema CDK, Sellick GS, Verbyla AP, Morris CA, Cullen NG, Pitchford WS (2005). Association of the exon 9 single nucleotide polymorphism of CAPN1 with beef tenderness. Young Scientists Pap., 4: 258-261.

Esmailizadeh KA, Bottema CDK, Sellick GS, Verbyla AP, Morris CA, Cullen NG, Pitchford WS (2008a). Effects of the myostatin F94L substitution on beef traits. J. Anim. Sci., 86: 1038-1046.

Esmailizadeh KA, Foladi MH, Mohammadabadi MR (2008b). Molecular dissection of beef tenderness using DNA technology. The 5th National Biotechnology Congress of Iran. Tehran, Iran, pp. 25-29.

Estelle j, Gill F, Vázquez MJ, Latorre R, Ramírez G, Barragán MC, Folch JM, Noguera MA (2002). A quantitative trait locus genome scan for porcine muscle fiber traits reveals overdominance and epistasis. J. Anim. Sci., 86: 3290-3299.

Fadiel A, Anidi I, Eichenbaum KD (2005). Farm animal genomics and Informatics: An update. Nucleic Acids. Res., 33(19): 6308–6318.

Freudenberg J, Propping P (2002). A similarity-based method for genome-wide prediction of disease-relevant human genes. Bioinformatics, 18: 110-115.

Frenking BA, Murphy SK, Wylie AA, Rhodes SJ, Keele JW, Leymaster KA, Jirtle RL, Smith TP (2002). Identification of the single base change causing the callipyge muscle hypertrophy phenotype, the only known example of polar overdominance in mammals. Genome. Res., 10: 1496-1506.

Gao Y, Zhang R, Hu X, Li N (2007). Application of genomic technologies to the improvement of meat quality of farm animals. Meat Sci., 77: 36-45.

Gerbens F, Van Erp AJ, Harders FL, Verburg FJ, Meuwissen TH, Veerkamp JH, Pas MFW (1999). Effect of genetic variants of the heart fatty acid-binding protein gene on intramuscular fat and performance traits in pigs. J. Anim. Sci., 77: 846-852.

Gerbens F, Jansen A, van Erp AJ, Harders F, Meuwissen THE, Rettenberger G, Veerkamp JH, Pas MFW (1998). The adipocyte fatty acid-binding protein locus: Characterization and association with intramuscular fat content in pigs. Mamm. Genome, 9: 1022–1026.

Harhay GP, Keele JW (2003). Positional candidate gene selection from livestock EST databases using gene ontology. Bioinformatics, 19: 249-255.

Hu ZL, Fritz ER, Reecy JM (2007). AnimalQTLdb: A livestock QTL database tool set for positional QTL information mining and beyond. Nucleic Acids. Res., 35: 604-609.

Kharrati Koopaei H, Pasandideh M, Esmailizadeh KA (2009). Genomic technologies and meat quality. The 2nd Agriculture Biotechnologies Conference, Kerman. Iran, pp. 156-160.

Karamichou E, Richardson RI, Nute GR, McLean KA, Bishop SC (2006). A partial genome scan to map quantitative trait loci for carcass composition, as assessed by X-ray computer tomography, and meat quality traits in Scottish Blackface Sheep. Brit. Soc. Anim. Sci., 82: 301-309.

Liu x, Zhang H, li H, li N, Zhang Y, Zhang Q, Wang S, Wang Q, Wang H (2008). Fine mapping quantitative trait loci for body weight and abdominal fat traits: Effects of marker density and sample size. Poult. Sci., 87: 1314-1319.

MacNeil MD, Grosz MD (2002). Genome-wide scans for QTL affecting carcass traits in Hereford x composite double backcross populations. J. Anim. Sci., 80: 2316-2324.

Pellegrini-Calace M, Tramontano A (2006). Identification of a novel putative mitogen-activated kinase cascade on human chromosome 21 by computational approaches. Bioinformatics, 22: 775-778.

Perez-Iratxeta C, Bork P, Andrade MA (2002). Association of genes to genetically inherited diseases using data mining. Nat. Genet., 31: 316-319.

Rosenvold K, Andersen HJ (2003). Factors of significance for pork quality a review. Meat Sci., 64(3): 219–237.

Sorensen B, Kuhan C, Teuscher F, Schneider F, Weselake R, Wegner J (2006). Diacylglycerol acyltransferase (DGAT) activity in relation to muscle fat content and DGAT1 genotype in two different breeds of Bos Taurus. Arch. Tierz., 49: 351-356.

Tabor HK, Risch NJ, Myers RM (2002). Candidate-gene approaches for studying complex genetic traits: Practical considerations. Nat. Res. Genet., 3: 391-397.

Tiffin N, Kelso JF, Powell AR (2005). Integration of text- and data-mining using ontologies successfully selects disease gene candidates. Nucleic Acids Res., 33: 1544-1552.

Wang Q, Ning L, Xuemei D, Zhengxing L, Hui L, Changxin W (2001). Single nucleotide polymorphism analysis on chicken extracelluar fatty acid binding protein gene and its associations with fattiness trait. Sci. China, 4(44): 429-434.

Wang Q, Li N, Li L, Wang Y (2006). Tissue Expression and Association with Fatness Traits of Liver Fatty Acid-Binding Protein Gene in Chicken. Poult. Sci., 85: 1890–1895.

Zhu M, Zhao S (2007). Candidate gene identification approach: Progress and challenges. Int. J. Biol. Sci., 3(7): 420-427.

4

Analysis and manipulation of the genome dynamic structure

Valentina Tosato* and Carlo V. Bruschi

Yeast Molecular Genetics Laboratory, International Centre for Genetic Engineering and Biotechnology, AREA Science Park – W, Padriciano 99, IT-34012 Trieste, Italy.

After the genomic era, during which DNA sequencing revealed genes and the post-genomic era, in which their functional analysis was implemented, the notion of a dynamic genome has become convincing. Indeed, since the early days of DNA transposition, new evidence has accumulated indicating a high level of intrinsic structural plasticity characterizing the genome. An ensemble of gross chromosomal rearrangements has been reported, together with their biological effects, some of which correlate to major cellular pathologies such as cancer. From microorganisms to human cells, the convoluted architecture of chromosomes has gained relevance not only from a descriptive point of view, but also as a potential multi-layered storage mechanism of genetic information. The higher-order structure of DNA, including hairpin turns, bending and curvature, as well as precise chromatin topology, could provide the metadata on super-information needed to explain the low number of inferred mammalian genes, perhaps conferring new scientific dignity to the infamous „junk DNA". In this view, genome dynamics, including the clustering of essential genes and the synteny of others, appears as the paramount cellular response to varying environmental conditions, resulting in massive differential regulation of gene expression. A deeper understanding of the various orders of complexity of genomic DNA structure has allowed the design of more sophisticated biochemical and biophysical tools for its analysis and manipulation, which, in turn, has yielded a better knowledge of the genome itself. This creative cycle is providing new generations of diagnostics and intelligent drugs of pharmacogenomic origin that exploit the ever changing, yet stable, genome dynamic structure.

Key words: Genome dynamics, secondary DNA structure, gross chromosomal rearrangements

Table of contents

INTRODUCTION

Historical facts

In 1912, the William Braggs father and son discovered the atomic structure of crystals, by X-ray diffraction patt-

ern, which provided the key for Watson, Crick, Franklin and Wilkins to visualize the DNA architecture just 40 years later. However, there was a then fashionable belief that the molecular structure of this macromolecule was a relatively stiff, static construction. Barbara Mc Clintock (Nobel Prize) had a prescient recognition of genomic plasticity prior to 1952, which took years to be generally accepted by the wider scientific community. By 1970 when Hamilton Smith isolated the first restriction enzyme, the idea that the genomic DNA could be manipulated and recombined *in vitro* began to be considered and assessed. In the eighties the plasticity of the genome was notable largely because it constituted a drawback in the taxonomic definition of the species of kinetoplastidae, such as *Trypanosoma* and *Crithidia*, the hyper-variability of few genetic loci impairing their classification by karyotyping. High frequency of genome rearrangements, due to repeated sequence elements, had already been described in archaea and other microbes like *Rhizobium* (Martinez et al., 1990; Flores et al., 2005). *Helicobacter pylori*, for another instance, is known to have such a high plasticity resulting in important differentiation among pathogenic strains. Real insight into the importance of genomic rearrangements however, came as a consequence of the enormous success of genome sequencing and functional analysis projects. Today, many researchers believe that DNA winds and unwinds or undergoes frequent melting of its double helical structure, generating single-strandedness at particular regions, then re-annealing to its normal status, an entropic phenomenon nicknamed "breathing". Among other things, this dynamic structure of the genome allows exogenous and endogenous DNA fragments to be integrated, regulates gene transcription by a topological control of the double-helix, and may lead to differentiation of undifferentiated cells, as well as to dedifferentiation (cancer) through gross chromosomal rearrangements (GCRs).

Mathematical models provide *in silico* predictions for chromatin organization, bendability and secondary structures

One of the most fascinating and fast-moving fields in genome dynamics studies is the mathematical prediction and interpretation of the biological significance of DNA sequences. Nonlinear models have been proposed to explain the complex dynamics of DNA bubbles occurring in denaturation, transcription and genetic recombination events (Komarova and Soffer, 2005). Particularly during recombination events, the rate of the movement of the crossover junction has been estimated by branch migration assays (Karymov et al., 2005; McKinney et al., 2005). Also, a stochastic differential equation has been applied to the dynamic feature of the B-Z DNA transition (Lim and Feng, 2005). The free energy spent by topoisomerase to bind and bend DNA has been calculated, computing the twist and the writhe of supercoiling (Barbi et

---

*Corresponding author. E-mail: tosato@icgeb.org.

al., 2005; Charvin et al., 2005). These models should help to elucidate how the genome movements following its entropic breathing may interact with the complex regulation of replication, repair and transcription mechanisms.

Here, we will focus on the detailed *in silico* analyses that have been performed on chromatin remodeling, DNA bending and genomic secondary structures. It is not yet completely clear, for example, how the repair mechanism (either homologous recombination - HRS - or non-homologous end joining - NHEJ) gains access to DNA within chromatin and how chromatin structure is restored after repair. Perhaps, physical disruptions of the double helix such as double strand breaks (DSBs) are detected through an altered chromatin topology, which transmits a signal to be sensed by the cell at a checkpoint. The checkpoint system then arrests the cell cycle until the break is repaired (for a review see Ehrenhofer-Murray 2004). Effective modification of histone H2A following a double strand break formed by the HO endonuclease was recently characterized at the molecular level in yeast (Shroff et al., 2004). This analysis suggests that a break of the chromosomal DNA leads to rearrangements of chromatin and re-positioning of the repair protein complex. Moreover, chromatin-remodeling factors are responsible for ploidy control. Indeed, a mutation in these genes results in an increase of genomic instability and it is usually associated with cancer induction (Vries et al., 2005). The majority of these experimental evidences are predicted by exhaustive comparative *in silico* analysis of chromatin remodeling. Genome-wide mapping predicts that chromatin accessibility and gene expression are controlled by histone modifications of regulatory elements such as enhancers, locus control regions (LCRs) and insulators. In favor of this prediction, islands of acetylation have been identified within promoters and highly transcribed regions (Roh et al., 2005). Unconventional chromatin assays (digital analysis of chromatin structure-DACS) coupled with sophisticated algorithms allow precise predictions of regulatory sequences and the accessibility of their chromatin (Sabo et al., 2004).

The *in silico* analysis of scaffold/matrix attachment sites (S/MARs), of the nucleosome-formation propensity, and of the presence of repeats within regions located upstream in human genes, brings into sight how much chromatin architecture may modify the transcription activity of the genome (Ganapathi et al. 2005). The analysis of chromatin features allows discrimination between housekeeping and tissue-specific genes, attributing to chromatin a super-informative function with respect to the genetic level. This super-information provided by chromatin is consistent with the effective low number of regulatory and specialized genes in the human genome. Unlike the case of polymorphism of the double helix (A, B or Z) structure, which is quite rare *in vivo*, the genomic DNA usually shows micro-polymorphisms in the B helical structure. The concept that DNA may be bent, especially in A-rich tracts, was proposed initially by Trifonov and

Curvature in the ORFs of the *B. subtilis* genome

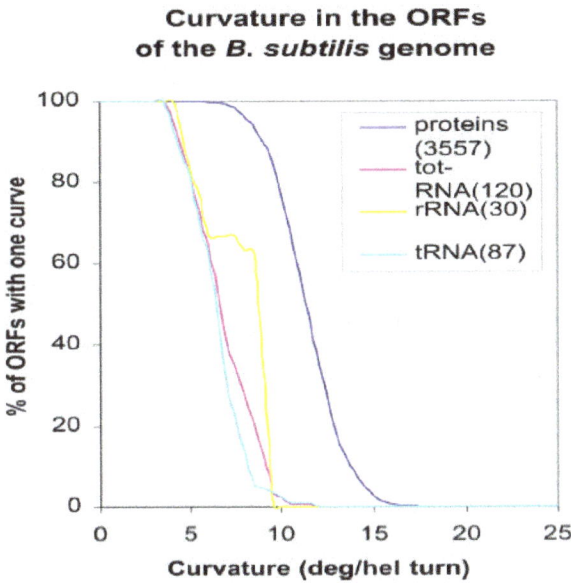

Figure 1 Percentage. of ORFs overlapping with curved motifs in the *B. subtilis* genome. The total number of protein, RNA, rRNA and tRNA ORFs is given in parenthesis.

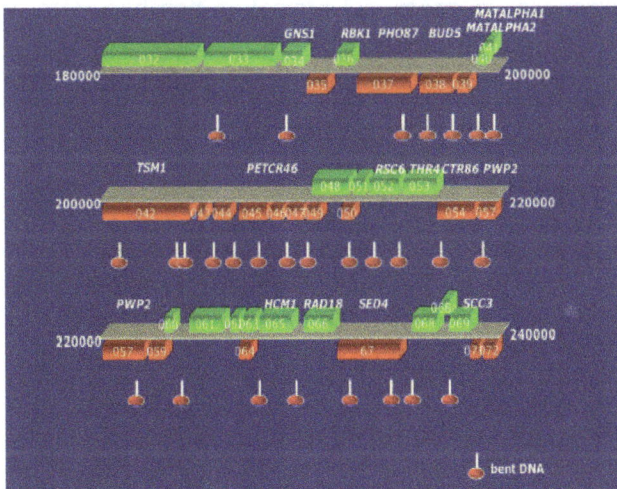

Figure 2. Position of significant (>14 degrees/ helical turn) curvature peaks along 60,000 nucleotides of the DNA sequence of yeast *Saccharomyces cerevisiae* chromosome III. The green and the red numbered boxes represent ORFs.

Sussman (1980) who observed that the natural anisotropy of the DNA molecule facilitates its smooth folding into chromatin. The curvature of DNA is measured in terms of angle of deflection between adjacent base pairs (degrees/helical turn). Several servers running bendability prediction programs have been developed to be applied to experimental systems, looking for a correlation between curvature and other features of the genomes (Vlahovicek et al., 2005).

As mentioned, AT-rich stretches in the "B" form usually correlate with high curvature. This is the case for all 16 centromeric DNA regions of *Saccharomyces cerevisiae* (Bechert et al., 1999). The structure of the centromere may correlate with the binding, assembly and the function of the chromosome determining element (CDE) protein complex, which regulates centromeric activity. To examine the effect of the A-T tracts or curvature, it is possible to simulate the molecular dynamics of the "B"-structure (McConnell and Beveridge 2001). Furthermore, there is a preference for curved segments to be distributed with respect to start and stop codons. Usually, intergenic regions such as promoters and terminators show a propensity to be curved. An example of that is reported in Figure 1 where only 6.2% of all ORFs contain one significant curvature (>14 degrees/helical turn) in the model bacterium *Bacillus subtilis* (Tosato et al., 2003).

An example of distribution of curvature around open reading frames in chromosome III of yeast is shown in Figure 2 where the exact DNA sequence and its corresponding coding and non-coding regions were initially identified following the genomic sequencing of the first eukaryotic chromosome from yeast (Oliver et al., 1992).

This non-uniform distribution of curvature correlates with a different propensity of specific regions of genomic DNA to spontaneously recombine. The promoter and terminator sequence are usually more prone to accept DNA integration by recombination events (Gjuracic et al., 2004). This phenomenon is significantly parallel to the nonrandom distribution of DSB regions along the chromosomes (Baudat and Nicolas, 1997). Hotspots for recombination via DSBs are known in all the chromosomes of yeast (Gerton et al., 2000) and many of them contain palindromic sequences (Nasar et al., 2000). The presence of secondary structures as palindromes may influence the recombination efficiency of a specific genomic region that thus becomes a fragile site in terms of breaks and recombination.

A decreasing gradient of large DNA hairpins from the origin towards the terC end of chromosomal replication as well as a high curvature in the intergenic regions characterizes the genome of *B. subtilis* (Tosato et al. 2003). Indeed, the lack of secondary structures around terC (with the exception of the two hairpins of the replication terminus) is in agreement with the low level of homologous recombination detected in this region (Chedin et al., 1998). This observation supports the idea that DNA secondary structures may trigger recombination (Lobachev et al., 1998). Higher order structures may also contain a different kind of super-information in pathogenic species. *Kinetoplastida* parasites, for example, are responsible for diseases severely affecting human health and retarding agriculture development in third world countries. *Kinetoplastida* species cause sleeping sickness (*Trypanosoma brucei*), Chagas (*Trypanosoma cruzei*) and leishmaniasis (*Leishmania* spp.). More than 400 million of people around the world are affected by

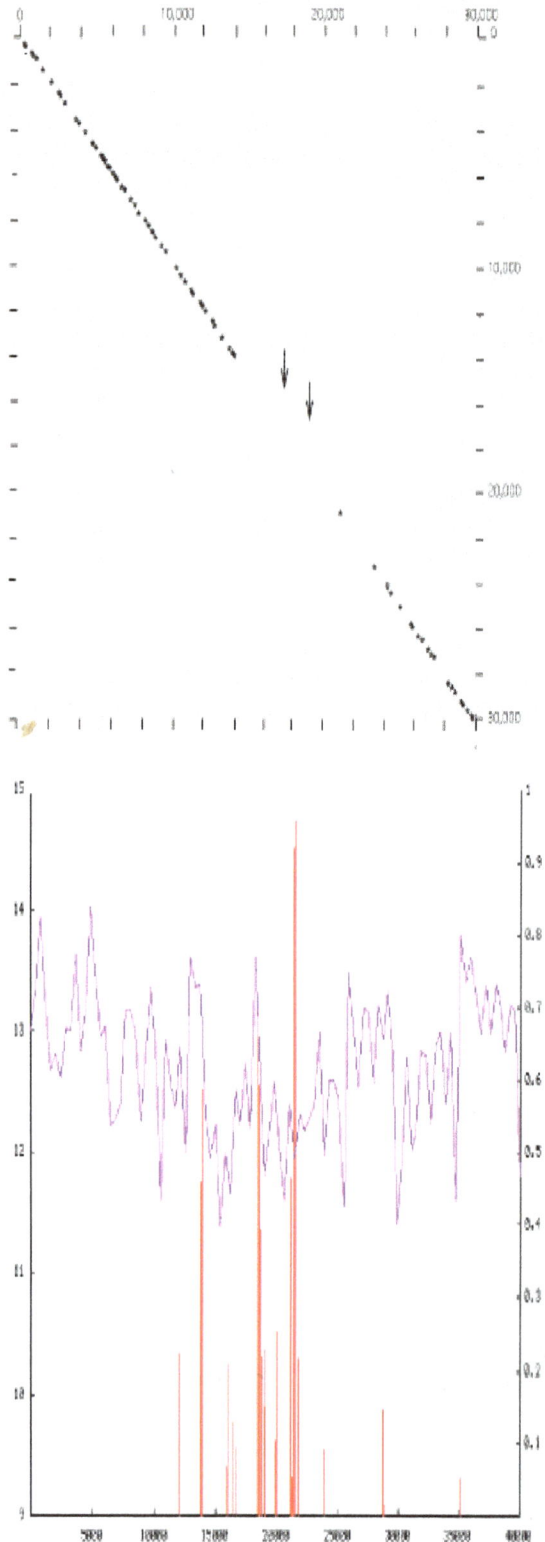

Figure 3. Distribution of hairpins, Top, and bendability, bottom, of the DNA sequence encompassing the switching point of *Lieshmania major* chromosome I. In the top panel, the arrows indicate the extremes of the switching point DNA and the dots the position of the hairpins. In the bottom panel, the wavy, violet line represents the G+C content while the vertical red bars indicate the predicted DNA curvature.

these pathologies, especially in the developing Countries. The parasites share a unique organization of protein-coding genes into long, strand-specific, polycistronic clusters and a lack of general transcription factors (Ivens et al., 2005). Large surface antigen families occur at non-syntenic chromosome-internal and sub-telomeric regions.

The regions between poly-cistronic clusters, where the direction of transcription switches from one DNA strand to the other, have been hypothesized to contain origins of replication and possibly also centromeres and promoters. The base skews of divergent strand-switches of *Trypanosoma* species show patterns analogous to those for bacterial origins of replication (Nilsson and Andersson, 2005), but they differ from those of *Leishmania major*. In *Leishmania* chromosomes, for example, there are several coding strand switch regions that show peculiar characteristics. An *in silico* analysis suggests that there is a trend towards decreased number and stability of hairpins in the direction of the switching points. Moreover, within a few kilobases on either side of the coding-strand switch locus, the predicted secondary structures disappear (Tosato et al., 2001). Some of the results are summarized in Figure 3. Other features of the switch regions are a high AT content and a strong intrinsic curvature. The functional meaning of these regions as bi-directional promoters has been later confirmed experimentally (Martinez-Calvillo et al., 2003). Whether the switching point coincides with a centromeric region has not yet been resolved. Using a telomere-associated chromosome fragmentation approach, it was shown that the region required for the mitotic stability of chromosome 3 of *T. cruzei* encompasses a transcript-tional switching point domain constituted of a GC-rich island. This region contains several retrotransposon-like insertions but, atypically, lacks the arrays of satellite repeats normally associated with centromeric regions (Obado et al., 1995).

Switching points are an example of junk DNA in lower eukaryotes, confirming the idea that the genetic code is only one of the keys we need to decipher the genome meaning. The super-information that is packed within higher-order structures, such as secondary DNA structures, bendability and chromatin topology, may contribute to the as yet unknown significance of this junk DNA.

New biochemical and biophysical tools for the analysis and manipulation of genomic DNA

The extremely rapid recent development of a large assortment of new genomic tools has frustrated their cataloguing and renders incomplete any review of the field. Nevertheless, we can summarize the most important new tools as examples of the technical sophis-tication reached by genomic analytical methodology in this field. In doing this, we must recognize that the notion of the dynamic genome has not yet influenced, in gene-ral, the strategies for discovery of new tools, neither their biochemical, or their biophysical characteristics.

Artificial chromosomes

Yeast and Bacterial Artificial Chromosome (YAC and BAC)-based technology has been extensively used in the subcloning, of large genomic DNA for the construction of genomic libraries essential to sequencing projects (Zhang and Win, 1997; Bruschi et al., 2006). Together with advanced gene-targeting techniques, the construction of "YAC mega-cassettes (YMCs)" has proven efficient for the disruption of entire mouse genetic loci, and their replacement with equivalent human DNA regions in murine embryonic stem (ES) cells. This was performed to create transgenic mice with specific modifications of genes, enabling them to produce only antigen-specific humanized antibodies (Ledermann, 2000). Although this approach offered new experimental opportunities, there are considerable technical limitations that make it difficult for routine use. The major obstacle is generally the low frequency of gene targeting in mammalian cell lines, ranging up to 10^{-6}/cell for most loci studied. The main reasons for this gene targeting inefficiency lay in: 1) the inherently refractory behaviour of the mammalian genome to mitotic homologous recombination; 2) the high frequency of competing repair pathways; and 3) the random DNA integration into chromosomal sites by illegitimate recombination, which outnumbers gene targeting by about 1000:1. Some enrichment in gene-targeted events could be obtained by using targeting vectors with both positive and negative selection markers or with "trap-vectors" containing modified selection markers which are expressed only in the case of vector integration into particular DNA sequences (promoter- and polyadenylation-trapping; Bradley et al. 1992). Another approach to increase the gene targeting frequency is based on the transient expression of the endogenous or foreign recombinational enzymes (Yanez and Porter, 1999; Scherbakova et al., 2000), resulting in a significant increase of targeted events.

However, retroviral targeting vectors have another limiting factor, namely the length of the DNA fragment homologous to the target locus. Due to the size limit that viral vectors can accommodate, the ideal length of DNA inserts ranges from 5 - 10 kb. This relatively small size of the homology between vector and target DNA influences not only the targeting frequency but also the size of the disruption/replacement that can be obtained (several tens of kb; Tsuda et al., 1997).

An alternative to retroviral targeting vectors could be provided by bacteriophage λ and P1 vectors, which can carry up to 90 kb of inserted DNA (Nehls et al. 1994; Brüggemann and Neuberger 1996). However, their cloning potential is still far below the size of many mammalian genes, which limits their use. Accommodation of even larger DNA fragments and their stable pro-

pagation in yeast *S. cerevisiae* was achieved by construction of yeast artificial chromosomes with megabase-long DNA inserts from several YAC clones bearing overlapping DNA inserts, through the exploitation of the efficient and accurate yeast recombination machinery (Silverman et al., 1990, Deunen et al., 1992). This characteristic of yeast also allows an introduction of subtle changes within cloned DNA and introduction of new selectable markers (retrofitting) for transfer of the modified YAC into mammalian cell lines. Many human loci have been transferred so far to murine ES cells using modified YAC vectors, and have been subsequently efficiently expressed in the transgenic mice tissues (Mendez et al.; Peterson, 1997). However, YACs were not used for improvement of gene targeting, but rather as vectors for the introduction of large human genomic DNA fragments into ES cell lines. To avoid heterogeneous expression of transgenic and endogenous loci, the cell line employed must have endogenous DNA region(s) silenced or, alternatively, the YAC-derived transgenic mouse need to be subsequently crossed with appropriate knock out strains. The main disadvantage of this approach lies in time-consuming process to obtain the desired genetic change in the mouse and in the necessarily prolonged culture passages that reduce dramatically the pluripotency of the ES cells, thus minimizing chances to produce transgenic animals. Within the category of artificial chromosomes, Mammalian Artificial Chromosomes (MACs) did not fare as well in the past, due to the difficulty to clone and perpetuate very large DNA molecules containing human centromeric regions that, *per se*, can reach the size of 1 megabase. These biotechnological instruments, therefore, were used only in particular, specific cases in which the genetic elements to be manipulated were already known and isolated from the whole genome. More recently though, mammalian artificial chromosome technology has been further improved and rendered almost as handy as BACs and YACs. Indeed, one class of MACs, the mammalian satellite DNA-based Artificial Chromosome Expression (ACE) systems can be replicated *de novo* in cell lines of different species and efficiently separated from the host chromosomes (Bunnell et al., 2005).

Chromosome knockout

A major manipulation of the genome of a eukaryotic cell is, without any doubt, the elimination of one of its chromosomes. This event dramatically changes genome homeostasis as well as overall chromosome dynamics during meiotic division and mitotic segregation. Historically, chromosome loss has been achieved in eukaryotes by treating cells with chemical and physical agents that interfere with spindle formation and dynamics, with the result that this kind of loss is random and occurs with additional toxic cellular effects (Esposito et al., 1982;

Howlett and Schiestl, 2000). Monosomic yeast strains have been obtained in cell division cycle (*cdc*) mutant strains defective in gene products playing an essential role in DNA synthesis and chromosome segregation (Bueno and Russel, 1992; Bruschi et al., 1995; Storici et al., 1995). Moreover, a specific chromosome loss achieved by 2-μ DNA plasmid integration was reported many years ago (Falco and Botstein, 1983). In another report, the *GAL1* promoter was integrated immediately upstream of a centromere. Transcription of *GAL1* interfered with the centromere function, thus inducing chromosome instability (Guacci and Kaback, 1991). However, these methods are very elaborate and yield contradictory results. The establishment of randomly monosomic mammalian cell lines obtained through the inactivation of DNA topoisomerase II, which plays an important role in mitotic chromosome disjunction (Clarke and Gimenez-Albian, 1999) has highlighted the importance of DNA topology and dynamics in the stability of the whole genomic architecture.

More recently, selective chromosome V loss by the deletion of the corresponding centromere in yeast diploid strain, using the standard EUROFAN knockout technology, has been reported (Zang et al., 2002). The experimental approach was to substitute the centromere region of interest with a linear, double-stranded DNA cassette containing a specific yeast selectable marker (the *LEU2* gene in that case) flanked by two 40-bp DNA sequences homologous to the target peri-centromeric regions. In this way, the substituted chromosome is replicated during DNA synthesis, but it is transferred casually to the daughter cell at each cell division, instead of systematically, because of the lack of the kinetochore-anchoring complex for the mitotic spindle fibers. The result is the segregation of a cell line lacking one chromosome. However, it has further been demonstrated that, either immediately or after a certain number of cell generations, the cells endoreduplicated the remaining homologous chromosome. This restores the normal euploidy, but fixes the Loss of Heterozygosity (LOH) for the markers previously in heterozygous configuration on the pair of chromosomes involved in the knock out.

Gross chromosomal rearrangements (GCRs)

The discovery of mutants with increased genomic instability suggested that chromosomal rearrangements normally are actively suppressed, probably by specific checkpoint genes (Kolodner et al. 2002). Transformation of yeast cells with a chromosomal fragmentation vector (CFV) resulted in the gain of a chromosomal fragment (CF) with or without the loss of the targeted chromosome, following DSB processing by break-copy duplication (Morrow et al., 1997). Later, it was demonstrated that a chromosomal DSB produced by the HO endonuclease could be repaired by break-induced-replication (BIR),

leading to non-homologous end joining (NHEJ)-mediated non-reciprocal translocation (Bosco and Haber, 1998). Recently, a *cre* site-specific recombination-based system has been developed to produce reciprocal translocations at pre-engineered *loxP* sites (Delneri et al., 2003). Finally, the HO system has been utilized for the production of a DSB on two chromosomes, resulting in reciprocal translocations by NHEJ (Yu and Gabriel, 2004). However, in these experimental yeast systems, cells needed to be previously engineered to obtain translocations at their modified chromosomal sites, and could not generate non-reciprocal translocations. Therefore, the technical difficulty of generating this type of gross chromosomal rearrangement (GCR) *in vivo* hinders the understanding of genome alterations and dynamics.

An experimental production of selectable translocants generated at desired chromosomal locations in wild-type yeast strains was developed (Tosato et al., 2005). Cells have been transformed with the Kan^R linear DNA cassette having the selectable marker flanked by two DNA sequences homologous to two different chromosomes. Using this BIT (Bridge Induced Translocation) system, induction of targeted non-reciprocal translocations in mitosis was achieved (Tosato et al., 2005). This *in vivo* approach to generate specific chromosomal translocations is the first step into the manipulative understanding of genome dynamics, to mimic the GCRs alterations responsible for many genetic diseases, including cancer.

Gene inactivation, DNA deletion, and integration technologies

Sophisticated genome manipulation requires the possibility of modifying any inter- or intragenic DNA sequence at will, without leaving large amounts of undesired vector DNA at the site of alteration. To this end, a long series of sophisticated vectors has been developed from the previous gene knockout plasmid systems, for example to integrate non-selectable foreign DNA at any desired genomic location in yeast, with a minimum amount of residual plasmid DNA. Some of these vectors have two mutated *Flp* recognition targets sequences (*FRTs*) of the endogenous 2-micron DNA plasmid for site-specific excision of the flanking *KanMX4* kanamycin-resistance gene. Outside the FRT boundaries, the plasmid carries multiple sites for subcloning of the DNA fragment to be integrated (Storici et al., 1999). Within the recyclable selection marker methodology, several other systems were available for gene disruption and replacement (Toh-e, 1995; Akada et al., 2002), as well as for epitope tagging of chromosomal genes (De Antoni and Gallwitz, 2000; Knop and Schiebel, 1997). However, none is capable of integrating at a specific locus, a desired DNA sequence having no directly detectable phenotype. The *FLP/FRT* system has now been improved by implementing a new advanced strategy for *in vivo* genomic DNA alterations. The new system, called STIK (Specific Targeted Integration of Kanamycin resistance-associated non-selectable DNA), allows the integration, at any genomic location, of DNA sequences that express no directly selectable phenotype, such as spacers, tags, nuclear localization signal sequence and any intergenic or otherwise heterologous sequence (Waghmare et al., 2003). The STIK system accomplishes this task by exploiting the temporary integration of the recyclable, positively selectable *KanMX4* marker. This selectable marker can be recycled by *Flp* site-specific excision between two identical *FRT*s, for the integration of further DNA fragments. This chaperone-like plasmid system provides for a new molecular tool to "stik" (integrate) any DNA fragment at any genome location in yeast strains. Moreover, the system can be extrapolated to other eukaryotic cells in which the *FLP/FRT* system functions efficiently.

Genome microarray imaging

One of the latest technologies in the field of genome analysis is the development of bioinformatics programs able to perform the manipulation of microarray images and the identification of known biological relationships among sets of genes (Zimmer et al., 2004). This technology lends itself to online consultation and interactive utilization for the re-ordering of microarray map data through genome image alignment programs that can be run or downloaded from http://coli.berkeley.edu/genomeimages/ and other URL sites.

From bacteria to eukaryotes: the distribution of genes as a consequence of evolution

Despite the idea that the highly expressed genes are positioned on the leading DNA strand to allow faster replication and transcription, today it is believed that, at least in bacteria, gene essentiality drives gene-strand bias, (Rocha and Danchin, 2003), while lethality can be provoked not only by deletion, but also by strand switching or scrambling of essential genes. In this view, the genetic organization in bacteria may represent a kind of chromosomal hyper-structure that regulates cellular segregation (Rocha et al., 2003). These signs of chromosomal organization are in apparent contrast to the loss of genetic order due to gross chromosomal rearrangements (Rocha, 2004). The idea that scrambling may be useful for adaptation and evolution partially justifies this possible conflict. In eukaryotes, frequent recombination events leading to massive chromosomal rearrangements, may be necessary in meiosis to generate differentiations, but represent a mistake in mitosis. Gene distribution and genomic evolution are driven by spontaneous segmental duplication or aneuploidization.

The size of the putative segments that undergo duplications varies from 41 kb to 268 kb (Koszul et al., 2006).

One of the proposed molecular mechanisms used to explain duplication, is a fork collapse-generated DSB followed by intra or inter-molecular BIR. These spontaneous events are then fixed in a cellular population by positive selection. Adaptive sweeps are associated with the fixation of these duplicated loci (Moore and Purugganan, 2003). Around 15% of genes in the human genome, and up to 20% in *Drosophila, S. cerevisiae* and *C. elegans* are believed to arise from duplication events. For these organisms, the average half-life of a duplicate gene is approximately 4 million years. A conservative estimate of the average rate of origin of new gene duplicates is on the order of 0.01 per gene per million years, with rates in different species ranging from about 0.02 down to 0.002. (Lynch and Conery, 2000). Genomic duplication is now accepted as the major evolutionary mechanism providing a dispensable copy of any gene upon which selection can act, shaping the modifications compatible with life by tests that do not eliminate the failing mutations. Moreover, the knowledge of the existence of tandem chromosomal duplications in pathogens (Brosh et al., 2000) will help in the monitoring the altered immunogenicity of few pathogenic strains and their use as vaccines while new strategies of genomic manipulations such as insertion-duplication mutagenesis may contribute to the identification of potential therapeutic targets (Opperman et al., 2003).

Homologous linear DNA molecules are bridging life to death via GCRs

Recombination is an essential step in meiosis to generate genomic variability, and genomic variability is necessary for speciation. Chromosomal rearrangements may have contributed to some of the speciation processes along the human and mouse lineages (Marques-Bonet and Navarro, 2005). Nevertheless, massive chromosomal rearrangements in meiosis may also lead to genetic diseases. In mitosis, where strict checkpoint mechanisms avoid excessive recombination among repeated elements and paralogs, the rearrangements are an occasional accident caused by unexpected Double Strand Breaks (DSBs). A replication fork barrier can generate the DSBs, following a chemical or physical damage (irradiation, methyl methanesulfonate-MMS exposure) or following an integration of an exo/endogenous DNA. In all these cases, the break should be immediately repaired; if not, the cell undergoes apoptosis. Different molecular sentinels sense the break and activate variegate salvage pathways like Homologous Recombination (HR), Non Homologous End Joining (NHEJ), Single Stand Annealing (SSA) and others. In all of these pathways, two distinct classes of genes are involved: DNA repair genes and checkpoint genes. Mutations in the first class of genes give rise to severe diseases such as the *Xeroder-*

a

b

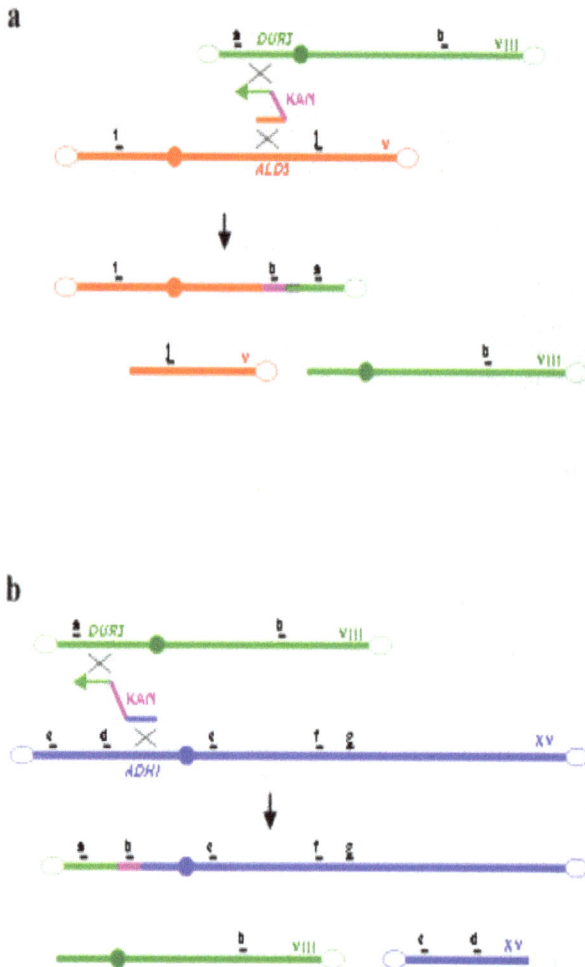

Figure 4. BIT generation of translocant chromosomes with two *Kan*^R DNA cassettes in the yeast *Saccharomyces cerevisiae.* a) selectable cassette with two ends homologous to the *ald5* and *dur3* loci on chromosomes V and VIII, respectively. b) Selectable DNA cassette with two ends homologous to the *dur3* and *adh1* loci on chromosome VIII and XV, respectively. Black letters indicate the positions of the strategic probes used for mapping the translocation by Southern hybridization. a = *apm2*, b = *crp1*, c = *cdc33*, d = *msh2*, e = *alg6*, f = *ade2*, g = *elg1*, h = kanamycin^R, i = *bud16*, j = *rad51*.

ma pigmentosum and the Hereditary-non-polyposis colorectal cancer. Alterations in the second class may lead, for instance, to Ataxia-telangiectasia, Nijmegen breakage syndrome, Bloom, Werner and Rothmund-Thomson syndromes. The two major pathways of repair are HR (which is more used in lower eukaryotes) and NHEJ (very frequent in mammalians, Kanaar et al., 1998). NHEJ occurs in mammals with a frequency of one event/10^2-10^4 cells vs 10^5-10^7 of HR. The NHEJ therefore represents a barrier to targeted integration in higher eukaryotes. It is believed that both pathways are active in the regulation of Gross Chromosomal Rearrangements, which are a natural consequence of DNA repair, since in the attempt to repair a DSB the cell generates random GCRs.

When a eukaryotic model cell such as yeast captures linear exogenous DNA with ends homologous to the genome, it integrates it by homologous recombination. If the homologous loci belong to two different chromosomes, a non-reciprocal translocation is generated with a frequency that depends upon the extension of the homology itself. For a tail of 40 nt, the frequency of integration is around 2%, while increasing the homology up to 70 nt, the frequency doubles becoming approximately 4.8%. This observation implies that an exogenous linear DNA may create chromosomal translocations exploiting the endogenous HR system of the cell. Thus, this bridge-induced translocation (BIT) event is responsible for the loss of heterozygosity that is usually associated with many types of cancer (Tosato et al. 2005, Figure 4). It has to be assessed whether these phenomena may also play a role during evolution. When a linear molecule of DNA enters into the nucleus, even if it shows homologous ends, generates a complex scenario of events which consists not only of targeted translocations, but also of ectopic integrations, non-specific translocations and intra-chromosomal deletions. Therefore, it can be assumed that the two main pathways of DNA repair, homologous recombination and the non-homologous end joining, are co-operating to avoid a persisting DNA break and that, in this way, their activity may cause severe genomic rearrangements.

The legacy of the genomic era is essential for vaccine development

Many infectious diseases that are endemic in developing Countries cannot be treated with an efficient vaccine. Drugs are usually available, but resistance is a recurrent event, especially in infections caused by protozoan parasites. The key obtained by deciphering the microbial genomes has also provided the possibility of developing new vaccines and drugs. Such genomic pioneering opened the curtain on a panorama of incredibly powerful technologies such as microarray, SAGE, cell-based drug design (CBDD) systems and more. When applied to parasites, these technologies highlighted new genes, relevant to the diseases, whose expression differs between the insect-borne stage and the human infectious stage (Duncan et al., 2004). Genome dynamics was observed by subtractive hybridization and microarray analysis also in *Yersinia pestis* (Wang et al., 2006; Zhou et al., 2004) and *Vibrio cholerae* (Dziejman et al., 2002) leading to the identification of the genes that correlate with bubonic/pneumonic plague and cholera. It was found that a surprising dynamics of the genome characterizes *Yersinia.* This is due to horizontal gene acquisition and genetic loss explaining the continuous bacterial evolution to deadly pathogenic lineages. By contrast, *V. cholerae* shows a high degree of conservation among different strains.

The advantage in terms of early diagnosis coming from genomic studies is well demonstrated for leishmaniasis. Knowledge of the genome (Ivens et al., 2005) allowed the development of a new PCR technique (gp63 PCR-RFLP) to discriminate among different *Leishmania* populations that are very close phylogenetically but not epidemiologically.

The elucidation of the dynamic of the microbial genomes helps in understanding the local sequences variations and the polymorphisms characterizing the different isolates. Therefore it is essential to characterize these genomic variations to design an appropriate vaccine. In rural areas of Tanzania the efficacy of different vaccines was tested on different patients bringing three subtypes of the HIV-1 strain (Arroyo et al., 2005). High-risk populations can provide an opportunity for the virus evolution, its recombination, and its adaptation to the host-specific genetic background (Herbinger et al., 2006). Sometimes the treatment of a disease with a vaccine may induce genomic changes and may help the persistence of the disease. One example is the wide spreading and reemergence of the endemic pertussis despite half a century of vaccination. Vaccination shifted the competitive balance between strains (Mooi et al., 2001). In particular the genes coding for two virulence factors, pertactin and pertussis toxin, varied with the adaptation of the microorganism to less react, acellular vaccines composed of purified *B. pertussis* proteins. DNA polymorphisms allowed the identification of three pertussis toxin variants. Five multi-locus sequence types (MLSTs) were found in different isolates of *B. pertussis* and their significant changes were observed after vaccination. The polymorphism in the pertussis toxin genes allowed the characterization of strains from widely separated geographic areas and their correlation with epidemics (van Loo et al., 2002).

The ability to manipulate genomes may be exploited to generate mutant collections without a revertible phenotype to be utilized as live attenuated vaccines. The manipulation of the genomic material bypasses by far the power of gamma-irradiation, long-term culture or chemical mutagenesis techniques. Technologies have been developed (Denise et al. 2004) to selectively delete a certain locus without leaving marker genes that mediate undesired antibiotic resistance to putative useful drugs. To understand the genome dynamics and the variations in gene expression of pathogens (also correlated with different stages of the parasites as for *Leishmania* spp.), several expression profilings were performed (Palacios et al., 2007; Leifso et al., 2007; Vora et al., 2005). They allowed the identification of strategies used by the pathogens for survival in the human host and the focusing on new candidate antigens for vaccine development. Few families that encode surface proteins in *Plasmodium falciparum* 3D7 (Daily et al., 2005) and several genes found in genomic regions with altered percentage of GC content in *Y. pestis* ((Lawson et al., 2006) are only few

among a multitude of putative targets suitable for vaccine development which have been highlighted through expression arrays. Therefore, from the epidemiological, diagnostic and technological point of view, exploitation of the genome dynamics potential helps in reading its informational message and understanding its meaning to a much deeper extent.

Conclusions

The enzymatic pathways that the cell activates in order to fix DNA injuries are not yet well understood, but represent the underlying molecular basis of genome plasticity. It is not completely clear, for example, if a crossing-over event of recombination or break-induced replication (BIR) machinery is the main mechanism responsible for the integration of a linear molecule into the genome. Because linear viral DNA integration as well as DNA transposition to naturally manipulate the genome had arisen during evolution, the elucidation of the molecular mechanism underlying this phenomenon will shed more light on how speciation occurred, and will foster the development of new drugs and vaccines. In this view, the dynamics of the genome structure represents still a crucial, unknown area of genetic information in which secondary DNA architecture warrants future investigations to better define its biological role.

REFERENCES

Akada R, Hirosawa I, Kawahata M, Hoshida H, Nishizawa Y (2002). Sets of integrating plasmids and gene disruption cassettes containing improved counter-selection markers designed for repeated use in budding yeast. Yeast 19: 393-402.

Arroyo MA, Hoelscher M, Sateren W, Samky E, Maboko L, Hoffmann O, Kijak G, Robb M, Birx DL, McCutchan FE (2005). HIV-1 diversity and prevalence differ between urban and rural areas in the Mbeya region of Tanzania. AIDS 19: 1517-1524.

Barbi M, Mozziconacci J, Victor JM (2005). How the chromatin fiber deals with topological constraints. Phys Rev E Stat Nonlin Soft Matter Phys. 71(3 Pt 1): 031910.

Baudat F, Nicolas A (1997). Clustering of meiotic double-strand breaks on yeast chromosome III. Proc Natl Acad Sci U S A 94: 5213-5218.

Bechert T, Heck S, Fleig U, Diekmann S, Hegemann JH (1999). All 16 centromere DNAs from *Saccharomyces cerevisiae* show DNA curvature. Nucleic Acids Res 27: 1444-1449.

Bosco G, Haber JE (1998). Chromosome break-induced DNA replication leads to nonreciprocal translocations and telomere capture. Genetics 150: 1037-1047.

Bradley A, Hasty P, Davis A, Ramirez-Solis R (1992). Modifying the mouse: design and desire. Biotechnology 10: 534-539.

Brosch R, Gordon SV, Buchrieser C, Pym AS, Garnier T, Cole ST (2000). Comparative genomics uncovers large tandem chromosomal duplications in *Mycobacterium bovis BCG Pasteur*. Yeast 17: 111-123.

Brüggemann M, Neuberger MS (1996). Strategies for expressing human antibody repertoires in transgenic mice. Immunol Today 17: 391-397.

Bruschi CV, Mcmilan JN, Coglievina M, Esposito M (1995). The genomic instability of yeast cdc6-1/cdc6-1 mutants involves chromosome structure and recombination. Mol. Gen. Genet 249: 8-18.

Bruschi CV, Gjuracic K, Tosato V. (2006). Yeast Artificial Chromosomes. In: Encyclopedia of Life Sciences. John Wiley & Sons, Ltd: Chichester, UK http://www.els.net/: doi:10.1038/npg.els.0006101.

Bueno A, Russell P (1992). Dual function of *CDC6*: a yeast protein required for DNA replication also inhibits nuclear division. EMBO J. 11: 2167-2176.

Bunnell BA, Izadpanah R, Lederbur Jr. HC, Perez CF (2005). Development of mammalian artificial chromosomes for the treatment of genetic diseases: Sandhoff and Krabbe diseases. Exp. Opin. Biol. Ther. 5:195-206.

Charvin G, Strick T, Bensimon D, Croquette V (2005). Topoisomerase IV bends and overtwists DNA upon binding. Biophys J. 89: 384-392.

Chedin F, Noirot P, Biaudet V, Ehrlich SD (1998). A five-nucleotide sequence protects DNA from exonucleolytic degradation by AddAB, the RecBCD analogue of *Bacillus subtilis*. Mol. Microbiol. 29: 1369-1377.

Clarke DJ, Gimenez-Abian JF (1999). Check points controlling mitosis. BioEssays 22:351-363.

Daily JP, Le Roch KG, Sarr O, Ndiaye D, Lukens A, Zhou Y, Ndir O, Mboup S, Sultan A, Winzeler EA, Wirth DF (2005). *In vivo* transcriptome of *Plasmodium falciparum* reveals overexpression of transcripts that encode surface proteins. J. Infect. Dis 191: 1196-203.

De Antoni A, Gallwitz D (2000). A novel multi-purpose cassette for repeated integrative epitope tagging of genes in *Saccharomyces cerevisiae*. Gene 246: 179-185.

Delneri D, Colson I, Grammenoudi S, Roberts IN, Louis EJ, Oliver SG (2003). Engineering evolution to study speciation in yeasts. Nature 422: 68-72.

Den Dunnen JT, Grootscholten PM, Dauwerse JD, Monaco AP, Walker AP, Butler R, Anand R, Coffey AJ, Bentley DR, Steensma HY, Van Ommen GJB (1992). Reconstruction of the 2.4 Mb human DMD-gene by homologous YAC recombination. Hum. Mol. Genet. 1: 19-28.

Denise H, Coombs GH, Mottram JC (2004). Generation of *Leishmania* mutants lacking antibiotic resistance genes using a versatile hit-and-run targeting strategy. FEMS Microbiol Lett 235: 89-94.

Duncan RC, Salotra P, Goyal N, Akopyants NS, Beverley SM, Nakhasi HL (2004). The application of gene expression microarray technology to kinetoplastid research. Curr. Mol. Med. 4: 611-621.

Dziejman M, Balon E, Boyd D, Fraser CM, Heidelberg JF, Mekalanos JJ (2002). Comparative genomic analysis of *Vibrio cholerae*: genes that correlate with cholera endemic and pandemic disease. Proc. Natl. Acad. Sci. U.S.A. 99: 1556-1561.

Ehrenhofer-Murray AE (2004). Chromatin dynamics at DNA replication, transcription and repair. Eur. J. Biochem. 271: 2335-2349.

Esposito MS, Maleas DT, Bjornstad KA, Bruschi CV (1982). Simultaneous detection of changes in chromosome number, gene conversion and intergenic recombination during mitosis of *S. cerevisiae*: Spontaneous and ultraviolet induced events. Curr. Genet. 6: 5-11.

Falco SC, Botstein D (1983). A rapid chromosome-mapping method for cloned fragments of yeast DNA. Genetics 105: 857-872.

Flores M, Morales L, Avila A, Gonzalez V, Bustos P, Garcia D, Mora Y, Guo X, Collado-Vides J, Pinero D, Davila G, Mora J, Palacios R. (2005). Diversification of DNA sequences in the symbiotic genome of *Rhizobium etli*. J. Bacteriol. 187: 7185-92.

Ganapathi M, Srivastava P, Das Sutar SK, Kumar K, Dasgupta D, Singh GP, Brahmachari V, Brahmachari SK (2005). Comparative analysis of chromatin landscape in regulatory regions of human housekeeping and tissue specific genes. BMC Bioinformat. 6: 126.

Gerton JL, DeRisi J, Shroff R, Lichten M, Brown PO, Petes TD (2000). Inaugural article: global mapping of meiotic recombination hotspots and coldspots in the yeast *Saccharomyces cerevisiae*. Proc. Natl. Acad. Sci. USA. 97: 11383-1390.

Gjuracic K, Pivetta E, Bruschi CV (2004). Targeted DNA integration within different functional gene domains in yeast reveals ORF sequences as recombinational cold-spots. Mol. Genet. Genomics 271: 437-446.

Guacci V, Kaback DB (1991). Distributive disjunction of authentic chromosomes in *Saccharomyces* cerevisiae. Genetics 127: 475-488.

Herbinger KH, Gerhardt M, Piyasirisilp S, Mloka D, Arroyo MA, Hoffmann O, Maboko L, Birx DL, Mmbando D, McCutchan FE, Hoelscher M. (2006) Frequency of HIV type 1 dual infection and HIV diversity: analysis of low- and high-risk populations in Mbeya Region, Tanzania. AIDS Res Hum Retroviruses 22: 599-606.

Howlett NG, Schlest RH (2000). Simultaneous measurement of the frequencies of intrachromosomal recombination and chromosome gain using the yeast DEL assay. Mutat. Res. 454: 53-62.

Ivens AC, Peacock CS, Worthey EA, Murphy L, Aggarwal G, Berriman M, Sisk E, Rajandream MA, Adlem E, Aert R, Anupama A, Apostolou Z, Attipoe P, Bason N, Bauser C, Beck A, Beverley SM, Bianchettin G, Borzym K, Bothe G, Bruschi CV, Collins M, Cadag E, Ciarloni L, Clayton C, Coulson RM, Cronin A, Cruz AK, Davies RM, De Gaudenzi J, Dobson DE, Duesterhoeft A, Fazelina G, Fosker N, Frasch AC, Fraser A, Fuchs M, Gabel C, Goble A, Goffeau A, Harris D, Hertz-Fowler C, Hilbert H, Horn D, Huang Y, Klages S, Knights A, Kube M, Larke N, Litvin L, Lord A, Louie T, Marra M, Masuy D, Matthews K, Michaeli S, Mottram JC, Muller-Auer S, Munden H, Nelson S, Norbertczak H, Oliver K, O'neil S, Pentony M, Pohl TM, Price C, Purnelle B, Quail MA, Rabbinowitsch E, Reinhardt R, Rieger M, Rinta J, Robben J, Robertson L, Ruiz JC, Rutter S, Saunders D, Schafer M, Schein J, Schwartz DC, Seeger K, Seyler A, Sharp S, Shin H, Sivam D, Squares R, Squares S, Tosato V, Vogt C, Volckaert G, Wambutt R, Warren T, Wedler H, Woodward J, Zhou S, Zimmermann W, Smith DF, Blackwell JM, Stuart KD, Barrell B, Myler PJ (2005). The genome of the kinetoplastid parasite, *Leishmania major*. Science 309: 436- 442.

Kanaar R, Hoeijmakers JH, van Gent DC. Molecular mechanisms of DNA double strand break repair. Trends Cell Biol. 1998 8: 483-489.

Karymov M, Daniel D, Sankey OF, Lyubchenko YL (2005). Holliday junction dynamics and branch migration: Single-molecule analysis. Proc Natl Acad Sci USA 102: 8186-8191.

Knop M, Schiebel E (1997). Spc98p and Spc97p of the yeast gamma-tubulin complex mediate binding to the spindle pole body via their interaction with Spc110p. EMBO J. 16: 6985-6995.

Kolodner RD, Putnam CD, Myung K (2002). Maintenance of genome stability in *Saccharomyces cerevisiae*. Science 297: 552-557.

Komarova NL, Soffer A (2005). Nonlinear waves in double-stranded DNA. Bull. Math. Biol. 67: 701-718.

Koszul R, Dujon B, Fischer G (2006). Stability of large segmental duplications in the yeast genome. Genetics 172: 2211-2222.

Lawson JN, Lyons CR, Johnston SA (2006). Expression profiling of *Yersinia pestis* during mouse pulmonary infection. DNA Cell Biol. 25: 608-616.

Ledermann B (2000). Embryonic stem cells and gene targeting. Exp. Physiol. 85: 603-613.

Leifso K, Cohen-Freue G, Dogra N, Murray A, McMaster WR (2007) Genomic and proteomic expression analysis of *Leishmania* promastigote and amastigote life stages: the *Leishmania* genome is constitutively expressed. Mol. Biochem. Parasitol. 152: 35-46.

Lim W, Feng YP (2005). Applying the stochastic difference equation to DNA conformational transitions: A study of B-Z and B-A DNA transitions. Biopolymers 78:107-20.

Lobachev KS, Shor BM, Tran HT, Taylor W, Keen JD, Resnick MA, Gordenin DA (1998) Factors affecting inverted repeat stimulation of recombination and deletion in *Saccharomyces cerevisiae*. Genetics 148: 1507-1524.

Lynch M, Conery JS (2000). The evolutionary fate and consequences of duplicate genes. Science 290: 1151-1151.

Marques-Bonet T, Navarro A (2005). Chromosomal rearrangements are associated with higher rates of molecular evolution in mammals. Gene 353:147-154.

Martinez E, Romero D, Palacios R (1990) The *Rhizobium* genome. Crit. Rev. Plant Sci. 9: 59-93.

Martinez-Calvillo S, Yan S, Nguyen D, Fox M, Stuart K, Myler PJ (2003) Transcription of *Leishmania major* Friedlin chromosome 1 initiates in both directions within a single region. Mol Cell 11:1291-1299.

McConnell KJ, Beveridge DL (2001). Molecular dynamics simulations of B'-DNA: sequence effects on A-tract-induced bending and flexibility. J. Mol. Biol. 314:23-40.

McKinney SA, Freeman AD, Lilley DM, Ha T (2005). Observing spontaneous branch migration of Holliday junctions one step at a time. Proc Natl Acad Sci U S A 102:5715-5720.

Mendez MJ, Green LL, Corvalan JRF, Jia XC, Maynard-Currie CE, Yang XD, Gallo ML, Louie DM, Lee DV, Erickson KL, Luna J, Roy CM, Abderrahim H, Kirschenbaum F, Noguchi M, Smith DH, Fukushima A, Hales JF, Finer MH, Davis CG, Zsebo KM, Jakobovits A (1997). Functional transplant of megabase human immunoglobulin loci recapitulates human antibody response in mice. Nature Genetics 15: 146-156.

Mooi FR, van Loo IH, King AJ (2001) Adaptation of *Bordetella pertussis* to vaccination: a cause for its reemergence? Emerg Infect Dis 7: 526-528.

Moore RC, Purugganan MD (2003). The early stages of duplicate gene evolution. Proc Natl Acad Sci U S A 100:15682-15687.

Morrow DM, Connelly C, Hieter P (1997). Break copy, duplication: a model for chromosome fragment formation in *Saccharomyces cerevisiae*. Genetics 147: 371-382.

Nasar F, Jankowski C, Nag DK (2000). Long palindromic sequences induce double-strand breaks during meiosis in yeast. Mol. Cell. Biol. 20: 3449-3458.

Nehls M, Messerle M, Sirulnik A, Smith AJH, Boehm T (1994). Two large Insert vectors, λPS and λKO, facilitate rapid mapping and targeted disruption of mammalian genes. Biotechniques 17: 770-775.

Nilsson D, Andersson B (2005) Strand asymmetry patterns in trypanosomatid parasites. Exp. Parasitol. 109: 143-149.

Obado SO, Taylor MC, Wilkinson SR, Bromley EV, Kelly JM (2005). Functional mapping of a trypanosome centromere by chromosome fragmentation identifies a 16-kb GC-rich transcriptional "strand-switch" domain as a major feature. Genome Res. 15:36-43.

Oliver SG, van der Aart QJ, Agostoni-Carbone ML, Aigle M, Alberghina L, Alexandraki D, Antoine G, Anwar R, Ballesta JP, Benit P, et al (1992). The complete DNA sequence of yeast chromosome III. Nature 357: 38-46.

Opperman T, Ling LL, Moir DT (2003). Microbial pathogen genomes – new strategies for identifying therapeutic and vaccine targets. Expert Opin Ther Targets 7: 469-473.

Palacios G, Quan PL, Jabado OJ, Conlan S, Hirschberg DL, Liu Y, Zhai J, Renwick N, Hui J, Hegyi H, Grolla A, Strong JE, Towner JS, Geisbert TW, Jahrling PB, Buchen Osmond C, Ellerbrok H, Sanchez-Seco MP, Lussier Y, Formenty P, Nichol MS, Feldmann H, Briese T, Lipkin WI (2007). Panmicrobial oligonucleotide array for diagnosis of infectious diseases. Emerg Infect Dis 13:73-81.

Peterson KR (1997). Production of transgenic mice with yeast artificial chromosomes. TIG 13: 61-66.

Rocha EP, Danchin A (2003) Essentiality, not expressiveness, drives gene-strand bias in bacteria. Nat Genet 34:377-378.

Rocha EP, Fralick J, Vediyappan G, Danchin A, Norris V. (2003) A strand-specific model for chromosome segregation in bacteria. Mol. Microbiol. 49: 895-903.

Rocha EP (2004). Order and disorder in bacterial genomes. Curr. Opin. Microbiol 7: 519-527.

Roh TY, Cuddapah S, Zhao K (2005). Active chromatin domains are defined by acetylation islands revealed by genome-wide mapping. Genes Dev. 19: 542-552.

Sabo PJ, Hawrylycz M, Wallace JC, Humbert R, Yu M, Shafer A, Kawamoto J, Hall R, Mack J, Dorschner MO, McArthur M, Stamato-yannopoulos JA (2004). Discovery of functional noncoding elements by digital analysis of chromatin structure. Proc. Natl. Acad. Sci. USA 101: 16837-16842.

Scherbakova OG, Lanzov, VA, Ogawa H, Filatov MV (2000). Overexpression of bacterial *RecA* protein in somatic mammalian cells increases the frequency of gene targeting. Mutat. Res. 459: 65-71.

Shroff R, Arbel-Eden A, Pilch D, Ira G, Bonner WM, Petrini JH, Haber JE, Lichten M (2004). Distribution and dynamics of chromatin modification induced by a defined DNA double-strand break. Curr. Biol. 14:1703-1711.

Silverman GA, Green ED, Young RL, Jockel JI, Domer PH and Korsmeyer SJ (1990). Meiotic recombination between yeast artificial chromosomes yields a single clone containing the entire BCL2 protooncogene. Proc. Natl. Acad. Sci. USA. 87: 9913-9917.

Storici F, Oberto J, Bruschi CV (1995). The *CDC6* gene is required for centromeric, episomal, and 2-μm plasmid stability in yeast *Saccharomyces cerevisaie*. Plasmid 34: 184-197.

Storici F, Coglievina M, Bruschi CV (1999). A 2-micron DNA-based marker recycling system for multiple gene disruption in the yeast *Saccharomyces cerevisiae*. Yeast 5: 271-283.

Toh-e A (1995). Construction of a marker gene cassette which is repeatedly usable for gene disruption in yeast. Curr. Genet. *27*:293-297.

Tosato V, Ciarloni L, Ivens AC, Rajandream MA, Barrell BG, Bruschi CV (2001). Secondary DNA structure analysis of the coding strand switch regions of five *Leishmania major* Friedlin chromosomes. Curr. Genet. 40: 186-94.

Tosato V, Gjuracic K, Vlahovicek K, Pongor S, Danchin A, Bruschi CV (2003). The DNA secondary structure of the *Bacillus subtilis* genome. FEMS Microbiol. Lett. 218: 23-30.

Tosato V, Waghmare SK, Bruschi CV (2005). Non-reciprocal chromosomal bridge-induced translocation (BIT) by targeted DNA integration in yeast. Chromosoma 114: 15-27.

Trifonov EN, Sussman JL (1980). The pitch of chromatin DNA is reflected in its nucleotide sequence. Proc Natl Acad Sci U S A. 77:3816-3820.

Tsuda H, Maynard-Currie CE, Reid LH, Yoshida T, Edamura K, Maeda N, Smithies O, Jakobovits A (1997). Inactivation of the mouse HPRT locus by a 203-bp retroposon insertion and a 55-kb gene-targeted deletion: establishment of new HPRT-deficient mouse embryonic stem cell lines. Genomics 42: 413-421.

van Loo IH, Heuvelman KJ, King AJ, Mooi FR (2002). Multilocus sequence typing of *Bordetella pertussis* based on surface protein genes. J Clin Microbiol 40: 1994-2001.

Vlahovicek K, Kaján L, Szabó G Jr, Tosato V, Bruschi CV, Pongor S (2005). Web servers for the prediction of curvature as well as other characteristics from DNA sequence. In Arseni Markoff (ed) Analytical Tools for Genes, Genomes and Sequence: The DNA Level, DNA Press, Eagleville, PA, pp. 215-236.

Vora GJ, Meador CE, Bird MM, Bopp CA, Andreadis JD, Stenger DA (2005). Microarray-based detection of genetic heterogeneity, antimicrobial resistance, and the viable but nonculturable state in human pathogenic *Vibrio spp.* Proc Natl Acad Sci U S A 102: 19109-19114.

Vries RG, Bezrookove V, Zuijderduijn LM, Kia SK, Houweling A, Oruetxebarria I, Raap AK, Verrijzer CP (2005). Cancer-associated mutations in chromatin remodeler hSNF5 promote chromosomal instability by compromising the mitotic checkpoint. Genes Dev. 19: 665-670.

Waghmare SK, Caputo V, Radovic S, Bruschi CV (2003). Specific targeted integration of kanamycin resistance-associated non-selectable DNA in the genome of the yeast *Saccharomyces cerevisiae*. BioTechniques 34: 1024-1033.

Wang X, Zhou D, Qin L, Dai E, Zhang J, Han Y, Guo Z, Song Y, Du Z, Wang J, Wang J, Yang R (2006). Genomic comparison of *Yersinia pestis* and *Yersinia pseudotuberculosis* by combination of suppression subtractive hybridization and DNA microarray. Arch Microbiol 186: 151-159.

Yanez RJ, Porter AC (1999). Gene targeting is enhanced in human cells overexpressing hRAD51. Gene Therapy 6: 1282-1290.

Yu X, Gabriel A (2004) Reciprocal translocations in *Saccharomyces cerevisiae* formed by nonhomologous end joining. Genetics 166: 741-751.

Zang Y, Garre` M, Gjuracic K, Bruschi CV (2002). Chromosome V loss due to centromere knockout or *MAD2*-deletion is immediately followed by restitution of homozygous diploidy in *Saccharomyces cerevisiae*. Yeast 19: 1-12.

Zhang HB, Wing RA (1997). Physical mapping of the rice genome with BACs. Plant Mol. Biol. 35: 115-127.

Zhou D, Han Y, Song Y, Tong Z, Wang J, Guo Z, Pei D, Pang X, Zhai J, Li M, Cui B, Qi Z, Jin L, Dai R, Du Z, Bao J, Zhang X, Yu J, Wang J, Huang P, Yang R (2004). DNA microarray analysis of genome dynamics in *Yersinia pestis*: insights into bacterial genome microevolution and niche adaptation. J. Bacteriol. 186: 5138-5146.

Zimmer DP, Paliv O, Thomas B, Gyaneshwar P, Kustu S (2004). Genome image programs: visualization and interpretation of *Escherichia coli* microarray experiments. Genetics 167: 2111-2119.

5

Molecular marker technology in cotton

Preetha, S* and Raveendren T. S

Centre for Plant Breeding and Genetics (CPBG), Tamilnadu Agricultural University, Coimbatore-641001, Tamilnadu, India.

Molecular markers have been a valuable tool in cotton breeding investigations. Various marker techniques used in cotton include Restriction Fragment Length Polymorphism (RFLP), Random Amplified Polymorphic DNA (RAPD), Amplified Fragment Length Polymorphism (AFLP), Simple Sequence Repeats (SSR), Inter Simple Sequence Repeats (ISSR) and Sequence Related Amplified Polymorphism (SRAP).They have large number applications like characterization of gene pool, DNA fingerprinting, phylogenetic analysis, molecular dissection of complex traits, and characterization of genome organization . Several challenges have been overcome in cotton genomic research and now genetic linkage maps of cotton have been developed based on both intraspecific (*intrahirsutum*) and interspecific (*Gossypium hirsutum x Gossypium barbadense*) population and the QTLs responsible for leaf shapes, plant trichomes, photopheriodism, stomatal conductance, disease resistance, yield and fibre quality traits have been mapped. This article would give an overview of genomic research on cotton.

Key words: Cotton, DNA markers, diversity studies, QTL mapping

CONTENTS

INTRODUCTION

Plant breeders select plants with desirable traits by looking at the phenotype. Most of these traits are polygenic with complex non-allelic and environmental interac

tions. Though biometrical genetics revealed the cumulative effects of the genetic loci involved in a quantitative trait, the specific locus involved could not be detected. If the quantitative traits could be resolved into individual genetic components by finding DNA marker closely linked to each trait, it would be easy to manipulate them efficiently and this would help to attain the desirable results

*Corresponding author E-mail:preethii1@rediffmail.com.

quickly and more precisely. Thus, DNA marker technology would provide a tool to the plant breeders to select desirable plants directly on the basis of genotype instead of phenotype.

Conventional breeding methods generally aim to improve agronomically relevant or otherwise interesting traits by combining characters present in different parental lines of cultivated species or their wild relatives. Transfer of genes from wild species is time consuming, not always successful and it is difficult to assess the introgression of *alien* genes. Moreover conventional methods to combine all the favorable traits from cultivars for development of superior varieties involve repeated back crossing, selfing and testing which are time consuming and less precise processes as compared to direct selection of plants based on molecular processes. Further conventional selection depends upon accurate screening methods and availability of lines with clear-cut phenotypic characters. With the use of molecular techniques it would be now possible to hasten the transfer of desirable genes among varieties and to introgress novel genes from related wild species into the local and popular genotypes which would accelerate the generation of new varieties. It would also be possible to establish genetic relationships between sexually incompatible species.

Morphological and physiological features of plants have been used to understand the genetic variation. Though morphological features are indicative of the phenotype, they are affected by environmental factors and growth practices. There are about 145 morphological markers identified in cultivated cotton but their utility in breeding programmes has remained limited because of their deleterious effect and the difficulties in accumulating multiple markers in a single genotype (Percy and Kohel, 1999). Therefore, more reliable markers such as protein, or more specifically, allelic variants of several enzymes so called isozymes were used. But as the number of isozymes available is limited and their expression is often restricted to a specific developmental stage of tissues, various DNA markers have come into use of the plant breeders. Tanksley (1983) listed five properties that distinguish molecular markers from morphological markers. These properties are (1) genotypes can be determined at the whole plant, tissue and/or cellular level, (2) a relatively larger number of naturally occurring alleles exists at many loci, (3) phenotypic neutrality i.e., deleterious effects are not usually associated with different alleles, (4) alleles at many loci are codominant, thus all possible genotypes can be distinguished and (5) few epistatic or pleiotrophic effects are observed.

DNA marker techniques used in cotton

Restriction fragment length polymorphism (RFLP)

RFLP was the first DNA marker that was used in crop improvement; the first true RFLP map in a crop plant (tomato) was constructed in 1986 with 57 loci (Bernatzky and Tanksley, 1986). It refers to variation among individuals in the lengths of DNA fragments produced by restriction enzymes that cut DNA at specific sites. RFLP analysis, in its original form, consists of DNA digestion with a restriction enzyme, separation of the restriction fragments by agarose gel electrophoresis, transfer of separated restriction fragments to a filter by southern blotting and detection by autoradiography. The differences in sizes of DNA fragments may result from base substitutions, additions, deletions, or sequence rearrangements within restriction enzyme recognition sequences.

The major disadvantage of this technique is that a large amount of DNA is required for each reaction and it is labour intensive to apply to numerous samples. Additionally, only one or a few loci are scanned with a single probe and restriction enzyme combination, but the markers are usually co-dominant. Despite the drawbacks of this technique it is well established and a significant volume of mapping work has been completed with RFLPs, including in cotton.

Meredith (1992) in a study of heterosis and varietal origins reported the first RFLP evaluation in upland cotton. Reinisch et al. (1994) found that 64% of cotton RFLPs is co-dominant. Detailed RFLP maps of cotton with 41, 5, 31, 24 and 17 linkage groups were developed by Reinisch et al. (1994), Shappley et al. (1996, 1998a, 1998b) and Ulloa and Meredith (2000) respectively. Wright et al. (1998) identified RFLP marker linked to resistance allele for bacterial blight pathogen. Moreover using a detailed RFLP map, genes affecting density of leaf and stem trichomes have been mapped by Wright et al. (1999). An effort for mapping the trait of low gossypol seed and high gossypol plant was made using 49 RFLP probes to trace introgression of parental DNA segment in trispecies hybrid and in three back cross generations (Vroh et al., 1999).

Random amplified polymorphic DNA (RAPD)

RAPD, the oldest PCR-based technique (Williams et al., 1990) relies on use of short, random PCR primers to amplify random portions of the genome. It has many advantages over RFLP technique such as non-radioactive detection, no prior sequence information for a genome is required, universal primers work in genome, very small amount of genomic DNA is needed, experimental simplicity and no need for expensive equipments beyond a thermocycler and a transilluminator (Rafalski, 1997). However it is often criticized for its lack of reproducibility (Jones et al., 1997). Several factors may influence reproducibility of RAPD profile within and between laboratories including DNA concentration, reproducibility of thermocycler profiles, primer quality and concentration, choice of DNA polymerase, and pipetting accuracy (Rafalski, 1997).

RAPD technique has been employed for marker assisted selection in the interspecific population involving *Gossypium sturtianum* and other species for getting glandless seed and glanded plant type (Mergeai et al., 1998; Vroh et al., 1998). Vroh et al. (1997) have used RAPD technique in recurrent selection programme utilizeing wild species. RAPD have greatly facilitated linkage mapping in cotton (Khan et al., 1999). Zhang et al. (2002) used RAPD along with SSR markers to construct a map containing 43 linkage groups to investigate the homeologous chromosome regions of the A and D subgenomes in the allotetraploid genome. A RAPD marker (R-6592) has been identified for male fertility restoration gene in cotton (Lan et al., 1999). This marker will serve as a potential tool in marker assisted selection for fertility restoration genes in cotton. Ulloa et al. (2000) utilized this technique to identify two putative QTLs for stomatal conductance. Besides, this has been utilized for DNA fingerprinting (Multani and Lyon, 1995; Krishna, 1998), assessing genetic diversity (Tatineni et al., 1996) and to establish relationship between genotypes of same and different species (Wajahtullah and Stewart, 1997).

Amplified fragment length polymorphism

AFLP technique was developed by Vos et al. (1995). It involves digesting DNA with two different restriction endonucleases (usually a four base and six base recognition sequence) and ligating adapters to the sticky fragment ends created by enzymes. PCR amplification is then performed, which increases reproducibility of the technique. Polymorphism is detected where there is an addition or loss of restriction site, or if a length polymorphism exists between two restriction endonuclease sites due to insertion or deletion of nucleotide sequences. Although AFLP markers have been reported to be co-dominant (Maughan et al., 1996), they are usually dominant, which reduces their information content. However, this is compensated for by the large number of loci that can be assayed with any given primer combination (often 40 – 50 bands per reaction).

AFLP technique has been employed for estimating genetic diversity in cotton by Vroh et al. (1999), Abdalla et al. (2001), Rana et al. (2004a) and Zhang et al. (2005). Altaf et al. (1997) developed a map of linkage groups using AFLP along with RAPD markers. Recently Lascape et al. (2003) and Zhang et al. (2005) have utilized this technique for map saturation.

Simple Sequence Repeats (SSR)

SSR or microsatellites are short, tandemly repeated DNA sequence motifs that consist of two to six nucleotide core units, and were initially described in humans (Litt and Lutty, 1989). They are highly abundant in eukaryotic genome but also occur in prokaryotes at lower frequencies. These small repetitive DNA sequences provide

the basis for PCR-based multi allelic, co-dominant genetic marker system. The high incidence of detectable polymorphism through changes in repeat numbers is caused by an intramolecular mutation mechanism called DNA slippage (Gupta et al., 1996).

The regions flanking the microsatellites are generally conserved and PCR primers relative to the flanking regions are used to amplify SSR- containing DNA fragments. The length of the amplified fragment will vary according to the number of repeated residues (Ellegien, 1993). SSRs are now considered as the marker of choice for self-pollinated crops with little intraspecific polymorphism (Roder et al., 1998). Furthermore, the reproducibility of SSRs is such that they can be efficiently used by different laboratories to produce consensus data, which makes them useful for genome mapping projects and results in their successful isolation and application within many plant species (Dietrich et al., 1996; Dib et al., 1996; Schmidt and Heslop, 1998).

In cotton, SSRs represent a new class of genetic markers which accelerated cotton genome mapping work. Liu et al. (2000) used 65 SSR primer pairs to amplify 70 marker loci localized to a specific cotton genome. The SSR markers identified in this study provide a framework that can be used with further conventional linkage mapping to other DNA markers to expand the genome-wide coverage of the cotton genetic map. A linkage map was constructed with 199 RAPD and SSR markers to assist in selection for stomatal conductance; two putative QTL for this difficulty to measure physiological trait were identified on two cotton linkage groups (Ulloa et al., 2000). Using SSR markers, 23 chromosomes have been covered with an average inter-loci distance of 4.9 cM (Frelichowski et al., 2006).

Inter Simple Sequence Repeats (ISSR)

In contrast to the SSR marker technique that amplifies with primers located on the flanking single copy DNA, microsatellites anchored primers that anneal to an SSR region can amplify regions between adjacent SSRs. The ISSR technique uses primers that are complimentary to a single SSR and anchored at either the 5' or 3' end with a one- to three-base extension (Zietkiewicz et al., 1994). The amplicons generated consist of regions between neighbouring and inverted SSRs. As a result, the high complex banding pattern obtained will often differ greatly between genotypes of the same species. Liu and Wendel (2001) reported ISSR as an easy and informative genetic marker system in cotton for revealing both inter and intraspecific variations. However, no ISSR based linkage map has been reported in cotton.

Sequence related amplified polymorphism (SRAP)

Li and Quiros (2001) introduced this new marker technique, which involves preferential amplification of ORFs by PCR. For this purpose two different primer pairs were used. One primer (forward), of 17 base pairs, contains a fixed sequence of 14 nucleotide rich in G and C in the 5'end and three selective bases in the 3'end.This primer amplifies preferentially exonic regions, which tend to be rich in these nucleotides. The second primer (reverse), with 19 base pairs, contains a sequence of 16 nucleotide rich in A and T in the 5'end and three selective bases in the 3'end.This primer preferentially amplifies intronic regions and regions with promoters, rich in these nucleotides. This technique combines simplicity, reliability, moderate throughput ratio and facile sequencing of selected bands. Further, it targets coding sequences in the genome and results in a moderate number of co-dominant markers. However, these techniques use no prior sequence information, and the markers generated are randomly distributed across the genome. In cotton this new marker technique is being used along with other markers (Lin et al., 2005 and He et al., 2007) for saturating the genome.

Status of cotton molecular marker technology

Cotton genomics is in its infancy and despite ongoing efforts by the cotton community, a high density molecular PCR-based map to facilitate gene discovery and marker aided selection in the public domain is lacking. Knowledge of cotton genomic structure remains at least ten years behind the well co-ordinated global efforts in *Arabidopsis*, rice, maize, *Medicago* and other model systems. Earlier molecular markers were not widely used in cotton as in other crop species because scientists faced several challenges in application of molecular markers to cotton genome. The first problem faced by researchers was developing a standard extraction protocol to produce high quality DNA suitable for molecular analysis. As cotton tissues contain high level of polysaccharides and polyphenolic compounds which interact with proteins and nucleic acids, it poses problem in extraction of high quality DNA for molecular analysis. The second challenge is the low level of polymorphism found within elite upland cultivars. Most genetic studies in cotton, both pedigree and marker based, have found a narrow genetic base in upland cotton (Wendel et al., 1992; Van Esbroeck et al., 1998; Gutierrez et al., 2002 and Saha et al., 2003). Methods for detection of sufficient polymorphism were lacking. Third complicating factor is the allotetraploid genome of the major cultivated cottons *G. hirsutum* and *G. barbadense* which means that most "single copy loci" will actually exist as two unlinked loci and the corresponding loci from each subgenome must be resolved. Moreover, polyploidy nature and chromosomal rearrangements affect the statistical interpretation of segregation data and linkage interferences (Doerge and Craig, 2000; Livingstone et al., 2000). Linkage analysis is complicated by extensive recombination per homologous chromosome and genetic redundancy associated

ciated with polyploidy (Brubaker et al., 1999; Reinisch et al., 1994). Some chromosomes may contain recombination hotspots, resulting in large genetic distances, whereas others contain a large portion of heterochromatin associated with low recombination frequencies. The interspecific hybrids result in a wide range of phenotypic segregation and fertility. F_1 hybrids are semisterile, and one third of the plants in F_2 and subsequent generation do not produce any bolls or seeds. Gametic and seed lethality is probable cause of segregation distortion and unmapped loci, which makes it difficult to develop a high resolution genetic map using interspecific hybrid derived population. Further genetic redundancy makes it difficult to develop polymorphism based DNA markers. The physical size of the cotton genome is relatively large in a range of 2702 to 2246 Mb (Arumuganathan and Earle, 1991; Michaelson et al., 1991). With a genome size of 2246 Mb, the average physical size of a cM in cotton is about 400 kb. The genetic map 5000 cM of cotton genome will require 3000 DNA probes at average 1 cM density and the physical genome of 2246 Mb will require 75,000 YACs/BACs) of average size 150 kb for 5 times coverage (Arumuganathan and Earle, 1991).

However, all these challenges have been overcome and genomic research on cotton has acquired a notable dimension with the development of cotton specific SSR markers. DNA extraction methods suitable for molecular analysis in cotton have been proposed by Paterson et al. (1993), Vroh et al. (1996), Saha et al. (1997) and Permingeat et al.(1998). Wang et al. (1993) and Krishna and Jawali (1997) have developed DNA isolation technique from single or half seed suitable for RAPD. Moreover, several thousand microsatellite markers are now available to saturate genetic maps of cotton. The first public microsatellite markers were developed at Brookhaven National Laboratory. They identified more than 500 microsatellite clones containing (GA)n and (CA)n repeats, the primer pairs were designated BNL. Liu et al. (2000) assigned several BNL microsatellite to the chromosomes using cytogenetic aneuploid stocks. An additional 150 (GA)n repeat loci designated CM were isolated at Texas A and M University (Connell et al., 1998; Reddy et al., 2000).

Reddy et al. (2001) using SSR-enriched genomic libraries identified 300 SSR markers (JESPR). Dow Agro Science seed company has reported more than 1,200 SSRs (Kumpatla et al., 2002). Nguyen et al. (2004) developed 390 SSRs rich in CA/GT repeats named as CIR. In addition to generation of repeat enriched libraries for development of SSR markers, sequence databases represent a valuable resource for identification of microsatellites. The rapid increase in cotton sequences (ESTs, BAC end sequences and STCs) deposited in public databases has also allowed the identification of microsatellite DNA containing regions from *G. hirsutum* and *G. arboreum* which would be used to develop markers. The abundance of expressed sequence tags (ESTs) in cotton

(Arpat et al., 2004) serves as an attractive potential source of microsatellite markers. Studies of Kantety et al. (2002) revealed that mRNA transcripts can also contain repeat motifs, and expressed sequence tags (ESTs) available makes this an attractive potential source of microsatellite markers. Cotton Genome Centre, University of California-Davis and Clemson University initiated large scale C-DNA analyses of cotton genome, which have provided an opportunity to use partially characterized cDNA clones for identifying expressed sequence tags containing SSR (EST-SSRs). Saha et al. (2003) have shown that SSR loci can be identified from data mining of growing number of cotton EST sequences in databases with an efficiency of 1 to 4.8% of sequenced ESTs. They identified over 18,000 SSR-ESTs in different plant species including cotton after surveying 739,258 ESTs in the publicly available Genbank databases, and developed EST-SSRs to study transcribed genes of cotton. Han et al. (2004) developed 544 SSR markers (NAU) from fibre ESTs from *G. arboreum*. Qureshi et al. (2004) utilized 9,948 ESTs belonging to *G. hirsutum* and designed 84 primer pairs (MGHES). Park et al. (2005) developed Simple Sequence Repeats (MUSS) and Complex Sequence Repeat (MUCS) primers from a total of 1232 ESTs. Apart from this BAC library assembled from *G. hirsutum* Acala 'Maxxa' with an average insert size of 137 kb (Tomkins et al., 2001) was utilized to design SSR primers (MUSB) by Frelichowski et al. (2006). Information on all these primers is available to public and can be downloaded at http://www.mainlab.clemson.edu/cmd/projects.

Applications of marker technology in cotton

Fingerprinting and diversity studies

Plant breeders desire their new varieties to be distinct, uniform and stable. In the past the ability to discriminate between varieties was heavily dependent on morphological traits. Lately, DNA markers have been employed as a promising method of finger printing. High resolution of molecular markers compared with other markers makes them a valuable tool for varietal and parental identifycation for protection of cotton breeders rights. DNA markers further add to repertoire of tools for determination of evolutionary relationship between *Gossypium* species and families. Molecular markers also allow an understanding of the relationship between chromosomes of related *Gossypium* sp. Similar to other crops, an understanding of the evolutionary and genomic relationships of cotton species and cultivars is critical for further utilization of extant genetic diversity in the development of superior cultivars (El-zik and Thaxton, 1989). Apart from this, molecular markers are employed for genetic diversity estimation in place of morphological markers as number of morphological descriptors in various crops is in vogue for characterization purpose. High polymorphism

independent of the effects related to environmental conditions and physiological stage of plant makes molecular markers a reliable tool for diversity studies.

Initially isozymes were applied to *Gossypium* as a taxonomical tool to distinguish differences at species level. Cherry et al. (1972) observed minor differences in isozyme banding pattern of A and B genome species whereas greater band variation between the more distantly related species in the C, D and E genome. Isozymes have been used to investigate the genetic diversity in *G. hirsutum* (Wendel et al., 1992). This study assayed 50 enzyme systems within wild *G. hirsutum* accession and upland cultivars. Thirty isozymes were found to be polymorphic with *G. hirsutum* but of these only 14 were polymorphic between upland cultivars and allelic diversity was minimal, which indicated a low level of diversity in upland cultivars. Wendel et al. (1994) used allozyme analysis to assess levels and patterns of genetic diversity in *Gossypium mustelinum* and its relationship to other tetraploid species. They inferred low level of genetic variation compared to other tetraploid species. Isozyme assays resolved little or no variability within species of real use and numbers are limited and their expression is often restricted to a specific developmental stage of tissues. So DNA markers have been employed for this purpose. Brubaker and Wendel (1994) examined genetic diversity in upland cultivars using RFLP markers. Their study found that despite surveying 205 loci from 23 upland cultivars, only 6 had unique, multilocus genotypes again suggesting limited diversity.

Multani and Lyon (1995) generated RAPD markers with 30 random primers and found that *Gossypium barbadense* accessions can be easily distinguished from *G. hirsutum* varieties by 104 markers. Cultivar specific markers were found to be consistent and have the potentiality to be used as genetic finger prints for varietal identification. RAPD marker technique has been used to differentiate cultivars resistant to aphids, mites and jassids (Geng et al., 1995) and as an alternative to cumbersome grow out test in identification of parents and F_1 hybrid cotton (Krishna, 1998). Martsinkovskaya et al. (1996) used RAPD markers in differentiating species and varieties using ribosomal intergenic sequences of genes Rrn 18 - Rrn 25. Tatineni et al. (1996) studied the diversity of 16 near homozygous elite cotton genotypes derived from inter-specific hybridization using 80 RAPD primers and generated 135 RAPD markers. Observations on morphological traits were also recorded. The results revealed that the clustering through taxonomic distance and genetic distance was found to be similar. Both procedures produced two clusters with one resembling *G. hirsutum* and the other resembling *G. barbadense*. They also identified several genotypes which are distant from typical *G. hirsutum* and *G. barbadense*. The level of polymorphism exhibited by these genotypes can be exploited to develop mapping population for tagging fibre quality genes. The study of Iqbal et al. (1997) clearly dis-

distinguished *G. hirsutum* lines from those of *G. arboreum* with 98% of the RAPD primers found polymorphic. Their study revealed that the intervarietal genetic relationship of several varieties was related to their centre of origin and also the narrow genetic base of most varieties.

Wajahatulla and Stewart (1997) studied the genomic affinity among some of the wild species by utilizing random primers. They reported that there are high intraspecific genetic similarities for *Gossypium nelsonii*, *Gossypium australe*, *Gossypium sturtianum* and *Gossypium bickii*. Genetic similarity of *Gossypium nandewarense* with two accessions of *Gossypium sturtianum* was high and median in placement, indicating that it should not be considered as a separate species. Genomic affinity among Australian *Gossypium* species was assessed by Wajahatullah et al. (1997). They concluded that *G. triphyllum* and *G. longicalyx* accounted for 20% of the total polymorphism. The intraspecific genome affinities were high compared to interspecific affinities. All accessions of a particular species were found to cluster in their respective groups except *G. nandewarense*, which got clustered along with *G. sturtianum* accessions.

Wang et al. (1997) analysed 25 short duration cotton cultivars of China using arbitrary primers. Based on similarity coefficients, they divided the cultivars into four groups. Results indicated that RAPD profiles were consistent to the pedigree of most of the cultivars evaluated. Gwoli et al. (1999) through RAPD analysis revealed close affinity between *G. sturtianum*, *G. nandewarense*, *Gossypium australe* and *Gossypium robinsonii*. The genetic similarity between the wild species (*G. sturtianum*, *Gossypium raimondii*, *Gossypium thurberii*) upland cottons and between upland cotton and BC$_3$ progenies of trispecies hybrids involving above species was assessed using AFLP markers (Vroh et al., 1999). The genetic similarity assessed was less (29.5 – 43.2%) in case of wild species and upland cotton and more (80%) in case of progenies derived from them based on 417 AFLP polymorphic fragments generated using five pairs of primers.

Pillay and Meyers (1999) utilized variation in ribosomal RNA genes (rDNA) to distinguish New and Old world cottons using AFLP markers to establish the extent of genetic diversity, they reported that AFLP technique differentiated two diploid species (*G. arboreum* and *Gossypium herbaceum*) from each other and as well as from tetraploid (*G. hirsutum* and *G. barbadense*) taxa. Zhen et al. (1999) distinguished the resistant and susceptible cotton cultivars for *Verticillium dahliae* using RAPD molecular marker technology. They classified the 25 varieties into four subgroups; group I including introduced cultivars resistant to *V. dahliae* and group II with cotton varieties from Shanxi and Liaoning Provinces. Resistant varieties of group III were derived from disease nursery under directional selection and had a complex genetic

basis, and the cultivar Taichang 121, a susceptible variety, was grouped in the last.

Genetic diversity of 31 available *Gossypium* species, three sub-species and one interspecific hybrid was studied by Khan et al. (2000). They screened these genotypes with 45 RAPD primers to distinguish the genotypes. The result indicated interspecific genetic relationship of several species as related to their centre of origin. The study further revealed a broader genetic base of most of species besides indicating the genetic relationship of *G. hirsutum* to standard cultivated *G. barbadense*, *G. herbaceum* and *G. arboretum*. They concluded that for constructing genetic linkage maps, RAPD markers are more efficient than morphological markers, isozymes and RFLP as RAPD detected the variation in closely related genotypes too. Liu et al. (2000) using SSRs determined the molecular variation among and within the converted day neutral BC$_4$F$_4$ (*G. hirsutum*) race stock accessions collected throughout the range of original domestication of *G. hirsutum* and postulated that the recovery of the primitive recurrent parent could be improved by marker-assisted backcrossing with SSR markers used in their study.

Abdalla et al. (2001) determined intra and inter specific genetic relationships of *G. herbaceum*, *G. arboreum*, *G. raimondii*, *G. hirsutum* and *G. barbadense* using AFLP technique. They showed that AFLP technique is useful for estimating genetic relationships across a wide range of taxonomic levels, and for analyzing the evolutionary and historical development of cotton cultivars at the genomic level. Genetic diversity of 30 *G. hirsutum* cotton cultivars collected from three main cotton growing regions of China was assessed through RAPD analysis (Shu et al., 2001). They reported the intervarietal genetic relationships between several cultivars and their association to the pedigree of the parents besides revealing the narrow genetic base in most of the cultivars evaluated. They also indicated that RAPD technique could reliably be used to determine the genetic relationship within diverse array of cotton genotypes. Lu and Meyers (2002) evaluated the level of genetic diversity of ten influential cotton varieties using RAPD markers. They were able to individually identify all the tested varieties by specific markers in genetic finger printing.

Kumar et al. (2003) studied genetic diversity of 30 elite cotton germplasm lines including 20 *G. hirsutum*, seven *G. arboreum*, one each of *G. herbaceum*, *G. thurberi* and *Gossypium klotzschianum* using RAPD markers and morphological characters. They reported one 1100 bp *G. arboreum* specific band. *G. klotzchianum* was reported to exhibit the least similarity coefficient of 0.5 with all other species studied. Clustering using morphological characters was found to be more or less similar to the clustering obtained through RAPD analysis. Results of study indicated narrow genetic base within *G. hirsutum* and *G. arboreum* genotypes. Rana and Bhat (2004a) employed AFLP and RAPD technique in sixteen diploid cotton (*G. herbaceum* and *G. arboreum*) cultivars for genetic diversity estimation and cultivar identification and they have found that AFLP markers are more efficient for genetic diversity estimation, polymorphism detection and cultivar identification. Rana and Bhat (2004b), in their study utilized five RAPD primers producing 46 polymorphic bands to quantify the magnitude of genetic diversity among and within 12 genotypes belonging to *G. hirsutum, G. barbadense, G. herbaceum* and *G. arboreum*. They revealed that proportion of variance components within and between genotypes was 1.71 and 9.12 respectively indicating that genotypic differences were more prominent. Within genotypic variation of the above mentioned species were 25, 43, 0, and 18%, respectively.

Tabar et al. (2004) used RAPD analysis to evaluate genetic diversity among commercial Indian cotton varieties belonging to *G. hirsutum* and *G. arboreum* and revealed that the intervarietal genetic relationships were related to their pedigree and tetraploids show much narrow genetic base than diploids. Rungis et al. (2005) utilized SSR markers for quantifying the level of polymorphism between *G. hirsutum* and *G. barbadense* within the species. Their investigation showed that though interspecific polymorphism was high, the level of polymorphism within *G. hirsutum* cultivars was low (~ 5%).SSR markers were employed to estimate genetic distance among selected genotypes (*G. hirsutum*) and its relationship with F_2 performance (Gutierrez et al., 2002). The results on genetic distance revealed a lack of genetic diversity among all the genotypes and it was a poor predictor of overall F_2 performance. However, when genotypes with maximum range of genetic diversity were present, it was a better predictor for some traits.

Bertini et al. (2006) generated SSR marker profile for *G. hirsutum* cultivars in order to discriminate the cultivars. Guo et al. (2007) revealed genetic affinity of D-genome species *G. raimondii and Gossypium gossypoides* based on two types of EST-SSR markers derived from *G. arboreum* and *G. raimondii*. Lascape et al. (2007) studied the allelic diversity of genotypes of cultivated and primitive land races of species *G. hirsutum, G. barbadense, Gossypium darwinii* and *Gossypium tomentosum*. Their results confirmed the proximity of *G. darwinii* to *G. barbadense* and one of the race *yucantanense* of *G. hirsutum* appeared to be the most distant from cultivated genotypes. The above studies have revealed lack of diversity in cultivated upland *G. hirsutum* cultivars. Van Esbroeck et al. (1998) pointed out that monoculture of some successful cultivars and their extensive use as progenitors in breeding programme has limited genetic diversity of cultivated cotton cultivars. According to Iqbal et al. (2001), one hypothesis which may explain the apparent lack of diversity in cultivated upland *G. hirsutum* was that one or more genetic bottlenecks may have occurred during the later stages of the development of *G. hirsutum*, possibly as a result of rigorous selection for

early maturity, including valuable alleles that confer resistance to insects, pathogens and environment adversities, variability would have been lost during this phase of domestication.

Construction of linkage map and QTL mapping

QTL mapping is based on the principle that genes and markers segregate *via.* chromosome recombination (called crossing-over) during meiosis thus allowing their analysis in the progeny (Paterson, 1996). Genes or markers that are close together or tightly linked will be transmitted together from parent to progeny more frequently than genes or markers that are located further apart. In a segregating population, the frequency of recombinant genotypes can be used to calculate recombination fractions, which may be used to infer the genetic distance between markers. Markers that have recombination frequency of 50% are described as unlinked and assumed to be located far apart on the same chromosome or on different chromosomes. Mapping functions (Kosambi mapping function and Haldane mapping function) are used to convert recombinant fractions into map units called Centi Morgans (cM) (Kearsey and Pooni, 1996).

Linkage maps are constructed from the analysis of many segregating markers. First step in linkage map construction is production of mapping population. The construction of linkage map requires segregating plant population consisting of 50-250 individuals (Mohan et al., 1997). Generally in self pollinated species, mapping population originates from parents that are both highly homozygous (inbred). In cross pollinated species, mapping population may be derived from a cross between a heterozygous parent and a haploid or homozygous parent (Wu et al., 1992). Several different mapping population viz., F_2s, backcross population, recombinant inbred lines (RIL), double haploids (DH) may be utilized, with each population type possessing advantage and disadvantage. The second step is to identify markers that reveal differences between parents i.e. polymorphic markers (Young, 1994). Once polymorphic markers have been identified, they must be screened across the entire mapping population including the parents which is known as marker genotyping of the population. The final step involves coding data for each marker on each individual of a population and conducting linkage analysis using computer programmes. Commonly used software programmes include Map maker/EXP (Lander et al., 1987) and Map manager QTX (Manly et al., 2001). Linkage between markers is usually calculated using odd ratios (i.e. the ratio of linkage versus no linkage). This ratio is more conveniently expressed as the logarithm of the ratio, and is called logarithm of odds (LOD) value or LOD score (Risch, 1992). LOD value of 3 between two markers indicates that linkage is 1000 times more likely than no linkage. LOD value may be lowered in order to detect a greater level of linkage.

Once linkage map has been constructed, QTL analysis is performed to identify/map genes. QTL analysis is based on the principle of detecting an association between phenotype and genotype of markers. Markers are used to partition the mapping population into different genotypic groups based on the presence or absence of a particular marker locus and to determine whether signifycant differences exists between groups with respect to the trait being measured (Tanksley, 1993; Young, 1996). A significant difference between phenotypic means of groups (either 2 or 3), depending on the marker system and type of population, indicates that the marker locus being used to partition the mapping population is linked to a QTL controlling the trait. Three widely used methods for detecting QTLs are single marker analysis, simple interval mapping and composite interval mapping (Liu, 1998; Tanksley, 1993).

Cotton linkage maps

The molecular information of a crop genome is usually presented in the framework of a genetic linkage maps wherein markers are placed into linkage groups in relation to one another, or to other morphological and physiological markers. Paterson (1996) described genetic map as a road map, reflecting the proximity of different land marks to one another, and molecular markers at defined places along each linear chromosome which enables the genetists to determine a particular gene of interest. Genetic linkage maps are useful tools for studying genome structure, organization evolution, to estimate gene effects of important agronomic traits, identifying introgression and for tagging genes of interest to facilitate Marker Aided Selection (MAS) and map based cloning. Genetic linkage maps saturated with poly-morphic molecular markers have already been generated for many species of economic and scientific interest. But the progress on cotton genome mapping is impeded by relatively larger genome size, inadequate DNA markers and polyploidy of widely cultivated tetraploid cotton.

A detailed molecular map of the cotton genome has been published by Reinisch et al. (1994) who used an interspecific cross of G. hirsutum race "palmeri" and G. barbadense accession K101 to assemble 705 RFLP loci into 41 linkage groups and 4675 cM, with average spaing between markers of about 7 cM. Since each gamete contains 26 chromosomes, at least 15 gaps exist in the map to bring overall genetic size of cotton genome to 5125 cM. Fourteen of the 26 chromosomes have been associated with linkage groups by using a series of monosomic interspecific substitution stocks developed by Stelly (1993). As F_2 mapping population of the above mentioned map were derived from undomesticated allotetraploid cotton races, their use in detecting and mapping agronomic traits are limited to some extent and could not totally represent the genome of cultivated cottons; however, they are valuable in revealing the organization and

evolution of cotton genome.

An F_2 population was derived from a cross between homozygous lines G. hirsutum cv. TM-1 and G. barbadense cv. 3-79 at USDA - ARS in Texas and segregating data of 171 F_2 individuals of this cross were obtained for 868 genetic markers (Yu et al., 1997). These markers have been mapped into 50 linkage groups spaning nearly 5000 cM of the cotton genome. By using diploid and aneuploid cottons, 21 linkage groups have been derived from the A sub-genome and 19 from the D-sub-genome. Eighteen of the 26 cotton gametic chromosomes were identified with linkage groups. A trispecific F_2 population was developed from three different cultivars to study inheritance patterns of segregating loci and to establish linkage groups among three genome species by Altaf et al. (1997). Using RAPDs and AFLPs, 11 linkage groups were identified that spanned 521.7 cM, with an average distance of 16.8 cM between markers with a large number of markers showing distorted segregation. Jiang et al. (1998) constructed a RFLP linkage map using 271 F_2 progeny from a cross of G. hirsutum x G. barbadense and the map included 261 RFLP markers in 27 linkage groups and spanned 3,767 cM with an average spacing of 14.4 cM between markers.

An intraspecific map of G. hirsutum has been developed (Shappley et al., 1998a) based on a cross between HS46 and MARCABUCAG8US-1-88. This map was based on segregation in 96 F_2:F_3 families and it consists of 120 RFLP loci placed into 31 linkage groups and covering 865 cM, or an estimated 18.6% of the cotton genome. Khan et al. (1998) used RFLP and RAPD marker to construct linkage map comprising 51 linkage groups, covering 6663 cM with 332 AFLPs, 91 RAPDs and 3 morphological traits. Brubaker et al. (1999) compared both allotetraploid cotton and its diploid progenitors with a set of RFLP markers. Ulloa and Meredith (2000) developed a RFLP genetic linkage map of cotton (G. hirsutum L.) from 119 $F_{2:3}$ progeny from the cross MD 5678ne x Prema. The linkage map comprises 81 loci mapped to 17 linkage groups with an average distance of 8 - 7 cM between markers covering 700.7 cM, or approximately 15% of the recombinational length of cotton genome.

Siu et al. (2000) identified 71 SSR markers involving 65 SSR primers and located them on different chromosomes of cotton. Zuo et al. (2000) constructed a map of 67 marker loci with F_2 population in the upland cotton cultivars. With the cultivar map, constructed using F_2 population by cultivar crosses in G. hirsutum L., it was almost impossible to cover whole genomes so researcher's interest turned towards developing species map. Since cultivated G. hirsutum L. cotton is characterized by high fibre production and G. barbadense in particular is of superior fibre qualities, it was thought that the species map developed from crossing them would be of great use in researching the genomes structure, organization and function and has a great role in identifying, detecting and mapping genes of critical agronomic traits including fibre traits.

Jiang et al. (2000) utilized 180 F_2 plants from an interspecific cross between a *G. hirsutum* genotype carrying four morphological mutants and a wild type *G. barbadense* to map 261 RFLP markers in 26 linkage groups which covered 3664 cM of genome with an average spacing of 14.1 cM between markers. The linear order of markers showed only small differences from the previously published map of Reinisch et al. (1994), usually associated with short distances between markers or polymorphism at different homeologous (duplicate) loci detected by individual probes. Ulloa et al. (2002) brought into light the fact that so far developed maps have covered only a relatively small part of cotton genome (11 ~ 18.6%) and more detailed maps covering a wide range of the genome is needed to map QTLs precisely and completely. They constructed a RFLP join map from four different mapping populations, covering 1502.6 cM and approximately 31% of the total recombination length of the cotton genome; however the usefulness of the join map in conventional breeding of cotton is yet to be evaluated.

Most of the above studies involved tentative (such as F_2) other than permanent population (such as RILS and DH) for mapping work. Zhang et al. (2002) used DH population which has following advantages: (i) it can be used for continuous research in plant genome and may be shared in several laboratories, (ii) smaller volume of population could be used to construct a precise linkage map relative to 3:1(for dominant loci) or 1:2:1(for co-dominant loci) segregated F_2 population, because the marker locus segregation is 1:1 in the DH population. They constructed a linkage map with 58 doubled and haploid plants of *G. hirsutum* cv.TM1and *G. barbadense* cv.Hai T124 cross. Among 624 marker (SSRs and RAPDs), 489 loci were assembled into 43 linkage groups covering 3,314.5 cM. Using monosomic and telodisomic genetic stocks, 17 linkage groups were associated with chromosomes of the allotetraploid genome and some of the unassociated groups were connected to corresponding A or D subgenomes. By assigning duplicated SSR loci in the chromosomes or linkage groups, chromosome pairs 1 and 15, 4 and 22, 10 and 20, 9 and 23 and 5 and 18 were identified as homoeologous chromosomes.

Lascape et al. (2003) published a map(BC$_1$ (*G. hirsutum* cv.Guazunchoz x *G. barbadense* VH8-4602) x *G. hirsutum* cv.Guazunchoz) which combines RFLP, AFLP and SSR markers, comprising 888 loci mapped on 26 long and 11 short linkage groups and spans a genetic length of 4,400 cM. Han et al. (2004) published a genetic map ((TM1 x Hai7124) x TM1) based on 99 SSR markers developed from fibre ESTs of *G. arboreum*. A total of 111 loci detected with these 99 EST-SSRs integrated with 511 SSR loci and two morphological marker loci assorted to 34 linkage groups containing 2 to 41 markers each, covering 5644.3 cM with an average inter-marker distance of 9.0 cM. Mei et al. (2004) developed a genetic map with an F_2 population derived from interspecific hy-

brids between *G. hirsutum* L. cv. Acala-44 and *G. barbadense* L. cv. Pima S-7. A total of 392 genetic loci, including 333 AFLPs, 47 SSRs and 12 RFLPs have been mapped into 42 linkage groups which span 3,287 cM and cover approximately 70% of the genome. Using chromosomal aneuploid interspecific hybrids and a set of 29 RFLP and SSR framework markers (Liu et al., 2000), they have assigned 19 linkage groups involving 223 loci to 12 chromosomes. They found that the genetic distance between mapped loci in the A subgenome chromosomes were often larger than those in homeologous chromosome in D subgenome which is probably caused by larger amount of repetitive DNA and heterochromatin in the A subgenome chromosomes.

Nguyen et al. (2004) developed a set of 204 SSR markers revealing 261 scorable segregating bands (BC$_1$ (*G. hirsutum* cv.Guazunchoz x *G. barbadense* VH8-4602) x *G. hirsutum* cv.Guazunchoz) which gave rise to 233 mapped loci. The 233 loci integrated to previously published map (Lascape et al., 2003) resulted in a genetic map comprising of 1,160 loci and 5,519 cM with an average distance of 4.8 cM between two loci. Rong et al. (2004) reported genetic maps for diploid (D) and tetraploid (AtDt) *Gossypium* genomes composed of sequence-tagged sites (STS) that foster structural, functional, and evolutionary genomic studies. The maps include, respectively, 2584 loci at 1.72-cM (approximately 600 kb) intervals based on 2007 probes (AtDt) and 763 loci at 1.96-cM (approximately 500 kb) intervals detected by 662 probes (D).They found no major structural changes between Dt and D chromosomes, but confirmed two reciprocal translocations between At chromosomes and several inversions. Further locus duplication patterns revealed all 13 expected homeologous chromosome sets.

Recently, sequence related amplified polymorphism (SRAP), a PCR based marker (Li and Quiros, 2001) has been used in linkage map construction. Lin et al. (2005) studied segregation of SRAP markers along with SSR and RAPD markers in sixty one F_2 progenies of 'Handan 208' (*G. hirsutum*) x 'Pima 90' (*G. barbadense*). A total of 566 loci have been assembled into 41 linkage groups and further twenty eight linkage groups were assigned to corresponding chromosomes by SSR markers with known chromosome locations. The map covered 5141.8 cM with a mean interlocus space of 9.08 cM. Park et al. (2005) reported the first fibre EST-microsatellite map constructed using a RIL population (*G. hirsutum* TM1 x *G. barbadense* Pima). The genetic map consists of 193 loci, including 121 new fiber loci not previously mapped which spanned 1277 cM providing approximately 27% genome coverage. Han et al. (2006) added 123 EST microsatellite loci to their back bone genetic map and developed a genetic map consisting of 907 loci and 5,060 cM, with an average between loci distance of 5.6 cM. Shen et al. (2005) constructed three genetic maps using F_2 population of three different crosses viz., (7235 x TM-

1), (HS 427-10 x TM-1) and (PD 6992 x SM 3) to tag QTS for fibre quality traits. Number of mapped loci were 86, 56 and 73 covering a length of 666.7, 557.8 and 588 cM, respectively.

A genetic linkage map with 70 loci (55 SSR, 12 AFLP and 3 morphological loci) was constructed using 117 F_2 plants obtained from a cross between two upland cotton cultivars Yumian1 and T586, which have relative high levels of DNA marker polymorphism and differ remarkably in fibre-related traits (Zhang et al., 2005). The linkage map comprised 20 linkage groups, covering 525 cM with an average distance of 7.5 cM between two markers, or approximately 11.8% of the recombinational length of cotton genome. Frelichowski et al. (2006) designed microsatellites from Acala 'Maxxa' BAC ends and utilized them to construct a map comprising 433 marker loci and 46 linkage groups with a genetic distance of 2,126.3 cM covering approximately 45% of the cotton genome and an average distance between two loci of 4.9 cM. Based on genome specific chromosomes identified in G. hirsutum, 56.9% of the coverage was located on the A subgenome while 39.7% was assigned to D subgenome in the genetic map, suggesting that the A subgenome may be more polymorphic and recombinationally active. Moreover the linkage groups have been assigned to 23 of the 26 chromosomes.

He et al. (2007) developed a genetic linkage map with an F_2 population derived from interspecific hybrids between G. hirsutum L. cv. Handan 208 and G. barbadense L. cv. Pima 90 using 834 SSRs, 437 SRAPs, 107 RAPDs, and 16 REMAPs (Retrotransposan Microsatellite Amplified Polymorphism). A total of 1029 genetic loci were mapped to 26 linkage groups that covered 5472.3 cM with an average between loci distance of 5.32 cM. Shen et al. (2007) constructed a SSR genetic linkage map consisting of 156 loci covering 1,024.4 cM using a series of recombinant inbred lines (RIL) developed from an F_2 population of an Upland cotton (G. hirsutum L.) cross 7235 and TM-1.

Genes mapped in cotton

In view of most measures of cotton quality and productivity of polygenic, QTL mapping is in high priority of many research programmes. Molecular markers hold promise for mapping many traits in cotton which have been introgressed from exotic germplasm, such as Verticillium wilt resistance and bacterial blight resistance, restoration of cytoplasmic male sterility (Weaver and Weaver, 1977) and improved fibre quality (Culp et al., 1979). Genetic mapping of 145 monogenic traits, characters and mutants has assembled 65 such marker genes into 18 linkage groups, with 13 being assigned into specific chromosomes (Percy and Kohel, 1999).

Information on QTL analysis has accumulated quickly, and will eventually help the manipulation of complex traits in cotton breeding (Tanksley, 1993). Jiang et al. (1998) using RFLP map of G. hirsutum x G. barbadense identi-

fied 14 QTLs for fibre related traits, and 10 QTLs were mapped on to D subgenome. Allele effects for 12 of the 14 were consistent with the difference between parents; with Gossypium barbadense alleles associated with long, strong, and fine fibers; and with G. hirsutum alleles associated with higher yield and early maturity. The two exceptions were fiber strength QTLs, where the G. barbadense alleles reduced fiber strength. In cotton several QTL studies have been done on inter-specific populations, but a few studies have used mapping populations derived from crosses within G. hirsutum (Shappley et al., 1998a). Shappley (1998a) investigated 19 agronomic and fibre traits and mapped 100 QTLs to 60 positions in 24 linkage groups. Their study revealed the influence of several QTLs on more than one trait. The most frequent association of QTL with multiple traits was for fibre traits related to maturity and fineness.

Jiang et al. (2000) used 180 F_2 plants from a cross of G. hirsutum and G. barbadense to map a total of 62 QTL for 14 different traits related to leaf morphology. They found that 38 (61.3%) QTLs mapped to the D-genome suggesting that the D-subgenome of tetraploid cotton has been subjected to relatively greater rate of evolution than the A subgenome, subsequent to polyploidy isolation. Twenty six QTLs were detected on nine linkage groups constructed from 119 $F_{2:3}$ progeny from a cross between MD 567 ne x Prema (Ulloa and Meredith, 2000). Two QTLs were detected for lint yield and three for lint percentage, explaining from 5 to 20% of the variation in each trait. Three QTLs for fibre strength (explaining 10.6 – 24.6% of the phenotypic variation), four for micronaire (explaining 6.2 – 21.7%), three for fibre strength (explaining 3.4 – 31.6%) and two for 2.5% span length (explaining 11.5 -44.6%) were detected. The clustering of QTLs positions on the linkage groups suggested that genes conferring fiber quality traits to be linked on the same chromosome(s).

Kohel et al. (2001) used an F_2 population derived from an interspecific cross between TM-1 (G. hirsutum) and 3-79 (G. barbadense) to map 13 QTLs. Four QTL influenceed bundle strength, three influenced fibre length and six influenced fibre fineness. These QTL collectively explained 30 to 60% of total phenotypic variation. The effect and mode of action of the individual loci characterized in the genetic background of TM-1 indicated that most of the QTLs for fibre quality properties appeared to be recessive, making marker assisted selection desirable and of great utility in fibre improvement. Paterson et al. (2003) mapped a total of 76 QTL for six fibre related traits (fibre length, length uniformity, elongation, strength, fineness and colour) under well watered Vs water-limited conditions. Using F_2 and F_3 populations derived from a cross between 7235 (G. anomalum introgression line) and TM-1 (genetic standard of upland cotton), Zhang et al. (2003) identified two QTLs for fibre strength; one was a major QTL, QTL_{FS1} explaining 30% of phenotypic variation mapped to chromosome 10. The QTL was detected in different locations such as Nanjing and Hainan in

China and Texas in USA. MAS revealed that DNA marker linked to this QTL could be used to increase fibre strength of commercial cultivars in early segregating breeding generations.

Mei et al. (2004) detected seven QTL for six fibre-related traits; five of these were distributed among A subgenome chromosomes, the genome donor of fibre traits. The detection of QTLs in both A subgenome in this study and D subgenome in a previous study. Jiang et al. (1998) suggest that fibre related traits are controlled by the genes in homoeologous genomes, which are subjected to selection and domestication. Out of111 EST-SSR markers derived from ESTs transcribed during fibre elongation in A genome *G.arboreum*, 72 were anchored to the At and 37 to Dt subgenome of allotetraploid cotton, a nearly 2:1 ratio (Han et al., 2004). Lin et al. (2005) reported 13 QTL associated with fibre traits, among which two QTL for fibre strength, four for fibre length and seven for micronaire value. These QTL were on nine linkage groups explaining 16.18 – 28.92% of the trait variation. Six QTLs were located in the A subgenome, six QTL in the D subgenome and one QTL in an unassigned linkage group.

Sarang et al. (2001) detected RFLP markers linked to leaf chlorophyll content, found that chlorophyll 'a' was correlated with dry matter under treatment with water-limited conditions and found one QTL for chlorophyll b on chromosome 14 associated with seed cotton yield. Investigation on BC_1 population of a cross TM-1 x Hai 7124 by Song et al. (2005) revealed thirty one QTLs, 10 for lobe length, 13 for lobe width, six for lobe angle and two for leaf chlorophyll content. The QTLs for leaf morphology were found to be distributed unevenly across the genome i.e. 29 QTLs were located on 15 of 34 linkage groups of their linkage map.

Zhang et al. (2005) used 117 $F_{2:3}$ plants from a cross between Yumian 1 and T586 to identify QTLs affecting lint per cent and fibre quality traits. Four QTLs for lint per cent, two QTLs for 2.5% span length, three QTLs for fibre length uniformity, three QTLs for fibre strength, two QTLs for fibre elongation and two QTLs for micronaire have been detected. Several QTLs affecting different fibre-related traits were detected within the same chromosome regions, suggesting that genes controlling fibre traits may be linked or the result of pleiotrophy. Shen et al. (2005) utilized three elite fibre lines of upland cotton (*G.hirsutum* L.) as parents and constructed three linkage maps to tag QTLs for fibre quality traits. A total of 39 QTLs (17 significant QTLs and 22 suggestive QTLs) affecting fibre traits were found in the three populations. Three common QTLs were identified from different genetic backgrounds.

Studies of Park et al. (2005) on genetic mapping of fibre loci using EST-derived microsatellites in an inter-specific recombinant inbred line population suggested that chromosomes 2, 3, 15 and 18 may harbour genes for traits related to fibre quality. Further investigation by Frelichowski et al. (2006) revealed that apart from the above mentioned chromosomes, loci on chromosome 12 may also affect

variation in fibre quality traits. In an investigation by Han et al. (2006), 62 and 61 EST-SSR markers (developed from *G. hirsutum*) were mapped on the At and Dt subgenome respectively essentially a 1:1 ratio. EST-SSR markers, developed from cotton developing fibres and ovule cells which were mapped on the D-subgenome revealed that there are some important genes for fibre development in the D-genome chromo-some. This may partially explain why QTL for fibre related traits are mapped in Dt genome chromosome in tetraploid cotton, even though D-genome species do not produce spinnable fibres (Jiang et al., 1998; Kohel et al., 2001; Mei et al., 2004). These results suggest there are complex evolutionary relationships between A, At, D and Dt genomes.

Abdurakhmonov et al. (2007) reported QTLs associated with lint percentage to be located on chromosomes 12, 18, 23 and 26. He et al. (2007) utilizing F_2 population (*G. hirsutum* L. cv. Handan 208 and *G. barbadense* L. cv. Pima 90), identified 4 QTLs for lint index, 8 for seed index, 11 for lint yield, 4 for seed cotton yield, 9 for number of seed per boll, 3 for fiber strength, 5 for fiber length, and 8 for micronaire value. A study on intra-hirsutum RIL population by Shen et al. (2007) using SSR markers revealed 25 major QTLS, 4 with large effects on fiber quality and 7 with large effects on yield components. In the present study chromosome D8 was densely populated with markers and QTLs, in which 36 SSR loci within a chromosomal region of 72.7 cM and 9 QTLs for 8 traits were detected.

Conclusion

From the above studies, comparison of QTLs revealed poor consistency among populations. Although some QTLs were found to be located on same chromosome in different populations, no common markers could prove that they were of the same QTL. Only a few stable and common QTLs have been reported up to now due to non-replicated experiments and difficulty in assignment of linkage groups. To identify stable QTLs we need to integrate different maps of interspecific population and for this it is important to work with a fixed population.

REFERENCES

(2002). Genetic distance among selected cotton genotypes and its relationship with F_2 performance. Crop Sci., 42: 1841-1847.

Abdalla AM, Reddy OUK, El-Zik KM, Pepper AE (2001). Genetic diversity and relationships of diploid and tetraploid cottons revealed using AFLP. Theor. Appl. Genet., 102: 222-229.

Abdurakhmonov IY, Buriev ZT, Saha S, Pepper AE, Musaev JA, Almatov A, Shermatov, SE, Kushanov FN, Mavlonov GT, Reddy, UK, Yu JZ, Jenkins, JN, Kohel RJ Abdukarimov A (2007). Microsatellite markers associated with lint percentage trait in cotton, *Gossypium hirsutum*. Euphytica, 156:141-156.

Altaf MK, Stewart JMCD, Wajahatullah MK, Zhang J (1997). Molecular and morphological genetics of a trispecies F_2 population of cotton. Proc. Beltwide Cotton Conf., pp. 448-452.

Arpat AB, Waugh M, Sullivan JP, Gonzales M, Frisch D, Main D, Wood

T, Leslie A, Wing RA, Wilkins TA (2004). Functional genomics of cell elongation in developing cotton fibres. Plant Mol. Biol. Rep., 54: 911-929.

Arumuganathan K, Earl ED (1991). Nuclear DNA content of some important plant species. Pl. Mol. Biol. Rep., 9: 208-218.

Bernatzky R, Tanksley SD (1986). Towards a saturated linkage map in tomato based on isozyme and random cDNA sequences. Genetics, 112: 887-898.

Bertini HC, Schuster I, Sediyama T, Barros, EGD, Moreira MA (2006). Characterization and genetic diversity analysis of cotton cultivars using microsatellites. Genetics and Molecular Biology, 29(2): 321-329.

Brubaker CL, Paterson AH, Wendel JF (1999). Comparative genetic mapping of allotetraploid cotton and its diploid progenitors. Genome, 42: 184-203.

Brubaker CL, Wendel JF (1994). Reevaluating the origin of domesticated cotton (Gossypium hirsutum L.) using nuclear restriction fragment length polymorphism (RFLPs). American. J. Bot., 81: 1309-1326.

Cherry JP, Katterman FRH, Endrizzi JE (1972). Seed esterases, leucine, and catalases of species of the genus Gossypium. Theor. Appl. Genet., 42: 218–226.

Connell JP, Pammi S, Iqbal MJ, Huizinga T, Reddy AS (1998). A high through put procedure for capturing microsatellites from complex plant genomes. Pl. Mol. Bio. Rep., 16: 341-349.

Culp TW, Harrel DC, Kerr T (1979). Some genetic implications in the transfer of high fibre strength genes to upland cotton. Crop Sci., 19: 481-484.

Dib C, Faure S, Fizamer CD, Samson N, Drouot A, Vignal P, Millaseau S, Marc J. Hazan, Seboun E, Lathrop M, Gyapay, Morissette GJ, Wellssenbach J (1996). A comprehensive genetic map of the human genome based on 5264 microsatellites. Nature, 380: 152-154.

Dietrich WF, Miller J, Steen R, Merchant MA, Damronboles D, Husain Z, Dredge R, Dely MJ, Ingalls KA, O'Connor TJ, Evans CA, De Angelis MM, Levinson DM, Kruglyak L, Goodmann N, Copeland NG, Jenkins NA, Hawkins TL, Stein L, Page DC, Lander, ES (1996). A comprehensive genetic map of the mouse genome. Nature, 380 : 149-152.

Doerge RW, Cairg BA (2000). Model selection for quantitative trait locus analysis in polyploids.Proc. Natl. Acad. Sci., USA., 97:7952-7956.

Ellegien H (1993). Genomic analysis with microsatellite markers. Ph.D Dissertation. University of Agricultural Science, Swedish (unpublished).

El-Zik KM, Thaxton PM (1989). Genetic improvement for resistance to pests and stresses in cotton. In: Integrated Pest Management Systems and Cotton Production. R.E. Frisbie, K.M. El-Zik and L.T. Wilson (eds.). John Wiley & Sons, NY, pp. 191-224.

Frelichowski JE, Palmer MB, Main D, Tomkins JP, Cantrell RG, Stelly DM, Yu J, Kohel RJ, Ulloa M(2006). Cotton genome mapping with new microsatellites from Acala 'Maxxa' Bac-ends. Mol. Gent. Genomics, 275:479–491.

Geng CD, Gong ZZ, Huang JQ, and Zhang ZC (1995). Identification of difference between cotton cultivars (G. hirsutum) using the RAPD method. Jiangsu J. Agric. Sci., 11: 21-24.

Guo WZ, Sang ZQ, Zhou BL, Zhang TZ (2007). Genetic relationships of D-genome species based on two types of EST-SSR markers derived from G. arboreum and G. raimondii. Plant Science, 172(4): 808-814.

Gupta PK, Balyan HS, Sharma PC, Ramesh B(1996). Microsatellites in plants: a new class of molecular markers. Current Sci. 70:45-53.

Gutierrez OA, Babu S, Saba S, Jenkins JN, Shoemaker DB, Cheathem C, McCarty Jr JC

Gwoli S, Xia CR,Bo WK, Hui LS, Fa ZJ, Hua GJ (1999). Analysis of genetic diversity of Australian species of Gossypium using RAPD. Acta Gossypi Sinica, 11: 65-69.

Han Z, Guo W, Song X, Zhang T (2004). Genetic mapping of EST-derived microsatellites from the diploid Gossypium arboreum in allotetraploid cotton. Mol. Genet. Genomics, 272: 308-327.

Han Z, Wang C, Song X, Guo W, Gou J, Li C, Chen X Zhang T(2006). Characteristics development and mapping of Gossypium hirsutum derived EST-SSRs in allotetraploid cotton. Theor. Appl. Genet., 112: 430-439.

He DH, Lin ZX, Zhang XL, Nie YC, Guo XP, Zhang YX, Li W (2007).

QTL mapping for economic traits based on a dense genetic map of cotton with PCR-based markers using the interspecific cross of Gossypium hirsutum Vs Gossypium barbadense. Euphytica 153(1-2): 181-197.

Iqbal, MJ, Aziz N, Saeed NA, Zafar Y, Malik KA (1997). Genetic diversity evaluation of some elite cotton varieties by RAPD analysis. Theor. Appl. Genet., 87: 934-940.

Iqbal, MJ, Reddy OUK, El-Zik KM, Pepper AE(2001). A genetic bottleneck in the evolution under domestication of upland cotton Gossypium hirsutum examined using DNA fingerprinting. Theor. Appl. Genet., 103: 547-554.

Jiang C, Wright RJ, Woo SS, DelMonte TA, Paterson AH (2000). QTL analysis of leaf morphology in tetraploid Gossypium. Theor. Appl. Genet., 100: 409-418.

Jiang CX, Wright RJ, El-Zik KM, Paterson AH (1998). Polyploid formation created unique avenues for response to selection in Gossypium (cotton). Proc. Natl. Acad. Sci. USA., 95: 4419–4424.

Jones CJ, Edwaeds KJ, Castaglione S, Winfield MO, Sela F, Van De Weil C, Bredemeijer G, Vosman B, Matthes M, Daly A, Brettschneider R, Bettni P, Buitti M, Maestri E, Malcevschi A, Marmiroli N, Aert R, Volckaert G, Rueda J, Linacero R, Vazquetz A, Karp A(1997). Reproducibility testing of RAPD, AFLP and SSR markers in plants by a network of European laboratories. Mol. Breed., 3: 381-390.

Kantety RV, La Raton M, Matthews DE, Sorrels ME (2002). Data mining for simple sequence repeats in the expressed sequence tags from barley, maize, rice, sorghum and wheat. Plant Mol. Bio. Rep., 48: 501-510.

Kearsey M, Pooni H (1996). The genetical analysis of quantitative traits. Chapman and Hall, London, pp. 81-90.

Khan MA, Zhang J, Stewart MCDJ (1998). Integrated molecular map based on a trispecific F_2 population on cotton. Proc. Beltwide Cotton Improvement Conference (ed. D.J. Herber and D.A. Richter), San Dieogo, CA, pp. 491-492.

Khan MA, Stewart JMCD, Zhang J, Myers GO, Cantrell RG (1999). Addition of new markers to trispecific cotton map. Proc. Beltwide Cotton Conf., pp. 439.

Khan SA, Hussain D, Askari E, Stewart JM, Malik KA, Zafar Y (2000). Molecular phylogeny of Gossypium sp. by DNA fingerprinting. Theor. Appl. Genet., 101: 931-938.

Kohel RJ, Yu J, Park YH, Lazo R (2001). Molecular mapping and characterization of traits controlling fibre quality in cotton. Euphytica, 121: 163-172.

Krishna TG (1998). DNA based markers for identification of inbreds and hybrid in cotton. Hybrid Crops Workshop, Pantnagar (UP), India, pp. 1-3.

Krishna TG, Jawali N (1997). DNA isolation from single or half seeds suitable for random amplified polymorphic DNA analysis. Anal. Biochem., 250: 125-127.

Kumar P, Singh K, Vikal Y, Radhawa LS, Chahal GS (2003). Genetic diversity studies of elite cotton germplasm lines using RAPD markers and morphological characteristics. Indian J. Genet., 63(1): 5-10.

Kumpatla SP, Horne EC, Shah MR, Gupta M, Thompson SA (2002). Development of SSR markers : towards genetic mapping in cotton (Gossypium hirsutum L.). 3rd International cotton genome initiative workshop, Nanjing, China. Cotton Sci., 14(1): 28.

Lan TH, Cook CG, Paterson AH (1999). Identification of a RAPD marker linked to male fertility restoration gene in cotton (Gossypium hirsutum L.). J.ournal of Agri. Genomics., 1: 1-5.

Lander ES, Green P, Abrahamson J, Barlow A, Daly MJ, Lincoln SE, Newburg L (1987). Mapmaker: An interactive computer package for constructing primary genetic linkage maps of experimental and natural population. Genomics, 1: 174-181.

Lascape JM, Dessauw D, Rajab M, Noyer JL, Hau B (2007). Microsatellite diversity in tetraploid Gossypium germplasm: assembling a highly informative genotyping set of cotton SSRs. Mol. Breed. ,19(1): 45-58.

Lascape JM, Nguyen TB, Thibivilliers S, Bojinov B, Courtois B, Cantrell RG, Burr B, Hau B (2003). A combined RFLP-SSR –AFLP map of tetraploid cotton based on a Gossypium hirsutum x Gossypium barbadense backcross population. Genome, 46: 612-626.

Li G, Quiros CF (2001). Sequence – related amplified polymorphism

(SRAP), a new marker system based on a simple PCR reaction : its application to mapping and gene tagging in *Brassica*. Theor. Appl. Genet., 103: 455-461.

Lin Z, He D, Zhang X, Nie Y, Guo X, Feng C, Stewart JMCD (2005). Linkage map construction and mapping QTL for cotton fibre quality using SRAP, ,SSR and RAPD. Plant Breeding, 124 : 180-187.

Litt M, Lutty JA (1989). A hypervariable microsatellite revealed by *in vitro* amplification of a dinucleotide repeat within the cardiac muscle actin gene. Am. J. Hum. Genet.., 44: p. 397.

Liu S, Saha S, Stelly D, Burr B, Cantrell RG(2000). Chromosomal assignment of microsatellite loci in cotton. J. Hered., 91: 326-332.

Liu B (1998). Statistical Genomics : Linkage, mapping and QTL analysis. CRC press, Boca Raton.

Liu B, Wendel JF (2001). Intersimple sequence repeat (ISSR) polymorphisms as a genetic marker system in cotton. Molecular Ecology Notes, 1: 205-208.

Livingstone KD, Churchill G, Jahn MK (2000). Linkage mapping in population with karyotypic rearrangements. J. Hered.., 91: 423-428.

Lu H, Meyers GO (2002). Genetic relationships and discrimination of ten influential upland cotton varieties using RAPD markers. Theor. Appl. Genet. 105:325–331.

Manly KF, Cudmore H, Robert, Meer JM (2001). Map manager QTX, cross-platform software for genetic mapping. Mamm. Genome, 12: 930-32.

Martsinkovskaya, AI, Moukhamedov RS, Abdukarimov AA (1996). Potential use of PCR amplified ribosomal intergenic sequence for differentiation of varieties and species of *Gossypium* cotton. Plant Mol. Bio. Rep., 14(1): 44-49.

Maughan PJ, Seghai MA, Maroof GR, Buss, Huestis GM (1996). Amplified fragment length polymorphism in soyabean : species diversity, inheritance and near –isogenic line analysis. Theor. Appl. Genet., 93: 392-401.

Mei M, Syed NH, Gao W, Thaxton PM, Smith CW, Stelly DM, Chen ZJ (2004). Genetic mapping and QTL analysis of fibre related traits in cotton. Theor. Appl. Genet., 108: 280-291.

Meredith WR(1992).RFLP association with varietal origin and heterosis. In: Proc. Beltwide Cotton Conf. (ed. D. Herber), Nashville, TN. pp. 607.

Mergeai G, Baudoin JP, Vroh BI (1998). Production of high gossypol cotton plants with low gossypol seed from trispecific hybrids including *Gossypium sturtianum*. World Cotton Research Conference-2, Athens, Greece, pp. 74.

Michaelson MJ, Price HJ, Ellison JR, Johnston JS (1991). Comparison of plant DNA contents determined by Feulgen microspectrophotometry and laser flow cytometry. American J. Bot., 78(2): 183-188.

Mohan M, Nair S, Bhagwat A, Krishna TG, Yano M, Bhatia CR, Sasaki T (1997). Genome mapping, molecular markers and marker-assisted selection in crop plants. Mol. Breed., 3: 87-103.

Multani DS, Lyon BR (1995). Genetic fingerprinting of Australian cotton cultivars with RAPD markers. Genome, 38: 1005-1008.

Nguyen TB, Giband M, Brottier P, Risterucci AM, Lascape JM (2004). Wide coverage of the tetraploid cotton genome using newly developed microsatellite markers. Theor. Appl. Genet., 109 : 167-175.

Park YH, Alabady MS, Ulloa M, Sickler B, Wilkins TA, Yu J, Stelly DM, Kohel RJ, El-Shiny OM, Cantrell RG (2005). Genetic mapping of new cotton fibre loci using EST-derived microsatellites in an interspecific recombinant inbred (RIL) cotton population. Mol. Genet. Genomics, 274: 428-441.

Paterson AH (1996). Making genetic maps. In : Genome mapping in plants (ed. A.H. Paterson), Academic Press, Austin, Texas, pp. 23-29.

Paterson AH, Brubaker CL, Wendel JF (1993). A rapid method for extraction of cotton genomic DNA suitable for RFLP or PCR analysis. Plant Mol. Biol. Rep., 11(2): 122-127.

Percy RG, Kohel RG(1999). Cotton qualitative genetics. In: Cotton (ed. C.W. Smith and J.T. Cothren), John Wiley & Sons, NY, pp. 319-360.

Permingeat HR, Romangnoli MV, Ruben H (1998). A simple method for isolating high yield and quality DNA from cotton (*Gossypium hirsutum* L.) leaves. Plant Mol. Biol. Rep., 16:1-6.

Pillay M, Meyers GO (1999). Genetic diversity assessed by variation in

ribosomal RNA genes and AFLP markers. Crop Sci., 39: 1881-1886.

Qureshi SN, Saha S, Kantety RV, Jenkins JN (2004). EST-SSR:a new class of genetic markers in cotton. J. Cotton Sci., 8: 112123.

Rafalski JA (1997). Randomly amplified polymorphic DNA (RAPD) analysis. In : DNA markers : Protocols, Applications and Overviews (eds. G.C.Anolles and P.M. Gresshoff), Wiley-Liss, Inc. USA, p. 364.

Rana MK, Bhat KV (2004a). A comparison of AFLP and RAPD markers for genetic diversity and cultivar identification in cotton. J. Plant Biochemistry and Biotechnology, 13: 19-24.

Rana MK, Bhat KV (2004b). Analysis of molecular variance in cotton (*Gossypium* sp.) using RAPD markers. Indian J. Genet., 64: 85-86.

Reddy OUK, Brooks TD, El-Zik KM, Pepper AE (2000). Development and use of PCR-based technologies for cotton mapping. Proc. Beltwide Cotton Res. Conf., 4-8 Jan, San Antonio, Texas, pp. 483.

Reddy OUK, Pepper AE, Abdurakhmonov I, Saha S, Jenkins JN, Brooks T, Bolek Y and El-Zik KM (2001). New dinucleotide and trinucleotide microsatellite marker resources for cotton genome research. J. Cotton Sci, 5: 103-113.

Reinisch MJ, Dong J, Brubaker CL, Stelly DM, Wendel JF, Paterson AH (1994). A detailed RFLP map of cotton, *Gossypium hirsutum x Gossypium barbadense* : chromosome organization and evolution in a disomic polyploid genome. Genetics, 138: 829-847.

Risch N(1992). Genetic linkage : Interpreting LOD scores. Science, 255: 803-804.

Roder MS, Korzun V, Wendehake K, Plaschke J, Tixier MH, Leroy P, Ganel MA(1998). A microsatellite map of wheat. Genetics, 149: 2007-2023.

Rong J, Abbey C, Bowers JE, Brubaker CL, Chang C, Chee PW, Delmonte TA, Ding X, Garza JJ, Marler BS, Park C, Pierce GJ, Rainey KM, Rastogi VK, Schulze SR, Trolinder NL, Wendel JF, Wilkins TA, Williams-Coplin TD, Wing RA, Wright RJ, Zhao X, Zhu L, Paterson AH (2004). A 3347-locus genetic recombination map of sequence-tagged sites reveals features of genome organization, transmission and evolution of cotton (*Gossypium*). Genetics, 166: 389-417.

Rungis D, Llewellyn D, Dennis ES, Lyon BR (2005). Simple sequence repeat (SSR) markers reveal low levels of polymorphism between cotton (*Gossypium hirsutum* L.) cultivars. Australian J. Agri. Res., 56: 301-307.

Saha S, Callahan F, Dollar D, Creech J (1997). Effect of lyophiliozation of cotton tissue on quality of extractable, DNA, RNA and protein. J. Cotton Sci., 1: 11-14.

Saha S, Karaca M, Jenkins JN, Zipf AE, Reddy UK, Kantey RV (2003). Simple sequence repeats as useful resources to study transcribed genes of cotton. Euphytica, 130: 355-364.

Sarang Y, Menz M, Jiang CX, Wright RJ, Yakir D, Paterson AH (2001). Genomic dissection of genotype x environment interactions conferring adaptation of cotton to arid conditions. Genome Res., 11: 1988-1995.

Schmidt T, Heslop HJs (1998). Genomes, genes and junk : the large-scale organization of plant chromosome. Trends in Plt. Sci., 3: 195-198.

Shappley ZW, Jenkins JN, Meredith WR, McCarty JC (1998b). An RFLP linkage map of upland cotton, *Gossypium hirsutum* L. Theor. Appl. Genet., 97: 756-761.

Shappley ZW, Jenkins JN, Watson CE, Kohler AL, Meredith WR (1996). Establishment of molecular markers and linkage groups in two F$_2$ population of upland cotton. Theor. Appl. Genet., 92: 915-919.

Shappley ZW, Jenkins JN, Zhu J, McCarty JC (1998a). Quantitative trait loci associated with agronomic and fibre traits of upland cotton. J. Cotton Sci., 4: 153-163.

Shen X, Guo W, Lu Q, Zhu X, Yuan Y, Zhang T (2007). Genetic mapping of quantitative trait loci for fiber quality and yield trait by RIL approach in Upland cotton. Euphytica, 155: 371-380

Shen X, Guo W, Zhu X, Yuan Y, Yu JZ, Kohel RJ, Zhang T (2005). Molecular mapping of QTLs for qualities in three diverse lines in upland cotton using SSR markers. Mol. Breed., 15: 169-181.

Shu B, Fenling K, Yao ZY, Mei ZG, Yuan ZQ, Gang WX (2001). Genetic diversity analysis of representative elite cotton varieties in three main cotton regions in China by RAPD and its relation with agronomic characteristics. Scientia Agricultura Sinica, 34: 597-603.

Siu L, Saha S, Stelly D, Burr B, Cantrell R (2000). Chromosomal assignment of microsatellite loci in cotton. J.Hered., 91(4):326-332.

Song XL, Wang K, Guo WZ, Han ZG, Zhang TZ (2005). Quantitative trait loci mapping of leaf morphological traits and chlorophyll content in cultivated tetraploid cotton. Acta Botanica Sinica, 47(11): 1382-1390.

Stelly DM (1993). Interfacing cytogenetics with the cotton genome mapping effort. In : Proc. Beltwide cotton improvement conference (ed. D.J. Herber and D.A. Richter), pp. 1545-1550.

Tabar MV, Chandrashekaran S, Rana MK, Bhat KV (2004). RAPD analysis of genetic diversity in Indian tetraploid and diploid cotton (Gossypium sp.). J. Plant Biochem.istry and, Biotechnol.ogy, 13: 81-84.

Tanksley SD (1983). Molecular markers in plant breeding. Plant Mol. Biol. Rep., 1 : 3-8.

Tanksley, S.D. (1993). Mapping polygenes. Annu. Rev. Genet., 27: 205-233.

Tatineni V, Cantrell RG, Davis DD (1996). Genetic diversity in elite cotton germplasm determined by morphological characteristics and RAPDs. Crop Sci., 36: 186-192.

Tomkins JP, Peterson DG, Yang TJ, Main D, Wilkins TA, Paterson, AH, Wing RA (2001). Development of genomic resources for cotton (Gossypium hirsutum): BAC library construction, preliminary STC analysis and identification of clones associated with fibre development. Mol. Breed., 8(3): 255-261.

Ulloa M, Cantrell RG, Percy RG, Lu Z, Zeiger E (2000). QTL analysis of stomatal conductance and relationship to lint yield in interspecific cotton. J. Cotton Sci., 4: 10-18.

Ulloa M, Meredith WR (2000). Genetic linkage map and QTL analysis of agronomic and fibre quality traits in an intraspecific population. J. Cotton Sci., 4: 161-170.

Ulloa M, Meredith WR (Jr), Shappley ZW, Kahler AL (2002). Genetic linkage maps from four F$_{2:3}$ populations and a join maps of Gossypium hirsutum L. Theor. Appld. Genet., 101: 200-208.

Van Esbroeck GA, Bowman DT, Calhoun DS, OL May (1998). Changes in the genetic diversity of cotton in the U.S. from 1970 to 1995. Crop Sci. 38:33-37.

Vos P, Hogers R, Blecker M, Reijans M, Van de Lee T, Hornes M, Fritjters A, Pot J, Peleman J, Kuiper M, M Zabeau. (1995). AFLP: a new technique for DNA fingerprinting. Nucleic Acids Res., 23: 4407-4414.

Vroh BI, Baudoin JP, Mergeai G (1998). Potentialities of DNA markers to monitor introgression of the glandless-seed and glanded plant trait from Gossypium sturtianum into upland cotton. World Cotton Research Conference-2, Athens, Greece, pp. 54.

Vroh BI, Harvenge L, Chandelier A, G Margaei, Jaredin PD (1996). Improved RAPD amplification of recalcitrant plant DNA by the use of activated charcoal during DNA extraction. Plant Breeding, 155(3): 205-206.

Vroh BI, Jardin PD, Mergeai G, Baudoin JP (1997). Optimisation and application of RAPD in a recurrent selection programme of cotton. Biotechnology, Agronomy, Society and Environment, 1(2): 142-150.

Vroh BI, Mergeai G, Baudoin JP, Jardin PD (1999). Breeding for "low-gossypol seed and high-gossypol plants" in upland cotton. Analysis of trispecies hybrids and backcross progenies using AFLP's and mapped RFLP's. Theor. Appl. Genet., 99: 1233-1244.

Wajahatullah MK, Stewart JM (1997). Genomic affinity among Gossypium subgenus sturtia species by RAPD analysis. Proc. Beltwide Cotton Conf., National Cotton Council , Memphis, TN, USA, pp. 452.

Wajahatullah MK, Stewart JM, Zhang J (1997). Use of RAPD markers to analyse genomic affinity among Australian Gossypium species. Special Report -Agricultural Experiment Station, Division of Agriculture, University of Arkansas, 183: 150-152.

Wang G, Wing R, Paterson AH (1993). PCR amplification of DNA extracted from single seeds, facilitating DNA-marker assisted selection. Nucl. Acids Res., 21: 2527.

Wang XY, Guo WZ, Zhang TZ, Pan JJ (1997). Analysis of RAPD fingerprinting on short-seasonal cotton cultivars in China. Acta Agronomica Sinica, 23: 669-676.

Weaver, D.B,. and J.B. Weaver JB. (1977). Inheritance of pollen fertility restoration in cytoplasmic male-sterile upland cotton. Crop Sci., 17: 497-499.

Wendel JF, Brubaker CL, Percival AE (1992). Genetic diversity in Gossypium hirsutum and the origin of upland cotton. Am.erican. J. Bot., 79: 1291-1310.

Wendel JF, Rowley R, Stewart J McD (1994). Genetic diversity in and phylogenetic relationships of the Brazilian endemic cotton, Gossypium mustelinum (Malvaceae). Plant Systematics and Evolution, 192:49-59.

Williams J, Kubelik A, Liviak JL, Rafalski JA, Tingey SV (1990). DNA polymorphism amplified by random primers are useful as genetic markers. Nucleic acid Res., 18: 6531-6535.

Wright R J, Thaxton PM, El-Zik KM, Paterson AH (1998).D-subgenome bias of Xcm resistance genes in tetraploid Gossypium suggests that polyploid formation has created novel avenues for evolution. Genetics, 149: 1987-1996.

Wright R J, Thaxton PM, El-Zik KM, Paterson AH (1999). Molecular mapping of genes affecting pubescence of cotton. The American Genetic Aassociation., 90: 215-219.

Wu K, Burnquist W, Sorrells ME, Tew T, Moore P, Tanksley SD (1992). The detection and estimation of linkage in polyploids using singledose restriction fragments. Theor. Appl. Genet., 83: 294–300.

Young ND (1994). Constructing a plant genetic linkage map with DNA markers. In: DNA-based markers in plants (eds. I.K.V. Ronald & L. Phillips), Kluwer, Dordrecht, Boston, London. pp. 39-57

Young ND (1996). QTLmapping and quantitative disease resistance in plants. Annu Rev Phytopathol , 34: 479–501.

Yu ZH, Park YH, Lazo GR, Kohel RJ (1997). Molecular mapping of the cotton genome. Proc. of 5th International Congress of Plant Molecular Biology, September 21-27, Singapore. p.147.

Zhang ZS, Xiao YH, Luo M, Li XB, Luo XY, Hou L, Li DM, Pei Y (2005). Construction of a genetic linkage map and QTL analysis of fibre related traits in upland cotton. Euphytica, 144: 91-99.

Zhang J, Guo W, Zhang T (2002). Molecular linkage map of allotetraploid cotton (Gossypium hirsutum L. x Gossypium barbadense L.) with a haploid population. Theor. Appl. Genet., 105: 1166-1174.

Zhang TZ, Yuan YL, Yu J, Guo WZ, Kohel RJ(2003). Molecular tagging of major QTL for fibre strength in upland cotton and its marked assisted selection. Theor. Appl. Genet., 106:262-268.

6

Agrobacterium-mediated transformation of plants: emerging factors that influence efficiency

Jelili T. Opabode

Department of Plant Science, Obafemi Awolowo University, Ile-Ife, Nigeria.

Despite production of fertile transgenic plants through transformation mediated by Agrobacterium tumefaciens, transformation efficiency is still low. Apart from plant genotype, Agrobacterium strains, plasmid vectors, virulence (vir) gene inducing compounds, medium composition and tissue specific factors, some other factors are becoming important for improving transformation efficiency of plant species. Sucrose treatment of explant increased T-DNA delivery in rice while desiccation improved the T-DNA delivery and stable transformation of sugarcane, maize, wheat and soybean. Silver nitrate suppresses the Agrobacterium growth and facilitates plant cell recovery that resulted in increased efficiency of transformation in wheat. Inclusion of thiol compounds, L-cysteine, dithiothreitol and sodium thiosulphate in co-cultivation medium increased transformation efficiency as high as 16.4% in soyabean. A temperature of 22^0C was found to be optimal for T-DNA delivery in tobacco. The optimal temperature for both T-DNA delivery and stable transformation was $23-25^0C$ for wheat and $\sim23^0C$ for maize. Surfactants Silwet 77, pluronic acid F68, Tween 20 enhanced T-DNA delivery in wheat. Evidence that Agrobacterium density, co-culture medium, antibiotic and selectable marker influence T-DNA delivery and integration and stable transformation of plants were also presented.

Key words: Agrobacterium, stable transformation, T-DNA delivery, T-DNA integration, transformation efficiency.

INTRODUCTION

Agrobacterium tumefaciens causes crown gall disease of a wide range of plants, especially members of the rose family such as apple, pear, peach, cherry, almond, raspberry and roses. The discovery of the bacterial origin of crown gall disease (Smith and Townsend, 1907) sparked a number of studies with understanding the mechanisms of oncogenesis in general and applied it to study of cancer disease in animals and humans as objectives. The elegant work of Binns and Thomashaw (1988) which revealed that A. tumefaciens is capable of transferring a particular DNA segment Transfer (T)-DNA of the tumour-inducing (Ti) plasmid into the nucleus of infected cells where it is subsequently integrated into the host genome, changed the objectives of research on A. tumefaciens to transformation of plants. Early realization of this goal was brighten with the report that the T-DNA contains two types of genes: the oncogenic genes, encoding for enzymes involved in the synthesis of auxins

and cytokinins and responsible for tumour formation; and the genes encoding for the synthesis of opines, a product resulted from condensation between amino acids and sugars, which are produced and excreted by the crown gall cells and consume by A. tumefaciens as carbon and nitrogen sources. Outside the T-DNA, are located the genes for the opine catabolism, the genes involved in the process of T-DNA transfer from the bacterium to the plant cell and for the bacterium-bacterium plasmid conjugative transfer genes (Zupan and Zambrysky, 1995).

Virulent strains of A. tumefaciens contain a large megaplasmid (more than 200 kb) that plays a key role in tumour induction and for this reason it was named Ti plasmid. The transfer is mediated by the co-operative action of proteins encoded by genes determined in the Ti plasmid virulence region (vir genes) and in the bacterial chromosome. The 30 kb virulence (vir) region is a region organised in six operons that are essential for the T-DNA

Table 1. *Agrobacterium*-mediated transformation of some dicots plants

Host plant	Strain plasmid	marker	Explant	TF(%)	Reference
Pigeon pea (*Cajanus cajan* L.)					
ICP787	LBA4404 (pdhdps-GUS)	*npt*II	CN	93.2	Thu et al.2002
Broad bean *Vicia faba* L.)					
Lobab lippoi	C58C1 (pArA4b)	none	IS	92.5	Jelenic et al.2000
Canola (*Brassica napus* L.)					
Westars	GV3850 (pBinmGFP5-ER)	*npt*II	H	17.0	Cardoza and Stewart 2003
Maplus	GV3850 (pNK55-Resy.KCS)	*npt*II	MP	25.0	Wang et al.2005
Chickpea (*Cicer arientum* L.)					
Semsen	AGL1 (pRM50)	*npt*II	CN	0.5	Sarmah et al 2004
Soybean (*Glycine max* L. Merill)					
Lambart	LBA4404 (pCAMBIA 1303)	*hpt*	CN	16.4	Olhoft et al 2003
Cotton (*Gossypium hirsutum*)					
Ekang 9	LBA4404 (pBin438)	*npt*II	EC	33.0	Wu et al.2005

TF-Transformation frequency; *npt*II-neomycin phosphotransferase;CN-Cotyledonary node;EC-embryonic calli ;MP-Mesopyhll protoplast; H-Hypocotyl; IS-Internodal segment

Table 2. *Agrobacterium*-mediated transformation of some monocot plants

Host plant	Strain (plasmid)	Marker	Explant	TF(%)	Reference
Banana(*Musa* spp.)					
Grand Nain (AAA)	LBA4404 (pBI141)	*npt*II	MCS	2.0	May et al. 1995
Barley (*Hordeum vulgare* L.)					
Winter (igri)	LBA4404 (pSBI: VG35PAT)	hpt	PC	2.2	Kumlehn et al.2006
Rice (*Oryza sativa* L.)					
Indica (basmati 370)	EHA101 (pIGI21Hm)	hpt	EC	22	Rashid et al.1996
Japonica (Taipei 309)	LBA4404 (pTOK233)	hpt	PCIE	3.0	Uze et al.1997
Rye (*Secale cereale* L.)					
Spring (L22)	AGLO (pJFnptII)	*npt*II	PCIE	3.5	Popelka and Altpeter,2003
Sugarcane (*Saccharium officinarium* L.)					
Ja60-5	LBA4404 (pBI141)	*hpt*	SC	0.94-1.15	Arencibia et al.,1998
Sorghum (*Sorghum bicolor* L.)					
C401	EHA101 (pPZP201)	*pmi*	IE	3.3	Gao et al. 2005
Pioneer 8505	EHA101 (pPZP201)	*pmi*	IE	2.8	Gao et al.2005
Maize (*Zea mays* L.)					
A188	EHA101 (pTF102)	Bar	FIIE	5.5	Frame et al.2002
A188	LBA4404 (pTOK233)	*hpt*	FIIE	11.8-30.6	Ishida et al.1996
Wheat (*Triticum aestivum* L.)					
Spring(Bobwhite)	ABI (pMON18365)	*npt*II	EC	10.5	Cheng et al.2003
Winter(Candenza)	AGLI (pAL151)	Bar	IE	1.7	Wu et al.2003

TF-Transformation frequency; *npt*II-neomycin phosphotransferase;
hpt-hygromycin phosphotranferase;pmi-phosphomannose isomerase
Bar-bialaphos-resistant gene;PCIE-Precultured immature embryo
EC-Embryogenic calluses;FIIE-Freshly isolated immature embryo;SC-suspension culture;IE-Immature embryo;MCS-Meritem corm slices
PC-pollen culture

transfer (virA, virB, virD, and virG) or for the increasing of transfer efficiency (virC and virE) (Zupan and Zambrysky, 1995; Jeon et al., 1998).

The initial results of the studies on T-DNA transfer process to plant cells demonstrate three important facts for the practical use of this process in plants transformation. Firstly, the tumour formation is a transformation process of plant cells resulted from transfer and integration of T-DNA and the subsequent expression of T-DNA genes. Secondly, the T-DNA genes are transcribed only in plant cells and do not play any role during the transfer process. Thirdly, any foreign DNA placed between the T-DNA borders can be transferred to plant cell, no matter where it comes from. These well established facts, allowed the construction of the first vector and bacterial strain systems for plant transformation (Rival et al., 1998; Opabode 2002)

The first record on transgenic tobacco plant expressing foreign genes appeared at the beginning of the last decade. Since that crucial moment in the development of plant science, a great progress in understanding the Agrobacterium-mediated gene transfer to plant cells has been achieved. However, Agrobacterium tumefaciens naturally infects only dicotyledonous plants and many economically important plants, including the cereals, remained accessible for genetic manipulation by other methods. For these cases, alternative direct transformation methods have been developed such as polyethyleneglycol-mediated transfer, microinjection, protoplast and intact cell electroporation and gene gun technology (Rival et al., 1998). However, Agrobacterium-mediated transformation has remarkable advantages over direct transformation methods, including preferential integration of defined T-DNA into transcriptionally active regions of the chromosome (Czernilofsky et al., 1986; Koncz et al., 1989, Le et al., 2001; Olhoft et al., 2004) with exclusion of vector DNA (Hiei et al., 1997; Fang et al., 2002), unlinked integration of co-transformed T-DNA (McKnight et al., 1987; Komari et al., 1996; Hamilton, 1997; Olhoft et al., 2004). The transgenic plants are generally fertile and the foreign genes are often transmitted to progeny in a Mendelian manner (Rhodora and Thomas, 1996).

Agrobacterium-mediated gene transfer into monocotyledonous plants was not possible until recently, when reproducible and efficient methodologies were established on rice, banana, corn, wheat, and sugarcane (Hiei et al., 1994; Cheng et al., 1998; May et al., 1995; Ishida et al., 1996; Enriquez-Obregon, 1998; Arencibia et al., 1998). Reviews on plant transformation using Agrobacterium tumefaciens and the molecular mechanisms involved in this process have been published during the last years (Hooykas and Schilperoort, 1992; Zupan and Zambrysky, 1995; Rival et al., 1998; Zupan et al., 2000; Cheng et al., 2004).

The transfer of T-DNA and its integration into the plant genome is influenced by several A. tumefaciens and plant tissue specific factors. These include plant genotype, explant, vectors-plasmid, bacteria strain, addition of vir-gene inducing synthetic phenolics compounds, culture media composition, tissue damage, suppression and elimination of A. tumefaciens infection after co-cultivation (Alt-morbe et al., 1989; Bidney et al., 1992; Hoekema et al., 1993; Hiei et al., 1994; Komari et al., 1996; Nauerby et al., 1997; Klee, 2000). Some of these factors are summarized in Tables 1 and 2 for selected plant species. Recently, some other factors have been found important in influencing the efficiency of Agrobacterium -mediated genetic transformation of crops. This review shall summarize those factors for further ptimization of existing transformation protocols and establishment of new ones for recalcitrant plant species.

OSMOTIC TREATMENT OF EXPLANT

After the explant is chosen, in vitro manipulation of the explant may be necessary to enhance competency of plant cells to T-DNA delivery, and to facilitate plant cell recovery after infection. Unlike biolistic-mediated transformation, osmotic treatment enhancement of Agrobacterium-mediated transformation largely depends upon species. Supplementation of co-culture medium with 68.5 gl^{-1} (200 m M) sucrose and 36 gl^{-1} (200 mM) glucose was extensively used in rice and maize transformation (Hiei et al., 1994; Zhao et al., 2001; Frame et al., 2002). However, the effect of osmotic medium on T-DNA delivery and stable transformation was not described. Uze et al., (1997) observed that plasmolysis with 65 gl^{-1} (292 mM) sucrose improved T-DNA delivery into precultured immature embryos rice. This treatment was extensively used to produce large numbers of transgenic plants for various projects (Ye et al., 2000; Lucca et al., 2001). However, osmotic treatment was not effective with precultured immature embryos of wheat (Uze et al., 2000). Osmotic treatment did not have a beneficial effect on T-DNA delivery in wheat (Cheng et al., 2003).

PRECONDITIONING, CO-CULTIVATION TIME AND A. TUMEFACIENS DENSITY

Optimizing the preconditioning time to 72 h and co-cultivation time with A. tumefaciens to 48 h provided an increase in the transformation efficiency from a baseline 4% to 25% in canola (Cardoza and Stewart, 2003). Zhang et al. (2000) reported that in Chinese cabbage, co-cultivation for 72 h yielded the highest transformation frequency. Co-cultivation of explants with A. tumefaciens has made possible the use of some explants, which were hitherto recalcitrant for transformation experiment. Canola was transformed by co-cultivation of mesophyll protoplast with a strain of A. tumefaciens carrying nptII and KCS genes (Wang et al., 2005). Similarly, high efficient transformation of cotton was achieved by co-cultured embryonic calli with A. tumefaciens (Wu et al., 2005). Hiei et al. (1997) reported that transformation of rice was possible when the Agrobacterium density was between 1.0×10^{6} and 1.0×10^{10} colony-forming units (cfu) ml^{-1}, and the optimal concentration was approximately 1.0×10^{10} cfuml^{-1} (Hiei et al., 1994). The same density of A. tumefaciens was successfully used later in maize (Ishida et al., 1996) and adopted by many other laboratories for various genotypes and explants in rice. A. tumefaciens densities higher or lower than 1.0×10^{10} cfuml^{-1} were evaluated systematically with N_6-based medium in maize (Zhao et al., 2001), transient GUS activity increased with higher A. tumefaciens density, but the callus initiation frequency was reduced and peak transformation frequency was achieved with A. tumefaciens at 0.5×10^{10} cfuml^{-1}. Similar results were reported with sorghum immature embryos (Zhao et al., 2000). Experiments with various explants of wheat showed that higher A. tumefaciens density could increase transient GUS expression, but was not correlated with higher stable transformation frequency (Cheng et al., 1997). With wheat suspension cells as a model system, an optimal A. tumefaciens density of around 0.5×10^{10} cfu ml^{-1} was identified. With higher or lower A. tumefaciens density, both transient and stable transformation decreased. A. tumefaciens density higher than 1×10^{10} cfu usually damaged the plant cells, and resulted in lower cell recovery that ultimately reduced the stable transformation frequency. Nevertheless, when a higher density of A. tumefaciens is necessary for recalcitrant explants or species, transformation frequency can be improved by a short inoculation time, gently rinsing the explants after inoculation with fresh inoculation medium as performed in dicot transformation, or addition of a bactericide agent such as silver nitrate in the co-culture medium (Zhao et al., 2000; 2001; Zhang et al., 2003).

Although efficient T-DNA delivery is a prerequisite for achieving efficient stable transformation in most cases, under many conditions increased T-DNA delivery has not resulted in increased stable transformation. For example, when surfactant was included in the inoculation medium for freshly isolated immature embryos of wheat, T-DNA delivery (as measured by transient gene expression) was increased, but stable transformation frequency was not improved. The likely reason for the lack of correlation between T-DNA delivery and stable transformation in this case was the detrimental effect of surfactant on plant cell/tissue recovery (Cheng et al., 1997). T-DNA delivery has correlated well with stable transformation frequency inoculation and co-culture conditions favour both T-DNA delivery and plant cell recovery. One example is the desiccation treatment post A. tumefaciens infection for precultured immature embryos or embryogenic calluses of wheat (Cheng et al., 2003). When T-DNA delivery is not rate-limiting for a given explant, adjust the transformation parameters to favour plant cell recovery has been an effective means of achieving efficient stable transformation.

DESICCATION OF EXPLANTS

A significant factor that enhances transformation of crop species is dessication of explants prior to, or post, A. tumefaciens infection. Arencibia et al. (1998) reported that air-drying sugarcane suspension cells prior to inoculation under laminar flow conditions for 15-60 min slightly improved T-DNA delivery and subsequently increased transformation efficiency, but the actual desiccation stringency was not defined in this report. Similarly, air-drying calluses derived from rice suspension cultures for 10-15 min increased the transformation efficiency 10-fold or more as compared to the control without air-drying (Urushibara et al., 2001). It is unclear to the investigators what factors were affected by air-drying, but it is possible that plasmolysis or wounding may be important. The effect of air-drying on other explants of rice such as embryonic calluses and precultured immature embryos was not evaluated. Using the same air-drying conditions, it was shown that air-drying precultured immature embryos and embryogenic calluses in wheat prior to inoculation did not have the same effect as in sugarcane and rice. However, Cheng et al. (2003) reported that desiccation of precultured immature embryos, suspension culture cells, embryonic calluses of wheat, and embrogenic calluses of maize greatly enhanced T-DNA delivery and plant tissue recovery after co-culture, leading to increased stable transformation frequency. This treatment was not only effective in monocot species, but also improved T-DNA delivery in recalcitrant dicot species such as soybean suspension cells based on our preliminary study (Cheng and Fry, 2000). Although the molecular mechanism of desiccation during co-culture remains unclear, it is known that desiccation suppresses the growth of Agrobacterium similar to the effect observed with silver nitrate. In addition, maize embryogenic calluses from the desiccation treatment recovered better than explants co-cultured under non-desiccation conditions (with H_2O), when co-culture plates were supplemented with 20 µM silver nitrate. Furthermore, osmotic treatments and abscisic acid (ABA) treatment before and during inoculation, and during co-culture, did not have the same effect on T-DNA delivery as the desiccation treatment.

ANTINECROTIC TREATMENTS

With respect to pretreatments, antinecrotic mixtures for pre-induction were shown to be important for reducing oxidative burst. Enrique-Obregon et al. (1998) treated merismatic spindle sections of sugarcane with a medium containing 15 mgl^{-1} (0.09 µM) ascorbic acid, 40 mgl^{-1}(0.33 µM) cysteine, and 2 mgl^{-1}(0.01µM) silver nitrate. An efficient transformation system was developed using this pretreatment in sugarcane. Transformed calluses were obtained only when the mixture of these antinecrotic compounds was added in their previous study (Enrique-Obregon et al., 1998). A similar protocol was applied to rice transformation using seedling explants (Enrique-Obregon et al., 1999). Explant viability was significantly improved when the plantlet explant were treated with this mixture of compounds. Inclusion of cysteine in the co-culture medium led to an improvement in both transient β-glucuronidase (GUS) expression in target cells and a significant increase in stable transformation frequency in maize. In Olhoft and Somers (2001) and Olhoft et al. (2003), T-DNA transfer into cotyledonary-node cells and genomic integration were increased through the inclusion of thiol compounds in the solid co-cultivation medium, resulting in an increased production of transgenic plants. Hygromycin B selection combined with the inclusion of the thiol compounds L-cysteine, dithiothreitol (DDT) and sodium thiosulphate in the co-cultivation medium, further improved the production of transgenic plants, with transformation efficiencies as high as 16.4% of independent Southern-positive T_o plants produced per explants treated (Olhoft et al., 2003). Inclusion of silver nitrate in co-culture medium enhanced stable transformation in maize (Armstrong and Rout, 2001; Zhao et al., 2001). Silver nitrate significantly suppresses the Agrobacterium growth during co-culture without compromising T-DNA delivery and subsequent T-DNA integration. The suppressed Agrobacterium growth on the target explants could facilitate plant cell recovery and result in increased efficiency of transformation (Cheng et al., 2003).

TEMPERATURE

The effect of temperature during co-culture on T-DNA delivery was first reported in dicot species. A temperature of 22°C was found to be optimal for T-DNA delivery in tobacco leaves (Dillen et al., 1997). However, in another report, co-culture at 25°C led to the highest number of transformed plants of tobacco, even though 19°C was optimal for T-DNA delivery (Salas et al., 2001). These results indicate that the optimal for stable transformation with a given species and explant. The optimal temperature for stable transformation should be evaluated with each specific explant and Agrobacterium strain involved (Salas et al., 2001). In monocots, the co-culture temperature for most of the crops ranged from 24 to 25°C, and in some cases, 28°C was used for co-culture (Rashid et al., 1996; Arencibia et al., 1998; Enriquez-Obregon et al., 1998; Hashizume et al., 1999). The effect of lower temperature ($\leq$ 23°C) on T-DNA delivery and stable transformation was also evaluated. Kondo et al. (2000) tested the effect of four temperatures, namely 18, 20, 22 and 24°C on T-DNA delivery with garlic calluses. The highest transient GUS expression was observed at 22°C, in which 64% of the total calluse showed GUS activity. The ratio of GUS-stained calluses decreased by 85% at 20°C and by 69% at 24°C. Higher transformation frequency was observed in maize immature embryo transformation at 20°C than at 23°C when using a standard binary vector (Frame et al., 2002).Transgenic maize plants have also been obtained from elite inbred lines PHP38 and PHN46 by co-culture of the immature embryos at 20°C followed by 28°C subculture (Gordon-Kamm et al., 2002). The effect of temperature on both transient and stable transformation

was extensively studied in other laboratories using suspension-cultured wheat (cv. Mustang) and maize (cv. BMS) cells as model systems. The optimal temperature for both T-DNA delivery and stable transformation was 23-25°C for wheat and ~23°C for maize (Rout et al., 1996).

SURFACTANTS

Including surfactants such as Silwet L77 and pluronic acid F68 in inoculation medium greatly enhanced T-DNA delivery in immature embryos of wheat (Cheng et al., 1997). Surfactants may enhance T-DNA delivery by aiding A. tumefaciens attachment and or by elimination of certain substances that inhibit A. tumefaciens attachment. The surfactant Silwet L77 was also shown to be useful to the success of the floral dip method of Arabidopsis thaliana transformation. Surfactant added to the inoculation medium may play a role similar to vacuum infiltration, facilitating the delivery of A. tumefaciens cells to closed ovules, the primary target for A. tumefaciens during in planta transformation of A. thaliana (Ye et al., 1999; Bechold et al., 2000; Desfeux et al., 2000).

INOCULATION AND CO-CULTURE MEDIUM

Medium component, sugar, plant growth regulators, and vir induction chemicals are also important factors that affect transformation frequency. The modified N6 medium (Chu et al., 1995) containing 2,4-dichlorophenoxyacetic acid (2,4-D) and casamino acids was shown to be suitable for co-culture in rice. Several laboratories with different genotypes and explants adopted a similar medium recipe. MS (Murashige and Skoog, 1962) or a modified MS-based medium was shown to be suitable for inoculation and co-culture in several report of rice transformation (Dong et al., 1996; Enriquez-Obregon et al., 1999; Mohanty et al., 1999; Luca et al., 2001). Ishida et al. (1996) reported transformation of maize immature embryo using LS-based (Linsmaier and Skoog, 1965) medium, and N6-based medium failed to generate transformed plants. With additional component added in the mixture such as silver nitrate. Zhao et al. (2001) showed that N6-based medium was also suitable for inoculation and co-culture of immature maize embryos, resulting in transgenic plants. Similarly, the addition of $CaCl_2$ in the medium increased transformation efficiency in barley (Kumlehn et al., 2006).

Reducing the salt strength in the inoculation and co-culture media was reported as beneficial for transformation of canola (Fry et al., 1987). Medium with reduced salts enhanced T-DNA delivery in wheat (Cheng et al., 1997). This treatment was used to regenerate stable transformed wheat plants from embryogenic callus with a superbinary vector in a recent study (Khanna and

Daggard, 2003). Medium with reduced salts also enhanced T-DNA delivery in maize (Armstrong and Rout, 2001), and half-strength MS salts in both inoculation and co-culture media have been used in maize transformation (Zhang et al., 2003). The impact of salt strength within the inoculation and co-culture medium on transient GUS expression was extensively assessed in barley with immature embryos as the target explants (Ke et al., 2002). One-tenth MS salt strength enhanced transient GUS expression 10-fold over full-strength salts. Furthermore, the distribution of cells expressing the GUS gene within each set of immature embryos was clearly altered, showing significantly more cells on the scutellar surface expressing GUS.

Chemicals such as acetosyringone for vir induction are recommended in most of crops transformation protocols (Hiei et al., 1994; Ishida et al., 1996; Cheng et al., 1997; Tingay et al., 1997; Zhao et al., 2000; Kumlehn et al., 2006). When acetosyringone was omitted, the level of transient GUS expression was low and stable transformed plants could not be regenerated in rice, onion or barley (Rashid et al., 1996; Hiei et al., 1997; Zheng et al., 2001; Kumlehn et al., 2006). However some explants of monocot species could be efficiently transformed without the aid of external vir induction chemicals for special treatment. For example, meristematic sections of sugarcane pretreated with an antinecrotic mixture (Enriquez-Obregon et al., 1999), and precultured immature embryos and embryogenic calluses of wheat co-cultured under desiccation conditions could be efficiently transformed (Cheng et al., 2003).

ANTIBIOTICS

Antibiotics such as cefotaxime, carbenecillin and timentin have been used regularly in Agrobacterium-mediated transformation of crops following co-culture to suppress or eliminate Agrobacterium (Cheng et al., 1996; Bottinger et al., 2001; Sunikumar and Rathore, 2001). Although cefotaxime worked will in Agrobacterium-mediated transformation of rice and maize initially, it was later found that cefotaxime at a concentration of 250 mgl[-1] (Ishida et al., 1996) had a detrimental effect to maize Hi II callus, Callus formation was greatly reduced when cefotaxime (50 or 250 mgl[-1]) was added in the callus induction medium, and consequently transformation frequency was reduced 3-fold compared to that with carbenicillin (100 mgl[-1]). Carbenicillin at 100 mgl[-1] was used for all the subsequent experiments (Zhan et al., 2001). Carbenicillin has been the antibiotic of choice in reports of Agrobacterium-mediated transformation of wheat and maize (Cheng et al., 1997, 2003; Zhang et al., 2003). On the other hand, 100 mg[-1] kanamycin was economical and improved the transformation efficiency in white spruce by enrichment of transformed tissue in bud-forming callus (Le et al., 2001) and increased the proportion of positively transformed shots during subculture on kanamycin containing medium in peanut and pigeon pea (Sharma and Anjaiah, 2000; Thu et al., 2003).

SELECTABLE MARKER

The most widely used selectable markers for transformation of crops are genes encoding hygromycin phosphotransferase (hpt), phosphinothricin acetyltransferase (pat or bar), and neomycin phosphotransferase (nptII). Use of these marker genes under the control of constitutive promoters such as the 35S promoter from cauliflower mosaic virus, the ubiquitin promoter from maize, works as efficiently for selection of Agrobacterium-transformed cells as for biolistics-mediated transformation. For Asparagus and banana, the npt II gene under the control of the nopaline synthase promoter has been used to successfully select stable transformants with kanamycin (May et al., 1995; Limanton-Grevet and Jullien, 2001). The positive selectable marker phosphomannose isomerase was first used for Agrobacterium-mediated transformation of sugar beet and was recently used to enhance transformation of sorghum (Joersbo et al., 1998; Lucca et al., 2001; Gao et al., 2005). To improve selectable marker genes for crops, Wang et al. (1997) inserted introns into the coding region of hpt as the strategy used in enhancing transgene expression in monocot species (Simpson and Filipowics, 1996). The intoduction of introns into the hpt not only improved transformation frequency in rice Agrobacterium-mediated transformation due to the elevated hpt expression, but also reduced copy numbers of the marker gene. Furthermore, inserting the introns into the marker gene also enabled better control of Agrobacterium growth during the transformation process (Wang et al., 1997). This modified selectable marker enhanced stable transformation with elite rice and barley cultivars as well (Upadhyaya et al., 2000; Wang et al., 2001). Glyphosate-insensitive plant 3-enolpyruvylshikimate-5-phosphate synthases (EPSPS) genes, the bacterial CP4 gene or a bacterial gene that degrades glyphosate, i.e. glyphosate oxidoreductase (GOX) gene, have been used in some laboratories to generate transgenic plants in wheat and maize with biolistics-mediated transformation approaches (Armstrong et al., 1995; Zhou et al., 1995; Russell and Fromm, 1997; Howe et al., 2002). One of these genes, CP4, has been successfully used in Agrobacterium transformation of wheat (Cheng et al., 2003; Hu et al., 2003).Transformation frequency was comparable to biolistics-mediated transformation in wheat (Hu et al., 2003) when a desiccation-based protocol was used.

CONCLUSION

Efficient transformation systems using readily available explants are in high demand for agronomically important plants. Though fertile transgenic plants have been generated from more than a dozen plants, yet the transformation frequency for most species is still low. In some cases, only a few transformed plants have been regenerated. Further optimizing the transformation parameter such as inoculation, co-culture condition and selectable marker could increase transformation frequency. Since indication that explant competency to Agrobacterium infection using techniques such as desiccation, antinecrotic mixture for pre-induction as well as plant growth regulation treatment is emerging. Understanding the molecular basis of several factors such as desiccation and antinecrotic treatments affecting both T-DNA delivery and stable transformation may facilitate application of these treatments to other species or transformation systems to further improved many published protocols.

ACKNOWLEDGEMENTS

The author thanks Nigerian Agricultural Biotechnology Programme (NABP) and USAID for a three-month fellowship at International Institute of Tropical Agriculture (IITA), Ibadan that made this review possible.

REFERENCES

Alt-Morbe J, Kithmann H, Schroder J (1989). Differences in induction of Ti-plasmid virulence genes virG and virD and continued control of vir D expression by four external factors. Mol. Plant-Microbe Interact 2: 301-308.

Arencibia AD, Carmona ERC, Tellez P, Chan MT, Yu SM, Trujillo LE, Oramas P (1998). An efficient protocol for sugarcane (Saccharum spp. L) transformation mediated by Agrobacterium tumefaciens. Transgenic Res.7: 213-222.

Armstrong CL, Rout JR (2001). A novel Agrobacterium-mediated plant transformation method. Int. Patent Publ. WOO1/09302 A2.

Bechold N, Jaudeau B, Jolivet S, Maba B, Vezon D, Voisin R, Pelletier G (2000). The maternal chromosome set is the target of T-DNA in planta transformation of Arabidopsis thaliana. Genetics 155:1875-1887.

Bidney D, Scelonge C, Martich J, Burus M, Sims L and Huffman G (1992). Microprojectile bombardment of plant tissues increased transformation frequency of Agrobacterium tumefaciens. Plant Mol. Biol. 18: 301-313.

Binns AN,Thomashaw MF (1988). Cell biology of Agrobacterium infection and transformation of plants. Annual Review of Microbiology 42: 575-606.

Bottinger P, Steinmetz A, Scheider O, Pickardt T (2001). Agrobacterium mediated transformation of Vicia faba. Mol. Breed. 8: 243-254.

Cadoza V, Stewart CN (2003). Increased Agrobacterium mediated transformation and rooting efficiencies in canola (Brassica napus L.) from hypocotyls segment explants. Plant Cell Rep. 21:599-604.

Chateau S. Sangwan, RS, Sangwan-Norreel, BS (2000). Competence of Arabidopsis thaliana genotypes and mutants for Agrobacterium tumefaciens-mediated gene transfer role of phytohormones. J. Exp. Bot. 51-1961-1968.

Cheng M, Fry JE (2000) An improved efficient Agrobacterium-mediated plant transformation method. Int. Patent publ. WO 0034/491

Cheng M, Fry JE, Pang S, Zhou I, Hironaka C, Duncan DRI, Conner TWL,Wang Y (1997). Genetic transformation of wheat mediated by Agrobacterium tumefaciens, Plant. Physiol. 115: 971-980.

Cheng M,Hu T, Layton JI, Liu C-N, Fry JE (2003). Desiccation of plant tissues post-Agrobacterium infection enhances T-DNA delivery and increases stable transformation efficiency in wheat. In Vitro Cell. Dev. Biol. Plant 39: 595-604l.

Cheng MI, Jarret RLI, Li ZI, Xing AI, Demski JW (1996). Production of fertile transgenic peanut (Arachis hypogea L.) plants using Agrobacterium tumefaciens. Plant Cell Rep. 15: 653-657.

Cheng M,Lowe BA,Spencer M,Ye X,Armstrong CL (2004) Factors influencing Agrobacterium-mediated transformation of monocotyledonous species.In Vitro cell. Dev.Biology-Plant 40: 31-45

Chu CC, Wang CC, Sun CS, Hsu C, Yin KC, Chu CY, Bi FY (1995). Establishment of an efficient medium for anther culture of rice through comparative experiments on the nitrogen sources. Sci, Sip 18: 659-668.

Czernilofsky AP, Hain R, Baker B, Wirtz U (1986). Studies of the structure and functional organization of foreign DNA integrated into the genome   of Nicotiana tabacum.DNA 5: 473-478.

Desfeux C, Clough SJ, Bent AF (2000). Female reproductive tissues are theprimary target of Agrobacterium-mediated transformation by the Arabidopsis floral-dip method. Plant, Physiol. 123: 859-904l.

Dillen W, De Clereq J, Kapila J, Zamnbre M, Van Montagu M, Angenon G (1997). The effect of temperature on Agrobacterium tumefaciens-method of gene transfer to plants. Plant J. 12: 1459-1462

Dong J, Teng W, Buchholz WGL, Hall TC (1996). Agrobacterium-mediated transformation of javanica rice. Mol. Breed. 2: 267-276

Enriquez-Obregon GA, Prieto-Samsonov DL, de la Riva GA,Perez MI, Selman-Housein G, Vazquz-Padron RI (1999). Agrobacterium-mediated Japonica rice transformation a procedure assisted by an antinecrotic treatment. Plant Cell Tiss. Organ Cult. 59: 159-168l.

Enriquez-Obregon GA, Vazquez-Padron RI, Prieto-Samsonov DL, de la

Riva GA, Selman-Housein G (1998). Herbicide-resistant sugarcane (Saccharum officinarum L.) plants by Agrobacterium-mediated transformation. Planta 205: 20-27.

Fang YD, Akula C, Altpeter F (2002). Agrobacterium-mediated barley (Hordeum vulgare L) transformation using green fluorescent protein as a visual marker and sequence analysis of the T-DNA::barley genomic DNA junction. J. Plant Physiol.159: 1131-1138.

Frame BR, Shou H, Chikwamba RK, Zhang ZI, Xiang CI, Fonger TM, Pegg SEK, Li B, Nettleton DS, Pei D, Wang K (2002). Agrobacterium tumefaciens-mediated transformation of maize embryos using a standard binary vector system. Plant Physiol. 129: 13-22.

Fry J, Barnason A, Horsch RB (1987). Transformation of Brassica napus with Agrobacterium tumefaciens based vectors. Plant Cell Rep. 6: 321-325.

Gao Z,Xie X,Ling Y,Muthukrishnan S,Liang HG (2005) Agrobacterium tumefaciens transformation using a mannose selection system.Plant Biotechnology Journal 3: 591-597.

Gordon-Kamm W, Dilkes BP, Lowe K, Hoerster G, Sun X, Ross M, Church KD, Bunde C, Farell J, Maddock S, Snyder J, Skyes L, Li Z, Woo YM. Bidney D, Larkins BA (2002). Stimulation of the cell cycle and maize transformation by disruption of the plant retinoblastoma pathway.Proc. Natl. Acad.Sci. USA. 99: 11975-11980.

Hamilton CM (1997). A binary-BAC system for plant transformation with high-molecular-weight DNA. Gene. 200: 107-116.

Hashizume F, Tsuchiya T, Ugaki M, Niwa Y, Tachibana N, Kowyama Y (1999). Efficient Agrobacterium-mediated transformation and the usefulness of a synthetic GFP reporter gene in leading varieties of japonical rice. Plant Biotechnol. 16: 397-401.

Hiei Y, Komari T, Kubo T (1997). Transformation of rice mediated by Agrobacterium tumefaciens. Plant Mol. Biol. 35: 205-218.

Hiei Y, Ohta S, Komari T, Kumashiro T. (1994). Efficient transformation of rice (Oryza sativa L.) mediated by Agrobacterium and sequence analysis of the boundaries of the T-DNA Plant Journ. 6: 271-282.

Hoekema A, Hirsch PR, Hooykaas PJ, Schilperpoort RA (1993).A binary plant vector strategy based on seperation of vir- and T-region of the Agrobacterium tumefaciens Ti-plasmid. Nature 303: 179-180.

Howe AR, Gasser CS, Brown SM, Padgette SR, Hart J, Parker G,Fromm ME, Armstrong CL (2002). Glyphosate as a selective agent for production fertile transgenic maize (Zea mays L.). Plant. Mol. Breed. 10: 153-164.

Hu T, Meltz S, Chay C, Zhou HP, Biest N, Chen G, Cheng M, Feng X, Radionenka M, Lu F, Fry JE (2003). Agrobacterium-mediated large scale transformation of wheat (Triticum aestivum L.).Plant Cell Rep. 21: 1010-1019.

Hooykaas PJJ, Shilperoort RA (1992). Agrobacterium and plant genetic engineering. Plant Molecular Biology 19: 15-38.

Ishida Y, Saito H, Ohta S, Hiei Y, Komari T, Kumashiro T (1996.) High efficiency transformation of maize (Zea mays L.) mediated by Agrobacterium tumefaciens. Nature Biotechnol. 14: 745-750

Jelenic S,Mitrikeski PT,Papes D,Jelaska S (2000). Agrobacterium – mediated transformation of broadbean Vicia faba L.Food Technology and Biotechnology 38: 167-172

Jeon G.A., Eum JS Sim WS (1998). The role of inverted repeat (IR) sequence of the virE gene _expression in Agrobacterium tumefaciens pTiA6. Molecules and Cells 8: 49-53.

Joersbo M, Donaldson I, Kreiber J, Peterson SG, Brunstedt J, Okkels FT (1998) Analysis of mannose selection used for transformation of sugar beet. Mol. Breed.4: 111-117.

Ke X-Y, McCormac AC, Harvey A, Lonsdale D, Chen D-F, Elliot MC (2002) Manipulation of discriminatory T-DNA delivery by Agrobacterium into cells of immature embryos of barley and wheat. Euphytica 126: 333-343.

Khanna HK, Daggard GE (2003). Agrobacterium tumefaciens-mediated transformation of wheat using a superbinary vector and a polyamine-supplemented regeneration medium. Plant Cell Rep.21: 429-436.

Klee H (2000). A guide to Agrobacterium binary Ti vectors. Trends in Plant Science 5: 446-451.

Komari T, Hiei Y, Saito Y, Murai N, Kumashiro T (1996). Vector caring two separate T-DNAs for co-transformation of higher plants mediated by Agrobacterium tumefaciens and segregation of transformants free from selective markers. Plant J. 10: 165-174.

Koncz C, Martini N, Mayerhofer R, Koncz-Kalman Z, Korber H, Redei GP (1989).High-frequency T-DNA mediated gene tagging in plants.Proc. Natl. Acad. Sci. 86: 8467-8471.

Kondo T, Hasegawa H, Suzuki M (2000). Transformation and regeneration of garlic (Allium sativum L.) by Agrobacterium-mediated gene transfer. Plant Cell Rep.19: 989-993.

Kumlehn J,Serazetdinora L,Hensel G,Becker D,Loerz H (2006). Genetic transformation of barley (Hordeum vulgare L.) via infection of androgenetic pollen culture with Agrobacterum tumefaciens.Plant Biotechnology Journal 4:251-258

Limanton-Grevet A, Jullien M (2001). Agrobacterium-mediated transformation Asparagus officinalis L.: Molecular and genetic analysis of transgenic plants. Mol. Breed. 7: 141-150.

Linsmaier EM, Skoog F (1965) Organic growth factor requirements of tobacco tissue cultures.Physiol. Plant. 18: 100-127.

Le VQ, Belles-Isles J, Dusabenyagusani M, Tremblay FM (2001). An improved procedure for production of white pruce (Picea glauca) transgenic plants using Agrobacterium tumefaciens. J. Exp. Bot. 52: 2089-2095.

Lucca P, Ye X, Potrykus I (2001). Effective selection and regeneration of transgenic rice plants with mannose as selective agent. Mol. Breed. 7: 43-49.

May GD, Afza R, Mason HS, Wiecko A, Novak FJ, Arntzen CJ (1995). Generation of transgenic banana (Musa acuminata) plants via Agrobacterium-mediated transformation. Bio/Technology 13: 486-492.

McKnight TD, Lillis MT, Simpson RB (1987). Segregation of genes transferred to one plant cell from two separate Agrobacterium strains Plant. Mol. Biol. 8: 439-445

Mohanty A, Sarma NP, Tyagi AK (1999). Agrobacterium-mediated high frequency transformation of an elite indica rice variety Pusa Basmati 1 and transmission of the transgene to R2 progeny. Plant Sci.147: 127-137.

Murashige T, Skoog F (1962). A revised medium for rapid growth and bioassays with tobacco tissue cultures. Physiol. Plant.15: 473-479.

Nauerby B, Billing K, Wyndaele R (1997). Influence of the antibiotic timentin on plant regeneration compared to carbernicillin and cefotaxime in concentration suitable for elimination of Agrobacterium tumefaciens. Plant Science 123: 169-177

Olhoft PM, Flagel LE, Donovan CM, Somers DA (2003). Efficient soybean transformation using hygromycin B selection in the cotyledonary-node method. Planta 216: 723-735

Olhoft PM, Somers DA (2001). L-cysteine increases Agrobacterium-mediated T-DNA delivery into soybean cotyledonary-node cells. Plant Cell Rep. 20: 706-711.

Olhoft PM, Flaye, LE, Sowers DA (2004). T-DNA locus structure in a large population of soyabean plant transform using the Agrobacterium-mediated cotyledonany-node methods. Plant Biotech. Journ. 2: 289-300.

Opabode,JT (2002). Factors influencing transformation of crops by Agrobacterium tumefaciens. A review seminar presented at Department of Plant Science, Obafemi Awolowo University, Nigeria, 23 March 2002.

Rashid H, Yokoi S, Toriyama K, Hinata K (1996). Transgenic plant production mediated by Agrobacterium in indica rice. Plant Cell Rep. 15: 727-730.

Rhodora RA, Thomas KH (1996). Agrobacterium tumefaciens mediated transformation of Japonica and Indica rice varieties. Planta 199: 612-617.

Riva GA, Gonzalez-Cabrera J, Vasqu-Padru J, Ayra-Pardo C (1998). Agrobacterium tumefaciens gene transfer to plant cell. Electronic Journal of Biotechnology (Online) 15 December 1998 Vol.2 no. 3. Available from http://www.ejbiotechnology.info/

Rout JR, Hironaka CM, Conner TW, DeBoer DL, Duncan DR, Fromm ME, Armstrong CL (1996) Agrobacterium-mediated stable genetic transformation of suspension cells of corn (Zea mays L.).38th Annual maize genetics conf. St Charles, IL, March 14-17.

Russell DA, Fromm ME (1997). Tissue-specific expression in transgenic maize of four endosperm promoters from maize and rice. Transgenic Res.6:157-168.

Salas MC, Park SH, Srivatanakul M, Smith RH (2001). Temperature influence on stable T-DNA integration in plant cells. Plant Cell Rep. 20: 701-705.

Sharma K.K. and Anjaiah V (2000). An efficient method for the production transgenic plants for peanut (Arachis hypogea L.) through Agrobacterium tumefaciens mediated genetic transformation. Plant Science 159:7-19

Sarmah BK, Moore A, Tate W,Molvig L, Morton RL, Rec DP, Chaise P, Chrispeel MJ, Higgins TJV (2004). Trangenic chickpea seeds expressing high levels of a bean α-amylase inhibitor. Molecular Breeding 14:73-82

Simpson GC, Filipowcz W (1996). Splicing of pre-cursors to mRNA in higher plants: mechanism, regulation and sub-nuclear organization of the spliceosomal machinery. Plant Mol. Biol. 32: 1-41.

Smith EF ,Towsend CO (1907). A plant tumour of bacterial origin. Science 25: 671-673.

Sunikumar G, Rathore KS (2001). Transgenic cotton: factors influencing Agrobacterium-mediated transformation and regeneration. Mol. Breed. 8: 37-52.

Thu TT, Mai TTX, Deade E, Farsi S, Tadesse Y, Angenum G, Jacobs M (2003). In vitro regeneration and transformation of pigeonpea (Cajanus cajan (L.) Mills P) Mol. Breed. 11: 159-168.

Tingay S, McElroy D, Kalla R, Fieg S, Wang M Brettel R (1997). Agrobacterium-mediated barley transformation. Plant J. 11: 1369-1376.

Torisky RS, Kovacs L, Avdiushko S, Newman JD, Hunt AG, and Collins GB (1997). Development of a binary vector system for plant transformation based on supervirulent Agrobacterium tumefaciens strain Chry5. Plant Cell Reports 17: 102-108.

Upadhyaya NM, Surin B,Ramm K, Gaudron J, Schunman PHD,Taylor W, Waterhouse PM, Wang MB (2000). Agrobacterium-mediated transformation of Australian rice cultivars Jarrah and Amaroo using modified promoters and selectable markers. Aust. J. Plant Physiol. 27: 201-210.

Urushibara S, Tozawa Y, Kawagishi-Kobayashi M, Wakasa K (2001). Efficient transformation of suspension-cultured rice cells mediated by Agrobacterium tumefaciens. Breed. Sci. 5: 33-38.

Uze M, Potrykus I, Sauter C (2000). Factors influencing T-DNA transfer from Agrobacterium to precultured immature wheat embryos (Triticum aestivum L.) Cereal Res. Commun. 28: 17-23.

Uze M, Wunn J, Pounti-Kaelas J, Potrykus I, Sauter C (1997). Plasmolysis of precultured immature embryos improves Agrobacterium mediated gene transfer to rice (Oryza sativa L) Plant Sci. 130: 87-95.

Wang M-B, Abbott DC, Upadhyaya NM, Jacobsen JV, Waterhouse PM (2001). Agrobacterium tumefaciens- mediated transformation of an elite Australian barley cultivar with virus resistance and reporter genes. Aust. J. Plant. Physiol. 28: 149-156.

Wang MB, Upadhyaya NM, Brettell RIS Waterhouse PM (1997). Intron-mediated improvement of a selectable marker gene for plant transformation using Agrobacterium tumefaciens. J. Genet. Breed 513: 25-334.

Wang YP,Sonntag K,Rudloff E,Han J (2005). Production of fertile transgenic Brassica napus by Agrobacterium-mediated transformation of protoplasts.Plant Breeding 124:1-5.

Wu J, Zhang X, Nie Y, Luu X (2005). High-efficiency transformation of Gossypium hirsutum embryonic calli mediated by Agrobacterium tumefaciens and regeneration of insect-resistant plants. Plant Breeding 124: 142-148.

Ye G-N, Stone D, Pang SZ, Creely W, Gonzalez K, Hinchee M (1999) Arabidopsis ovule is the target for Agrobacterium in planta vacuum infiltration transformation. Plant J. 19: 249-257.

Ye X, Al-Babili S, Kloti A, Zhang J, Lucca P, Beyer P, Potrykus I (2000) Engineering the provitamin A (β-carotene) biosynthetic pathway into (carotenoid-free) rice endosperm. Science 287: 303-305.

Zhang FL, Takahata Y, Watanabe M, XU JB (2000). Agrobacterium mediated transformation of cotyledonary explants of chined cabbage (Brassica campestris L. ssp. pekinensis). Plant Cell Rep. 19: 569-575.

Zhang W, Subbarao S, Addae P, Shen A, Armstrong C, Peschke V, Gilbertson L (2003). Cre/lox mediated gene excision in transgenic maize (Zea mays L.) plants. Theor. Appl. Genet. 107: 1157-1168.

Zhao Z-Y, Cai T, Tagliani L, Miller M, Wang N, Pang H, Rudert M, Schroeder S, Hondred D, Seltzer J, Pierce D (2000). Agrobacterium-mediated sorghum transformation. Plant Mol. Biol. 44: 789-798.

Zhao Z-Y, Gu W, Cai T, Tagliani L, Hondred D, Bond D, Schroeder S, Rudert M, Pierce D (2001). High throughput genetic transformation mediated by Agrobacterium tumefaciens in maize.Mol. Breed. 8: 323-333.

Zheng SJ, Khrustaleva L, Henken B, Jacobsen E, Kik C, Krens FA (2001). Agrobacterium tumefaciens-mediated transformation of Allium cepa L: the production of transgenic onion and shallots.Mol. Breed. 7: 101-115.

Zhou H, Arrowsmith JW, Fromm ME, Hironaka CM, Taylor ML, Rodriguez D,Pajeau ME, Brown SM, Santino CG, Fry JE (1995). Glyphosate-tolerant CP4 and GOX genes as a selectable marker in wheat transformation. Plant Cell Rep. 15: 159-163.

Zupan J,Muth TR,Draper O,Zambryski P (2000). The transfer of DNA from Agrobacterium tumefaciens into plants: A feast of fundamental insights. The Plant Journal 23: 11-28

Zupan JR Zambryski PC (1995). Transfer of T-DNA from Agrobacterium to the plant cell. Plant Physiology 107: 1041.1047.

7

Cassava Biotechnology, a southern African Perspective

Murunwa Makwarela and Christine Rey*

School of Molecular and Cell Biology, University of the Witwatersrand, Private Bag 3, WITS 2050, Johannesburg, South Africa.

The pre-requisite for any cassava (Manihot esculenta Crantz) transformation program that proposes to develop improved plants is the availability of a reliable regeneration system. Presently many laboratories that prioritize cassava research are able to reliably regenerate plants from a range of cultivars. Unfortunately, some cultivars are still either recalcitrant or resisting attempts to induce useful levels of embryogenesis from their tissues.The review gives a brief account on the different uses of cassava, its introduction into southern Africa and the region's current cassava disease complex with a particular focus in South Africa. Different cassava regeneration and gene transfer systems are also discussed.We conclude by presenting future prospects in southern African cassava biotechnology.

Key words: Cassava, CMD, Cassava biotechnology.

INTRODUCTION

This review focuses on a brief account on the different uses of cassava, its introduction into southern Africa and the region's current cassava disease complex with a particular focus on South Africa is examined. We will also discuss the different cassava regeneration and gene transfer systems currently residing within five laboratories in Europe and America forming an advanced cassava transformation group. Those laboratories are ILTAB, CIAT, ETH, Ohio State University and Wageningen University. We will conclude by sharing future prospects in cassava biotechnology in order to make the technology more widely applicable in the southern African region.

Cassava (*Manihot esculenta* Crantz) is a vegetatively propagated root crop used as a staple throughout the tropics and sub tropics. It is the fourth most important and cheapest staple food crop after rice, wheat and maize in developing countries, providing food for over 600 million people (Schöpke *et al.*, 1993a). Cassava, otherwise known as tapioca, yucca, manioc or mandioca is an outcrossing, monecious member of the family *Euphorbiaceae*. Considered as allopolyploid (2n=36), it is a highly heterozygous, semi-woody, perennial shrub varying from 1-4 m in height depending on the cultivar, and produces between 3 and 36 storage roots per plant. Storage roots on a fresh mass basis contain between 20% and 36% starch and approximately 77% on a dry mass basis (Gray, 2003). Cassava is propagated vegetatively usually via lignified stem cuttings. After

planting, new roots are produced and axillary buds sprout to form the shoot system. About two months later, photosynthates produced by the established leaf canopy are diverted to root system where the excess energy is converted to starch and stored in the parenchyma of greatly thickened storage roots, generally referred to as tubers. It can also be cultivated in association with several other crops in most African countries, as discussed by Nweke (1994).

Although still a subject of debate, its centre of origin is generally believed to be southern border of the Amazon basin (Allem, 2002). It was introduced into Africa in the Congo River delta by the Portuguese in the 15th century (Jones, 1959), and its cultivation spread rapidly to many agro-ecologies including East Africa through Madagascar and Zanzibar and later to Asia. Cassava was introduced into Mozambique by the Portuguese in the 17th century, and was adopted as a food crop by the Tonga tribesmen, in eastern Transvaal now Mpumalanga Province, Swaziland and northern Natal (Daphne, 1980). Cultivation of cassava by neighbouring tribes started gradually and it appeared, therefore, that the cultivation of cassava in SA was related to the major tribal movements of the 1830s and 1860s (Daphne, 1980). Cultivation continued to increase throughout the 20th century, most noticeably in Africa where colonial powers often encouraged its cultivation as a famine reserve.

Large scale cassava production in SA was impaired by a taste preference for maize, but in the late 1970s there was a renewed interest in cassava, and extensive yield trials were conducted throughout sub-tropical regions of

*Corresponding authors E-mail: chrissie@gecko.biol.wits.ac.za.

Kwazulu-Natal Province and Northern Province (now called Limpopo Province) under a range of environments (Daphne, 1980). In Mozambique, cassava is the second most important staple food after maize, which more than 50% of the population depend on. The crop plays a big role as a food security crop, and is mostly produced through subsistence farming. It is grown mostly in the northern and coastal regions of the country. The main cassava production provinces are Nampula, Zambézia, Cabo Delgado and Inhambane. Across the country, roots are eaten fresh or dried (flour), and leaves as a vegetable. Yields in Mozambique are consistently low (8-9 tonnes per hectare) compared with yield potential of 70 ton/ha (FAO, 2003). Low yields are attributed to both abiotic and biotic factors.

Cassava is widely consumed as a porridge, which is prepared from dried and pounded roots, but eaten in a very wide range of forms in different parts of the African continent. Cassava is reported to be consumed in 28 different forms in Cameroon, alone (Kokora Nicole, pers. comm., 2002). In SA, there are a number of cassava-processing methods consisting of drying and pounding of the roots to produce porridge known as 'Xigema', and cooking of leaves to produce a condiment known as 'Mathapi' (Diana Sikulane pers. comm. 2002). Cassava is also consumed as a snack food in various parts of the continent. Varieties used as snack food are 'sweet' types, low in cyanic acid, which can be boiled and eaten or even consumed raw. In certain regions, the leaves, which contain appreciable quantities of protein and vitamins, are used as a major component of the diet to provide supplementary protein, vitamins and minerals to complement the carbohydrate rich staple (Lacanster and Brooks, 1983).

CASSAVA DISEASE COMPLEX IN SOUTHERN AFRICA

The most important diseases affecting cassava production in Southern Africa are Cassava brown streak disease (CBSD) and Cassava mosaic disease (CMD). CBSD is caused by cassava brown streak virus (CBSV) while CMD is caused by several whitefly-transmitted begomoviruses. Both diseases cause enormous losses to cassava production. In Mozambique, CBSD was reported for the first time in 1997 when farmers were reporting major losses of cassava caused by root rotting, which by 1998 were considerable, and had infected large areas of coastal Nampula province in the northern part of the country. As a result many farmers started to turn to alternative crops. Farmers in Zambézia Province, south of Nampula, also had the same problem of cassava root rotting. A survey carried out by Natural Resources Institute (NRI) in 1999 in the affected areas of Nampula, Zambézia and Cabo Delgado Provinces identified the

disease as caused by CBSV. CBSD is still confined to the north of the country, mainly in the coastal regions.

A study was conducted by Berry and Rey (2001) in six countries in southern Africa, namely Angola, Mozambique, South Africa, Swaziland, Zambia and Zimbabwe. It was found that African cassava mosaic virus (ACMV) occurred in five of the six countries (except Angola), East African cassava mosaic virus (EACMV) was present in five countries (except Zambia) and South Africa cassava mosaic virus SACMV) was detected only in South Africa and Swaziland. In addition, their report for the first time in southern Africa implicated the appearance of the Ugandan variant virus (EACMV-Ug), which was found in mixed infections with other cassava-infecting begomoviruses.

There is evidence that mixed infections by cassava mosaic Geminiviruses (CMGs) are more damaging than single infections, as reported in studies in Uganda and Cameroon (Fondong et al., 2000; Pita et al., 2001). Symptom severity is associated with the magnitude of yield loss (Thresh et al., 1994) and some yield loss models are derived from the relationships between disease severity and disease incidence.

CMGs are disseminated in the stem cuttings used routinely for vegetative propagation. They are also transmitted by the whitefly, Bemisia tabaci Gennadius. Dissemination by stem cuttings accounts for the occurrence of the disease in areas where there is little or no spread by the whitefly vector (Calvert and Thresh, 2002). The distribution of immigrant whiteflies and of plants newly affected by CMD is influenced by the direction of the prevailing wind and by the effects of wind turbulence around and within stands. The incidence of whiteflies and CMD is highest at the crop margins, especially along the windward and leeward edges and environmental gradients have been observed when whitefly populations decrease with increasing distance from the field boundaries. Incidence is also increased by breaks or discontinuities in the crop canopy, which facilitate the alighting and establishment of viruliferous vectors (Calvert and Thresh, 2002).

Obvious benefits are realised by decreasing the losses caused by CMD and this can be achieved by a reduction in the incidence and/or severity of the disease. Various approaches to control are possible, however, the main attention has been given to the use of resistant varieties and phytosanitation, involving the use of CMD-free planting material and the removal (rouging) of any additional diseased plants that occur (Legg and Thresh, 2003). Farmers occasionally use insecticides in attempts to restrict the spread of CMD by controlling the whitefly vector. However, the use of insecticides on cassava or other tropical crops has received little attention from researchers in Africa and this approach is unlikely to be effective. It is also inappropriate considering the costs that are involved and the risk to farmers, consumers and the environment (Calvert and Thresh, 2002).

Reported losses, as indicated by evidence, are more qualitative than quantitative. Trials to assess the effect of CMD on cassava yield have provided differing results ranging from virtually no loss to almost total loss (Thresh et al., 1994).

GENETIC ENGINEERING OF CASSAVA

Biotechnology is fast proving to be a valuable tool for genetic improvement of plants. However, the prerequisites for efficient exploitation of biotechnology are the development of reliable transformation systems, efficient tissue culture regeneration methods, transformation and selection methods of transgenic plants. Genetic engineering is a powerful tool that complements traditional breeding and can extend the genetic pool of useful gene sources beyond the species (Fregene and Puonti-Kaerlas, 2002). Transgene technology also offers the advantage of precisely transferring single or even quantitative traits without the problems of linkage encountered in traditional breeding.

In cassava improvement programs, the limiting factors in production and utilization are among others, inadequate resistance to viruses, bacteria and insect pests difficulties in the production of novel value-added products from cassava, poor starch characteristics, rapid post-harvest deterioration of tubers, and a ubiquitous problem limiting marketability of the crop, which hinders the development of medium to large scale processing centres. Additional problems still unresolved by traditional breeding are, low protein content of cassava products and the presence of cyanogenic compounds in the tubers (Puonti-Kaerlas, 1998).

Conventional breeding for the agronomic improvement of cassava is frustrated by the crop's inherent heterozygous nature, inbreeding depression and the polygenic and recessive nature of many desirable traits. Therefore, genetic engineering has been identified as a powerful tool to overcome these limitations. Traditional breeding of cassava is difficult as few natural resistance genes have been found in sexually compatible germplasm. The allotetraploid nature of cassava that leads to polymorphisms after crossing, its high outcrossing nature and its low fertility linked to inbreeding depression restrict the use of traditional breeding (Puonti-Kaerlas, 1998).

Genetic engineering, therefore, presents an alternative to traditional plant breeding. Using the techniques of molecular biology, a single gene that codes for a desired trait, such as insect resistance, increased protein content, or tolerance to drought is isolated and then combined with a promoter sequence that will allow the gene to be expressed. This combination of genes is then introduced directly into the plant genome (Chrispeels and Sadava, 2003). To improve cassava by genetic engineering, an essential prerequisite as earlier stated, is the develop ment of an efficient regeneration and transformation procedures.

Cassava in vitro regeneration

Plant cells are generally considered to be totipotent, thus being able to regenerate whole plants from single cells in vitro. The ability to regenerate in vitro is, however, often limited to certain tissues and developmental stages, and the requirements for transformation and regeneration competence may not always be compatible. Furthermore, a method for efficient transfer and stable integration of the transgenes into plant genomic DNA is essential for transformation, as well as a means for identifying and selecting transformed cells (Fregene and Pounti-Kaerlas, 2002).

The main constraint is usually not the delivery of foreign DNA to the regenerable cells, but the recovery of the transformed cells. Finally, the introduced genes must be correctly expressed in the primary transgenic plants and transmitted stably to their progenies (Zhang, 2000). As cassava is vegetatively propagated, the transgenes can be fixed already at the level of primary transgenic plants, and stable inheritance is of concern only when the transgenic plants are to be incorporated in breeding programs.

Plant regeneration through tissue culture can be accomplished using one of the three methods, namely, meristem culture, somatic embryogenesis and organogenesis. Figure 1 illustrates how all three methods are used in regeneration and recovery of transgenic cassava plants. Of the different explants used for regeneration, meristems are the tissue of choice as they represent 'growth centres' of plants (Fregene and Pounti-Kaerlas, 2002; Zhang et al., 2001). Therefore, this system is easy, fast and relatively genotype-independent. Applications of this system include germplasm preser-vation, micropropagation, transformation and eliminating virus and other pathogens from plant materials. In cassava, meristems can be induced to form multiple shoots on cytokinin-containing medium. Most of the shoots are derived from pre-existing auxiliary meristems, but also de novo formation of new meristems and shoots occurs (Konan et al., 1997). Transient and stable expression of both GUS and luciferase have been demonstrated in meristems and meristem-derived somatic embryos and multiple shoot clusters after particle bombardment (Puonti-Kaerlas et al., 1997).

Somatic embryogenesis is the production of embryo-like structures from somatic cells. The somatic embryo is a bipolar structure which is independent and not vascularly attached to the tissue of origin. This system is now the most commonly used regeneration method in cassava. In cassava, somatic embryogeneis is restricted to meristematic and embryonic tissues. Somatic embryogenesis can only be induced on a limited number of explants such as cotyledons of zygotic embryos

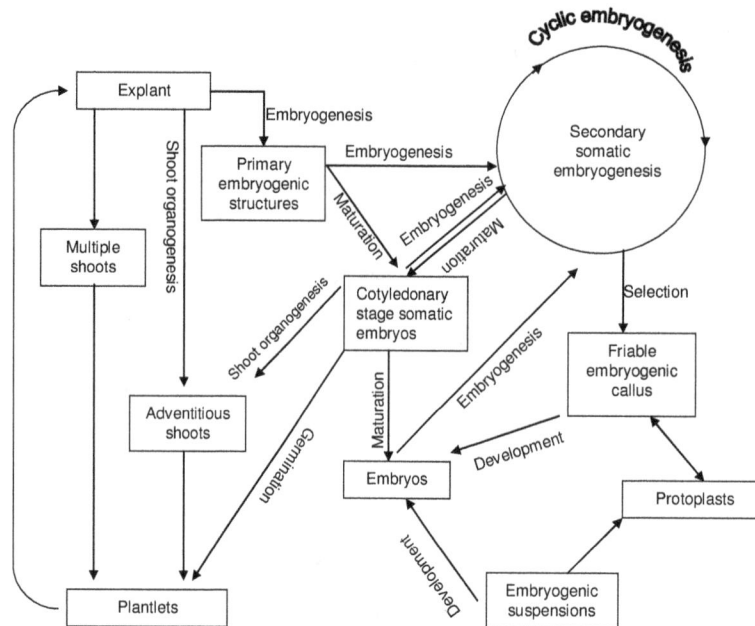

Figure 1. Schematic representation of different regeneration steps in cassava (adapted from Zhang 2000).

(Stamp and Henshaw, 1982) and immature leaf lobes, (Stamp and Henshaw, 1987; Taylor *et al.*, 1996). Other tissues include meristems and shoot tips (Puonti-Kaerlas, 1998), anthers and immature inflorescences. Primary somatic embryos can be induced to produce secondary somatic embryos by further sub-culturing on auxin-containing medium (Stamp and Henshaw, 1987). By constant sub culturing of somatic embryos, a cyclic embryogenesis system can be established either in liquid or solid medium, where the embryos rarely pass the 'torpedo' stage, until transferred to germination medium.

Cassava has proven to be recalcitrant to plant regeneration from protoplasts. Cassava protoplast isolation and culture have been performed by various laboratories (Szabados *et al.*, 1987), but regeneration was observed only in one instance and has not been repeated. Recently, protoplasts isolated from FECs and embryogenic suspensions of cassava cv. TMS60444 were found to divide and develop readily into callus after culture in a medium supplemented with 0.5 mg/l NAA and 1 mg/l zeatin (Sofiari *et al.*, 1998). After 2 months of culture, about 60% of the callus had a friable embryogenic nature. One disadvantage of this system is the long span (≥20 weeks) from explant to suspension culture and to regenerated plantlets, which may result in somaclonal variation and loss of regeneration capacity, the most common problem associated with suspension cultures. Another limiting factor is the production of FEC,

which is strongly genotype-dependent (Zhang, 2000).

A method for regeneration of cassava plants through somatic embryogenesis has been available since 1982 (Stamp and Henshaw, 1982). However, the use of embryogenic structures generated by this culture system as target tissues for genetic transformation via *Agrobacterium* (Calderón, 1988; Schöpke *et al.*, 1997) and electroporation (Luong *et al.*, 1995) has yielded at best only chimeric embryos. Taylor *et al.* (1996) developed an alternative regeneration system in which clusters of embryogenic cells are suspended in liquid medium. These suspension cells are far more suitable for genetic transformation protocols with regard to accessibility or regenerable cells and selection procedures. The first reports on successful regeneration of transgenic cassava plants have been published only in the second half of the 1990s (Li *et al.*, 1996; Raemakers *et al.*, 1996). The current status of cassava transformation is summarized in Table 1.

Cassava genetic transformation

Plant transformation is performed using a wide range of tools such as *Agrobacterium* Ti plasmid vectors, microprojectile bombardment, microinjection, chemical (PEG) treatment of protoplasts and electroporation of

Table 1. A summary of methods used in genetic engineering cassava programs.

Target tissue	Regeneration mode	Gene transfer system	Selection	Transgenic tissue	Analysis	Reference
Somatic embryos	Somatic embryogenesis	Eletroporation	-	Chimeric embryos	Transient GUS expression	Luong et al. (1995)
Somatic cotyledons	Shoot organogenesis	Agrobacterium	Hygromycin geneticin	Transgenic plants	Southern, Northern	Li et al. (1996)
Embryogenic suspension	Somatic embryogenesis	Particle bombardment	Paramomycin	Transgenic plants	Southern	Schöpke et al. (1996)
Embryogenic suspension	Somatic embryogenesis	Particle bombardment	Luciferase	Transgenic plants	Southern	Raemakers et al. (1996) Schöpke et al. (1997a)
Embryogenic suspension	Somatic embryogenesis	Particle bombardment	-	Chimeric suspensions	Transient gene expression	Munyikwa et al. (1998)
Embryogenic suspension	Somatic embryogenesis	Particle bombardment	Luciferase and phosphinotricin	Transgenic plants	Southern, Northern	González et al. (1998)
Embryogenic suspension	Somatic embryogenesis	Agrobacterium	Paramomycin	Transgenic plants	Southern	Sarria et al. (2000)
Somatic cotyledons	Somatic embryogenesis	Agrobacterium	Basta	Transgenic plants	Southern	Zhang et al. (2000)
Somatic cotyledons	Shoot organogenesis	Particle bombardment	Hygromycin	Transgenic plants	Southern, RT-PCR	Zhang and Puonti-Kaerlas (2000)
Somatic cotyledons	Shoot organogenesis	Particle bombardment	Mannose, hygromycin	Transgenic plants	Southern, Northern, RT-PCR	Zhang and Puonti-Kaerlas (2000)
Embryogenic suspension	Somatic embryogenesis				RT-PCR	Zhang et al. (2001)
Embryogenic suspension	Shoot organogenesis, Somatic embrygenesis	Agrobacterium	Mannose, hygromycin	Transgenic plants	Southern, Northern, RT-PCR	Zhang et al. (2001)

Adapted from Fregene and Pounti-Kaerlas (2002).

protoplasts. Of the above-mentioned methods for delivering foreign DNA into plant cells, the most used are Agrobacterium-mediated gene transfer and particle bombardment (Fregene and Puonti-Kaerlas, 2002). Pathogenecity of different Agrobacterium strains is highly variable and genotype dependent (Chavariaga-Aquirre et al., 1993; Sarria et al., 1993).

The naturally evolved unique ability of Agrobacterium tumefaciens to precisely transfer defined DNA sequences to plant cells has been very effectively utilized in the design of a range of Ti plasmid-based vectors. Agrobacterium-based DNA transfer system offers many unique advantages in plant transformation including:

The simplicity of Agrobacterium gene transfer makes it a relatively inexpensive vector.
A precise transfer and integration of DNA sequences with defined ends.
A linked transfer of genes of interest with the transformation marker.
The higher frequency of stable transformation with many single copy insertions.
Reasonably low incidence of transgene silencing and lastly.
The ability to transfer large T-DNA (>150 kb).

Cotyledons from somatic embryos of cassava cv. MCol22 have been the target for Agrobacterium-mediated transformation of cassava (Li et al., 1996). Their study showed that regeneration of transgenic shoots was achieved via organogenesis from somatic embryo cotyledon explants after co-cultivation with A. tumefaciens and selection on hygromycin or geneticin. Agrobacterium strains LBA4404 (pTOK233) and LBA4404 (pBin9GusInt) gave the highest transient transformation rates. Schöpke et al. (1993a) reported the use of plasmids for transformation that contained an uidA-intron gene controlled by different versions of the Cauliflower mosaic virus (CaMV) 35S promoter and the htp or nptII genes as selectable markers. Optimal transient expression of uidA was observed when cotyledon pieces were co-cultivated for 4 days with Agrobacterium strain LBA4404. In another study by González et al. (1998), Agrobacterium-mediated transformation was applied to introduce the uidA-intron and nptII genes into cassava tissue derived from embryogenic suspension cultures.

Although the success of Agrobacterium vectors has been paramount, the technique continues to have problems and limitations. Resultantly, its inability to infect monocots inspired researchers to develop an alternative delivery system. Particle bombardment is a procedure in which microscopic gold or tungsten particles coated with genetically engineered DNA are explosively accelerated into plant cells. This technique has become the second most widely used vehicle for plant genetic transformation after Agrobacterium - mediated transformation system

(Gray and Finer, 1993). Several distinct particle guns have been used including the Biolistic PDS 1000/He (Kirkert, 1993), which is the only commercially available device. The most attractive of the non-commercial devices is the particle inflow gun (Finer et al., 1992), which is based on a flowing helium device described by Takeuchi et al. (1992), since it can be fabricated from a steel plate with readily available parts and offers performance on par with the Biolistic PDS 1000/He (Brown et al., 1994). Despite its crude nature, this technique requires careful preparation and administration, and quite often tissue regeneration. Target cells in the front line are usually destroyed, but projectiles normally penetrate those cells just behind without killing them. Some of those cells that would have survived the bombardment incorporate the DNA from the microprojectiles into the genome and begin to express the gene product of the foreign DNA.

The use of different marker genes is necessary for identification as stable transformation frequencies are low. The most commonly used visual markers are GUS-encoded by the uidA gene (Jefferson, 1987), the luciferase genes from the firefly Phonitus pyralis (Ow et al., 1986), soft coral Renilla reniformis (Mayerhofer et al., 1995) and green fluorescent protein (Chalfie et al., 1994). The most selectable marker genes encode for resistance to aminoglycoside antibiotics (Fraley et al., 1983), hygromycin (van den Elzen, 1985) and phosphinotrin (Wohllenben et al., 1988).

The effectiveness of microprojectile-mediated system was shortly demonstrated successfully by scientists after the discovery in transforming monocots, the first of which was Black Mexican Sweet corn (Fromm et al., 1990). Raemakers et al. (1996) investigated the effect of different bombardment and culture parameters on transient and stable expression of the firefly luciferase gene (luc) after particle bombardment of cassava embryogenic suspension cultures. Continuous selection and subculture of light-emitting tissue eventually resulted in cultures consisting totally of transformed tissue. Differentiation and maturation of somatic embryos occurred on an MS-based medium supplemented with a complex mixture of organic components in addition to 4.14 μM picloram and 0.43 μM adenine sulfate. Different promoters (35S, e35S, 4Oe35S, UBQ1) fused to the uidA gene were bombarded into cassava leaves of cv. Señorita with a pneumatic particle gun (Schöpke et al., 1993b) in order to study their efficiency in cassava tissue. Transient gene expression was measured 24 h after bombardment with fluorometric GUS assay using methylumbelliferone glucorinide (MUG) as a substrate. Higher activities measured with MUG assays corresponded to larger diameters of blue spots. Puonti-Kaerlas et al. (1997) used a particle inflow gun to investigate the efficiency of shoot meristem transformation in cassava. After bombardment with a particle preparation containing a range of sizes, particles were found to have lodged in the first and second cell layers and even deeper. Using the *uidA* gene resulted in 50% of bombarded meristems showing 2-8 blue spots per meristem after GUS assays. Bombardment with the *luc* gene allowed screening for gene expression in living tissues and thus eliminating non-expressing plants. In a different study, Schöpke *et al.* (1997b) established and optimized conditions for particle bombardment of tissue derived from embryogenic suspension cultures of cassava cv. TMS60444. The optimal conditions for particle bombardment parameters were found to be 1100 psi pressure, 1.0 µm particle size, 2 bombardments /sample, and an osmotic treatment with 0.1 M sorbitol and 0.1 M mannitol. Observing these conditions resulted in an average number of 1350 blue spots/cm^2 of bombarded sample of embryogenic-derived tissue.

Positive selection is a new concept for the selection of transgenic plant cells (Joersbo *et al.*, 1998). The transgenic cells are selected by the addition of a compound e.g. mannose, which is converted by the transformed cells into a compound inducing a positive response, for example, growth or shoot formation. Non-transgenic cells stay alive without shoot formation, which means that neighbouring cells are not exposed to toxic selections from dying cells. Simultaneously, cells containing the transgene can utilize a component in the medium which results in growth or differentiation and non-transformed cells remain unaffected, therefore having no detrimental effect in transgenic cells (Zhang, 2000).

Calderón (1988) was the first to describe transformed callus lines of cassava. He infected cassava leaf and stem pieces, and embryogenic callus with *Agrobacterium* containing plasmids with the coding sequences for neomycin phosphotransferase II (*npt II*), phosphinotricin acetyltransferase (*bar*), or β-glucuronidase (*uidA*). Southern blot analysis with one callus line demonstrated the stable integration of T-DNA into the cassava genome.

Mechanisms of genetic engineering for virus resistance

Plant virus diseases are major constraints in the cultivation of a wide range of economically important crops worldwide. The application of genetic transformation for increased resistance to the major cassava viruses is a major priority. Strategies for the management of viral diseases normally include control of vector population using insecticides, use of virus-free propagating material, appropriate cultural practices and use of resistant cultivars. However, each of the above methods has its own drawback. Sanford and Johnson (1985), working with bacteriophages described the concept of pathogen-derived resistance (PDR), which was later demonstrated by Abel *et al.* (1986). PDR resistance strategies have proved effective in other crops

(Beachy, 1997) and are being developed against both the major viral diseases of cassava, namely, cassava common mosaic disease (CsCMD) and African cassava mosaic disease (CMD). Beachy's illustration was in tobacco, by introduction of the coat protein (CP) of tobacco mosaic virus (TMV) into tobacco and observed TMV resistance in the transgenic plants. Many viral host resistance genes have now been isolated and are used in transgenic plants to provide protection against viral infection. In a number of crops, transgenics resistant to an infective virus have been developed by introducing a sequence of the viral genome in the target crop by genetic transformation (Dasgupta *et al.*, 2003). These different crops include maize, ice, wheat, apricot, grape, papaya, pepper, tomato, sugar-beet and peanut to mention, but a few. Virus-resistant transgenics have been developed in many crops by introducing either viral coat protein (CP) or replicase (Rep) gene encoding sequences. Resistance obtained by using CP is conventionally called coat protein mediated resistance (CPMR). In this case, resistance has been shown to be due to an inherent plant response, known as post-transcriptional gene silencing (PTGS). It has been reported that gene silencing can be induced by plant virus infections in the absence of any known homology of the viral genome to host genes and this silencing may occur at the transcriptional or post-transcriptional level (Covey *et al.*, 1997). Therefore it seems possible that plants can naturally escape virus infection in a post-transcriptional manner. Because of the essential nature of the viral movement protein (MP) for intercellular movement of plant viruses, movement protein sequence has also been used for achieving viral resistance (Okeese and Pinto, 2003). Other pathogen-derived approaches include the use of satellite RNA and defective-interfering viral genomic components.

Fauquet *et al.* (1993) employed *Agrobacterium*-mediated transformation of *N. benthamiana*, which can be infected by both ACMV and CsCMV to study the expression of the viral CPs and their ability to provide protection against the respective viruses. Plants transgenic for the ACMV-CP gene were shown to contain low levels of mRNA corresponding to the coding sequence of the ACMV-CP gene. When the CP-positive plants with ACMV resulted in some degree of resistance at a virus concentration of 20-100 ng/ml.

The application of plant viral replicase-associated genes for the transformation of host plants, which leads to the generation of plant lines resistant to the donor virus, is termed replicase-mediated resistance. It has been shown to be effective in several cases (Palukaitis and Zaitlin, 1997). Hong and Stanley (1996) found that integration and expression of the ACMV AC1 (Rep) gene driven by the enhanced 35S promoter in *N. benthamiana*, imparted elevated resistance to infection by this virus. Sangaré *et al.* (1999) showed expression of mutated AC1 gene to delay symptom apparition and severity and to reduce virus movement. In addition, truncated versions of the replication-associated protein gene of geminiviruses are also able to provide protection against viral infection. Transgenic *N. benthamiana* plants expressing the N-terminal 210 amino acids of *Tomato yellow leaf curl Sardinia virus* C1, could delay virus accumulation after inoculation (Chatterjii *et al.*, 2001). Resistance was associated with the concentration of the defective viral mRNA and it was protein at the time of infection and correlated with a substantial reduction of viral DNA replication. Similarly, transgenic *N. benthamiana* expressing N-terminal 160 amino acids of the replication-associated C1 protein of *Tomato leaf curl New Delhi virus* inhibited homologous viral DNA accumulation (Chatterjii *et al.*, 2001). In addition to the sequences representing diverse functional viral proteins, defective or truncated versions of these genes, either expressed in sense or antisense can confer resistance. These types of resistance have been found to operate completely at the RNA level and are referred to as RNA-mediated virus resistance (van der Vlugt *et al.*, 1992).

Movement proteins (MP) are essential for cell-to-cell movement of plant viruses. These proteins have been shown to modify the gating function of plasmodesmata, thereby allowing the virus particles or their nucleoprotein derivatives to spread to adjacent cells (Dasgupta *et al.*, 2003). The MP confers resistance based on the competition between wild-type virus-encoded MP and the preformed dysfunctional MP to bind the plasmodesmatal sites (Lapidot *et al.*, 1993). The cell-to-cell MP and the nuclear shuttle protein NSP genes have also been used to confer resistance to begomoviruses. A recent report by Freistas-Astúna *et al.* (2001) demonstrated that tobacco plants transformed with ToMoV MP gene behaved biologically as if the resistance was RNA-mediated (recovery phenotype), but exhibited some characteristics at the molecular level that are typical of protein-mediated resistance (low, but detectable levels of MP mRNA and protein after being challenged with ToMoV).

Some viral infections, in particular those involving tombusviruses and carmoviruses, are associated with the accumulation of defective interfering (DI) RNAs (Keese and Pinto, 2003). These RNAs contain sequences essential for their replication by the helper virus, but have incomplete coding regions. They often reduce accumulation of the helper virus and may result in amelioration of symptoms. With the development of infectious full-length viral clones, artificial DI RNAs and DNAs can now be generated and tested. In a study conducted by Rubio *et al.* (1999), *N. benthamiana* was transformed with a DNA cassette designed to transcribe DI RNA from Tomato bushy stunt virus (TBSV). Self-cleaving sequences were added to the termini so that transcripts were competent to be replicated by the helper virus. Subsequent viral challenge showed resistance to TBSV and closely related tombusviruses. In another study by von Arnim *et al.* (1993), incorporation of sub ge-nomic DNA A and B conferred resistance to ACMV in *N. benthamiana*.

CONCLUSION

Plant virus diseases, including CMD continue to cause severe constraints on the productivity of a wide range of economically important crops worldwide. The application of genetic transformation for increased resistance to cassava begomoviruses is a major priority. In SA, interest in cassava production has increased astronomically with Cassava Starch Manufacturing Company (CSM) taking the lead in Dendron, Limpopo Province and Barberton, Mpumalanga Province. CSM is a privately owned commercial company that belongs to Mr. Jim Casey. The company has dedicated more than 2000 ha of land to cassava production. In addition, CSM has contracted small-scale farmers to produce the crop for processing purposes. The main markets for the starch are food, textile, paper, corrugated cardboard and mining industries in SA. There are plans to extend cassava production in several districts in Mpumalanga and KwaZulu-Natal provinces, also using contracted small-scale rural farmers.

In SA, cassava transformation capability is still being developed amid a concern about declining cassava yields due to infection by begomoviruses. It is considered that cassava transformation technology can be transferred to the African environment when suitable institutes and individuals have been identified in the continent. One such laboratory, ILTAB has established a program by which African scientists are trained in cassava biotechnology and return to their home institutions to help transfer skills and capacity building. However, the issue of staff turnover due to lack of resources and proper funding can frustrate such a program. Currently, USAID through CIAT and ILTAB are funding a program that will enable Agricultural Research Council (ARC) scientists in SA train elsewhere in developed laboratories on cassava transformation aiming at improved starch qualities and resistance against CMD. South Africa could test for efficacy of transgenic cassava as issues concerning intellectual property rights (IPR) and biosafety implications are already legislated and in place. These issues can be a bottleneck in many African countries as recently experienced by ILTAB while trying to deploy a field test on their cassava replicase transgenic event. Successful application of transgenic technologies in cassava will depend not only on technical advances, but also on successful transfer of knowledge, tools and expertise to southern African countries wherein cassava has both important socioeconomic and industrial roles. In order for the cassava farmers to benefit from cassava biotechnology advances already achieved by well established overseas laboratories, e.g. ILTAB, Wageningen University, ETH, Ohio State University and CIAT, there has to be a strong technology transfer programme that will support local scientists in the transfer of already existing cassava transformation systems.

REFERENCES

Abel PP, Nelson RS, De B, Hoffmann N, Rogers SG, Fraley RT, Beachy RN (1986). Delay of disease development in transgenic plants that express the tobacco mosaic virus coat protein gene. Sci. 232: 738-743.

Allem AC (2002). The origins and taxonomy of cassava. In: Hillocks RJ, Thresh JM, Bellotti AC (eds.) Cassava: Biology, Production and Utilization. CAB International, Wallingford, Oxon. pp.1-16.

Beachy RN (1997). Mechanisms and applications of pathogen-derived resistance in transgenic plants. Current Opinions in Biotechnol. 8: 215-220.

Berry S, MEC Rey (2001). Molecular evidence for diverse populations of cassava-infecting begomoviruses in southern Afr. Archives of Virol. 146: 1795-1802.

Brown DCW, Tian L, Buckley DJ, Lefebvre M, McGrath A, Webb J (1994). .Development of a simple particle bombardment device for gene transfer into plant cells. Plant Cell Tissue Organ Culture 37: 47-53.

Calderon-Urrea A (1988). Transformation of Manihot esculenta (cassava) using Agrobacterium tumefaciens and expression of the introduced foreign genes in transformed cell lines. MSc thesis, Vrije University, Brussels, Belgium

Calvert LA, Thresh JM (2002). The viruses and virus diseases of cassava. In: Hillocks RJ, Thresh JM, Bellotti AC (eds.) Cassava: Biol. Production and Utilization. CAB Int. Wallingford, Oxon pp.237-260.

Chalfie M, Tu Y, Euskirchen G, Ward WW, Prasher DC (1994). Green fluorescent protein as a marker gene for gene expression. Sci. 263: 802-805.

Chatterji A, Beachy R, Fauquet CM (2001). Expression of the oligomerization domain of the replication-associated protein (Rep) of Tomato leaf curl New Delhi virus interferes with DNA accumulation of heterologous geminiviruses. J. Biol. Chem. 276: 25631-25638

Chatterjii A, Beachy RN, Fauquet CM (2001). Expression of the oligomerization domain of the replication-associated protein (Rep) of tomato leaf curl: New Delhi virus interferes with DNA accumulation of heterologous geminiviruses. Journal of Biology Chemistry 276: 2 5631–2 5638.

Chavariaga-Aquirre P, Schöpke C, Sangare A, Fauquet CM, Beachy RM (1993). Tranformation of cassava (Manihot esculenta Crantz) embryogenic tissues using Agrobacterium tumefaciens. In: Roca WM, Roca AM (eds.) Proceedings of the First International Scientific Meeting of the Cassava Biotechnol. Network. Cartegena de Indias, Colombia, CIAT Working Document 123, pp.222-228

Chrispeels M, Sadava D (2003). Plants, Genes, and Crop Biotechnology. Jones and Bartlett, Sudbury, Massachusetts p. 360

Covey SN, Al-Kaff N, Langara A, Turner DS (1997). Plants combat infection by gene silencing. Nature 385: 781-782.

Daphne PE (1980). Cassava, A South African venture. Optima 1: 61-68.

Dasgupta I, Malathi VG, Murkhejee SK (2003). Genetic engineering for virus resistance. Current Science 84(3): 341-354.

FAO (2003). Cassava production statistics 2002. http://www.fao.org

Fauquet C, Schöpke C, Sangare A, Chavarriaga P, Beachy RN (1993). Genetic engineering technologies to control and their application to cassava viruses. In: Roca, Thro AM (eds.). Proceedings of the First Int. Sci. Meeting of the Cassava Biotechnol. Network. Cartegena de Indias, Colombia, CIAT Working Document 123, pp.190-207.

Finer JJ, Vain P, Jones MW, McMullen M (1992). Development of the particle inflow gun for DNA delivery to plan cells. Plant Cell Reports 11: 323-328.

Fondong VN, Pita JS, Rey MEC, de Kochko A, Beachy RN, Fauquet CM (2000). Evidence of synergism between African cassava mosaic virus and a new double-recombinant geminivirus infecting cassava in Cameroon. J. General Virol. 81: 287-297.

Fraley RT, Rogers SG, Horsch RB, Sanders PR, Flick S, Adams SP, Bittner ML, Brand LA, Fink CL, Fry JS, Galluppi GR, Goldberg SB,

Hoffman NL, Woo SC (1983). Expression of bacterial genes in plant cells. Proceedings of Nat. Academy of Sci. USA 80: 4803-4807.

Fregene M and Puonti-Kaerlas J, (2002). Cassava Bbiotechnology. In: Hillocks RJ, Thresh JM, Bellotti AC (eds.) Cassava: Biol., Prod.and Utilization.. CAB International, Wallingford, Oxon pp.179-207.

Freistas-Astúa J, Polston JE, Hiebert E (2001). Further characterization of the resistance of the resistance in tobacco lines expressing the Tomato mottle virus BC1 movement protein gene. Annals, 3rd Int. Geminivirus Symposium, Norwich, UK. pp. 33.

Fromm ME, Morrish F, Armstrong C, Williams R, Thomas J, Kein TM (1990). Inheritance and expression of chimeric genes in the progeny of transgenic maize plants. Biotechnol. 8: 833-839.

González AE, Schöpke C, Taylor N, Beachy RN, Fauquet CM (1998). Regeneration of transgenic cassava plants (Manihot esculenta Crantz) through Agrobacterium-mediated transformation of embryogenic suspension cultures. Plant Cell Reports 17: 827-831.

Gray DJ, Finer JJ (1993). Editorial introduction. Special section: development and operation of five particle guns for introduction of DNA into plant cells. Plant Cell Tissue Organ Culture 33: 219.

Gray V, (2003). Modelling cassava crop production under semi-arid subtropical conditions in South Afr. J. Sci. 99: 321-331.

Hong Y and Stanley J (1996). Virus resistance in Nicotiana benthamiana conferred by African cassava mosaic virus. Replication-associated protein (AC1) transgene. Molecular Plant Microbe Interactions 9: 219-225.

Jefferson RA (1987). Assaying chimeric genes in plants: the GUS gene fusion system. Plant Mol. Biol. Reports 5: 387-405.

Joersbo M, Donaldson I, Kreiberg K, Guldager, Peterson S, Brunstedt J, Okkels FT (1998). Analysis of mannose selection used for transformation of sugar beet. Mol. Breeding 4: 111-117.

Jones WO (1959). Manioc in Africa. Standard University Press, Stanford, California, p.315.

Keese PK, Pinto YM (2003). Molecular crop protection using recombinant DNA techniques. In: Plant Virol. in sub-Saharan Afr. IITA. Pp 476-496

Keese PK, Pinto YM (2003). Molecular crop protection using recombinant DNA techniques. In: IITA Publication. Mol. crop protection using recombinant DNA techniques. pp. 476-491.

Kikkert JR (1993). The biolistic PD-1000/He device. Plant Cell Tissue Organ Culture 33: 221-226.

Konan NK, Schöpke C, Carcamo R, Beachy RN, Fauquet C (1997). An efficient mass propagation system for cassava (Manihot esculenta Crantz) based on nodal explants and axillary bud-derived meristems. Plant Cell Rep 16: 444-449.

Lacanster PA and Brooks JE (1983). Cassava leaves as human food. Econ. Bot. 37: 331-348.

Lapidot M, Gafny R, Ding B, Wolf S, Lucas WJ, Beachy RN (1993). A Dysfunctional Movement Protein of Tobacco Mosaic Virus That Partially Modifies The Plasmodesmata and Limits Virus Spread in Transgenic Plants. The Plant J. 4: 959-970.

Legg JP and Thresh JM (2003). Cassava virus diseases in Africa. In: Proceedings of the First International Conference on Plant Virology in Sub-Saharan (4–8 June 2001, Ibadan, Nig.), IITA, Ibadan, Nig., pp 517-522.

Li HQ, Sautter C, Potryus I, Pounti-Kaerlas J (1996). Genetic transformation of cassava (Manihot esculenta Crantz). Nature Biotechnol. 14: 736-740.

Luong HT, Shewry PR and Lazzeri PA (1995). Transient gene expression in cassava somatic embryos by tissue electroporation. Plant Sci. 107: 105-115.

Munyikwa TRI, Raemakers KCJM, Schreuder M, Kok R, Schippers M, Jacobsen E, Visser RGF (1998). Pinpointing towards improved transformation and regeneration of cassava (Manihot esculenta Crantz). Plant Sci. 135: 87-101.

Nweke, F.I., Poulson, R, Strauss J. (1994). Cassava production trends in Africa. In: Ofori, F. and Hahn, S.K. (eds.) Tropical Roots Crops in a Developing Economy. Proceedings of the 9th Sympossium of the International Society for Tropical Roots Crops, 20-26 October, 1991, Accra, Ghana. Govt of Ghana, pp. 311-321.

Ow DW, Wood KV, DeLuca M, de Wet JR, Helinski DR, Howell SH (1986). Transient and stable expression of the firefly luciferase gene in plant cells and transgenic plants. Sci. 234: 856-859.

Paukaitis P, Zaitlin M (1997). Replicase-mediated resistance to plant virus
disease. Advances in Virus Research 48: 349-377.

Pita JS, Fondong VN, Sangaré A, Otim-Nape GW, Ogwal S, Fauquet CM (2001). Recombination, pseudorecombination and synergism of geminiviruses are determinant keys to the epidemic of severe cassava mosaic disease in Uganda. J. General Virol. 82: 655-665.

Puonti-Kaerlas J (1998). Cassava biotechnology. Biotech Genetic Engineering Reviews 15: 329-361.

Puonti-Kaerlas J, Frey P, Potrykus I (1997). Deveopment of meristem gene transfer techniques for cassava. In: Thro AM, Akaroda MO (eds.) Procedures of Third International Meeting. Cassava Biotechnol. Network. Afr. J. Root and Tuber Crops 2: 175-180.

Raemakers CJJM, Sofiari E, Taylor N, Henshaw G, Jacobsen E, Visser RGF (1996). Production of transgenic cassava (Manihot esculenta Crantz) plants by particle bombardment using luciferase activity as selection marker. Molecular Breeding 2:339-349.

Rubio T, Borja M, Scholthof HB, Feldstein PA, Morris TJ, Jackson AO (1999). Broad-spectrum protection against tombusvirus elicited by defective interfering RNAs in transgenic plants. J. Virol. 73: 5070-5078.

Sanford JC, Johnson SA (1985). The concept of parasite-derived resistance: deriving resistance genes from the parasites own genome. J. Theoretical Biol. 115: 395-405.

Sangáre A, Deng D, Fauquet CM, Beachy RN (1999). Resistance to African cassava mosaic virus conferred by mutant of the putative NTP-binding domain of the Rep gene (AC1) in Nicotiana benthamiana. Mol Biol Rep 5: 95-102.

Sarria R, Ocamp C, Ramirez H, Hershey C, Roca WM (1993). Genetics of esterase and glutamate oxaloacetate transaminase isozymes in cassava (Manihot esculenta Crantz). In: Roca WM, Roca AM (eds.) Proceedings of the First International Scientific Meeting of the Cassava Biotechnology Network. Cartegena de Indias, Colombia, CIAT Working Document 123, pp.75-80.

Sarria R, Torres E, Angel F, Chavarriaga P, Roca WM (2000). Transgenic plants of cassava (Manihot esculenta) with resistance to Basta obtained by Agrobacterium-mediated transformation. Pant Cell Reports 19: 339-344.

Schöpke C, Chavarriaga P, Fauquet C, Beachy RN (1993a). Cassava tissue culture and transformation: Improvement of culture media and the effect of different antibiotics on cassava. In: Roca W, Thro AM (eds.). Proceedings of the First International Scientific Meeting of the Cassava Biotechnology Network. Cartegena de Indias, Colombia, CIAT Working Document 123, pp.140-145.

Schöpke C, Franche C, Bogusz D, Chavarriaga P, Fauquet C, Beachy RN (1993b). Transformation in Cassava (Manihot esculenta Crantz). In: Bajaj (ed.) Biotechnology in Agriculture and Forestry, Vol 23. Plant Protoplasts and Genetic Emgineering IV Springer Verlag Berlin Heidelberg pp. 273-289.

Schöpke C, Taylor N, Cárcamo R, González de Schöpke AE, Konan NK, Marmey P, Henshaw GG, Beachy RN and Fauquet C (1997a). Stable transformation of cassava (Manihot esculenta Crantz) by particle bombardment and by Agrobacterium. African Journal of Root and Tuber Crops 2(1&2): 187-193.

Schöpke C, Taylor N, Carcamo R, Konan NK, Marmey P, Henshaw GG, Beachy RN, Fauquet C (1996). Regeneration of transgenic cassava plants (Manihot esculenta Crantz) from microbombarded embryogenic suspension cultures. Nature Biotechnol. 14: 731-735.

Schöpke C, Taylor NJ, Cárcamo R, Beachy RN, Fauquet C (1997b). Optimization of parameters for particle bombardment of embryogenic suspension cultures of cassava (Manihot esculenta Crantz) using computer image analysis. Plant Cell Reports 16: 526-530.

Sofiari E, Raemakers CJJM, Bergervoet JEM, Jacobsen E, Visser RGF (1998) Plant regeneration from protoplasts isolated from friable embryogenic callus of cassava. Plant Cell Reports 18(1-2): 159-165.

Stamp JA and Henshaw GG (1982). Somatic embryogenesis in cassava. Z. Pflanzenphysiol. 105: 183- 187.

Stamp JA and Henshaw GG (1987). Secondary somatic embryogenesis and plant regeneration in cassava. Plant Cell Tissue and Organ Culture 10: 227-233.

Szabados L, Hoyos R and Roca W (1987). In vitro somatic embryogenesis and plant regeneration of cassava. Plant Cell Reports 6: 248-251.

Takeuchi Y, Dotson M, Keen NT (1992). Plant transformation: a simple particle bombardment device based on flowing helium. Plant Mol. Biol. 18: 835-839.

Taylor NJ, Edwards M, Kiernan RJ, Davey CDM, Blakesay D and Henshaw GG. (1996). Development of friable embryogenic callus and embryogenic suspension culture systems in cassava (Manihot esculenta Crantz). Nature Biotechnol. 14: 726-730.

Thresh JM, Otim-Nape GW and Jennings DL (1994). Exploiting resistance to African cassava mosaic virus. Aspects of Appl. Biol. 39: 51-60.

van den Elzen P, Townsend J, Lee KY, Bedbrook J (1985). A chimeric hygromycin resistance gene as a selectable marker in plant cells. Plant Mol. Biol. 5: 299-302.

Van der Vlugt RAA, Ruiter RK, Goldbach R (1992). Evidence for sense RNA-mediated resistance to PVY^{N}in tobacco plants transformed with the viral coat protein cistron. Plant Mol. Biol. 20: 631-639.

Von Armin, A., Frishmuth, T, Stanley, J (1993). Detection and possible function of African cassava mosaic virus DNA B gene products. Virol. 192: 264-272.

Wohllenben W, Arnold W, Broer I, Hillemann D, Strauch E, Peuhler A (1988). Nucleotide sequence of the phosphinothricin N-acetyltransferase gene Streptomyces viridochromogenes Tü494 and its expression in Nicotiana tabacum. Gene 70: 25-38.

Zhang P (2000). Studies on cassava (Manihot esculenta Crantz) transformation: towards genetic improvement. Ph.D. thesis, Swiss Federal Institute of Technology Zürich, Switzerland, Diss. ETH No. 13962.

Zhang P, Phansiri S, Puonti-Kaerlas J (2001). Improvement of cassava shoot organogenesis by the use of silver nitrate in vitro. Plant Cell Organ and Tissue Cult. 67: 47-54.

Zhang P, Puonti-Kaerlas J (2000). PIG-mediated cassava transformation using positive and negative selection. Plant Cell Rep. 19: 939-945.

8

Integrating molecular tools with conventional breeding strategies for improving phosphorus acquisition by legume crops in acid soils of Sub-Saharan Africa

Maureen Fonji ATEMKENG[1]*, Teboh Jasper MUKI[2], Jong-Won PARK[3] and John JIFON[3]

[1]Leguminous Crop Program, Institute of Agricultural Research for Development (IRAD), P.O. Box 2067, Yaounde, Cameroon.
[2]School of Plant, Environmental, and Soil Sciences, Lousiana State University – AgCenter, 238 Sturgis Hall, Baton Rouge, LA, 70803, USA.
[3]Texas AgriLife Research, Texas A&M University System, 2415 E Business 83, Weslaco, Texas 78596, USA.

Leguminous crops are key components of low input agricultural cropping systems, and play an important role in ensuring food security in many societies in Sub-Saharan Africa (SSA). However, legume crop productivity in SSA is frequently limited by mineral nutrient deficiencies (particularly phosphorus, P). A common remedy for P deficiency is the application of P-fertilizers or in the case of low input cropping systems the reliance on symbiotic relations between crops and beneficial soil bacteria (rhizobia) and fungi (mycorrhizas). More recently, identification of legume species and genotypes with high efficiencies of P uptake and P use has been the focus of improvement programs using conventional breeding techniques. Due to inherent time limitations in conventional breeding approaches, progress in improving legume P uptake and P use efficiencies has been slow. Advances in attaining this goal could be by integrating molecular tools with conventional improvement strategies. A consideration of molecular and physiological mechanisms underlying differences in P uptake and P use efficiencies can result in more precise targeting of genetic variation and improvement through marker-assisted selection and other conventional techniques. This article discusses the potential for improving legume crop P uptake and P use efficiency in low-P, acid soils of SSA by integrating physiological and genomic tools, with conventional crop improvement in acid soils.

Key words: Legume, phosphorus uptake, breeding, comparative genomics, crop improvement, Sub-Saharan Africa.

INTRODUCTION

Land degradation and soil fertility depletion are among the major causes of low agricultural productivity which in turn leads to food insecurity and natural resource destruction in many parts of Sub-Saharan Africa (SSA) (Sanchez, 2010). These factors have important economic and social implications, and are responsible for the gap between food production and population growth in SSA, as well as low productivity leading to poor nutrition and health. Therefore, a significant investment in advanced technologies that allow crop plants to adapt to inherent soil limitations to productivity as a complement to the conventional approach of adapting the production environment (soil) to meet crop requirements are needed to sustain and improve yields. This approach is particularly relevant in the smallholder cropping systems that dominate food production in SSA. Intercropping with legume crops is a key feature of these smallholder cropping systems, and legume crops play an important role not only in the overall productivity of these systems, but also in human health and nutrition (Kamanga et al., 2010).

Fabaceae (Legume family) is one of the most species-rich and economically important plant families because

*Corresponding author. E-mail: atemaureen@yahoo.com.

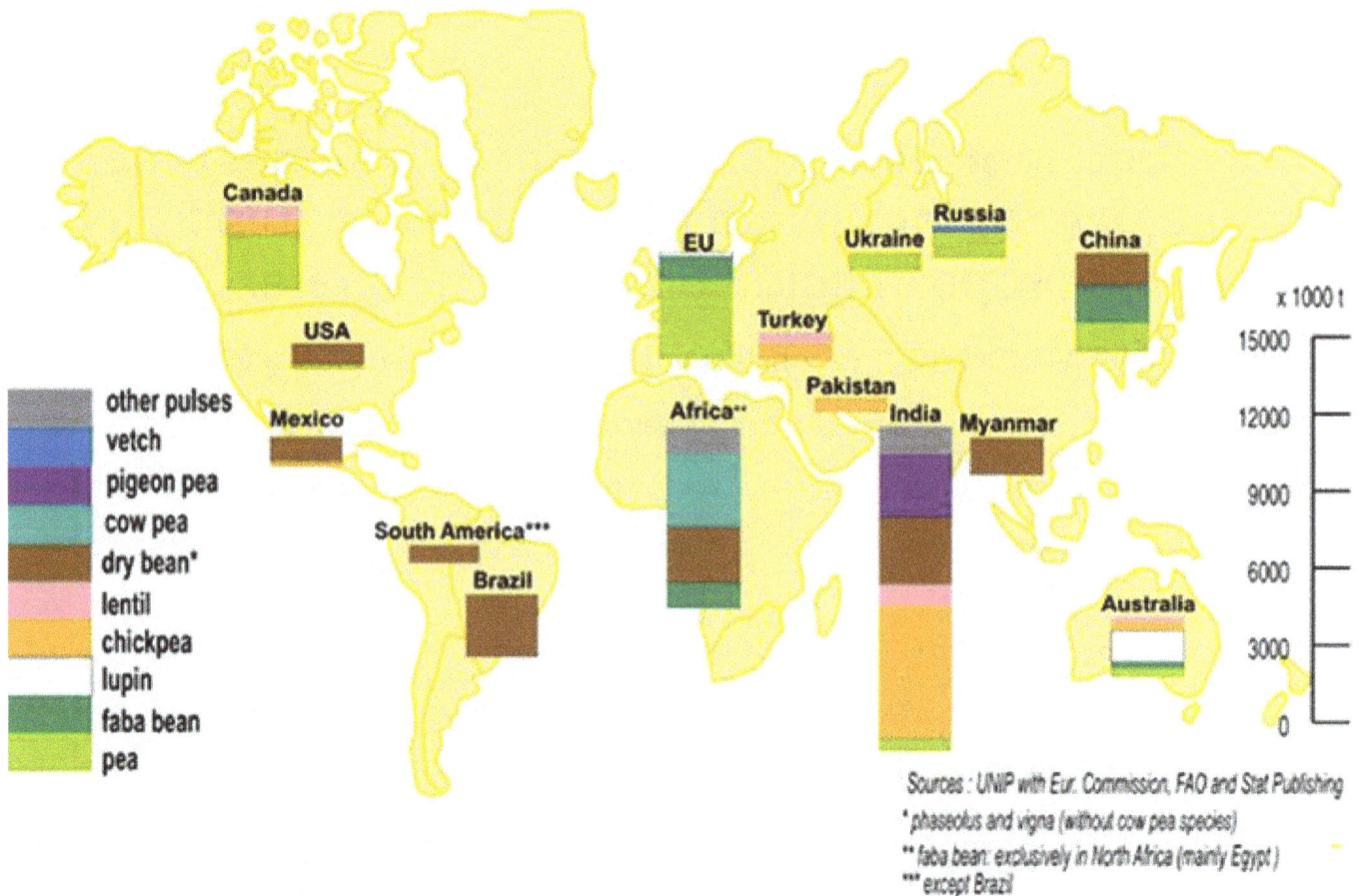

Figure 1. Market and consumption of grain legumes in the world. There are some grain legumes like cowpea that are produced uniquely in Africa.

they are an important source of food, fodder, energy and cover crops. Some of the most important commercial legume species in SSA (Figure 1) include the common bean *(Phaseolus vulgaris)*, cowpea (*Vigna unguiculata*), soybeans (*Glycine max*), peas (*Pisum sativum*), groundnut (peanut, *Arachis hypogaea*), and alfalfa (*Medicago sativa*). Their ability to fix atmospheric nitrogen (N) makes them key components of low-input cropping systems as a sustainable approach for improving soil N status (Ghosh et al., 2006). In this regard, they are ideal crops in intercropping production systems. Legume seeds and foliage also constitute an inexpensive and, in some cases, the only source of high-quality proteins in many low-income societies (Vance et al., 2000; Schneider, 2002; Xinshen et al., 2003).

However, in most parts of SSA, and the tropics in general, low phosphorus (P) availability is a major constraint for legume crop establishment, nitrogen fixation, growth, and productivity (Hague et al., 1986; Sanchez et al., 1997; Bouhmouch et al., 2005; Sanchez, 2010). Unlike N, which some crops can obtain through

symbiotic biological fixation, P must be supplied from mainly commercial fertilizers, plant and animal manures, industrial and domestic wastes, and soil parent materials (Brady and Weil, 2001; Havlin et al., 2005). Highly weathered acid soils that fix large quantities of P dominate arable lands in SSA (Table 1, Scherr, 1999), and include Aridisols (36.8%), Alfisols (21.0%), Ultisols and Oxisols (22.6%), Entisols (14.1%), while Inceptisols and Vertisols cover ca. 2.5 and 2.1%, respectively (Hague et al., 1986; Aubert and Tavernier, 1972). The widespread P deficiency in these soils is evidenced by the positive crop response to fertilizer P additions (Hague et al., 1986; Smit et al., 2009). However, in many instances, most of the applied P is rapidly fixed by Fe- and Al-oxides in these soils, or bound in sparingly soluble P pools not immediately for plant uptake (Sample et al., 1980). Additionally, soil erosion and P removal in harvested crops also contribute to the P deficiency problem. Maintaining a sufficient supply of plant available P in arable lands is, therefore, a key prerequisite for optimal productivity and quality.

Table 1. Global estimates of soil degradation in agricultural land (from Scherr, 1999).

Agricultural land (million hectares)	Region							
	Africa	Asia	South America	Central America	North America	Europe	Oceania	World
Total	187	536	142	38	236	287	49	1475
Degraded	121	206	64	28	63	72	8	562
Percent	65	38	45	74	26	25	16	38

In Africa, highly weathered acid soils that fix large quantities of Phosphorus dominate arable lands.

Low soil solubility and mobility of P, and high P sorption in soils make plant available P, usually as $H_2PO_4^-$ or HPO_4^{2-} orthophosphate ion (Marschner, 1995; Havlin et al., 2005) one of the major limitations to plant growth. In most arable soils, as well as in nutrient-poor grassland and forest soils, over 70% of phosphate is present in highly unavailable organic forms (Macklon et al., 1994) and only about 25% of the P applied in fertilizers in tropical soils is recovered by crops (Baligar and Bennett ,1986). Because of low available P in many soils, plants have evolved a variety of P uptake and utilization strategies including: (1) formation of mycorrhizal associations between roots and symbiotic fungi to increase soil exploration and uptake of water and immobile nutrients, notably P, (2) increasing the length and density of root hairs to increase the effective absorptive area of roots, and reduce the diffusive pathway distance for P to reach the root surface, (3) modifications in root architecture and branching patterns to thoroughly explore the soil, (4) exudation of organic acids, H^+ and phosphatases to solubilize and release organic and inorganic P from the soil, and (5) increasing the concentration of phosphate transporters in the root cell plasma membranes (Kochian, 2000; Lambers et al., 1998; Marschner, 1995; Havlin et al., 2005). These varied mechanisms for increase P uptake by plants have led to genetic variation among species for P uptake and P use efficiency. Such variation presents a unique opportunity for breeding programs to further improve on these traits and thereby ensure sustainable productivity of many cropping systems in P-deficient soils.

Farmers in SSA routinely incorporate legume crops in farming systems as an intercrop or in rotation to improve associated crop yields (Jemo et al., 2006; Wendt and Atemkeng, 2004; Niang et al., 2002). Until recently, little attention had been given to the P nutrition of legume crops despite the pivotal role that P plays in legume crop growth and productivity. In an intercrop study on low P soils including maize and Faba bean, Li et al. (2007) used permeable and impermeable root barriers to demonstrate that superior yields of intercropped maize compared to monocropping resulted from its uptake of P mobilized by acidification of the rhizosphere through Faba bean root release of organic acids and protons.

These enhancements in P nutrition seem to be greater when symbiotic associations between crop and microbes exist. For instance, Jemo et al. (2007) demonstrated that inoculation of Macuna (*Macuna pruriens*) with Bradyrhizobia and arbuscular mycorrhizal fungi on acid soils significantly increased P-uptake and P-use efficiency. Numerous recent studies have now confirmed that useful genetic variation for P uptake and P-use exists among various economically important legumes species. For instance, large genotypic differences P acquisition, P-use efficiency and important traits for P uptake (root length and root hair densities) have recently been reported among cowpea genotypes (Krasilnikoff et al., 2003; Saidou et al., 2007; Abaidoo et al., 2007). Similarly, genotpic differences in P acquisition and P-use efficiency have also been demonstrated among pigeonpea [*Cajanus cajan* (L.) Millsp.] genotypes, with short-duration genotypes accumulating more P, producing more total dry matter, and producing more dry matter per unit of absorbed P than the medium- and long-duration genotypes (Vesterager et al., 2006). These authors also reported positive correlations between P uptake and P-use efficiency. These findings of considerable genetic variability in P uptake-efficiency among legume genotypes have made it possible for conventional legume improvement programs to select and breed for this important trait as a means to improve productivity. Modern molecular techniques such as marker-assisted selection (MAS) can greatly expedite this process and ultimately assist in narrowing the gap between food production and demand in SSA.

Only recently, have researchers begun to use molecular genetics as a priority tool to enhance P uptake (Muchhal et al.,1996) with biochemical, physiological, and morphological plant response to P stress the subject of many recent reviews (Vance et al., 2003). This review presents opportunities, gaps and potentials of integrating molecular biological tools with traditional crop improvement efforts, to understand and exploit underlying genetic mechanisms of P-deficiency adaptation for food legume improvement in acid soils in SSA. We have therefore: (1) highlighted the importance of P in legume production, it's deficiency, and strategies to improve availability in acid soils (2) elaborated on the importance of microbial P turnover in preventing P sorption and enhancing P availability in acid soils (3) surveyed the literature on molecular biological tools and materials important in breeding legumes for adaptation to low-P (4) presented a brief overview of plant biology research carried out on some model legumes [common bean (*P. vulgaris* L.*)*, barrel medic (*M. truncatula* Gaertna) and white lupine (*Lupinus albus*)] with respect to P-stress adaptation and (5) concluded with a future outlook. This summary will enable legume plant breeders to use QTL markers in selecting and developing P-efficient legume plants, which would in turn, benefit low-input agricultural systems and enhance environmentally friendly cropping in intensively cultivated systems.

PHOSPHORUS AVAILABILITY IN SSA SOILS

A predominance of high activity Al and Fe in most SSA soils is a result of excessive weathering of basic cations (Ca and Mg) due to frequent heavy rainfall that usually exceeds evapotranspiration, with a resultant drop in soil pH commonly below 5.5. Among the trio of major plant nutrients, P is commonly the most limiting in these soils (Buehler et al., 2002), existing in inorganic and organic forms. Because of its particular chemistry, orthophosphate (HPO_4^{2-}, $H_2PO_4^-$) which is the preferred inorganic form taken up by plants reacts readily and forms relatively low soluble, high energy bonds with positively charged Al and Fe at low pH and Ca at high pH (Kideok and Kubicki, 2004). Even when total soil P may be high, >80% still exists in unavailable forms to plants (Rengel and Marschner, 2005). Rhizosphere pH is further reduced when N-fixing legumes take up cations and excrete H^+ from roots to maintain internal electroneutrality. Therefore since legume nodules are sinks for P with a concentration higher than that of other organs and require much ATP for nitrogenase functioning (Gniazdowska et al., 1998) different strategies have to be employed to enhance P uptake and use efficiency (Vance et al., 2003).

PLANT STRATEGIES FOR IMPROVING P-AVAILABILITY IN ACID SOILS

A variety of adaptive strategies to improve P acquisition and use involve changes in root morphology and architecture (Yan et al., 2004; Beebe et al., 2006; Ochoa et al., 2006) root proliferation and elongation (Bates and Lynch, 2000) as well as changes in shoot and flower development (Bucciarelli et al., 2006). Some plants may excrete hydrogen ions (H^+) to acidify the rhizosphere which could result from surplus uptake of nutrient cations or rather, from light induced photosynthesis observed

with cowpea seedlings (Rao et al., 2002). Phosphorus-stressed plants also tend to allocate a greater proportion of biomass to root dry matter compared to P-sufficient plants (López-Bucio et al., 2003). Some P-stressed plants exude organic acids (Wang et al., 2007) and phosphatase enzymes (Tomasi et al., 2008) into the rhizosphere, or produce cluster roots (Lynch and Brown, 2001) enhance symbiotic relationships with soil organisms (Brundrett, 2002) such as vesicular-arbuscular mycorrhizas (VAM) (Kaeppler et al., 2000). But at what cost are root exudation and root development for example, to the overall yield of the legume crop? For inherently P deficient or P depleted soils, excretion of organic ions can significantly reduce plant yields costing up to 70% of the C source (Johnson et al., 1996a). Kuzyakov and Domanski (2000 and 2002) reported that of the total C allocated below-ground, 7 to 13% is ultimately found in roots, 2 to 5% exuded and 7 to 14% used up in root respiration for maintenance, root growth and ion uptake. This is an important consideration in selecting P adaptive traits because P stress and sugar signaling are related (Müller et al., 2007) whereby, a proportionate increase in the amount of sugars is translocated to roots (Hernández et al., 2007) despite reduced demand of photosynthate and higher sugar levels (Morcuende et al., 2007).

MICROBIAL P TURNOVER

Albeit methodological challenges for assessing microbial turnover of soil organic P, the importance of biological process in enhancing soil P availability to crops, in subsistence farming especially, is well documented (Tiessen and Shang, 1998; Buehler et al., 2002; Oberson and Joner, 2005; Steffens et al., 2009). For example, the incorporation of P into microbial cells prevents P sorption thereby maintaining it in easily mineralizeable form. Since food legumes supply large amounts of organic residues above and below ground, and organic matter boosts soil microbial activity, the microbiologically-driven processes in soil P dynamics are enhanced, and the microbial P pool increased in legume cropping systems. However, as more genes regulating arbuscular mycorrhizal fungi (AMF) and plant P uptake remain uncharacterized (Tesfaye et al., 2007), there is still a paucity of research findings to determine that myorrhization of legumes significantly contributes to P uptake. Yan et al. (2006) showed no variation in P uptake amongst mycorrhizal and non-mycorrhizal soybean genotypes. Howeveer, breeding legumes to establish symbiotic relationships with Rhizobia and AMF in low pH soils, for the purpose of enhancing P uptake in SSA, has the dual advantage of enhancing soil N fertility following the formation of N-fixing root nodules of host plant roots.

In acid soils, nodulation of a range of pasture and crop

legumes (e.g. subclover, lucerne, white clover, pea, cowpea, bean, etc.) is reduced, mainly because of sensitivity of early nodulation events, such as attachment, root hair curling and initiation of infection thread formation (Vlassak and Vanderleyden, 1997). *Nod* gene expression, more notable in acid-sensitive than in acid-tolerant rhizobial strains, may account for these deleterious acidity-related effects (McKay and Djordjevic, 1993). Several of the genetically identified symbiotic genes have been cloned (Geurts et al., 2005) facilitated by the development of model systems (*Medicago* and *Lotus*) for which efficient molecular tools became available. These genetic tools have facilitated the cloning of orthologous genes from pea and soybean (Kaló et al., 2005). Improvements in the symbiotic relationship between rhizobia and legumes may be brought about by introducing the association of legume-rhizobia and plant growth-promoting rhizobacteria, legume-rhizobia and helper bacteria or legume-rhizobia and "arbuscular mycorrhizas" AM. Legume-AM symbiosis is an exquisite, highly regulated interaction between the legume host and AM, requiring coordinated expression of genes from two vastly different organisms. A complete review of AM symbiosis is beyond the scope of this paper, but can be referenced to several important and recent reviews (Kuster et al., 2007; Oldroyd et al., 2005). With advances in the development of molecular tools, genes responding to AM can be characterized, with the ultimate aim of identifying genetic strategies that regulate AM symbioses and P acquisition.

MOLECULAR TOOLS AND PLANT BREEDING

Conventional phenotypic selection for specific traits requires the evaluation of the trait from multiple environments over several years; this is often very expensive, time consuming, and labor intensive (Yuan et al., 2002). On the other hand, molecular marker technology is a powerful tool for selecting specific traits (Babu et al., 2003). Recently, a number of studies on quantitative trait loci (QTL) analysis relating root morphology and physiology to P nutrition in plants were reported (Beebe et al., 2006; Li et al., 2007; Chen et al., 2009; Cichy et al., 2009; Li et al., 2009). The first QTL descriptive analysis began by relating P stress to root weight in field-grown maize (Reiter et al., 1991) and subsequently to root architecture traits such as root hair length, and lateral root branching and length (Zhu et al., 2005a, b). Meanwhile, QTL analysis of root traits and P efficiency in legumes started much later.

Quantitative trait loci mapping of yield and quality components, and the components of other physiologically or biochemically complex pathways, can provide crop breeders with a better understanding of the basis for genetic correlations between economically important traits (linkage and/or pleiotropic relationships between gene blocks controlling associated traits; for example, flowering time and biomass; inflorescence size and inflorescence number). This can facilitate more efficient incremental improvement of specific individual target traits like P-acquisition ability. It has been shown in maize and common beans, that the root architecture is closely related to the crop's P-acquisition efficiency under various P levels (Lynch, 2001; Rubio et al., 2003; Zhu et al., 2005a, b). The studies with recombinant inbred lines (RILs) of maize and common beans identified QTLs affecting root development under P-stressed condition (Zhu et al., 2005; Beebe et al., 2006; Ochoa et al., 2006; Zhu et al., 2006). These findings provide a foundation for molecular marker development in breeding to develop new varieties with enhanced tolerance to P-stressed conditions. Therefore, RILs provide a powerful and useful genetic tool for mapping and marker development required for the selection of a new variety with the improved genetic traits of interest (Zhuang et al., 2002; Andaya and Mackill, 2003; Liu et al., 2008). The RILs are a population of homozygotes generated by a cross between two inbred parental lines followed by repeated selfing or crossing between progenies (Broman, 2005). Each RIL is a new inbred line and harbors a unique mosaic combination of two parental chromosomes which is a useful source for the genetic mapping (Broman, 2005).

Doubled Haploid (DH) technology allows breeders to generate a population of homozygous progenies in a single generation from heterozygous parents thus saving cost, and time needed for the generation of genetic materials for further analysis such as genetic mapping (Baenziger et al., 1989; Murigneux et al., 1993; Smith et al., 2008). The technology is often designed to segregate for several traits of importance at the same time and superior individuals from the populations progressed in breeding programs where developed markers can be applied directly. Double Haploidy lines can be generated either by anthers culture or by crossing with other species whose genomes are excluded from the embryos followed by chemical treatment to duplicate the haploid genome (Collins and Sadasivaiah, 1972; Devaux, 1988; Wan et al., 1989; Devaux et al., 1993; Thomas et al.,1997). This process results in a complete pair of homozygous genomes in each DH line in a single generation, which is not only more efficient but also reliable and predictable than conventional self producing segregating progeny (Choo et al., 1985; Bernardo, 2003; Bonnet et al., 2005; Smith et al., 2008). To accelerate variety breeding in order to meet consumer demands, the use of DH technology and molecular markers in practical breeding has been promoted for varied agronomic and end-use quality traits in several crops including barley, wheat, rapeseed, oat and rye (Manninen et al., 2004; Marwede et al., 2004; Tuvesson et al., 2007; Amar et al., 2008).

Because the practical use of markers is not evident in all crops due to limited access to trait-linked markers, there is a global call for collaboration and technology transfer in the improvement of DH protocols in recalcitrant crops.

The functional genomics approach such as transcriptomics and metabolomics provides opportunities to discover gene(s) whose expression level change against biotic and abiotic stress. The RNA interference (RNAi) approach that abolishes or reduces the expression level of a target gene at the RNA level is useful to identify the function of the identified target gene (Miki and Shimamoto, 2004; McGinnis et al., 2005; Li et al., 2006). Hernandez et al. (2007) investigated the changes of transcripts under the P stress condition and discovered genes whose expression levels are up- or down-regulated. The functionality of those genes for the P efficiency can be investigated by RNAi that adopts an intrinsic cellular surveillance system that protects an organism from the invasive or parasitic genetic materials such as virus and transposable elements Hannon, 2002). In addition, the functional genomics approach can be used to develop gene-specific molecular markers applicable to the selection of a new variety with improved P efficiency.

As molecular genetics advances, a vast amount of molecular data will be available for breeders to incorporate into the conventional breeding program. The use of molecular genetic information together with MAS will improve the breeding program especially for the crop traits that are difficult to improve by conventional approach (Dekkers and Hospital, 2002). Furthermore, these molecular marker tools can also be used in ways that allow more effective discovery and exploitation of the evolutionary relationships between organisms, through comparative genomics (Devos et al., 2000).

INFORMATION AND MATERIAL RESOURCES

The completion of the *Arabidopsis thaliana* genome sequence in 2002 (Arabidopsis Genome Initiative, 2000) followed by progress on genome sequencing for rice in 2002 (Yu et al., 2002) caused much excitement among researchers. These landmark efforts were followed by advances in characterization of genomes of other crops including corn (*Zea mays* L.), wheat (*Triticum aestivum* L.), soybean, and barril medic (Ware et al., 2002; Lunde et al., 2003; Young et al., 2003). There is therefore a rich library of research information that allows plant researchers to explore new paradigms to address fundamental and practical questions in a multidisciplinary manner. The new genetic tools for studying abiotic stress tolerance implored by the International Centre for Tropical Agriculture and the International Centre for Research in the Semi-Arid Tropics present opportunities

to identify and manipulate gene blocks contributing to within-species differences building on the examples of drought tolerance in bean (Broughton et al., 2003) and P-acquisition ability (Kaeppler et al., 2000) in maize, respectively. The success (Figure 2) of these approaches requires (1) heritable genetic variation for the trait of interest (2) effective screening procedures for efficient detection of these genetic differences (at least once, and under conditions that ultimately are relevant to farmers' fields in the breeding program's target environment) (3) adequate levels of marker polymorphism, and (4) potential parents of mapping populations that differ in both the trait(s) of interest (in economically important levels) and in marker genotype (at least in the vicinity of gene blocks contributing to the traits of interest). In what follows, genetic variation for P-stress tolerance and effective screening procedures are further discussed.

P-stress tolerance variability

Abounding research has established an association of genetic variability for various traits with enhanced P-acquisition ability albeit some circumstantial evidence (Kaeppler et al., 2000; Subbarao et al., 1997a, b; Wissuwa et al., 1998). Genotypic differences have been detected between pigeon pea cultivars for producing P solubilizing root exudates (Ishikawa et al., 2002; Subbarao et al., 1997a) and some between genotypes of groundnut and pigeon pea with apparent relative abilities to access and take up Fe- and/or Al-bound P (Subbarao et al., 1997a, b). Similarly, substantial genetic differences have been detected for root growth in chickpea (Ali et al., 2002) and pearl millet (Krishna et al., 1985), as well as for both P-use efficiency (Bationo et al., 2002) and response to mycorrhizal colonization (Krishna et al., 1985) in pearl millet. A reverse genetic tool, TILLING (Targeted Induced Local Lesions In Genomes), has been developed (Perry et al., 2003) that allows the identification of all alleles of a gene of interest from large ethylmethane sulfonate-mutagenized populations (McCallum et al., 2000). The identification of alleles with a weak or wild type-like phenotype can then be used, in addition to the knockout phenotype, to obtain insight into the function of a gene of interest. In some cases (e.g. Kaeppler et al., 2000; Wissuwa et al., 1998) QTLs associated with enhanced P uptake have already been mapped.

Available and reliable screening procedures for accurate phenotyping

There exist reliable screening procedures for P uptake in controlled conditions (Kaeppler et al., 2000; Subbarao et al., 1997a, b). Similarly, a system permitting rapid assessment of root volume on large numbers of plants

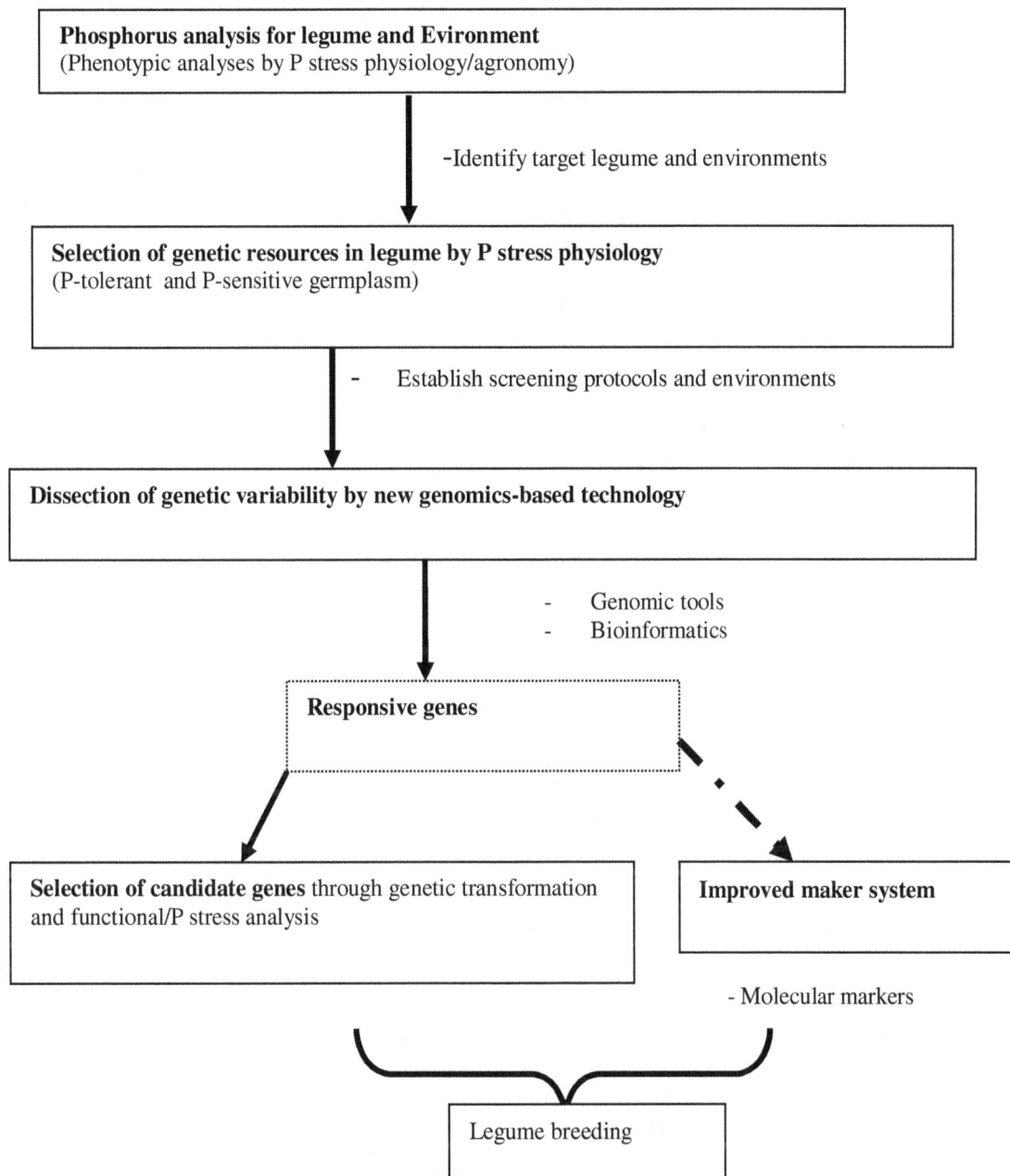

Figure 2. The top-down (phenotype to gene) approach for developing P- stress tolerance in food legumes. Research carried out on the physiology and agronomy of food legumes permits the analysis of P-stress and selection of promising genetic material. This research combined with genomic tools leads to subsequent identification of candidate genes for P-stress tolerance in food legume that can be exploited by legume breeders.

under field conditions has been described (Van Beem et al.,1998) and evaluated for improving drought tolerance in maize (Mugo et al., 1999). Additional procedures that could be used in phenotypic characterization of factors contributing to genetic variation in P-acquisition exist. Among this set of procedures, it is evident that some can be used to assess genetic variation in P-acquisition.

Others may be relevant to the specific causes of non-availability of P in a particular target environment, can be used on large enough numbers of mapping progenies, with high enough heritability, to permit QTL detection in mapping populations that are segregating for the trait and having adequate marker polymorphism, as has already been done for rice (Wissuwa et al., 1998) and maize

(Kaeppler et al., 2000).

MODEL LEGUME RESEARCH ON P-STRESS ADAPTATION

Phosphorus stress studies on legumes have mainly focused on, common bean, white lupine, and to a lesser extent barrel medic, and soybean (*G. max*). In what follows, a brief review on plant biology research related to P-stress adaptation is presented for common bean, white lupine and barrel medic.

Common bean *P. vulgaris* l

Common bean, the most important food legume worldwide has information on genetic variability for the capacity to produce grain in low soil P conditions (Broughton et al., 2003; Ochoa et al., 2006). The genetics of inheritable traits conferring low soil P tolerance in bean has also been reported in Africa (Kimani et al., 2007). For promising genomics approaches, the ability to transform a crop is the preferred method for providing "proof of concept" (Meagher, 2002; Wang et al., 2003) since transformation allows confirmation of the function of candidate genes. For crops such as common bean, where a reliable transformation system is lacking, the 'proof of concept" approach cannot be applied directly, and testing has to be conducted in other species that have high transformation efficiencies. If tests are successful, the candidate gene(s) can be transferred using conventional breeding methods combined with MAS.

An account of bean improvement from classical to molecular breeding for both abiotic and biotic stresses has been provided by Miklas et al. (2006). In addition, the QTL identification approach is now used to analyze P stress tolerance and adaptation in common bean and *Arabidopsisis* (Beebe et al., 2006; Ochoa et al., 2006). By employing the composite interval mapping approach, Beeble et al. (2006) used RILs from a cross between two bean genotypes with contrasting total P accumulation in low P conditions, to identify QTLs for P accumulation and associated root architectural traits in common bean. They found a total of 26 individual QTLs. The P accumulation QTLs often coincided with QTLs for basal root development, indicating that basal roots appear to be important in P acquisition. Similarly, Ochoa et al. (2006) generated RILs from a cross of two common bean accessions with contrasting root architecture traits for adventitious roots and identified 19 QTLs for adventitious root formation, screening 86 F5:7 RILs under P stress and P-sufficient conditions. Furthermore, QTL analysis applied to RIL of a cross of G19833 and DOR 364 common beans showed that root hair formation and root

organic acid exudation are important traits for marker-assisted selection and breeding of P stress tolerance and adaptation (Yan et al., 2004). Moreover, several thousand Expressed Sequence Tags (ESTs) derived from P-stressed common bean roots have been characterized (Ramírez et al., 2005) and a limited number (575) is registered in the NCBI database: http://www.ncbi.nlm.nih.gov/ dbEST. It is also worth mentioning that Hernàndez et al. (2007) have completed a P stress root transcriptome survey in common bean, identifying some 125 genes responsive to P stress. This information will go a long way to facilitate genetic improvement for low P adaptation in low-input agricultural systems.

Barrel medic (*M. truncatula*)

Bucciarelli et al. (2006) reported that barrel medic, a model legume for plant biology research responds to soil P deficiency by delaying (1) leaf development and leaf expansion along the main and axillary shoots; (2) axillary shoot emergence and elongation, resulting in stunted plants; and (3) timing and frequency of flower emergence. It was also observed that P-stressed barrel medic formed shorter petioles and shorter blade lengths relative to plants in P-sufficient conditions. These morphological changes require more supporting evidence to attribute overall delay in whole plant development as a P stress response. However, the plastic nature of plant morphological traits (Beebe et al., 2006; Ochoa et al., 2006) coupled with the lack of a standardized approach to describe plant growth and phenotypic responses to P stress (Bucciarelli et al., 2006), makes result comparisons from different laboratories difficult. Changes in root architecture are often associated with plant adaptation to P stress. However, Bucciarelli et al. (2006) reported no root architecture differences between barrel medic plants grown under P-sufficient and P-deficient conditions until 28 days after planting, when lateral root length and number of P-limited plants showed a decline. On the contrary, alfalfa (*M. sativa*) roots show changes in architecture when grown under P stress. Genetic regulation of root architecture changes due to P stress within and among species is not fully understood and thus, offers a fruitful area of emphasis for future research. According to Tesfaye et al. (2007), the sequencing of the genomes of *Medicago* and lotus will soon be completed.

Given the conserved synteny among legume genomes, using positional cloning, it should be possible to identify specific genes that contribute to QTLs affecting adaptation to P stress in *Medicago*. Phytochromes have also been proven to play a role in legume adaptation to P stress. It is noteworthy that barrel medic, bean and lupine plants exposed to P stress have increased density and length of root hairs. Some 40 genes are suggested to be involved in root hair development (Grierson et al. 2001). Amongst these, a key gene, 1-aminocyclopropane-1-carboxylic acid oxidase, in ethylene biosynthesis is over represented in the ESTs derived from P-stressed roots of Medicago, lupine and bean (Graham et al., 2006). This indicates that ethylene production and/or plant responsiveness to ethylene plays a role in root adaptation to P deficiency.

White lupine (*Lupinus albus*)

White lupine is a non-mycorrhizal species that adapts to P deficient soil. Lupine displays a highly synchronous suite of molecular and biochemical adaptations to P stress by developing proteoid (cluster) roots, increasing organic acid exudation, and enhancing the expression of many genes, such as secreted acid phosphatase (LaSAP1) and Pi transporters LaPT1 (Uhde-Stone et al., 2003a, b; Vance et al., 2003,). A recent bioinformatic analysis of legume gene indices (Medicago, Glycine, Phaseolus, and Lupinus) queried for genes overrepresented in P-stressed tissue revealed the annotation of several putative transcription factor genes, including WRKY, MYB, and zinc finger family of genes (Graham et al., 2006). In addition, leaves and root tissues showed non-overlapping sets of transcription factor genes (Wu et al., 2003). Tesfaye et al. (2007) have also observed the PHR1 imperfect palindromic consensus sequence motif within the 5' upstream region of many P stress-induced genes, including the white lupine LaPT1 and LaSAP1 (Liu et al., 2001; Müller et al., 2007). Also noteworthy is that contigs encoding orthologs of PHR1 occur in the common bean, barrel medic, and soybean gene indices. Auxin (principally indole-3-acetic acid) has been implicated in the regulation of many aspects of plant growth and root development, including P stress-induced proteoid (cluster) root development (Gilbert et al., 2000). More so, exogenous application of auxin to P-sufficient white lupine mimics proteoid cluster root formation as seen under P-deficient conditions (Gilbert et al. 2000). Auxin transport inhibitors added to P-deficient plants dramatically reduced the formation of cluster roots. Phosphorus deficiency stress is also known to stimulate ethylene production in lupine and other plant species (Gilbert et al. 2000), and Arabidopsis (Ma et al., 2003). Stimulation of ethylene production results in an increase in root hair density and length (Grierson et al., 2001; López-Bucio et al., 2003) characteristic of lupine plants exposed to P stress. These results strongly suggest that cluster root development in response to P deficiency in white lupine is controlled by auxin and ethylene availability. The role of cytokinins in root growth and P deficiency stress is not resolved. In P-stressed lupine proteoid roots, CKX gene expression showed a 3- to 5-fold increase in expression (Vance et al., 2003). Moreover, application of cytokinin to P-deficient white

lupine inhibits proteoid root formation, and kinetin content is increased in proteoid roots (Neumann et al., 2000). Aloni et al. (2006) have proposed a mechanism for lateral root initiation in P-sufficient plants that involves the interaction of auxin, cytokinin, and ethylene. They propose that factors that modulate root tip cytokinin production allow ethylene and auxin to increase at lateral root initiation sites, giving rise to new laterals.

CONCLUSION

Legume contribution of soil nutrients in cropping systems is known as well as legume production constraints especially in acid soils where Al toxicity and P deficiency are common. With the availability of a wide array of genomic and bioinformatic research platforms, P stress research is advancing toward an exciting phase geared towards signal transduction, regulation of developmental plasticity, gene function, and increased efficiency of use. Information from the literature indicates that for some food legumes, genetic variation and molecular tools already exist that can permit plant breeders to enhance the P-acquisition component of efficiency in low-nutrient environments. Where not yet already in place, these genetic tools can be expected to become available in the very near future. To promote sustainable agriculture, plant breeders are expected to identify mechanisms in plants that improve P acquisition and exploit these P stress adaptations to produce plants that rely on low energy consumption, and are efficient in acquiring and utilizing soil P. Most importantly, comparative genomics utilizing the integrated information from different plants is expected to provide a common language to aid knowledge transfer among different species, knowing well that improving soil P availability in legumes enhances the practice of economical and environmentally friendly agriculture.

ACKNOWLEDGEMENTS

The authors want to thank Dr. Collins Kimbeng and Dr. Brenda Tubaña both of Louisiana State University for providing useful comments and suggestions. Support from USDA-NIFA grant No. 2010-34402-20875 through Texas A and M, Vegetable and Fruit Improvement Center's Designing Foods for Health Program is greatly appreciated.

REFERENCES

Abaidoo RC, Okogun JA, Kolawole GO, Diels J, Randall P, Sanginga N (2007). Evaluation of cowpea genotypes for variations in their contribution of N and P to subsequent maize crop in three agro-ecological zones of West Africa. In: Advances in Integrated Soil Fertility Management in sub-Saharan Africa: Challenges and

Opportunities. Bationo A, Waswa B, Kihara J, Kimetu J (eds). Springer, Dordrecht, Netherlands, pp. 401-412.

Ali MY, Krishnamuthy L, Saxena NP, Rupela OP, Jagdish K, Johansen C (2002). Scope for genetic manipulation of mineral nutrition in chickpea. In: Food Security in Nutrient-Stressed Environments: Exploiting Plants' Genetic Capabilities. Adu-Gyamfi JJ (ed). Proceedings of an International Workshop 27–30 September, 1999, Patancheru, India. Developments in Plant and Soil Sciences volume 95. Kluwer Academic Publishers, Dordrecht, The Netherlands.

Aloni R, Aloni E, Langhans M, Ullrich CI (2006). Role of cytokinin and auxin shaping root architecture: regulating vascular differentiation, lateral root initiation, root apical dominance and root gravitropism. Ann. Bot., 97: 883-893.

Amar S, Becker HC, Mollers C (2008). Genetic variation and genotype x environment interactions of phytosterol content in three doubled haploid populations of winter rapeseed. Crop Sci., 48: 1000-1006.

Andaya VC, Mackill DJ (2003). QTLs conferring cold tolerance at the booting stage of rice using recombinant inbred lines from a Japonica x indica cross. Theor. Appl. Genet., 106: 1084-1090.

Arabidopsis Genome Initiative (2000). Analysis of the genome sequence of the flowering plant Arabidopsis thaliana. Nature, 408: 796-815.

Aubert G, Tavernier R (1972). Soil survey. In: Soils of the humid tropics. National Academy of Sciences, Washington D.C., pp. 17-34.

Babu RC, Nguyen BD, Chamarerk V, Shanmugasundaram P, Chezhian P, Jeyaprakash Ganesh P, Palchamy SK, Sadasivam A, Sarkarung S, Wade LJ, Nguyen HT (2003). Genetic analysis of drought resistance in rice by molecular markers: Association between secondary traits and field performance. Crop Sci., 43: 1457–1469.

Baenziger PS, Wesenberg DM, Smail VM, Alexander WL, Schaeffer GW (1989). Agronomic performance of wheat doubled-haploid lines derived from cultivars by anther culture. Plant Breed., 103 (2): 101–109.

Baligar VC, Bennett OL (1986). NPK-fertilizer efficiency. A situation analysis for the tropics. Fert. Res., 10: 147–164.

Bates TR, Lynch JP (2000). The efficiency of Arabidopsis thaliana (Brassicaceae) root hairs in P acquisition. Am. J. Bot., 87: 964–970.

Bationo A, Henao J, Anand KK (2002). Phosphorus use efficiency as related to sources of P fertilizers, rainfall, soil, crop management and genotypes in the West African semi-arid tropics. In: Food Security in Nutrient-Stressed Environments: Exploiting Plants' Genetic Capabilities . Adu-Gyamfi JJ (ed) .Proceedings of an International Workshop 27–30 September 1999, Patancheru, India. Kluwer Academic Publishers, Dordrecht, The Netherlands. Dev. Plant Soil Sci., vol. 95.

Beebe SE, Rojas-Pierce M, Yan X, Blair MW, Pedraza F, Muñoz F, Tohme J, Lynch JP (2006). Quantitative trait loci for root architecture traits correlated with P acquisition in common bean. Crop Sci., 46: 413–423.

Bernardo R (2003). Parental selection, number of breeding populations, and size of each population in inbred development. Theor. Appl. Genet.. 107: 1252-1256.

Bonnet DG, Rebetzke GJ, Spielmeyer W (2005). Strategies for efficient implementation of molecular markers in wheat breeding. Mol. Breeding, 15: 75-85.

Bouhmouch I, Souad-Mouhsine B, Brhada F, Aurag J (2005). Influence of host cultivars and Rhizobium species on the growth and symbiotic performance of Phaseolus vulgaris under salt stress. J. Plant Physiol., 162: 1103–1113.

Brady NC, Weil RR (2001). The Nature and Properties of Soils Prentice Hall. 13ed., p. 960.

Broman KW (2005). The genomes of recombinant inbred lines. Genetics, 169: 1133-1146.

Broughton WJ, Hernández G, Blair M, Beebe S, Gepts P, Vanderleyden J (2003). Beans (Phaseolus spp.): Model food legumes. Plant Soil. 252: 55–128.

Brundrett MC (2002). Coevolution of roots and mycorrhizas of land plants. New Phytol., 154: 275–304.

Bucciarelli B, Hanan J, Palmquist D, Vance CP (2006). A standardized method for analysis of Medicago truncatula phenotype development. Plant Physiol., 142: 207–219.

Buehler S, Obeson A, Rao IM, Friesen DK, Frossard E (2002). Sequential phosphorus extraction of a ^{33}P-labeled oxisol under contrasting agricultural systems, Soil Sci. Soc. Am. J., 66: 868–877.

Chen JY, Xu L, Cai YL, Xu J (2009). Identification of QTLs for phosphorus utilization efficiency in maize (Zea mays L.) across P levels. Euphytica, 167: 245–252.

Choo TM, Reinbergs E, Kasha KJ (1985). Use of haploids in breeding barley. Plant Breed. Rev., 3: 219-252.

Cichy KA, Blair MW, Mendoza CHG, Snapp SS, Kelly JD (2009). QTL analysis of root architecture traits and low phosphorus tolerance in an Andean bean population. Crop Sci., 49: 59–68.

Collins GB, Sadasivaiah RS (1972). Meiotic analysis of haploid and doubled haploid forms of Nicotiana otophora and N. tabacum. Chromosoma, 38(4): 387-404.

Dekkers JCM, Hospital F (2002). The use of molecular genetics in the improvement of agricultureal populations. Nat. Rev. Genet., 3: 22-32.

Devaux P (1988). Comparison of anther culture and the Hordeum bulbosum method for the production of doubled haploids in winter barley. Plant Breed., 100: 181-187.

Devaux P, Kilian A, Kleinhofs A (1993). Anther culture and Hordeum bulbosum-derived barley doubled haploids: Mutations and methylation. Mol. Gen. Genet., 241(5-6): 674-679.

Devos KM, Pittaway TS, Reynolds A, Gale MD (2000). Comparative mapping reveals a complex relationship between the pearl millet genome and those of foxtail millet and rice. Theor. Appl.Genet., 100: 190–198.

Geurts R, Fedorova E, Bisseling T (2005). Nod factor signalling genes and their function in the early stages of Rhizobium infection. Curr. Opin. Plant Biol., 8: 346-352.

Ghosh PK, Manna MC, Bandyopadhyay KK, Ajay TAK, Wanjar RH, Hati KM, Misra AK, Acharya CL, Subba Rao A (2006). Interspecific interaction and nutrient use in soybean/sorghum intercropping system. Agron. J., 98: 1097-1108.

Gilbert GA, Knight JD, Vance CP, Allan DL (2000). Proteoid root development of P-deficient lupin is mimicked by auxin and phosphonate. Ann. Bot. (Lond), 85: 921–928.

Gniazdowska A, Mikulska M, Rychter AM (1998). Growth, nitrate uptake and respiration rate in bean root under phosphate deficiency. Biol. Plantarum, 41: 217-226.

Graham MA, Ramirez M, Valdes-Lo´pez O, Lara M, Tesfaye M, Vance CP, Hernandez G (2006). Identification of candidate P stress induced genes in Phaseolus vulgaris through clustering analysis across several plant species. Funct. Plant Biol., 33: 789–797.

Grierson CS, Parker JS, Kemp AC (2001). Arabidopsis genes with roles in root hair development. J. Plant Nutr. Soil Sci., 164: 131–140.

Hague I, Nnadi LA, Mohamed-Saleem MA (1986). Phosphorus management with special reference to forage legumes in sub-Saharan Africa. In : Potentials of forage legumes in farming systems of sub-Saharan Africa. Haque I, Jutzi S, Neate PJH (eds). Proceedings of a workshop held at ILCA, Addis Ababa, Ethiopia, 16-19 September, 1985. ILCA, Addis Ababa.

Hannon GJ (2002). "RNA interference." Nature, 418(6894): 244-251.

Havlin JL, Tisdale SL, Beaton JC, Nelson WL (2005). Soil Fertility and Fertilizers: An Introduction to Nutrient Management. 7ed, Pearson Prentice Hall. Upper Saddle River, New Jersey, p. 515.

Hernández G, Ramírez M, Valdés-López O, Tesfaye M, Graham MA, Czechowski T, Schlereth TA, Wandrey M, Erban A, Cheung F, Wu HC, Lara M, Town CD, Kopka J, Udvardi MK, Vance CP (2007). Phosphorus stress in common bean: Root transcript and metabolic responses. Plant Physiol., 144: 752–767.

Ishikawa S, Adu-Gyamfi JJ, Nakamura T, Yoshihara T, Wagatsuma T (2002). Genotypic variability in P solubilizing activity of root exudates by crops grown in low-nutrient environments. In: Food security in nutrient-stressed environments: exploiting plants' genetic capabilities. Adu-Gyamfi JJ (ed). Proceedings of an International Workshop 27-30 September 1999, Patancheru, India.. Kluwer Academic Publishers, Dordrecht, The Netherlands. Dev. Plant Soil Sci., vol. 95.

Jemo M, Abaidoo RC, Nolte C, Tchienkoua M, Sanginga N,Horst WJ (2006). P benefits from grain-legume crops to subsequent maize grown on acid soils of southern Cameroon. Plant and Soil, 284: 385-397.

Jemo M, Nolte C, Nwaga D (2007). Biomass production, N and P uptake of Mucuna after Bradyrhizobia and arbuscular mycorrhizal fungi inoculation, and P-application on acid soil of Southern Cameroon. In : Advances in Integrated Soil Fertility Management in sub-Saharan Africa: Challenges and Opportunities. Bationo A,Waswa B, Kihara J, Kimetu J (eds) . Springer, Dordrecht, Netherlands, pp. 855-863.

Kaeppler SM, Parke JL, Mueller SM, Senior L, Stuber C, Tracy WF (2000). Variation among maize inbred lines and detection of quantitative trait loci for growth at low P and responsiveness to arbuscular mycorrhizal fungi. Crop Sci., 40: 358-364.

Kaló P, Gleason C, Edwards A, Marsh J, Mitra RM, Hirsch S, Jakab J, Sims S, Long SR, Rogers J, Kiss GB, Downie JA, Oldroyd GE (2005). Nodulation signaling in legumes requires NSP2, a member of the GRAS family of transcriptional regulators. Science, 308: 1786-1789.

Kamanga BCG, Waddington SR, Robertson MJ, Giller KE (2010). Risk analysis of maize-legume crop combinations with smallholder farmers varying in resource endowment in central Malawi. Exp. Agric., 46(1): 1-21.

Kideok KD, Kubicki JD (2004). Molecular orbital theory study on surface complex structures of phosphates to iron hydroxides: Calculation of vibrational frequencies and adsorption energies. Langmuir, 20: 9249-9254.

Kimani MJ, Kimani MP, Kimenju WJ (2007). Mode of inheritance of common bean (Phaseolus vulgaris L.) traits for tolerance to low soil phosphorus (P). Euphytica, 155: 225–234.

Kochian LV (2000). Molecular Physiology of Mineral Nutrient Acquisition, Transport and Utilization. In: Biochemistry and molecular biology of plants. Buchanan B, Gruissem W, Jones J (eds). pp. 1204-1249.

Krasilnikoff G, Gahoonia T, Nielsen NE (2003). Variation in phosphorus uptake efficiency by genotypes of cowpea (Vigna unguiculata) due to differences in root and root hair length and induced rhizosphere processes. Plant Soil, 251: 83-91.

Krishna KR, Shetty KG, Dart PJ, Andrews DJ (1985). Genotype dependent variation in mycorrhizal colonization and response inoculation of pearl millet. Plant Soil, 86: 113-125.

Kuster H, Vieweg MF, Manthey K, Baier MC, Hohnjec N, Perlick AM (2007). Identification and expression of symbiotically activated legume genes. Phytochem., 68: 8–18.

Kuzyakov Y, Domanski G (2000). Carbon input into the soil - Review. J. Plant Nutr. Soil Sci., 163: 421-431.

Kuzyakov Y, Domanski G (2002). Model for rhizodeposition and CO_2 efflux from planted soil and its validation by 14C pulse labeling of ryegrass. Plant Soil, 239: 87-102.

Lambers H, Chapin FS, Pons TL (1998). Plant Physiological Ecology. Springer, New York, p. 540.

Li L, Li SM, Sun JH, Zhou LL, Bao XG, Zhang HG, Zhang FS (2007). Diversity enhances agricultural productivity via rhizosphere phosphorus facilitation on phosphorus-deficient soils. PNAS, 104(27): 11192-11196.

Li XP, Gan R, Peng-Li L, Yuan-Yuan M, Li-Wen Z, Zhang R, Wang Y, Wang NN (2006). Identification and functional characterization of a leucine-rich repeat receptor-like kinase gene that is involved in regulation of soybean leaf senescence. Plant Mol. Biol., 61(6): 829-844.

Li JZ, Xie Y, Dai AY, Liu LF, Li ZC (2009). Root and shoot traits responses to phosphorus deficiency and QTL analysis at seedling stage using introgression lines of rice. J. Genet. Genom., 36: 173-183.

Liu G, Bernhardt JL, Jia MH, Wamishe YA, Jia Y (2008). Molecular characterization of the recaminant inbred line population derived from a Japonica indica rice cross. Euphytica, 159: 73-82.

Liu J, Uhde-Stone C, Li A, Vance CP, Allan D (2001). A phosphate transporter with enhanced expression in proteoid roots of white lupin (Lupinus albus L.). Plant Soil, 237: 257-266.

López-Bucio J, Cruiz-Ramirez A, Herrera-Estrella L (2003). The role of nutrient availability in regulating root architecture. Curr. Opin. Plant Biol., 6: 280–287.

Lunde CF, Morrow DJ, Roy LM, Walbot V (2003). Progress in maize gene discovery: A project update. Funct. Integr. Genom., 3: 25–32.

Lynch JP, Brown KM (2001). Topsoil foraging-an architectural adaptation of plants to low phosphorus availability. Plant Soil, 237: 225–237.

Ma Z, Baskin TI, Brown KM, Lynch JP (2003). Regulation of root elongation under P stress involves changes in ethylene responsiveness. Plant Physiol., 131: 1381–1390.

Macklon AES, Mackie-Dawson LA, Sim A, Shand CA, Lilly A (1994). Soil P sources, plant growth and rooting characteristics in nutrient poor upland grasslands. Plant Soil, 163: 257-266.

Manninen O, Tanhuanpää P, Tenhola-Roininen T, Kiviharju E (2004). Doubled haploids and genetic mapping in barley, rye and oat. In: Genetic variation for plant breeding. Vollmann J, Grausgruber H, Ruckerbauer R (eds). Publisher BOKU – University of Natural Resources and Applied Life Sciences,Vienna, Austria, p. 30.

Marschner H (1995). Mineral nutrition of higher plants. 2nd Ed. Academic Press, London, p. 889.

Marwede V, Schierholt A, Möllers C, Becker HC (2004). Genotype x environment interactions and heritability of Tocopherol contents in canola. Crop Sci., 44: 728-731.

McCallum CM, Comai L, Greene EA, Henikoff S (2000). Targeted screening for induced mutations. Nat. Biotechnol., 18: 455-457.

McGinnis K, Chandler V, Cone K, Kaeppler H, Kaeppler S, Kerschen A, Pikaard C, Richards E, Sidorenko L, Smith T, Springer N, Wulan T (2005). Transgene-induced RNA interference as a tool for plant functional genomics. Methods Enzymol., 392: 1-24.

McKay IA, Djordjevic MA (1993). Production and excretion of Nod metabolites by Rhizobium leguminosarum bv. trifolii are disrupted by the same environmental factors that reduce nodulation in the field. Appl. Environ. Microbiol., 59: 3385–3392.

Meagher RB (2002). Post genomics networking of biotechnology for interpreting gene function. Curr. Opin. Plant Biol., 5: 135-140.

Miki D, Shimamoto K (2004). Simple RNAi vectors for stable and transient suppression of gene function in rice. Plant Cell Physiol., 45(4): 490-495.

Miklas PN, Kelly JD, Beebe SE, Blair MW (2006). Common bean breeding for resistance against biotic and abiotic stresses: From classical to MAS breeding. Euphytica, 147: 105–131.

Morcuende R, Bari R, Gibon Y, Zheng W, Pant BD, Bläsing O, Usadel B, Czechowski T, Udvardi MK, Sttitt M, Scheible WR (2007). Genome-wide, reprogramming of metabolism and regulatory networks of Arabidopsis in response to phosphorus. Plant Cell Environ., 30: 85–112.

Muchhal US, Pardo JM, Raghothama KG (1996). Phosphate transporters from the higher plant Arabidopsis thaliana. Proc. Nat. Acad. Sci. USA, 93: 10519–10523.

Mugo SN, Banziger M, Edmeades GO (1999). The effects of divergent selection for root capacitance in maize. Agron. Abstr., p. 67.

Müller R, Morant M, Jarmer H, Nilsson L, Nielsen TH (2007). Genome-wide analysis of the Arabidopsis leaf transcriptome reveals interaction of phosphate and sugar metabolism. Plant Physiol., 143: 156–171.

Murigneux A, Barloy D, Leray P, Beckont M (1993). Molecular and morphological evaluation of doubled haploid lines in maize. I. Homogeneity within DH lines. Theor. Appl. Genet., 86: 837-842.

Neumann G, Massonneau A, Langlade N, Dinkelaker B, Hengeler C, Römheld V, Martinoia E (2000). Physiological aspects of cluster root function and development in P- deficient white lupin (Lupinus albus L.). Ann. Bot. (Lond), 85: 909–919.

Niang AI, Amadalo B, de Wolf Y, Gathumbi SM (2002). Species screening for short-term planted fallows in the highlands of western Kenya. Agroforest. Syst., 56: 145–154.

Oberson A, Joner EJ (2005). Microbial turnover of P in soil. In: Organic P in the Environment . Turner BL, Frossard E, Baldwin DS (eds). CAB International, Wallingford, UK, pp. 133–164.

Ochoa IE, Blair MW, Lynch JP (2006). QTL analysis of adventitious root formation in common bean under contrasting P availability. Crop Sci., 46: 1609-1621.

Oldroyd GED, Harrison MJ, Udvardi M (2005). Peace talks and trade deals. Keys to long-term harmony in legume microbe symbioses. Plant Physiol., 137: 1205–1210.

Perry JA, Wang TL, Welham TJ, Gardner S, Pike JM, Yoshida S, Parniske M (2003). A TILLING reverse genetics tool and a web-accessible collection of mutants of the legume Lotus japonicus. Plant Physiol., 131: 866-871.

Ramírez M, Graham MA, Blanco-López L, Silvente S, Medrano-Soto A, Blair MW, Hernaández G, Vance CP, Lara M (2005). Sequencing and
analysis of common bean ESTs. Building a foundation for functional genomics. Plant Physiol., 137: 1211–1227.

Rao TP, Yano K, Iijima M, Yamauchi A, Tatsumi J (2002). Regulation of rhizosphere acidification by photosynthetic activity in cowpea (Vigna unguiculata L. Walp.) seedlings Ann Bot., 89(2): 213-220.

Reiter RS, Coor JG, Sussman MR, Gabelman WH (1991). Genetic analysis of tolerance to low-phosphorus stress in maize using restriction fragment length polymorphisms. Theor. Appl. Genet., 82: 561–568.

Rengel Z, Marschner P (2005). Nutrient availability and management in rhizosphere: Exploiting genotypic difference. New Phytol., 168: 305-312.

Rubio G, Liao H, Yan X, Lynch JP (2003). Topsoil foraging and its role in plant competitiveness for phosphorus in common bean. Crop Sci., 43: 598–607.

Saidou AK, Abaidoo RC, Singh BB, Iwuafor ENO, Sanginga N (2007). Variability of cowpea breeding lines to low phosphorus tolerance and response to external application of Phosphorus. In: Advances in Integrated Soil Fertility Management in sub-Saharan Africa: Challenges and Opportunities. Bationo A, Waswa B, Kihara J, Kimetu J (eds). Springer, Dordrecht, Netherlands, pp. 413-422.

Sample EC, Soper RJ, Racz GJ (1980). Reactions of phosphate fertilizers in soils. In: Khasawneh FE, Sample EC, Kamprath EJ (eds) The role of phosphorus in agriculture. ASA, CSSA, SSSA. Madison, WI, pp. 263–310.

Sanchez P (2010). Tripling crop yields in tropical Africa. Nat. Geosci., 3: 299-300.

Sanchez PA, Shepherd KD, Soule MJ, Place FM, Buresh RJ, Izac AN, Mokwunye AU, Kwesiga FR, Ndiritu CG, Woomer PL (1997). Soil Fertility Replenishment in Africa: An investment in natural resource capital. In: Replenishing Soil Fertility in Africa. Buresh RJ, Sanchez PA, Calhoun FG (eds). ASA, SSSA, Madison, WI, pp. 1–46.

Scherr S J (1999). Soil degradation, a threat to developing-country food security by 2020? Food, Agriculture, and the Environmental Discussion Paper 27. International Food Policy Research Institute. Washington, DC.

Schneider AVC (2002). Overview of the market and consumption of pulses in Europe. Brit. J. Nutr., 88: 243-250.

Smit AL, Bindraban P, Schröder JJ, Conijn JG, Meer HG (2009). Phosphorus in agriculture: global trends and developments. Plant Res. Int. (Report 282).

Smith JSC, Hussain T, Jones ES, Graham G, Podlich D, Wall S, Williams M (2008). Use of doubled haploids in maize breeding: Implications for intellectual property protection and genetic diversity in hybrid crops. Mol. Breed., 22: 51-59.

Steffens D, Leppin T, Schubert S (2009). Organic soil phosphorus is plant-available but is neglected by routine soil-testing methods. The Proceedings of the International Plant Nutrition Colloquium XVI, Department of Plant Sciences, UC Davis.

Subbarao GV, Ae N, Otani T (1997a). Genotypic variation in iron-, and aluminum-phosphate solubilizing activity of the pigeonpea root exudates under P deficient conditions. Soil Sci. Plant Nutr., 43: 295–305.

Subbarao GV, Ae N, Otani T (1997b). Genetic variation in acquisition, and utilization of P from iron-bound P in pigeonpea. Soil Sci. Plant Nutr., 43: 511–519.

Tesfaye M, Liu J, Allan DL, Vance CP (2007). Genomic and genetic control of phosphate stress in legumes. Plant Physiol., 144: 594–603.

Thomas J, Chen Q, Howes N (1997). Chromosome doubling of haploids of common wheat with caffeine. Genome, 40(4): 552-558.

Tiessen H, Shang C (1998). Organic matter turnover in tropical land use systems. In: Carbon and Nutrient Dynamics in Natural and Agricultural Tropical Ecosystems. Bergström L, Kirchmann L (eds). CAB International, Wallingford, UK, pp. 1-14.

Tomasi N, Weisskopf L, Renella G, Landi L, Pinton R, Varanini Z, Nannipieri P, Torrent J, Martinoia E, Cesco S (2008). Flavonoids of white lupin roots participate in P mobilization from soil. Soil Biol. Biochem., 40: 1971-1974.

Tuvesson S, Dayteg C, Hagberg P, Manninen O, Tanhuanpää P, Tenhola-Roininen T, Kiviharju E, Weyen J, Förster F, Schondelmaier J, Lafferty J, Marn M, Fleck A (2007). Molecular markers and doubled haploids in European plant breeding programmes. Euphytica, 158: 305–312.

Uhde-Stone C, Gilbert G, Johnson JMF, Litjens R, Zinn KE, Temple SJ, Vance CP, Allan DL (2003a). Acclimation of white lupin to P deficiency involves enhanced expression of genes related to organic acid metabolism. Plant Soil, 248: 99–116.

Uhde-Stone C, Zinn KE, Ramirez-Yáñez M, Li A, Vance CP, Allan DL (2003b). Nylon filter arrays reveal differential gene expression in proteoid roots of white lupin in response to P deficiency. Plant Physiol., 131: 1064–1079.

Vance CP, Graham PH, Allan DL (2000). Biological nitrogen fixation. Phosphorus: A critical future need. In: Nitrogen fixation: From molecules to crop productivity. Pedrosa FO, Hungria M, Yates MG, Newton WE (eds). Kluwer Academic Publishers, Dordrecht, The Netherlands, pp. 506–514.

Vance CP, Uhde-Stone C, Allan DL (2003). P acquisition and use: Critical adaptations by plants for securing a nonrenewable resource. New Phytol., 157: 423–447.

Van Beem J, Smith ME, Zobel RW (1998). Estimating root mass in maize using a portable capacitance meter. Agron. J., 90: 566–570.

Vlassak KM, Vanderleyden J (1997). Factors influencing nodule occupancy by inoculant rhizobia. Crit. Rev. Plant Sci., 16: 163–229.

Vesterager JM, Nielsen NE, Høgh-Jensen H (2006). Variation in phosphorus uptake and use efficiencies between pigeonpea genotypes and cowpea. J. Plant Nutr., 29: 1869-1888.

Wan Y, Petolino JF, Widholm JM (1989). Efficient production of doubled-haploid plants through colchicine treatment of anther-derived maize callus. Theor. Appl. Genet., 77: 889-892.

Wang W, Vinocur B, Altman A (2003). Plant response to drought, salinity and extreme temperatures: Towards genetic engineering for stress tolerance. Planta, 218: 1–14.

Ware DH, Jaiswal P, Ni J, Yap IV, Pan X, Clark KY, Teytelman L, Schmidt SC, Zhao W, Chang K, Cartinhour S, Stein LD, McCouch SR (2002). Gramene, a tool for grass genomics. Plant Physiol., 130: 1606–1613.

Wendt JW, Atemkeng MF (2004). Soybean, cowpea, groundnut, and pigeonpea response to soils, rainfall, and cropping season in the forest margins of Cameroon. Plant Soil, 263: 121-132.

Wissuwa M, Yano M, Ae N (1998). Mapping of QTLs for P-deficiency tolerance in rice (Oryza sativa L.). Theor. Appl. Genet., 97: 777–783.

Wu P, Ma L, Hou X, Wang M, Wu Y, Liu F, Deng XW (2003). Phosphate starvation triggers distinct alternations of genome expression in Arabidopsis roots and leaves. Plant Physiol., 132: 1260–1271.

Xinshen D, Dorosh P, Rahman SM, Meijer S, Rosegrant M, Yanoma Y, Li W (2003). Market opportunities for African agriculture: An examination of demand-side constraints on agricultural growth. DSGD Discussion Paper No. 1. International Food Policy Research Institute, Washington DC, USA.

Yan X, Liao H, Beebe SE, Blair MW, Lynch JP (2004). QTL mapping of root hair and acid exudation traits and their relationship to P uptake in common bean. Plant Soil, 265: 17–29.

Yan X, Liao H, Tang J, Wang X, Li J, Tu P, Cheng F, Cao G, Liu J (2006). Increasing phosphorus efficiency and production of grain legumes in China and Africa. McKnight Foundation's Collaborative Crop Research Program Annual Report. December, 2006.

Young ND, Mudge J, Ellis TH (2003). Legume genomes: More than peas in a pod. Curr. Opin. Plant Biol., 6: 199–204.

Yu J, Hu S, Wang J, Wong KSG (2002). A draft sequence of the rice genome (Oryza sativa L. ssp. indica) Science, 296 (5565): 79–92.

Yuan J, Njiti VN, Meksem K, Iqbal MJ, Triwitayakorn K, Kassem MA, Davis GT, Schmidt ME, Lightfoot DA (2002). Quantitative trait loci in two soybean recombinant inbred line populations segregating for yield and disease resistance. Crop Sci., 42: 271-277.

Zhu JM, Kaeppler SM, Lynch JP (2005a). Mapping of QTL controlling root hair length in maize (*Zea mays* L.) under phosphorus deficiency. Plant Soil, 270: 299-310.

Zhu JM, Kaeppler SM, Lynch JP (2005b). Mapping of QTLs for lateral root branching and length in maize (*Zea mays* L.) under differential phosphorus supply. Theor. Appl. Genet., 111: 688-695.

Zhu J, Mickelson SM, Keappler SM, Lynch JP (2006). Detection of quantitative trait loci for seminal root traits in maize (*Zea mays* L.) seedlings grown under differential phosphorus levels. Theor. Appl. Genet., 113: 1-10.

Zhuang JY, Ma WB, Wu JL, Chai RY, Lu J, Fan YY, Jin MZ, Leung H, Zheng KL (2002). Mapping of leaf and neck blast resistance genes with resistance gene analog, RAPD and RFLP in rice. Euphytica, 128: 363-370

9

Programmed cell death or apoptosis: Do animals and plants share anything in common

Nishawar Jan, Mahboob-ul-Hussain and Khurshid I. Andrabi*

Department of Biotechnology, the University of Kashmir-1900 06 (J&K), India.

Plants, animals and several unicellular eukaryotes use programmed cell death (PCD) for defense and developmental mechanisms. While cell death pathways in animals have been well characterized, relatively little is known about the molecular mechanism of such a strategy in plants. Although, very few regulatory proteins or protein domains have been identified as conserved across all eukaryotic PCD forms, still plants and animals share many hallmarks of PCD, both at cellular and molecular levels. Morphological and biochemical features like chromatin condensation, nuclear DNA fragmentation, and participation of caspase like proteases in plant PCD appear to be similar across the eukaryotic kingdom and in conformity with the process in metazoans as well. Transgenic expression of mammalian anti- and pro-apoptotic proteins in plants has been shown to influence the regulatory pathways of cell death activation and suppression, indicating the existence of functional counterparts of such genes in plants, several of which have now been cloned and characterized to various extents. This suggests that despite differences, there may be a fair level of functional similarity between the mechanistic components of plant and animal apoptosis. Although genome scan of Arabidopsis thaliana seems to rule out the existence of major mammalian apoptotic counterparts in plants, the identification of caspase like proteins and other structural homolgs (metacaspases) together with mildly conserved apoptotic players like Bax-1 inhibitor may seemed to suggest some degree of common grounds both in execution and in the regulation of the cell death phenomenon. The overall review of the available data pertaining to mechanism of PCD in plants primarily supports an ancestral relationship with animal apoptosis rather than any common executional or regulational strategies. The establishment of mechanistic details of the phenomenon in plants is certain to throw up many surprises to necessitate a fresh review of this intriguing phenomenon. Metacaspases and Paracaspases having been ruled out to possess caspase activity is the beginning for this surprise to unfold.

Key words: Programmed cell death, apoptosis, caspases.

Table of content

INTRODUCTION

Programmed cell death (PCD) describes a physiological and pathological process of cell deletion that plays an important role in maintaining tissue homeostasis (Wertz and Hanley 1996). It is a highly regulated cellular suicide process essential for growth and survival in all eukaryotes. The origin of the phenomenon seems to be as old as the very first cell, because cellular homeostasis and preventing self-destruction would have been impossible to accomplish without such machinery (Ameisen, 2002). Therefore, this apparatus appears to have existed in all cells from the very origin. It has, indeed, been recognized in several prokaryotes and unicellular eukaryotes and related to numerous phenomena. Only later, during evolution of multicellular organisms, PCD is believed to have 'fine tuned' for purposes such as the social control of cell members (Gray, 2004). Multicellular organisms use the physiological mechanisms of cell death to regulate developmental morphogenesis and remove infected, mutated or damaged cells from healthy tissues (Vaux and Korsmeyer, 1999). This phenomenon is characterized in detail, especially in animal apoptosis systems, by a stereotypical set of morphological and biochemical changes such as condensation or shrinkage of the cell, reorganization of the nucleus, membrane blebbing, formation of apoptotic bodies (Kerr et al., 1972), and chromatin condensation (Earnshaw, 1995; O'Brien et al., 1998). This process finally results in activation of certain endonucleases, leading to the fragmentation of chromatin in multiples of 180 bp nucleosomal units, a process known as DNA laddering (Earnshaw, 1995; Fath et al., 1999; McCabe et al., 1997; Wang et al., 1996b; Wyllie, 1980). Most, but not all, of the above apoptotic features are commonly observed during PCD in a wide range of eukaryotic organisms.

In plants, PCD occurs during development, such as during xylogenesis, embryogenesis, parenchyma formation, several plant reproductive processes, seed development and leaf senescence (Pennel et al., 1997; Gray, 2004). In addition, PCD is well documented in relation to manifestation of hypersensitive response (HR) caused by the interaction between the host plant and an incompatible pathogen (Hatsugai et al., 2004). This hypersensitive response (HR) is thought to directly kill invaders and / or to interfere with their acquisition of nutrients (Heath, 2000). In contrast to animal system, signaling pathways and molecular mechanism of PCD are largely unknown in plants. Although a number of morphological and biochemical changes such as cell shrinkage, blebbing of the plasma membrane, condensation and fragmentation of the nucleus, and inter-nucleosomal cleavage of DNA, which are commonly observed during animal apoptosis, appear to be conserved in plant cells undergoing PCD, very little is known about the execution process that leads to cell death in plants.

In this review we provide a brief insight into some of the comparative features of PCD in plants and animals. Additionally, this article attempts to review some of the peculiar and specific features and regularities of apoptosis in plants.

MORPHOLOGICAL HALLMARKS OF PCD

The morphological features of PCD have been intensively studied in animals. PCD in animal systems is reported to result in the disassembly of cells involving the condensation, shrinkage and fragmentation of cytoplasm and nuclei into several sealed packets (often called apoptotic bodies), which are then phagocytosed, by the neighboring cells or the macrophages. Thus, there are no remnants of the cell corpses left. Nuclear fragmentation is preceded by chromatin condensation and marginalization in the nucleus. Fragmentation of DNA at the nucleosome linker sites then takes place and the fragmented oligonucleosomal bits are reported to be 180 bp (Danon and Gallois, 1998). Fragmentation is effected by endonucleases such as NUC 1, DNaseI and DNaseII (Peitsch et al., 1993), which are present in the nucleus and are activated by Ca^{2+} and Mg^{2+} but inhibited by Zn^{2+} and by several caspase activated nucleases such as CAD (Caspase activated DNase) or DFF40 (DNA fragmentation factor 40KDa) (Enari et al., 1998; Halenbeck et al., 1998; Liu et al., 1998). The DNA fragments can be cytochemically determined by Terminal deoxynucleotidyl transferase mediated dUTP Nick End Labelling (TUNEL) of DNA at 3'-OH group. When all these events are combined to result in a distinct morphological expression then PCD is termed as apoptosis (Cohen 1993). In other words, apoptosis is a distinct form of PCD (D'Silva et al., 1998; Danon and Gallois, 1998). However there are others who consider apoptosis and PCD as one and the same (White, 1996; Miller and Marx, 1998; Hengartner, 1998; Chinnaiyan and Dixit, 1996).

In the last few years and due to new interest in a possible apoptosis like phenomenon existing in plants, morphological changes have been investigated during plant PCD. Only some of the hallmarks are similar to those reported in animal apoptosis. Condensation & shrinkage of the cytoplasm and nucleus have been described in carrot cell culture, after cell death induced by heat shock (McCabe et al., 1997). The DNA processing reported earlier for the animal PCD is believed to exist in the dying cells of plant as well (Wang et al., 1996a; Wang et al., 1996b; Mc Cabe et al., 1998; Orzaez and Granell, 1997a). In plants, DNA ladders have been reported during development represented by death of monocot aleurone layer (Wang et al., 1996a) and endosperm (Young et al., 1997), senescence of petal, carpel tissue and leaves (Orzaez et al., 1997a; Orzaez et al.1997b; Yen et al., 1998) or during anther development (Wang et al., 1999) as well as during death induced by different stresses such as: cold (Koukalova et al., 1997), nutrient deprivation (Callard et al., 1996), salt or D-mannose stresses (Katsuhara et al., 1997; Stein et al., 1999), UV radiattion (Danon et al., 1998), pathogens or a pathogen toxin (Navarre et al., 1999; Ryerson et al., 1996; Wang et al., 1996b).In aleuronic cells of grass species such as barley, in dying root cap cells and in tobacco cells subjected to HR, nuclear condensation and shrinkage as well as oligo-

*Corresponding author. E-mail: andrabik@kashmiruniversity.ac.in.

nucleosome sized DNA fragments have been recognized through the presence of 3'-OH group detected by TUNEL experiments (Pennel and Lamb, 1997; Mittler and Lam, 1997). Although in case of animals, one of the major hallmarks of PCD is the fragmentation of DNA at the nucleosome linker sites into oligonucleosomal bits of precisely 180 bp, the major problem relating to nuclear changes in plant PCD is that there is no consistency regarding the size of DNA fragments during DNA fragmentation: fragments of more or less 50Kb in some cases (Mittler and Lam, 1997) and as small as 0.14 Kb in others (Cohen, 1993; O'Brien et al., 1998). It is believed that the activation of some endonucleases leads to 50 Kb DNA fragments followed latter by a different set of endonucleases causing the production of oligonucleosomal length of DNA fragments (Pandey et al., 1994). The first type of cleavage is believed to be the result of the release of chromatin loops and is observed in almost all cases of apoptosis and the subsequent nucleosomal laddering occurs less often and is considered to be not essential for apoptosis (Oberhammer et al., 1993).

The enzymes involved in nuclear dismantling in plants are still poorly known. Several DNase activities and nuclease genes have been recognized to be up regulated in different models of plant PCD (Sugiyama et al., 2004). However, molecular and biochemical evidences on their involvement in the cell death has been reported only in some of them. Recent work reported the induction of the activity of a 28 KDa endonuclease (p28) activity in victorin (Tada et al., 2001) treated oat leaves and this preceded the DNA laddering and heterochomatin condensation. The p28 activity also markedly increased in parallel with the rate of DNA fragmentation and cell death (Tada et al. 2001). In addition to p28, an inducible nuclease, p24 (24 kDa) and four constitutive nucleases, p22 (22 kDa), p31 (31 kDa), p33 (33 kDa) and p35 (35 kDa), have been detected in oat cell lysates using an in-gel assay for nuclease activity (Tada et al., 2001; Kusaka et al., 2004). A Mg^{2+} dependent nucleolytic activity has been identified in the intermembrane space of mitochondria responsible for the generation of 30Kb DNA fragments in Arabidopsis (Ito and Fukuda, 2002). ZEN1, a Zn^{2+} dependent endonuclease, has been directly impli-cated in the degradation of the nuclear DNA in Zinnia tracheary elements (Ito and Fukuda 2002). ZEN1 is localized to vacuoles, which collapse before DNA is degraded (Obara et al., 2001). However, ZEN1 activity does not produce the characteristic DNA laddering shown by the nucleases executing DNA fragmentation in apoptotic animal cells (Enari et al., 1998; Halenbeck et al., 1998; Liu et al., 1998). Based on the biochemical differences of ZEN1 and the nucleases involved in animal apoptosis, it has been proposed that plants and animals have evolved independent systems of nuclear DNA degradation during cell death. In contrast with tracheary elements, the tissues undergoing PCD in cereal grains show the characteristic DNA laddering indicative of inter-nucleosomal fragmentation of DNA

(Dom'ınguez et al., 2001; Young and Gallie, 1999; Young and Gallie, 2000; Dom'ınguez et al., 2004; Wang et al., 1996a), which is a hallmark of apoptosis in animal cells (Earnshaw, 1995). Recently, a Ca^{2+}/Mg^{2+} endonuclease localized in the nucleus wheat aleuron cells undergoing PCD has been identified which is detected prior to DNA laddering (Dom'ınguez et al., 2004). A cell-free system used to analyze nucleus degeneration in nucellar cells in wheat grains (Domı'nguez and Cejudo, 2006), shows that a different wheat tissue, the nucellus, which undergoes PCD at early stage of grain development (Dom'ınguez et al., 2001), presents a nucleus localized nuclease with identical cation requirements, but with a different electrophoretic mobility than the aleuron nuclease. These results suggest that both animal apoptosis (Samejima et al., 2005) and plant PCD involves more than one nuclease. Nuclear extracts from such cells have been shown to be capable of triggering DNA fragmenation in both plant and human nuclei, demonstrating that similar features of nucleus degradation could be shared between plant and animal cells.

Apoptotic bodies have not been shown to form during plant cell death. These bodies may be absent in plant PCD because they are functionally irrelevant due to the absence of possible phagocytosis by adjacent cells in the presence of cell wall. Instead, the plant pathway might involve autolysis. Although cells that die as part of the HR typically exhibit features of an oncotic cell death, which is characterized by the retention of a dead protoplast containing swollen organelles (Jones, 2000), many other plant cell suicide programs include cellular disassembly via autophagy and/or autolysis. The degree of processing of dead and dying cells ranges from that apparently limited to nucleus or nuclear DNA to complete autolysis that includes the extracellular matrix. Degradation of nucleus and nuclear DNA has been evaluated in several recent investigations of plant PCD. The results are consistent with earlier work in a variety of systems. They includes reports of nuclear blebbing and fragmentation (Schussler and Longstreth, 2000; Yamada et al., 2000; Filonova et al., 2000), and the detection of oligosomal DNA ladders (Xu and Hanson 2000; Yamada et al., 2000; Filonova et al., 2000; Delorme et al., 2000) and labelled fragmented DNA in nucleus (Jordan et al., 2000; De Jong et al., 2000; Yamada et al., 2000; Filonova et al., 2000; Simeonova et al., 2000).

Autophagy has been observed as engulfment and degradation of nucleus and other organelles by provacuoles, vacuoles and other autophagic organelles derived from leucoplast (Filonova et al., 2000). Autolysis does not require engulfment and contributes to the degradation of organelles and soluble cellular components. Unlike autophagy, autolysis can continue after cell death, as occurs during treachery element differentiation. In most cases, autolysis and autophagic mechanisms cooperate to yield cellular disassembly, such as that occurring during embryo suspensor death (Filonova et al., 2000). So if using a strict morphological definition, the term apoptosis-like phenomenon in plants should be used instead of apoptosis since some of the terminal hallmarks of apoptosis are absent.

Basic executors of PCD

The real effector molecules of animal PCD are the cysteine aspartate specific proteases (caspases) and granenzymes. The former are the conserved cysteine proteases, while the latter are serine proteases, both specifically cleave after the aspartate residues of many proteins. The studies in Caenorhabditis elegans (C. elegans) identified two genes ced-3 and ced-4 required for apoptosis in the worm, if either gene is inactivated by mutation, the 131 cell deaths that normally happen during the development of the worm (which has only about 1000 cells when mature) fail to occur (Ellis et al., 1991). Remarkably, the mutant worms with 131 extra cells have a normal life span, showing that in this organism apoptosis is not essential for either life or normal ageing. By contrast, more complex animals cannot survive without apoptosis: mutations that inhibit apoptosis in the fruit fly Drosophila melanogaster, for example, are lethal early in development (White et al., 1994) as are mutations in mice that inhibit apoptosis mainly in the developing brain (Kuida et al., 1996).The protein encoded by the ced-3 gene was found to be very similar to a human protein called interleukin-1-converting enzyme (ICE) (Yuan et al., 1993). ICE is an intracellular protein cleaving enzyme (a protease) that cuts out interleukin-1, a signalling protein that induces inflammation, from a larger precursor protein (Nicholson, and Thornberry, 1997).The similarity between the CED-3 and ICE proteins was the first indication that the death programme depends on protein cleavage (proteolysis). Till date 14 to 15 different caspases that play a role in inflammation (group1 caspases) and apoptosis (group 2 caspases) have been identified in animals (Rudel, 1999). All these are believed to share a fair level of sequence homology and similarity in sequence specificity (Rudel, 1999; Nicholson and Thornberry, 1997; Thornberry and Lazebnik, 1998). Up till now, the caspase family in animals is composed of 12 different proteases classified in 3 phylogenetic groups [Interleukin 1β converting Enzyme (ICE), ICH1 and cysteine protease 32 (CPP32)]. All these caspases have in common a highly conserved catalytic site, a stringent substrate specificity to cleave after an aspartic acid residue and requirement for at least 4 amino acids N terminal to cleavage site (Garcia-Calvo, 1998). It is possible to classify these caspases on the basis of their affinity for different substrates including two tetrapeptides in particular: DEVD (ICH1 and CPP32) and YVAD (ICE caspases). Corresponding caspase activity can be blocked with same peptide substrate coupled with aldehyde (CHO: reversible inhibitor) or methyl ketone radical [Chlorometylketone (CMK), flouromethylketone (FMK): irreversible inhibitor).

Caspases are made as a large, inactive precursor (procaspase), which is itself activated by cleavage at aspartic acids, usually by another caspase (de Murcia et al., 1994; Figure1). In apoptosis, caspases are thought to be activated in an amplifying proteolytic cascade, cleaving one another in sequence. Once activated, the effector caspases ultimately cleave numerous substrates, thereby causing the typical morphological features of apoptosis (Kumar, 2007; Timmer and Salvesen, 2007). They cleave proteins supporting the nuclear membrane (lamins) for example, thereby helping to dismantle the nucleus; they cleave protein constituents of the cell skeleton and other proteins involved in the attachment of the cell to their neighbors, thereby helping the dying cell to detach and round up making it easy to ingest; they cleave a protein Inhibitor of Caspase Activated DNase (ICAD of CAD-ICAD complex) that normally holds the CAD- a DNA degrading enzyme in an inactive form, freeing the DNase to cut up the DNA in the cell nucleus (Enari 1998). The other important substrates include PARP [poly(ADP-ribose)polymerase], DNA dependent protein kinase (DNA PK), Serum response element binding protein (SRE/BP), p21(CDKN1A)-activated kinase 2 (PAK2), 70KDa components of U1Sn-RNP, procaspases and so on (Rudel, 1999). PARP is among the first target proteins shown to be specifically cleaved by caspases to a signature of 89 KDa apoptotic fragments during cell death (Lazebnik et al., 1994). It is believed to be involved in the regulation of the repair of DNA strand breaks and in cell recovery from DNA damage (de Murcia et al., 1994).

Cell death in plants exhibits morphological features comparable to caspase mediated apoptosis in animals, suggesting that plant cell death is executed by (Caspase like) proteases. The recent characterization of cell death associated plant proteases with aspartate specific cleavage activity demonstrates the involvement in plant PCD of proteolytic activities functionally resembling animal caspases. The result of a study carried out show induction of YVADase activity whereas no DEVDase activity was detected (del Pozo and Lam, 1998). Surprisingly, both inhibitor peptides (DEVD and YVAD) were efficient in blocking the HR and YVADase activity. Encouragingly, none of the classical protease activity could suppress the hypersensitive response or YVADase activity. This is cited as an evidence for the presence of caspase like plant proteases that participate in hypersensitive response cell death. Different results were found during the plant response to UV-C radiation where both caspase inhibitors could prevent DNA digestion detected by TUNEL reaction and where UV-C induced DEVDase activity but no YVADase activity was found. Danon et al. (2004) and Lincoln et al. (2002) reported that the heterologous expression of Baculovirus p35 protein, a broad range caspase inhibitor that can effect-tively suppress PCD in animals, blocked AAL (Arternaria alternata) toxin induced cell death in transgenic tomato plants and provided protection against the pathogen Arternaria alter-

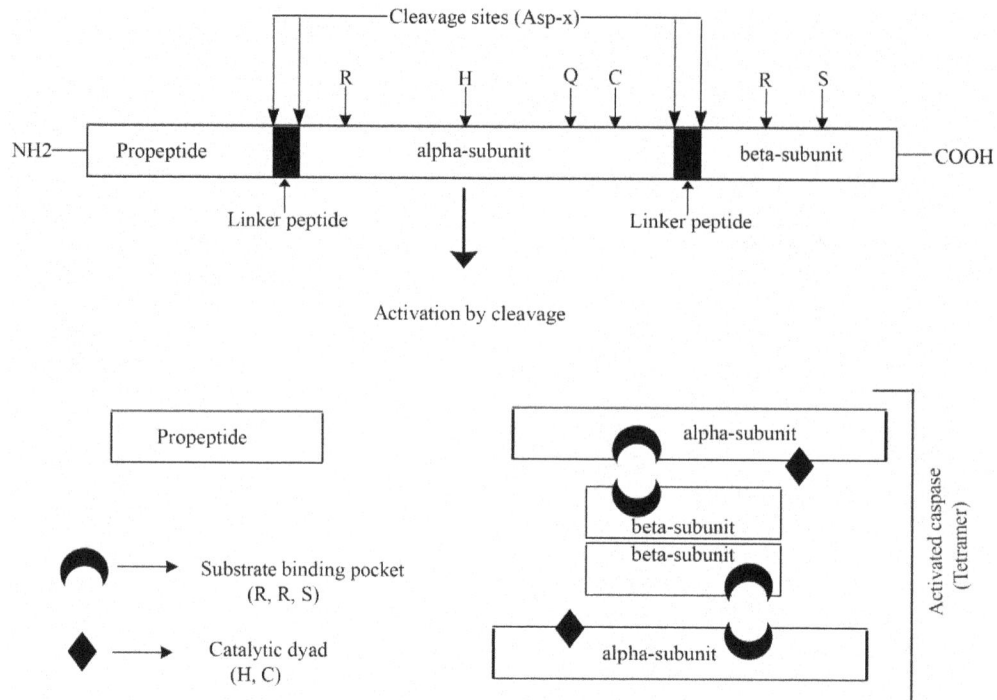

Figure1. Diagrammatic representation of the activation of human caspase 1: Caspases are activated by proteolytically removing the propeptide and the linker peptides from the inactive caspase and cleaving the rest of the caspase into larger α-subunit and smaller β-subunit. These subunits assemble into active tetramer containing two active sites (Nicholson, 1999; Grutter, 2000)

nate. Additionally, p35-expressing transgenic tomato plants displayed partial inhibition of cell death associated with non-host hypersensitive response cell death upon bacteria and virus challenge (Del Pozo and Lam, 2003). Because p35 shows a high degree of specificity towards caspases and it shows a little or no inhibitory activity towards other proteases, these physiological inhibitor studies support an important role for caspase like proteases during cell death in plants (Del Pozo and Lam, 2003). In addition, natural caspase substrates such as bovine and plant PARP are cleaved by plant proteases at caspase cleavage sites. Exogenous (bovine) PARP is endoproteolytically cleaved by extracts from fungus-infected cowpea (*Vigna unguiculata*) plants that were developing a HR but not by extracts from noninfected leaves. This cleavage activity inhibited by caspase-3 inhibitor (Acetyl-DEVD-CHO) but not by caspase-1 inhibitor (Acetyl-YVAD-CHO) (D'Silva et al., 1998). Interestingly, a polypeptide (GDEVDGIDEV) mimicking the PARP caspase-3 cleavage site (DEVD-G) partially inhibited PARP cleavage, whereas a modified peptide in which the essential Asp was replaced by Ala (GDEVAGIDEV) did not affect PARP cleavage. This cleavage activity was also inhibited by other Cys protease inhibitors (E-64, IA, and N-ethylmaleimide). Inhibitors to other types of proteases (Ser-, metallo-, Asp proteases, and calpain) were without effect in this system. In these experiments, PARP cleavage eventually yielded four different fragments of 77, 52, 47, and 45 kD (D'Silva et al., 1998). Cleavage of endogenous (plant) PARP (116 KDa) reacting with a PARP antibody occur during menadione-induced PCD in tobacco protoplasts, and this cleavage of PARP and induction of DNA fragmentation has been shown to be inhibited by caspase-1 (Acetyl-YVAD-CHO) and caspase-3 (Acetyl-DEVD-CHO) inhibitors (Sun et al., 1999). Also in heat shock induced PCD in tobacco suspension cells, endogenous PARP was cleaved, yielding a 89 kDa fragment (Tian et al., 2000). This is similar to the cleavage of PARP described in animal apoptotic cells. In both mammals and plants, two different types of PARP exist, and both types are presumably involved in DNA repair. The Arabidopsis PARP-1 shows high homology to human PARP-1 including a conserved caspase-3 recognition site (DSVD-N). In plants, PARP genes have been cloned from *Zea mays* and *Arabodopsis thaliana* and PARP activity has been identified in few species (O'Farrell, 1995).

Although there have been numerous efforts to identify proteinases that exhibit caspase activities, plant caspases have remained unidentified (Woltering et al., 2002). Recent work, however, has unraveled this mystery. Vacuolar processing enzyme (VPE) has been shown to be a protease that exhibits caspase-1 activity and is essential for virus-induced hypersensitive cell death (Hatsugai et al., 2004). Direct evidence has been reported for the involvement of VPEs in plant cell death (Sanmartin et al., 2005; Hatsugai et al., 2004; Hatsugai et al., 2006; Kuro-

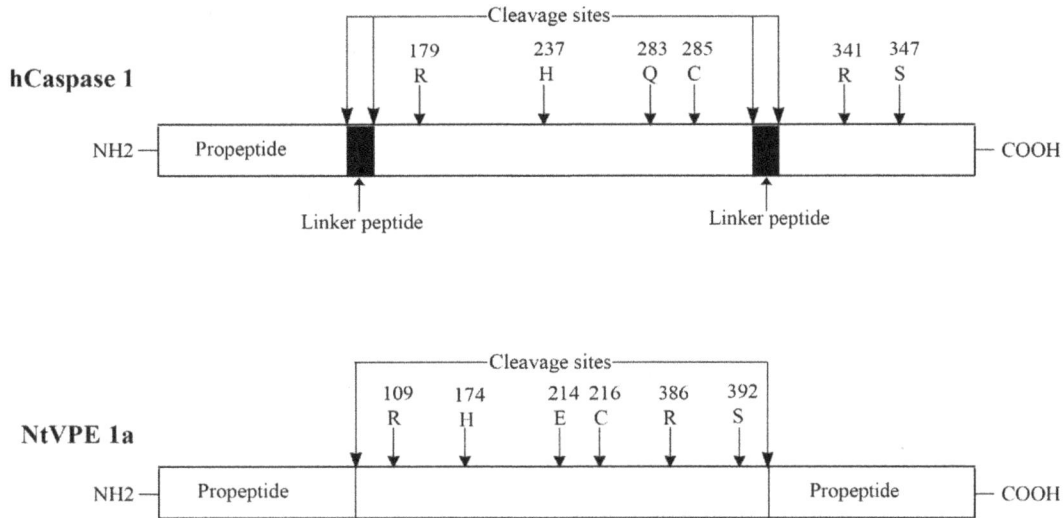

Figure 2. Schematic representation of active site residues among the human caspase1 (hCaspase1) and Nicotiana tabacum VPE (NtVPE1a). Histidine 237 and Cystein 285 that are involved in catalysis in hcaspase1 correspond to H174 and C216 of NtVPE1a and Argenine 179, Argenine 341, Serine 347 involved in substrate binding in hcaspase1 correspond to R109, R386 and S392 in case of NTVPE1a. Caspase1 and plant VPEs are activated by proteolytically removing the propeptides and the linker peptides from the inactive caspases which is then converted into active caspase.

yanagi et al., 2005; Hara-Nishimura et al., 2005; Lam, 2005). Hara-Nishimura and coworkers (Hatsugai et al., 2004) have used a temperature sensitive N-TMV tobacco plant pathogen system to identify the proteases responsible for caspase like activity. The temperature-sensitive N-TMV (tobacco mosaic virus) tobacco plant–pathogen system allows massive cell death to be synchronized. At 30ºC, TMV can systemically infect tobacco plants because induction of cell death and defence gene expression is completely suppressed. Upon shifting the temperature to 23ºC, cell death appears throughout the infected plant. A biotinylated caspase inhibitor (biotin-XVAD-FMK) was used to identify the proteins complexing with this inhibitor. The inhibitor when infiltrated into tobacco leaves before temperature shift, effectively blocked cell death and specifically bound to protein fractions of 40 and 38 KDa. Using antibodies against the intermediate and mature forms of VPE these fractions were recognized as two forms of VPE, indicating that the caspase like activity was performed by VPE. Infiltration of leaves with YVAD-CHO as well as specific VPE inhibitor (ESEN-CHO) abolished lesion formation. In addition, in VPE silenced plants hypersensitive cell death response to TMV was greatly suppressed. These results clearly demonstrate the caspase like activity of VPE and its involvement in TMV induced cell death. Cell death is accompanied by an increase in YVADase activity but not by DEVDase activity (Lam and Del Pozo, 2000). This suggests the involvement of VPEs in cell death and shows that VPEs are among the targets of caspase inhibitors in plants. Arabidopsis has four VPE genes (αVPE, βVPE, γVPE, δVPE) which are separated into seed type and vegetative

type (Yamada et al. 2005; Nakaune et al., 2005; Kinoshita et al., 1995). VPE is an asparaginyl endopeptidase (Hara-Nishimura et al., 1998; Hara-Nishimura et al., 2003). It cleaves peptide bonds on the C-terminal side of Asparagine residues exposed on molecular surface of proprotein precursors to generate the respective mature proteins (Hara-Nishimura et al., 1991; Hara-Nishimura et al., 1987; Hara-Nishimura et al., 1993). However, it also has been shown that VPE cleaves peptide bonds on the C-terminal side of aspartic acid residues (Becker et al., 1995; Hiraiwa et al., 1999). VPE recognizes aspartic acid when it is part of the YVAD sequence of a caspase-1 substrate, but does not necessarily recognize other aspartic acid residues (Hatsugai et al., 2004), similar to caspase-1 (Stennicke et al., 1998; Earnshaw et al., 1999). VPE, like caspases, is a cysteine protease. Although VPE is not related to the caspase family or the metacaspase family, VPE and caspase-1 share several enzymatic properties (Figure 2). Two residues of the catalytic dyad in VPE (histidine and cysteine; Figure 2) (Hiraiwa et al., 1999) are comparable to His237 and Cys285 of the catalytic dyad in human caspase-1 (Nicholson, 1999; Cohen, 1997). The QACRG pentapeptide of the active site of caspase-1 is similar to the E(A/G)CES pentapeptide of the active site of VPEs. A similar comparison was also done for human caspase-8 and Arabidopsis γVPE (Sanmatin et al., 2005). In addition, each of three essential amino acids, Arg179, Arg341 and Ser347, which form the substrate-binding pocket of caspase-1 (Wilson et al., 1994; Nicholson 1999) are conserved in VPEs. This is the case for all of the more than 20 VPEs that are currently in databases. The substrate binding pocket of VPE might be similar to the substrate-binding pocket of caspase-1. Both VPE (Hiraiwa et al., 1997; Hiraiwa et al., 1999; Kuroyanagi et al., 2002) and caspase-1 (Cohen, 1997) are subjected to self-catalytic conversion / activation from their inactive precursors.

With the sequencing of the complete genome of the model plant A. thaliana (Arabidopsis Genome Initiative, 2000), these caspase like activities have steered an intensive but frustrating search for caspase genes within plants. At the end of 2000, distant caspase relatives were discovered in silico in plants, the metacaspases that contain some of the structural features that are characteristic of the animal caspases (Uren et al., 2000; Vercammen et al., 2004). The Arabidopsis thaliana genome contains nine metacaspases. The function(s) and substrate specificity of the metacaspases from plants have not yet been investigated. The increased expre-ssion of one of the tomato metacaspases during infection with the necrotrophic pathogen Botrytis cinerea suggests a possible role for plant metacaspases in cell death (Hoeberichts et al., 2003). The recent findings by Peter Bozhkov and colleagues (Bozhkov et al., 2004) also indicate that a plant metacaspase might be involved in cell death. These authors studied proteolytic activity during embryogenesis in Norway spruce (Picea abies). Concomitant with massive cell death during shape remodelling, an increase in VEIDase activity (equivalent to activity of human caspase-6) was observed. Treatment with VEIDase inhibitor VEID-fluoromethylketone (VEID-FMK) inhibited cell death and prevented normal embryo development. The authors used a range of other caspase substrates but, apart from IETD- 7-amino-4-methyl-coumarin, these were cleaved poorly. The VEIDase activity was sensitive to pH, ionic strength and Zn^{2+} comparable to human caspase-6. The substrate specificity of the Norway spruce VEIDase appears similar to that of the yeast metacaspase YCA1 (Madeo et al., 2002), suggesting that the plant VEIDase involved in cell death is a metacaspase. This group also showed that silencing of a metacaspase gene (EMBL database Accession no. AJ534970) reduced VEIDase activity and cell death, and inhibited embryonic pattern formation (Suarez et al., 2004). These findings suggested that plant metacaspases were among the targets of the human caspase inhibitors and perhaps metacaspases functionally resemble animal caspases. But later in vitro experiments have shown that mcII-Pa (type II metacaspase gene) had Arg but not Asp specificity (Bozhkov et al. 2005). Because knocking down mcII-Pa not only disrupted cell death but also blocked embryonic differrentiation, it was speculated that mcII-Pa might be primarily involved in cellular differentiation rather than in cell death. Possibly, mcII-Pa regulates the actin reorganization observed during cellular differentiation (Smertenko et al., 2003), like mammalian caspases do in the cytoskeletal rearrangements during apoptosis (Mashima et al., 1999). In Arabidopsis, mere constitutive

over expression or disruption of meta-caspase genes does not lead to an obvious phenotype (Vercammen et al., 2006; Belenghi et al., 2007) and, thus, a role for metacaspases in cell death or other processes has not been identified yet. Redundancy may exist between the various members of this family, or additional factors may be necessary to activate ectopically expressed metacaspases.

The proteins responsible for the activation of the executors of PCD: The adaptors

The controlled activation of caspase precursors (Zymogens) is achieved by adaptor proteins that bind to them through shared motifs. Tumour Necrosis Factor (TNF) receptors superfamily or apoptogenic cofactors released by the mitochondria can be mentioned as examples of adaptors. Caspases-8 is activated when death effector domains (DEDs) in its pro domain bind to the C-terminal DED in adaptor molecule Fas-associated death domain (FADD); similarly Caspase-9 is activated after the association of Caspase Recruitment Domain (CARD) in its prodomain with the CARD in another cofactor protein, Apoptosis Protease Activating Factor-1 (APAF-1) (Vaux et al. 1999; Thornberry et al. 1998). In the worm Caenorhabditis elegans, Ced-4 acts as the adaptor molecule.

Database searches have identified several motifs of similarities between Ced-4, APAF-1 and proteins encoded by resistance genes regulating HR in plants. The conserved domain has been coined as NB-ARC (Van der Biezen and Jones, 1998; Rojo et al., 2004).

Intracellular controls: The regulators

A decision to die should not be taken lightly and so it is not surprising that the death programme is regulated in complex ways. A major class of intracellular regulators is the B-cell leukemia/lymphoma 2 (Bcl-2) family of proteins, which like caspase family, has been conserved in evolution from Worms to humans (Adams and Cory 1998). ced-9 (ced for cell death abnormal) gene in C. elegans encodes such a protein: if it is inactivated, most of the cells in the developing worm die and worm, therefore, dies early in development, but if ced-4 is also inactivated so that apoptosis cannot occur, the worm and all of the cells live (Kerr et al., 1972). Ced-9 prevents caspase activation by binding to adaptor Ced-4 (Vaux and Korsmeyer, 1999) in the worms. So, it seems that the only reason any cells in the developing worm live is that ced-9 normally keeps the death programme suppressed in these cells. ced-9 gene is similar to the humans bcl-2 gene. Fifteen bcl-2 family members have been identified so far in the mammals. Some such as Bcl-2, Bcl-XL, Bcl-W etc. suppress apoptosis (anti apoptotic): others such as BCL2 associated x protein (Bax), BCL2 antagonist/ killer (Bak), BCL2-related ovarian killer (Bok), BCLxL/BCL2 associated death promoter (Bad) and BH3 interacting domain death agonist (Bid) promote it (proapoptotic) (Adams and Cory, 1998). Some of these proteins can bind to each other: when an apoptosis suppressor forms a complex with an apoptosis promoters, each protein inhibits the others function. The ratio of suppressor to promoters help determine a cell's susceptibility to apoptosis (Merry and Korsmeyer, 1997).

It is now an established fact that mitochondria which are called the powerhouses of the cell, not only generate energy for cellular activities but also play an important role in cell death (Green et al., 1998; Shah et al., 2000) in animals. They release several death promoting factors such as cytochrome C (Cyt-C) (which contribute to caspase activation), Apaf-1, Apoptosis inducing Factor (AIF), procaspase-3, Ca^{2+} & reactive oxygen species (ROS) in response to various stimuli. Different mechanisms have been suggested to explain the release of apoptogenic factors from mitochondria, induced by pro-apoptotic proteins (Bernardi et al., 2001; Kroemer and Reed, 2000). The first involves Bax that could simply oligo-merise in outer mitochondrial membrane (OMM) to form a channel. Alternatively, Bax, in association with either the voltage-dependent anion channel (VDAC) or truncated Bid (tBid), could promote the formation of pores allowing the passage of soluble proteins. Alternative models have been suggested in which, during early stages of apoptosis, the inner mitochondrial membrane (IMM) plays a key role. The first one implies that water and solutes enter the mitochondrial matrix, inducing swelling of mitochondria (Bernardi et al., 2001; Kroemer and Reed, 2000). This process is mediated by either VDAC or the opening of a permeability transition pore (PTP) (Desagher and Martinou, 2000). The PTP may be defined as a voltage-dependent, cyclosporin A (CsA)–sensitive, high conductance inner membrane channel. The molecular structure of PTP is still unknown, although evidence suggests that it may be formed of several components, including matrix cyclophilin D, the outer membrane VDAC, the inner membrane adenine nucleotide translocase (ANT), peripheral benzodiazepin receptor and Bcl-2, hexokinase bound to VDAC, and intermembrane creatine kinase (Bernardi, 1999; Zoratti and Sza`bo, 1995). Recent evidence indicates that the mitochondria-associated hexokinase plays an important role in the control of apoptosis in mammals (Downward, 2003; Birnbaum, 2004; Majewski et al., 2004). Hexo-kinase is an integral component of the PTP through its interaction with porin or the voltage-dependent anion channel (VDAC) (Wilson, 2003), and hexokinase binding to the VDAC interferes with the opening of the PTP, thereby inhibiting cytochrome c release and apoptosis (Pastorino et al., 2002; Azoulay-Zohar et al., 2004).Thus, detachment of hexokinase from the mitochondria potentiates, and its over expression inhibits mitochondrial dys-function and cell death induced by various stimuli (Gottlob et al., 2001; Bryson et al., 2002; Majewski et al., 2004). Recent studies have shown that cyclophilin D, another compo-

nent of the PTP, is a key factor in the regulation of PTP function and that cyclophilin D–dependent mitochondrial permeability transitions are required to mediate some forms of necrotic cell death but not apoptotic cell death (Baines et al. 2005; Nakagawa et al., 2005). However, these observations do not exclude the possibility that certain forms of apoptosis are mediated by the mitochondrial permeability transitions, because some forms of apoptosis are significantly inhi-bited by cyclosporin A, a specific inhibitor of cyclophilin activity (Green and Kroemer, 2004). Additionally, cyclophilin D–overexpressing mice exhibited an increase in apoptotic heart muscle cells (Baines et al. 2005). Furthermore, in cancer cells, cyclophilin D over expression suppresses apoptosis via the stabilization of hexokinase II binding to the mitochondria (Machida et al., 2006). The pore open–closed transitions are highly regulated by multiple effectors at discrete sites. Factors affecting PTP can be subdivided into matrix and membrane effectors. The former include both openers (Ca^{2+}, phosphate, oxidizing agents, ¯OH and atractylate) and inhibitors [CyclosporinA (CsA)], Adenosine diphospate H^+, bongkrekate and reducing agents]. Among the latter, a high (inside-negative) membrane potential tends to stabilize the PTP in a closed conformation, whereas depolarisation by different uncouplers determines its aperture. PTP is also regulated by quinones, which prevent Ca^{2+}-dependent pore opening. In addition, the Pore is regulated by Bcl-2 proteins & intracellular ATP levels (Green and Reed, 1998; Adams and Cory, 1998). The Bcl-2 proteins are membrane spanning and have at least one of the four Bcl-2 homology (BH) domains. It is shown that the proapoptotic Bax interacts with VDAC & ANT & brings about a conformational alteration to form a megachannel leading to the release of Cyt-C (Martinou, 1999; Bernadi et al., 2001). The pore conductance of VDAC has been shown to increase in the presence of Bax in artificial membranes and this increase is blocked by Bcl-XL (Shimizu et al., 2000). Bax and Bim interact with VDAC and lead to the release of Cyt-C, whereas Bcl-XL blocks this release (Shimizu et al., 1999; Sugiyama et al., 2002). The permeabilization of the IMM to solutes with molecular mass up to 1.5 kDa, caused by the aperture of the PTP results in the complete dissipation of mitochondrial electrical potential. Consequently, the high concentration of solutes present in the matrix induces an osmotic swelling that could ultimately lead to OMM rupture and the consequent release of proteins from the intermembrane mitochondrial space (IMS) (Kroemer and Reed, 2000).

Cyt-C, the most investigated protein involved in caspase activation, binds the scaffolding protein, named apoptotic protease activating factor-1 (Apaf-1), leading to an ATP- or dATP-dependent conformational change that induces Apaf-1 oligomerisation (van Gurp et al., 2003). This high molecular mass complex, called the apoptosome, is assembled by binding Cyt-C and Apaf-1 with procaspase-9 through the interaction between their caspase recruitment domains (CARDs). Procaspase-9 activity is greatly enhanced in the apoptosome that, in turn, proteolytically activates caspase-3, finally resulting in the morphological and biochemical changes associated with apoptosis (Kaufmann and Hengartner 2001). The most common hallmark used to identify the involve-ment of plant mitochondria in PCD is the release of Cyt-C. The release of Cyt-C from mitochondria has been detected in different plant systems, in which PCD was induced. In particular, the release of Cyt-C precedes the appearance of PCD symptoms and has been recognized in A. thaliana cells treated with mannose, where the effect is also associated to endonuclease activation (Stein and Hansen, 1999), and in maize cells infected by Agrobacterium sp. (Hansen, 2000). In addition, harpin (a bacterial proteinaceous elicitor)–induced HR in tobacco cells is associated with an alteration of mitochondrial functions (Xie and Chen, 2000). The initial steps of cell death are accompanied by an oxidative burst, depletion of ATP, collapse of the mitochondrial electrical potential and release of Cyt-C. A strong stimulation of the expression of the alternative oxidase (AOX) and small heat-shock proteins (HSPs) has also been described (Krause and Durner, 2004). Consistent with this, induction of PCD in A. thaliana cell cultures by ceramide, protoporphyrin IX and an elicitor of HR (AvrRpt2) leads to the dissipation of mitochondrial electrical potential followed by morphological changes and Cyt-C release (Yao et al., 2004). By analogy with animal mitochondria, several authors have correlated the detected release of Cyt-C to the activity of PTP, on the basis of the inhibitory effect exhibited by CsA (Balk and Leaver, 2001; Lin et al., 2005; Tiwari et al., 2002). This contention seems to be confirmed by the observation that Nitric Oxide-induced programmed death in Citrus sinensis cell cultures is also prevented by CsA (Saviani et al., 2002). Plant mitochondria have the main components probably involved at the contact sites of OMM and IMM, e.g. ANT, VDAC (Godbole et al., 2003) and cyclophilin (Yokota et al., 2004), Hexokinase (Kim et al., 2006). A functional genomic screen to assess the functions of various signaling genes in Nicotiana benthamiana revealed that a tobacco rattle virus (TRV)–based virus induced gene silencing (VIGS) of a hexokinase gene, Hxk1, induced the spontaneous formation of necrotic lesions in leaves (Kim et al., 2006). Hxk1 was associated with the mitochondria, and its expression was stimulated by various cell death–inducing stresses. VIGS of Hxk1 resulted in apoptotic cell death in leaves, indicating that depletion of mitochondrial hexokinases activated programmed cell death (PCD). Conversely, overexpression of the mitochondria-associated Arabidopsis hexokinases, HXK1 and HXK2, conferred enhanced resistance to oxidative stress–induced cell death. Finally, the exogenous addition of recombinant Hxk1, but not Hxk1DN, which lacks the membrane anchor, inhibited clotrimazole (CTZ)/H2O2–induced Cyt-C release from mitochondria. These results suggest a

direct link between plant hexokinases and the PCD process. In any case, the opening of this channel would determine the entry into mitochondria of osmotically active solutes and water. This would cause a mitochondrial swelling with the consequent rupture of the OMM and release of Cyt-C.

A further model refers to the non-swelling mechanism involving the OMM. In this mechanism a crucial role is performed by VDAC, which interacts with Bax, forming a pore through which Cyt-C is released (Lam et al. 2001). The first evidence derives from a study in which the over-expression of mammalian Bax gene in tobacco plants causes hypersensitive-like lesions and induces defence genes (Lacomme and Santa Cruz, 1999). Recent experimental findings seem to corroborate this mechanism, suggesting that VDAC can play a crucial role in PCD pathway being a conserved element in both plants and animals (Godbole et al., 2003; Swidzinski et al., 2004). In agreement, VDAC expression increases during HR, senescence and heat-induced PCD in A. thaliana cells (Lacomme and Roby, 1999; Swidzinski et al., 2004). This evidence indicates a putative dual role for VDAC, as a component of PTP or as a channel that interacts with Bax.

In animal cells, the significance of the release of Cyt-C is the subsequent assembly of the apoptosome complex that is followed by the activation of the executioner caspases. Although evidences are lacking for the formation of apoptosome in plant cells, sequence alignments have revealed significant similarities among regions of C. elegans cell death gene that encodes a protease activating factor (Ced-4), human Apaf-1 and several plant resistance genes. Although these resistance genes do not contain CARD but may be assumed to function as controlling adaptors in plant protein complexes which are activated during HR (Vander Biezen and Jones 1998). Thus, the subsequent fate of Cyt-C is still problematic, because the formation of the complex like the apoptosome, is largely speculative. If an apoptosome- like complex exists in plants, it may interact with caspase-like proteases (Metacaspases, VPE) by analogy with the system in animal cells. In addition, a further evidence supporting the mitochondrial involvement in plant PCD is provided by the reports of strong increase in HSP during harpin HR in A. thaliana cells (Krause and Durner, 2004). HSPs are considered to partially suppress apoptosis in animal cells, by preventing Cyt-C release and disrupting the apoptosome. Plant HSPs are considered to accomplish comparable effects (Hoeberichts and Woltering, 2002).

It has been demonstrated that Bax inhibitor1 (BI-1) protein inhibits Bax induced apoptosis in mammalian cells and when ectopically expressed in yeast (Xu and Reed 1998). BI-1 contains six or seven predicted trans-membrane domains. As an integrate membrane protein, the localization of BI-1 is found to be similar to Bcl-2 exhibiting a nuclear envelope and endoplasmic reticulum association pattern (Xu and Reed, 1998). Moreover BI-1 has been isolated as one of the candidate suppressors of TNF-related apoptosis inducing ligand (TRAIL), an apoptosis-inducing member of the Tumour Necrosis Factor (TNF) (Burns and El-Deiry, 2001). The fundamental features of PCD are believed to be conserved throughout metazoans and plants (Pennel and Lamb, 1997). In support of this, a study carried out on tobacco shows that the expression of Bax, which is a mammalian pro-apoptotic protein, triggered cell death in tobacco leaf (Lacomme and Santa Cruz, 1999). Moreover, over-production of animal cell death suppressors Bcl-XL and Ced-9 conferred enhanced resistance to UV-b and paraquat treatment and salt, cold and wound stresses in tobacco plants (Mitsuhara et al., 1999; Qiao et al., 2002). However, recent research demonstrated that homologous plant genes for cell death suppressors such as *bcl-2*, *Bcl-XL* from humans and *ced-9* from *C. elegans* are not found in *Arabidopsis* genome whose sequence has been presented as the first complete genomic sequence of higher plants (Lam et al., 2001). Recently, it has been reported that a gene encoding a homolog of Bax inhibitor (*BI-1*) from *Arabidopsis* inhibits mammalian Bax action *in planta* (Kawai-Yamada et al., 2001). This is the first report on the direct contribution to plant cell death of a plant originating gene that is a homolog of animal cell death related gene. An Arabidopsis homolog *AtBI-1*(*Arabidopsis* Bax inhibitor 1) was identified from the genome sequencing project. The identity level is 37.5% AtBI-1 shares 41% amino acid identity with mammalian BI-1(mBI-1). Plant homologs of *BI-1* gene for several plant species including *Oryza sativa* (*OsBI-1*), *A. thaliana* (*AtBI-1*) have been cloned and characterized to various extents (Bolbuc and Brisson, 2002; Bolduc et al., 2003; Eichmann et al., 2004; Huckelhoven et al., 2003; Kawai et al., 1999; Kawai-Yamada et al., 2004; Matsumara et al., 2003; Sachez et al., 2000). Ectopic overexpression of OsBI-1 leads to the elimination of cell death caused by Bax protein in Budding yeast *S. cerevisiae* (Kawai et al., 1999). Intriguingly enough, however, when AtBI-1 was transfected into the mammalian cell systems, it did not suppress Bax induced cell death in the human cells. Infact, AtBI-1 induced cell death comparable to Bax (Yu et al., 2002). The possibility exists that AtBI-1 might directly damage the mitochondrial structure causing Cyt-C release. However, co-transfection of the cells with both human BI-1 and AtBI-1 crippled cell death, suggesting preferably a dominant-negative mechanism in which AtBI-1 induced apoptosis is minimized by overexpressed mBI-1 (Yu et al., 2002). It may, thus, be speculated that AtBI-1 competitively interacts with endogenous mBI-1 or BI-1 target protein, interfering with its function & thereby triggering cell death. In this regard, *in vitro* binding of BI-1 with Bcl-2 but not with Bax has been demonstrated (Xu and Reed, 1998). Thus, it remains unclear how BI-1 suppresses Bax's function given that yeast and plants contain no obvious Bcl-2 homologs (Lam et al., 2001).

Interestingly, down regulation of a tobacco BI-1 homolog using an antisense RNA approach resulted in accelerated cell death of tobacco BY-cells upon carbon starvation (Bolduc and Brisson 2002). Down regulation of rice BI-1 in cultured rice cells upon challenge with a fungal elicitor from *Megnaporthe grisea* was concomitant with the progression of cell death and, conversely, overexpressed rice BI-1 can improve cell survival against the elicitor (Matsumara et al., 2003). Another study found that decreased BI-1 expression correlated with chemical-induced resistance of barley to the infection of a biotrophic fungal pathogen powdery mildew (*Blumeria graminis*), and overexpression of barley BI-1 at a single-cell level induces hyper susceptibility and could reverse the fungal resistance that is conferred by the loss of MLO, a negative regulator of cell death and defense response in barley (Huckelhoven et al., 2003). Although these observations support the idea that BI-1 homologs of yeast and plants have an anti-PCD function, the physiological importance of BI-1 and the impact of its loss of function in plants are still unclear at the whole plant level as clear genetic evidence is absent. However, a study carried out more recently (Watanabe and Lam, 2006) identified and characterized two independent Arabidopsis mutants with T-DNA insertion in the AtBI-1 gene. The phenotype of atbi1-1 and atbi1-2, with a C-terminal missense mutation and a gene knockout, respectively, is indistinguishable from wild-type plants under normal growth conditions. However, these two mutants exhibit accelerated progression of cell death upon infiltration of leaf tissues with a PCD-inducing fungal toxin fumonisin B1 (FB1) and increased sensitivity to heat shock-induced cell death. Under these conditions, expression of AtBI1 mRNA has been shown to be upregulated in wild-type leaves prior to the activation of cell death, suggesting that increase of AtBI1 expression is important for basal suppression of cell death progression. Over-expression of AtBI-1 transgene in the two homozygous mutant backgrounds rescued the accelerated cell death phenotypes. Together, these results provide direct genetic evidence for a role of BI-1 as an attenuator for cell death progression triggered by both biotic and abiotic types of cell death signals in Arabidopsis. Plant homologs of the animal anti-apoptotic defender against Apoptotic Death 1 (DAD1) gene have been reported in the cells of *A. thaliana* (Gallois et al., 1997), pea (Orzaez and Grannel, 1997b) and rice (Tanaka et al., 1997). DAD 1 has been discovered in hamster cells where the corresponding mutant cell line dies via apoptosis (Kusaka et al. 2004). The suppressor function of this protein was further suggested in *C. elegans* via it's over expression protects some of the cells destined to die by apoptosis during development (Sugimoto et al., 1995). At least two A. *thaliana* homologues have been found and transformation of the mutant hamster cell line with one of them, demonstrates that the function of the protein is conserved between plants and

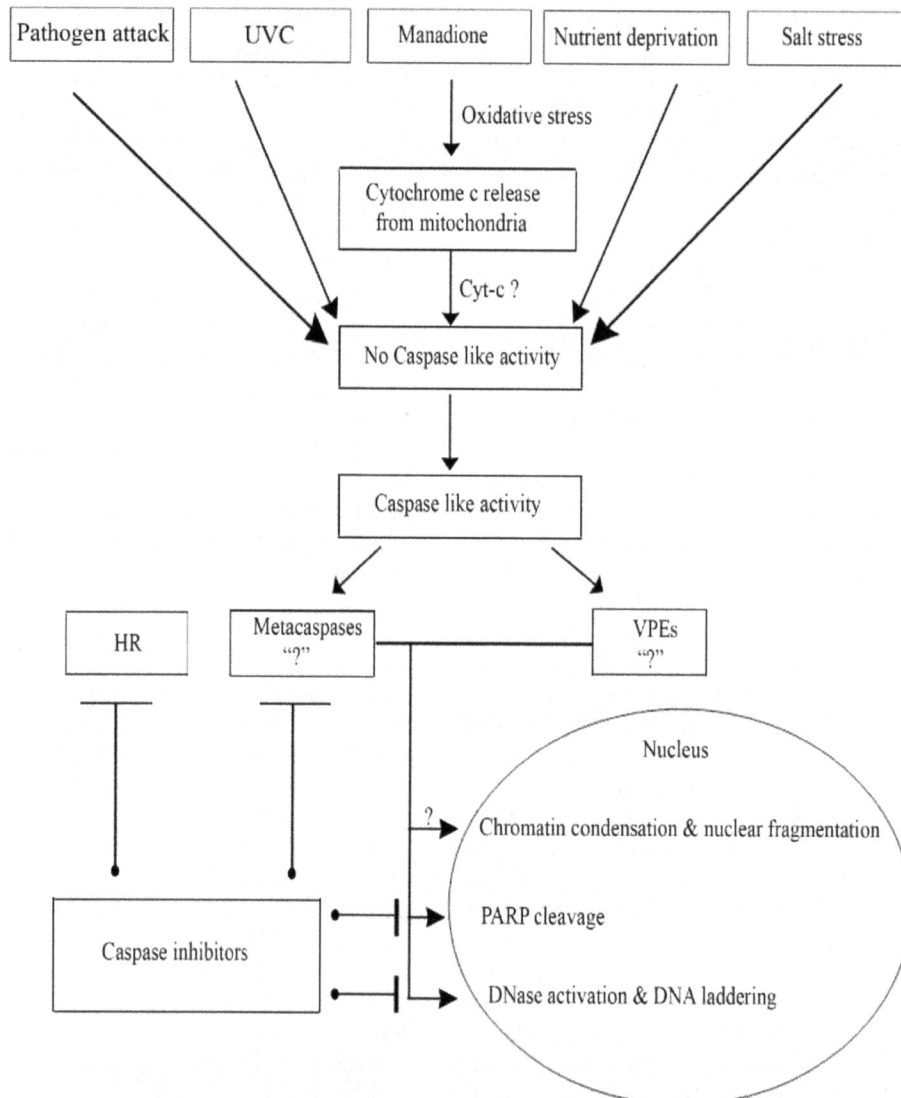

Figure 3. Chosen examples of treatments triggering different animal or plant PCD hallmarks in plant cell**?** Indicates events that have been shown in animals and have not been detected in plants and **"?"** Indicates arbitrary functional assignment in plants.

animals (Gallois et al., 1997). Other experiments have shown that DAD 1 is a subunit of mammalian Oligosaccaryl transferase complex (Kelleher et al., 1997). In mice, overexpression of DAD 1 does not have any protecting function against apoptosis but affects cell cycle (Hong et al., 1999). All these results together make the role of this gene in animal apoptosis unclear, whereas its putative suppressor role in plants is still not established.

CONCLUSIONS AND FUTURE PERSPECTIVES

There are likely to be inherent differences in the operational mechanism of PCD between plants and animals; there is also the possibility for the involvement of different operational mechanisms of PCD in different plant cell types, that is, more than one pathway of PCD is likely to

be operative in plants, while in animals there seems to be only one programme. No plant system is yet described which shows all features common to animal PCD. The comparison of plant PCD and apoptosis in animals show important differences regarding morphological changes occurring in the dying cells and enzymes involved in the process. However, the final execution of the process, DNA fragmentation and nuclear disorganization, has similarities in animal and plant cells, suggesting that it might have evolved from a common ancestor. Extensive studies have provided evidence that PCD in plants and animals share components that include caspase-like activity (Lam et al., 2001; Hatsugai et al., 2004; Suarez et al., 2004; Sanmartin et al., 2005; Kuroyanagi et al., 2005; Hara-Nishimura et al., 2005; Lam, 2005; Hatsugai et al., 2006) and these caspase-like activities could be inhibited with caspase inhibitors but not caspase-unrelated protease inhibitors (del Pozo and Lam, 1998). Furthermore, the caspase inhibitors have been shown to abolish these PCDs (del Pozo and Lam, 1998; De jong et al., 2000). The existence of plant proteases that recognize and process the natural caspase substrate PARP apparently at caspase recognition site (Sun et al., 1999; Tian et al., 2000; O'Farrell, 1995) and the functionality of natural caspase substrates in plants (Danon et al., 2004 and Lincoln et al., 2002) substantiates the involvement of caspase like activity in plant programmed cell death (Figure 3)

VPEs and metacaspases appear to be the prime candidates that seem to be responsible for the caspase like activities observed. However, until now, the role of metacaspases in cell death still remained enigmatic, and both up- and down-regulation of metacaspases have yielded conflicting data. However, such approaches bear the risk that a constitutive perturbation of genes that are essential for normal cellular homeostasis leads to over-interpretation. Alternative routes toward unraveling the function of metacaspases could involve the identification of their substrates by using technologies that allow direct characterization of *in vivo* protein processing on a proteome-wide scale (Gevaert et al., 2006). Knowing the degradome specificity of metacaspases could reveal their role in cellular and developmental processes, including cell death. Overproduction of the cleavage fragments and/or of uncleavable mutant proteins would help elucidate the functional consequences of substrate cleavage by metacaspases. The many intriguing similarities with PCD in animals will need to be rigorously tested to demonstrate that they are conserved, and are derived from a common ancestral origin. In spite of significant progress in our understanding of plant PCD in recent years, its control mechanism remains unclear. Deployment of reverse genetic approaches such as PTGS/RNAi strategies (Wang and Waterhouse, 2002) and knockout screens using T-DNA or transposon insertion collections (Bouche and Bouchez, 2001), coupled with global expression approaches such as micro-array analysis should help speed up the first essential step of identifying the important players involved in plant cell death activetion. This approach would be complementary to forward genetic approaches that are revealing new regulator(s) that may not have counterparts in other organisms.

ACKNOWLEDGEMENTS

Fellowship grant to Nishawar Jan [F.No.10-2(5)/2003(II).E.U.II] from University Grants Commission, India is Gratefully acknowledged. The authors also wish to thank Dr. Riyaz .A. Pandit for his help with graphics.

REFERENCES

Adams JM, Cory S (1998). Bcl-2 protein family: the arbiters of cell survival. Sci. 281:1322–1326.

Ameisen JC (2002). On the origin, evolution & nature of programmed Cell Death: a timeline of four billion years. Cell. Death. Differ. 9: 367–393.

Azoulay-Zohar H, Israelson A, Abu-Hamad S, Shoshan-Barmatz, V (2004). In self-defence: Hexokinase promotes voltage dependent anion channel closure and prevents mitochondria-mediated apoptotic cell death. Biochem. J. 377: 347–355.

Baines CP, Kaiser RA, Purcell NH, Blair NS, Osinska H, Hambleton MA, Brunskill EW, Sayen MR, Gottlieb RA, Dorn GW, Robbins J, Molkentin JD (2005). Loss of cyclophilin D reveals a critical role for mitochondrial permeability transition in cell death. Nature. 434: 658–662.

Balk J, Leaver CJ (2001). The PET1-CMS mitochondrial mutation in sunflower is associated with premature programmed cell death and cytochrome c release. Plant Cell. 13: 1803–1818.

Becker C, Shutov AD, Nong VH, Senyuk VI, Jung R, Horstmann C, Fischer J, Nielsen NC, Muntz K (1995). Purification, cDNA cloning and characterization of proteinase B, an asparagine-specific endopeptidase from germinating vetch (Vicia sativa L.) seeds. Eur. J. Biochem. 228: 456-462.

Belenghi B, Romero-Puertas MC, Vercammen D, Brackenier A, Inzé D, Delledonne M, Van Breuseqem F (2007). Metacaspase activity of Arabidopsis thaliana is regulated by S-nitrosylation of a critical cysteine residue. J. Biol. Chem. 282: 1352–1358.

Bernardi P (1999). Mitochondrial transport of cations: channels, exchangers, and permeability transition. Physiol. Rev. 79:1127–1155.

Bernardi P, Petronilli V, Di Lisa F, Forte M (2001). A mitochondrial perspective on cell death. Trends Biochem. Sci. 26: 112–117.

Birnbaum MJ (2004). On the Interaction between hexokinase and the mitochondrion. Dev. Cell. 7: 781–782.

Bolduc N, Brisson LF (2002). Antisense down regulation of NtBI-1 in tobacco BY-2 cells induces accelerated cell death upon carbon starvation. FEBS Lett. 532: 111–114.

Bolduc N, Ouellet M, Pitre F, Brisson LF (2003). Molecular characterization of two plant BI-1 homologues which suppress Bax-induced apoptosis in human 293 cells. Planta. 216: 377–386.

Bouche N, Bouchez D (2001). Arabidopsis gene knockout: phenotype wanted. Curr. Opin. Plant Biol. 4: 111–117.

Bozhkov PV, Filanova LH, Suarez MF, Helmersson A, Smertenko AP, Zhivotovsky B, Von Arnold S (2004). VEIDase is a principal caspase-like activity involved in plant programmed cell death and essential for embryonic pattern formation. Cell Death Differ.11: 175–182.

Bozhkov PV, Suarez MF, Filonova LH, Daniel G, Zamyatnin AA, Rodriguez-Nieto Jr. S, Zhivotovsky B, Smertenko A (2005). Cysteine protease mcII-Pa executes programmed cell death during plant organogenesis. Proc. Natl. Acad. Sci. USA. 102: 14463–14468.

Bryson JM, Coy PE, Gottlob K, Hay N, Robey RB (2002). Increased hexokinase activity, of either ectopic or endogenous origin, protects renal epithelial cells against acute oxidant-induced cell death. J. Biol. Chem. 277: 11392–11400.

Burns TF, El-Deiry WS (2001). Identification of Inhibitors of TRAIL-Induced Death (ITIDs) in the TRAIL sensitive colon carcinoma cell line, SW480, using a genetic approach. J. Biol. Chem. 276: 37879-37886.

Callard D, Axelos M, Mazzolini L (1996). Novel marker for late phases of the growth cycle of Arabidopsis thaliana cell-suspension cultures are expressed during organ senescence. Plant Physiol. 112: 705–715.

Chinnaiyan AM, Dixit VM (1996). The cell-death machine. Current Biol. 6: 555–562.

Cohen GM (1997).Caspases: the executioners of apoptosis. Biochem. J. 326: 1-16.

Cohen JJ (1993). Apoptosis. Immunol. Today 14: 126–130.

D'Silva I, Poirier GG, Heath MC (1998). Activation of cysteine proteases in cowpea plants during the hypersensitive response: a form of programmed cell death. Exp. Cell Res. 245: 389–399.

Danon A, Rotari VI, Gordon A, Mailhac N, Gallois P (2004). Ultraviolet-C Overexposure Induces Programmed Cell Death in Arabidopsis, Which Is Mediated by Caspase-like Activities and Which Can Be Suppressed by Caspase Inhibitors, p35 and Defender against Apoptotic Death. J. Biol. Chem. 279: 779–787.

Danon A, Gallois P (1998). UV C radiation induces apoptotic like changes in *Arabidopsis thaliana*. FEBS Lett. 437: 131–136.

De Jong AJ, Hoeberichts FA, Yakimova ET, Maximova E, Woltering EJ (2000). Chemical-induced apoptotic cell death in tomato cells: involvement of caspase-like proteases. Planta. 211: 656-662.

de Murcia G, Ménissier de Murcia J (1994). Poly (ADPribose) polymerase: a molecular nick-sensor. Trends Biochem. Sci. 19: 172–176.

del Pozo O, Lam E (1998). Caspase and programmed cell death in the hypersensitive response of plants to pathogens. Current Biol. 8: 1129–1132.

del Pozo O, Lam E (2003). Expression of the baculovirus p35 protein in tobacco affects cell death progression and compromises N gene mediated disease resistance response to tobacco mosaic virus. Mol. Plant–Microbe Interact. 16: 485–494.

Delorme VGR, McCabe PF, Kim DJ, Leaver CJ (2000). A matrix metalloproteinase gene is expressed at the boundary of senescence and programmed cell death in cucumber. Plant Physiol. 123: 917-927.

Desagher S, Martinou JC (2000). Mitochondria as the central control point of apoptosis. Trends Cell Biol. 10: 369–377.

Dom´ınguez F, Moreno J, Cejudo FJ (2001). The nucellus degenerates by a process of programmed cell death during the early stages of wheat grain development. Planta. 231: 352–360.

Dom´ınguez F, Moreno J, Cejudo FJ (2004). A gibberellin-induced nuclease is localized in the nucleus of wheat aleurone cells undergoing programmed cell death. J. Biol. Chem. 279: 11530–11536.

Dom´ınguez F, Cejudo FJ (2006). Identification of a nuclear-localized nuclease from wheat cells undergoing programmed cell death that is able to trigger DNA fragmentation and apoptotic morphology on nuclei from human cells. Biochem. J. 397: 529–536.

Downward J (2003) Metabolism meets death. Nature 424:896–897.

Earnshaw WC (1995). Nuclear changes in apoptosis. Curr. Opin. Cell Biol. 7: 337–343.

Earnshaw WC, Martins LM, Kaufmann SH (1999). Mammalian caspases: structure, activation, substrates, and functions during apoptosis. Annu. Rev. Cell Biol. 68: 383-424.

Eichmann, R, Holger S, Kogel K-H Hu¨, ckelhoven R (2004). The barley apoptosis suppressor homologue Bax inhibitor-1 compromises nonhost penetration resistance of barley to the inappropriate pathogen Blumeria graminis f. sp. tritici. Mol. Plant Microbe Interact. 17: 484–490.

Ellis RE, Yuan JY, Horvitz HR (1991). Mechanisms and functions of cell death. Annu. Rev. Cell Biol. 7: 663–698.

Enari M, Sakahira H, Yokoyama H, Okawa K, Iwamatsu A, Nagata S (1998). A caspase-activated DNase that degrades DNA during apoptosis. Nature (London). 391: 43–50.

Fath A, Bethke PC, Jones RL (1999). Barley aleurone cell death is not apoptotic: characterization of nuclease activities and DNA degradation. Plant J. 20: 305–315.

Filonova LH, Bozhkov PV, Brukhin VB, Daniel G, Zhivotovsky B, von Arnold S (2000). Two waves of programmed cell death occur during formation and development of somatic embryos in the gymnosperm, Norway spruce. J. Cell Sci. 113: 4399-4411.

Gallois P, Makishima T, Hecht V, Despres B, Laudié M, Nishimoto T, Cooke R (1997). An Arabidopsis thaliana cDNA complementing a hamster apoptosis suppressor mutant. Plant J. 11: 1325–1331.

Garcia-Calvo M, Peterson EP, Leiting B, Ruel R, Nicholson DW, Thornberry NA (1998). Inhibition of human caspase by peptide-based and macromolecular inhibitors. J. Biol Chem. 273: 32608–32613.

Gevaert K, Van Damme P, Ghesquière B, Vandekerckhove J (2006). Protein processing and other modifications analyzed by diagonal peptide chromatography. Biochem. Biophys. Acta.1764: 1801–1810.

Godbole J, Varghese A, Sarin MK, Mathew (2003). VDAC is a conserved element of death pathways in plant and animal systems. Biochim. Biophys. Acta.1642: 87– 96.

Gottlob K, Majewski N, Kennedy S, Kandel E, Robey RB, Hay N (2001). Inhibition of early apoptotic events by Akt/PKB is dependent on the first committed step of glycolysis and mitochondrial hexokinase. Genes Dev. 15: 1406–1418.

Gray J (2004). Paradigms of the evolution of programmed cell death. In 'Programmed Cell Death in Plants'. (Eds J Gray) (CRC Press: Boca Raton). pp. 1–25.

Green DR, Kroemer G (2004) The pathophysiology of mitochondrial cell death. Sci. 305: 627–629.

Green DR, Reed JC (1998). Mitochondria & apoptosis. Sci. 281: 1309–1312.

Grutter MG (2000). Caspases: key players in programmed cell death. Curr. Opinion Struc. Biol.10: 649–655.

Halenbeck R, MacDonald H, Roulston A, Chen TT, Conroy L, Williams, LT (1998). CPAN, a human nuclease regulated by the caspase-sensitive inhibitor DFF45. Curr. Biol. 23: 537–540.

Hansen G (2000). Evidence for Agrobacterium-induced apoptosis in maize cells. Mol. Plant Microbe Interact. 13: 649–657.

Hara-Hishimura I, Takeuchi Y, Inoue K, Nishimura M (1993). Vesicle transport and processing of the precursor to 2S albumin in pumpkin. Plant J. 4: 793-800.

Hara-Nishimura I (2003) Asparaginyl endopeptidase. In 'Handbook of Proteolytic Enzymes'. (Eds AJ Barrett, ND Rawlings, JF Woessner) (Academic Press). pp. 846-849.

Hara-Nishimura I, Hatsugai N, Nakaune S, Kuroyanagi M, Nishimura M (2005). Vacuolar processing enzyme: an executor of cell death. Curr. Opin. Plant Biol. 8: 404–408.

Hara-Nishimura I, Inoue K, Nishimura M (1991). A unique vacuolar processing enzyme responsible for conversion of several proprotein precursors into the mature forms. FEBS Lett. 294: 89-93.

Hara-Nishimura I, Kinoshita T, Hiraiwa N, Nishimura M (1998) Vacuolar processing enzymes in protein-storage vacuoles and lytic vacuoles. J. Plant Physiol. 152: 668-674.

Hara-Nishimura I, Nishimura M (1987). Proglobulin processing enzyme in vacuoles isolated from developing pumpkin cotyledons. Plant Physiol. 85: 440-445.

Hatsugai N, Kuroyanagi M, Nishimura M, Hara-Nishimura I (2006). A cellular suicide strategy of plants: vacuole-mediated cell death. Apoptosis.11: 905–911.

Hatsugai N, Kuroyanagi M, Yamada K, Meshi T, Tsuda S, Kondo M, Nishimura M, Hara-Nishimura I (2004) A plant vacuolar protease, VPE, mediates virus-induced hypersensitive cell death. Sci. 305: 855-858.

Heath MC (2000). Hypersensitive response-related death. Plant Mol. Biol. 44: 321-334.

Hengartner MO (1998). Apoptosis: Death by crowd control. Sci. 281:1298–1299.

Hiraiwa N, Nishimura M, Hara-Nishimura I (1997). Expression and activation of the vacuolar processing enzyme in Saccharomyces cerevisiae. Plant J. 12: 819-829.

Hiraiwa N, Nishimura M, Hara-Nishimura I (1999). Vacuolar processing enzyme is self-catalytically activated by sequential removal of the C-terminal and N-terminal propeptides. FEBS Lett. 447: 213-216.

Hoeberichts FA, ten Have A, Woltering EJ (2003). A tomato metacaspase gene is upregulated during programmed cell death in Botrytis cinereainfected leaves. Planta. 217: 517–522.

Hoeberichts FA, Woltering EJ (2002). Multiple mediators of plant programmed cell death: interplay of conserved cell death mechanisms and plant-specific regulators. Bioessays 25: 47–57.

Hong NA, Kabra NH, Hsieh SN, Cado D, Winoto A (1999). In *vivo* overexpression of Dad1, the defender against apoptotic death-1, enhances cell proliferation but does not protect against apoptosis. J. Immunol. 163:1888–1893.

Huckelhoven R, Dechert C, Kogel KH (2003). Overexpression of barley BAX inhibitor 1 induces breakdown of mlo-mediated penetration resistance to Blumeria graminis. Proc. Natl. Acad. Sci. USA. 100:5555–5560.

Ito J, Fukuda H (2002). ZEN1 is a key enzyme in the degradation of nuclear DNA during programmed cell death of tracheary elements. Plant Cell. 14: 3201–3211.

Jones AM (2000). Does the plant mitochondrion integrate cellular stress and regulate programmed cell death? Trends Plant Sci. 5: 225-230.

Jordan ND, Franklin FCH, Franklin-Tong VE (2000). Evidence for DNA fragmentation triggered in the self-incompatibility response in pollen of Papaver rhoeas. Plant J. 23: 471-479.

Katsuhara M (1997) Apoptosis-like cell death in barley roots under salt stress. Plant Cell Physiol. 38:1091–1093.

Kaufmann SH, Hengartner MO (2001). Programmed cell death: alive and well in the new millennium. Trends Cell Biol. 11: 526–534.

Kawai M, Pan L, Reed JC, Uchimiya H (1999). Evolutionally conserved plant homologue of the Bax inhibitor-1 (BI-1) gene capable of suppressing Bax-induced cell death in yeast. FEBS Lett. 464: 143–147.

Kawai-Yamada M, Jin L, Yoshinaga K, Hirata A, Ucimiya H (2001). Mammalian Bax-induced plant cell death can be down-regulated by over expression of Arabidopsis Bax inhibitor-1(AtBI-1). Proc.Natl.Acad.Sci. USA. 98: 12295-1300.

Kawai-Yamada M, Ohmori Y, Uchimiya H (2004). Dissection of Arabidopsis Bax inhibitor-1 suppressing Bax-, hydrogen peroxide-, and salicylic acid-induced cell death. Plant Cell. 16: 21–32.

Kelleher DJ, Gilmore R.(1997). DAD1, the defender against apoptotic cell death, is a subunit of the mammalian oligosaccharyl-transferase. Proc.Natl.Acad.Sci. USA. 94: 4994–4999.

Kerr JFR, Wyllie AH, Currie AR (1972). Apoptosis: a basic biological phenomenon with wide-ranging implications in tissue kinetics. Br. J. Cancer. 26: 239–257.

Kim M, Lim JH, Ahn CS, Park K, Kim GT, Kim WT, Pai HS (2006). Mitochondria-associated Hexokinases play a role in the control of Programmed Cell Death in Nicotiana benthamiana .The Plant Cell. 18: 2341–2355.

Kinoshita T, Nishimura M, Hara-Nishimura I (1995). Homologues of a vacuolar processing enzyme that are expressed in different organs in Arabidopsis thaliana. Plant Mol. Biol. 29: 81-89.

Koukalova B, Kovarik A, Fajkus J, Siroky J (1997). Chromatin fragmentation associated with apoptotic changes in tobacco cells exposed to cold stress. FEBS Lett. 414: 289–292.

Krause M, Durner J (2004). Harpin inactivates mitochondria in Arabidopsis suspension cells. Mol. Plant Microbe Interact. 17: 131–139.

Kroemer G, Reed JC (2000). Mitochondrial control of cell death. Nat. Med. 6: 513–519.

Kuida K, Zheng TS, Na SQ, Kuan CY, Yang D, Karasuyama H, Rakic P, Flavell RA (1996). Decreased apoptosis in the brain and premature lethality in CPP32-deficient mice Nature. 384: 368–372.

Kumar S (2007) Caspase function in programmed cell death. Cell Death Differ. 14: 32–43.

Kuroyanagi M, Yamada K, Hatsugai N, Kondo M, Nishimura M, Hara-Nishimura I (2005). Vacuolar processing enzyme is essential for mycotoxin induced cell death in Arabidopsis thaliana. J. Biol. Chem. 280: 32914–32920.

Kuroyanagi M, Nishimura M, Hara-Nishimura I (2002). Activation of Arabidopsis vacuolar processing enzyme by self-catalytic removal of an auto-inhibitory domain of the C-terminal propeptide. Plant Cell Physiol. 43: 143-151.

Kusaka K, Tada Y, Shigemi T, Sakamoto M, Nakayashiki H, Tosa Y, Mayama S (2004) Coordinate involvement of cysteine protease and nuclease in the executive phase of plant apoptosis. FEBS Lett. 578: 363–367.

Lacomme C, Roby D (1999). Identification of new early markers of the hypersensitive response in Arabidopsis thaliana. FEBS Lett. 459: 149–153.

Lacomme C, Santa cruz S (1999). Bax induced cell death in tobacco is similar to hypersensitive response. Proc. Natl. Acad. Sci. USA. 96: 7956-7961.

Lam E (2005). Vacuolar proteases livening up programmed cell death. Trends Cell Biol. 15:124-127.

Lam E, Del Poza O (2000). Caspase-like protease involvement in the control of plant cell death. Plant Mol. Biol. 44: 417–428.

Lam E, Kato N, Lawton M (2001). Programmed Cell Death, mitochondria and the plant hypersensitive response. Nature. 411: 848-853.

Lazebnik YA, Kaufmann SH, Desnoyers S, Poirier GG, Earnshaw WC (1994). Cleavage of poly(ADPribose)polymerase by a proteinase with properties like ICE. Nature. 371: 346–347.

Lin JS, Wang Y, Wang GX (2005). Salt stress-induced programmed cell death via Ca^{2+} -supercript stop-mediated mitochondrial permeability transition in tobacco protoplasts. Plant Growth Regul. 45: 243–250.

Lincoln JE, Richael C, Overduin B, Smith K, Bostock R, Gilchrist DG (2002). Expression of the antiapoptotic baculovirus p35 gene in tomato blocks programmed cell death and provides broad-spectrum

resistance to disease. Proc. Natl. Acad. Sci. USA. 99: 15217-15221.

Liu X, Li P, Widlak P, Zou H, Wo X, Garrard WT, Wang X (1998). The 40-kDasubunit of DNA fragmentation factor induces DNA fragmentation and chromatin condensation during apoptosis. Proc. Natl. Acad. Sci. USA. 95: 8461–8466.

Machida K, Ohta Y, Osada H (2006). Suppression of apoptosis by cyclophilin D via stabilization of hexokinase II mitochondrial binding in cancer cells. J. Biol. Chem. 281: 14314–14320.

Madeo F, Herker E, Maldener C, Wissing S, Lachelt S, Herlan M, Fehr M, Lauber K, Sigrist SJ, Wesselborg S, Frohlich KU (2002). A caspase-related protease regulates apoptosis in yeast. Mol. Cell 9: 911–917.

Majewski N, Noqueira V, Bhaskar P, Coy PE, Skeen JE, Gottlob K, Chandel NS, Thompson CB, Robey RB, Hay N (2004). Hexokinase-mitochondria interaction mediated by Akt is required to inhibit apoptosis in the presence or absence of Bax and Bak. Mol. Cell 16: 819–830.

Martinou J-C (1999). Apoptosis: Key to the mitochondrial gate. Nature. 399: 411–412.

Mashima T, Naito M, and Tsuruo T (1999) Caspase-mediated cleavage of cytoskeletal actin plays a positive role in the process of morphological apoptosis Oncogene. 18: 2423–2430.

Matsumara H, Nirasawa S, Kiba A, Urasaki N, Saitoh H, Ito M, Kawai-Yamada M, Uchimiya H , Terauchi R (2003). Overexpression of Bax inhibitor suppresses the fungal eliciterinduced cell death in rice (Oryza sativa L.) cells. Plant J. 33: 425–434.

McCabe PF, Levine A, Meijer PJ, Tapon NA, Pennell RI (1997). A programmed cell death pathway activated in carrot cells cultured at low cell density. Plant J. 12: 267–280.

Merry DE, Korsmeyer SJ (1997). Bcl-2 gene family in the nervous system. Annu. Rev. Neurosci. 20: 245–267.

Miller LJ, Marx J (1998) Apoptosis. Sci. 281:1301.

Mitsuhara I, Malic AM, Miura M, Ohashi Y (1999). Animal cell death suppressors Bcl-XL and Ced-9 inhibit cell death in tobacco plants. Curr. Biol. 9: 775-778.

Mittler R, Lam E (1997). Characterization of nuclease activities and DNA fragmentation induced upon hypersensitive response cell death and mechanical stress. Plant Mol. Biol. 34: 209–221.

Nakagawa T, Shimizu S, Watanabe T, Yamaguchi O, Otsu K, Yamagata H, Inohara H, Kubo T, Tsujimoto Y (2005). Cyclophilin D-dependent mitochondrial permeability transition regulates some necrotic but not apoptotic cell death. Nature. 434: 652–658.

Nakaune S, Yamada K, Kondo M, Kato T, Tabata S, Nishimura M, Hara-Nishimura I (2005). A novel-type VPE, dVPE, is involved in seed coat formation at the early stage of seed development. Plant Cell 17: 876-887.

Navarre DA, Wolpert TJ (1999). Victorin induction of an apoptotic/senescence-like responses in oats. Plant Cell. 11: 237–249.

Nicholson DW (1999). Caspase structure, proteolytic substrates, and function during apoptotic cell death. Cell Death Differ. 6: 1028-1042.

Nicholson DW, Thornberry NA (1997). Caspases: The killer proteases. Trends Biochem. Sci. 22: 299–306.

O'Brien IEW, Baguley BC, Murray BG, Morris BAM, Ferguson IB (1998). Early stages of the apoptotic pathway in plant cells are reversible. Plant J. 13: 803–814.

O'Farrell M (1995) ADP-ribosylation reactions in plants. Biochimie. 77: 486–491.

Obara K, Kuriyama H, Fukuda H (2001). Direct evidence of active and rapid nuclear degradation triggered by vacuole rupture during progra-mmed cell death in Zinnia. Plant Physiol. 125: 615–626.

Oberhammer F, Wilson JW, Dive C, Morris ID, Hickman JA, Wakeling AE, Walker PR, Sikorska M (1993). Apoptotic death in epithelial cells: Cleavage of DNA to 300 and or 50kb fragments prior to or in the absence of internucleosomal fragmentation. EMBO J. 12: 3679–3684.

Orzaez D, Granell A (1997a). DNA fragmentation is regulated by ethylene during carpel senescence in Pisum sativum. Plant J. 11: 137–144.

Orzaez D, Granell A (1997b). The plant homologue of the defender against apoptotic death gene is down-regulated during senescence of flower petals. FEBS Lett. 404: 275–278.

Pandey S, Walker PR, Sikorska M (1994) Separate pools of endonuclease activity are responsible for internucleosomal and high molecular mass DNA fragmentation during apoptosis. Biochem. Cell Biol. 72: 625–629.

Pastorino JG, Shulga N, Hoek JB (2002). Mitochondrial of hexokinase II inhibits Bax-induced cytochrome c release and apoptosis. J. Biol. Chem. 277: 7610–7618.

Peitsch MC, Polzar B, Stepham H, Crompton T, Mac Donald HR, Mannherz HG. Tshopp J (1993). Characterization of the endogenous deoxyribonuclease involved in nuclear DNA degradation during apotosis (Programmed cell death). Embo J. 12: 371–377.

Pennel RI, Lamb C (1997). Programmed cell death in plants. Plant Cell. 9: 1157–1168.

Qiao I, Mitsuhara I, Yozaki Y, Sakano K, Gotoh Y, Miura M, Ohashi Y (2002). Enhanced resistance to salt, cold & wound stresses by overproduction of animal cell death suppressors Bcl-xl and Ced-9 in tobacco cells-Their possible contribution through improved function of organelle. Plant Cell Physiol. 43: 992-1005.

Rojo E, Martin R, Carter C, Zouhar J, Pan S, Plotnikova J, Jin H, Paneque M, Sanchez-Serrano JJ, Baker B (2004). VPE gamma exhibits a caspase-like activity that contributes to defense against pathogens. Curr. Biol. 14: 1897-1906.

Rudel T (1999). Caspase inhibitors in prevention of apoptosis. Herz. 24: 236–241.

Ryerson DE, Heath MC (1996). Cleavage of nuclear DNA into oligonucleosomal fragments during cell death induced by fungal infection or by abiotic treatments. Plant Cell. 8: 393–402.

Samejima K, Earnshaw WC (2005). Trashing the genome: the role of nucleases during apoptosis. Nat. Rev. Mol. Cell Biol. 6: 677–688.

Sachez P, deTorres Zebala M, Grant M (2000). AtBI-1, a plant homologue of Bax inhibitor-1, suppresses Bax-induced cell death in yeast and is rapidly upregulated during wounding and pathogen challenge. Plant J. 21: 393–399.

Sanmartin M, Jaroszewski L, Raikhel NV, Rojo E (2005). Caspases. Regulating death since the origin of life. Plant Physiol. 137: 841-847.

Saviani EE, Orsi CH, Oliveira JFP, Pinto-Maglio CAF and Salgado I (2002). Participation of the mitochondrial permeability transition pore in nitric oxide-induced plant cell death. FEBS Lett. 510: 136–140.

Schussler EE, Longstreth DL (2000). Changes in cell structure during the formation of root aerenchyma in Sagittaria lancifolia (Alismataceae). Amer. J. Bot. 87: 12-19.

Shah NK, Taneja TK, Hasnain SE (2000). Mitochondria can power cells to life and death. Resonance. 5: 74–84.

Shimizu S, Ide T, Yanagida T, TsujimotoY (2000). Electrophysiological study of a novel large pore formed by Bax and the voltage-dependent anion channel that is permeable to cytochrome c. J. Biol. Chem. 27: 12321-12325.

Shimizu S, Narita M, Tsujimoto Y (1999). Bcl-2 family proteins regulate the release of apoptogenic cytochrome c by the mitochondrial channel VDAC. Nature. 399: 483–487.

Simeonova E, Sikora A, Charzynska M, Mostowska A (2000). Aspects of programmed cell death during leaf senescence of mono- and dicotyledonous plants. Protoplasma. 214: 93-101.

Smertenko AP, Bozhkov PV, Filonova L, von Arnold S, Hussey PJ (2003). Re-organisation of the cytoskeleton during developmental programmed cell death in Picea abies embryos Plant J. 33: 813–824.

Stein JC, Hansen G (1999). Mannose induces an endonuclease responsible for DNA laddering in plant cells. Plant Physiol. 121: 71–79.

Stennicke HR, Salvesen GS (1998). Properties of the caspases.Biochim. Biophys. Acta. 1387: 17-31.

Suarez M, Filonova L, Smertenko A, Savenkov E, Clapham D, von Arnold S, Zhivotovsky B, Bozhkov P (2004). Metacaspase-dependent programmed cell death is essential for plant embryogenesis. Curr. Biol. 14: R339–R340.

Sugimoto A, Hozak RR, Nakashima T, Nishimoto T, Rothman JH (1995). Dad-1, an endogenous programmed cell death suppressor in Caenorhabditis elegans and vertebrates. EMBO J. 14: 4434–4441.

Sugiyama M, Ito J, Aoyagi S, Fukuda H (2004). Endonucleases. Plant Mol. Biol. 44: 387–397.

Sugiyama T, Shimizu S, Matsuoka Y, Yoneda Y, Tsujimoto Y (2002). Activation of mitochondrial voltage-dependent anion channel by a proapoptotic BH3-only protein Bim. Oncogene. 21: 4944–4956.

Sun YL, Zhao Y, Hong X, Zhai ZH (1999). Cytochrome c release and caspase activation during menadione induced apoptosis in plants. FEBS Lett. 462: 317–321.

Swidzinski JA, Leaver CJ, Sweetlove LJ (2004). A proteomic analysis of plant programmed cell death. Phytochem. 65: 1829–1838.

Tada Y, Hata S, Takata Y, Nakayashiki H, Tosa Y , Mayama S (2001). Induction and signaling of an apoptotic typified by DNA laddering in the defence response of response oats to infection and elicitors. Mol. Plant Microbe Interact. 14: 477–486.

Tanaka Y, Makishima T, Sasabe M, Ichinose Y, Shiraishi T, Nishimoto T, Yamada T (1997). Dad-1: A putative programmed cell death suppressor gene in rice. Plant Cell Physiol. 38: 379–383.

Thornberry NA, Lazebnik Y (1998). Caspases enemies within. Science. 281: 1312–1316.

Tian RH, Zhang GY, Yan CH, Dai YR (2000). Involvement of poly (ADP-ribose) polymerase and activation of caspase-3-like protease in heat shock-induced apoptosis in tobacco suspension cells. FEBS Lett. 474: 11–15.

Timmer JC, Salvesen GS (2007). Caspase substrates. Cell Death Differ. 14: 66–72.

Tiwari BS, Belenghi B, Levine A (2002). Oxidative stress increased respiration and generation of reactive oxygen species, resulting in ATP depletion, opening of mitochondrial permeability transition, and programmed cell death. Plant Physiol. 128: 1271–1281.

Uren AG, O'Rourke K, Aravind L, Pissabaro MT, Sheshagiri S, Koonin EV, Dixit VM (2000). Identification of paracaspases and metacaspases: Two ancient families of caspase-like proteins, one of which plays a key role in MALT Lymphoma. Mol. Cell. 6: 961–967.

Van der Biezen EA, Jones JDG (1998). The NB-ARC domain: a novel signalling motif shared by plant resistance gene products and regulators of cell death in animals. Curr. Biol. 8: 226–227.

van Gurp M, Festjens N, van Loo G, Saelens X, Vandenabeele P (2003). Mitochondrial intermembrane proteins in cell death. Biochem. Biophys. Res. Commun. 304: 487–497.

Vaux DL, Korsmeyer SJ (1999). Cell death in development. Cell. 96: 245-254.

Vercammen D, Belenghi B, van de Cotte B, Beunens T, Gavigan JA, De Rycke R, Brackenier A, Inzé D, Harris JL, Van Breusegem F (2006). Serpin1 of Arabidopsis thaliana is a suicide inhibitor for metacaspase 9. J. Mol. Biol. 364: 625–636.

Vercammen D, van de Cotte B, De Jaeger G, Eeckhout D, Casteels P, Vandepoele K, Vandenberghe I, Van Beeumen J, Inzé D, Van Breusegem F (2004). Type II metacaspases Atmc4 and Atmc9 of Arabidopsis thaliana cleave substrates after arginine and lysine. J. Biol. Chem. 279: 45329–45336.

Wang H, Li J, Bostock RM, Gilchrist DG (1996b). The apoptosis: a functional paradigm for programmed plant cell death induced by a host-selective phytotoxin and invoked during development. Plant Cell. 8: 375–391.

Wang M, Hoekstra S, van Bergen S, Lamers GEM, Oppedijk BJ, van der Heijden MW, de Priester W, Schilperoort RA (1999). Apoptosis in developing anthers and the role of ABA in this process during androgenesis in Hordeum vulgare L. Plant Mol. Biol. 39: 489–501.

Wang M, Oppedijk BJ, Lu X, van Duijn B, Schilperoort RA (1996a). Apoptosis in barley aleurone during germination and its inhibition by abscisic acid. Plant Mol. Biol. 32: 1125–1134.

Wang M-B, Waterhouse PM (2002) Application of gene silencing in plants. Curr. Opin. Plant Biol. 5: 146–150.

Watanabe N, Lam E (2006). Arabidopsis Bax inhibitor-1 functions as anattenuator of biotic and abiotic types of cell death. Plant J. 45: 884–894.

Wertz IE, Hanley MR (1996). Diverse molecular provocation of programmed cell death. Trends Biochem. Sci. 21: 359–364.

White E (1996) Life, death, and the pursuit of apoptosis. Genes Dev. 10: 1–15.

White K, Grether ME, Abrams JM, Young L, Farrell K, Steller H (1994). Genetic control of programmed cell death in Drosophila. Science. 264: 677-683.

Wilson JE (2003). Isozymes of mammalian hexokinase: Structure, subcellular localization and metabolic function. J. Exp. Biol. 206: 2049–2057.

Wilson KP, Black JAF, Thomson JA., Kim EE, Griffith JP, Navia MA, Murcho MA, Chambers SP, Aldape RA, Raybuck SA (1994). Structure and mechanism of interleukin-1 beta converting enzyme. Nature. 370: 270-274.

Woltering EJ, van der Bent A, Hoeberichts FA (2002). Do plant caspases exist? Plant Physiol. 130: 1764-1769.

Wyllie AH (1980). Glucocorticoid-induced thymocyte apoptosis isassociated with endogenous endonuclease activation. Nature 284: 555–556.

Xie Z, Chen Z (2000). Harpin-induced hypersensitive cell death is associated with altered mitochondrial functions in tobacco cells. Mol. Plant Microbe Interact. 13: 183–190.

Xu Q, Reed JC (1998). Bax inhibitor-1, a mammalian apoptosis suppressor identified by functional screening in yeast. Mol. Cell. 1: 337-346.

Xu Y, Hanson MR (2000). Programmed cell death during pollination-induced petal senescence in petunia. Plant Physiol. 122: 1323-1333.

Yamada K, Shimada T, Nishimura M, Hara-Nishimura I (2005). A VPE family supporting various vacuolar functions in plants. Physiol. Plant. Special Issue 123: 369-375.

Yamada T, Marubashi W, Niwa M (2000). Apoptotic cell death induces temperature-sensitive lethality in hybrid seedlings and calli derived from the cross of Nicotiana suaveolens x N. tabacum. Planta. 211: 614-622.

Yao N, Eisfelder BJ, Marvin J, Greenberg JT (2004). The mitochondrion—an organelle commonly involved in programmed cell death in Arabidopsis thaliana. Plant J. 40: 596–610.

Yen CH, Yang CH (1998). Evidence for programmed cell death during leaf senescence in plants, Plant Cell Physiol. 39: 922–927.

Yokota S, Okabayashi T, Yokosawa N, Fuji N (2004). Growth arrest of epithelial cells during measles virus infection is caused by upregulation of interferon regulatory factor 1. J. Virol. 78: 4591–4598.

Young TE, Gallie DR (1999). Analysis of programmed cell death in wheat endosperm reveals differences in endosperm development between cereals. Plant Mol. Biol. 39: 915–926.

Young TE, Gallie DR (2000). Regulation of programmed cell death in maize endosperm by abscisic acid. Plant Mol. Biol. 42: 397–414.

Young TE, Gallie DR, Demason DA (1997). Ethylene mediated programmed cell death during maize endosperm development of wild-type and shrunken2 genotypes, Plant Physiol. 115: 737–751.

Yu L-H, Kawai-Yamada M, Naito M, Watanabe K, Reed JC, Uchimiyaa H (2002). Induction of mammalian cell death by a plant Bax inhibitor. FEBS Lett. 512: 308-312.

Yuan JY, Shaham S, Ledoux S, Ellis HM , Horvitz HR (1993). The C. elegans Cell Death Gene ced-3 Encodes a Protein Similar to Mammalian Interleukin-1β-Converting Enzyme. Cell. 75: 641–652.

Zoratti M, Szabo I (1995). The mitochondrial permeability transition. Biochim. Biophys. Acta 1241: 139– 176.

10

The Challenge of Mosquito Control Strategies: from Primordial to Molecular Approaches

Subbiah POOPATHI* and Brij Kishore TYAGI

Centre for Research in Medical Entomology (Indian Council of Medical Research), Ministry of Health and Family Welfare, Govt of India, Chinna Chokkikulam, Madurai- 625002, Tamil Nadu, India.

Mosquito control programs worldwide have been evaluating the feasibility to implement biological control strategies by using *Bacillus sphaericus* (*Bs*) and/or *B. thuringiensis* serovar *israelensis* (*Bti*). A comprehensive review is presented here to assess the potentiality of biological control agents in mosquito control operation. Vector control is primordial and very essential means for controlling transmission of filariasis, malaria, Japanese encephalitis and dengue in human society. Over the last few decades, there is growing realization that alternate methods to synthetic chemical control needs to be studied and perfected. In the last decade the bacilli based mosquito larvicides popularly known as biological larvicides are becoming more popular in vector management program the world over. The toxicity to mosquito larvae is due to crystal toxins encoded by specific genes. The major advantages of these biolarvicides are reduced application cost, safety to environment, human beings, animals and other non-target organisms. This special review paper explores the importance of bacterial toxin in controlling vector mosquitoes and the tactics for managing resistance to the mosquitocidal bacteria which include rotating different mosquitocidal strains and using genetic engineering to produce new combinations of toxins. This paper also provides a focus on continuous research toward identification of novel mosquitocidal toxins suitable for use if resistance to existing toxins.

Key word: Vector control, synthetic chemicals, biopesticides, mode of action, resistance, bioengineering.

Table of Contents

1.0   INTRODUCTION

Mosquitoes transmit some of the world's worst life threatening and debilitating parasitic and viral diseases including malaria (*Anopheles*), filariasis (*Culex, Mansonia*

---
*Correspondence author. E-mail:
Subbiahpoopathi@rediffmail.com;
Telegram: MOSQUITO.

Present address: Division of Microbiology and Immunology,

Vector Control Research Centre (Indian Council of Medical Research), Ministry of Health and Family Welfare, Govt of India, Medical Complex, Indira Nagar, Pondicherry–605 006, India.

and some *Anopheles* spp.,) and dengue fever (principally *Aedes aegypti*). These diseases are on the rise in many tropical and subtropical areas (WHO 1999). Therefore, approaches to reduce the incidence of vector-borne diseases by controlling mosquito populations are highly warranted. In this review paper, efforts have been made to assess the development of mosquito control strategies switching over from synthetic pesticides, use of bio-control agents (entomophagous bacteria, fungi, microsporidians, predators and parasites) to genetic engineering techniques against disease transmitting mosquito vectors. Vector control is an essential and effective means for controlling transmission of vector-borne diseases, especially in areas where resistance in parasite to drugs is growing. Unlike insecticides, bio-control agents are host specific, safer to the environment, find easy application in the field, are cost-effective in production, lack infectivity and pathogenicity in mammals including man and has little evidence of resistance development in target mosquito species. Many populations of mosquito vectors of diseases have developed resistance to synthetic organic insecticides, used mostly during the last half of the 20th century. Thus interest in alternate strategies as well as in integrated control grew increased (WHO 1982a,b, 1986). Since single method of control is ineffective to produce the desired results, emphasis should be laid on the comprehensive mosquito control strategies including the use of insecticides, bio-control agents and environmental management. This present paper reviews the primordial and current biotechnological approaches to tackle the rising emergence of mosquito vectors.

2.0   VECTOR CONTROL STRATEGIES

I.   Chemical- based control measures
II.   Non-chemical based control measures
III. Biological control agents

2.1   Chemical-based control measures

2.1.1   Insecticide-impregnated paint

This technology has been in use for decades in developed countries. The Vernacide is an example of insecticide-impregnated paint effective against mosquitoes (*Culex quinquefasciatus*) (Das et al., 1986a). This paint is safer than conventional water dispersible powder (WDP) formulation of adulticides, and can be conveniently employed in public places where pest free conditions are desirable for considerably longer time.

2.1.2   Insecticide impregnated bednets

Insecticide-impregnated fabrics have been used effectively against mosquito-bites (Jambulingam et al., 1989). The impregnated synthetic pyrethroid insecticide (deltamethrin) ropes, bed nets and curtains in the human dwellings were found to be promising against *Anopheles*

and *Culex* species (Sharma et al., 1989; Poopathi and Rao 1995, Poopathi 2001a). Mosquito nets treated with a water-dispersible tablet formulation of deltamethrin (K-O TAB) was evaluated against malaria vectors and found to be effective (Bhatt et al., 2005). Olyset nets are permethrin insecticide impregnated bednets, recommended by WHO, and are currently are in use in rural malaria endemic areas (Tami et al., 2004).

2.2   Non-chemical based control measures

2.2.1   Surface active agents (SAA)

These nonionic, biologically degradable chemicals on application to mosquito breeding habitats spread spontaneously to form a monomolecular film on the water surface and exhibits mortality of mosquito larvae. Arosurf as SAA has been used effectively to control *Culex quinquefasciatus, Anopheles stephensi* and *Aedes aegypti* in different breeding habitats (cesspits, cesspools, drains and wells) (Das et al., 1986b). Arosurf in combination with fast acting and residual larvicide (fenthion) enables better coverage of breeding habitats and long effective life of the film in inaccessible habitats like marshes and lagoons.

2.2.2   Expanded polystyrene beads (EPS)

A synthetic non-biodegradable product of very light thermocol beads can remain effective on treated water surface. Curtis and Minjas (1985) and Sharma et al. (1985) have shown tremendous potential of the EPS in abating breeding of *Cx. quinquefasciatus* and *An. stephensi*, in latrine pits in Tanzania and in overhead tanks in India, respectively. These beads are non-polluting and socially acceptable. Its long-life, non-hazardous nature and one time application renders it cost-effective and feasible.

2.2.3   Insect growth regulators (IGRs)

Insect growth regulators (IGRs), the third generation insecticides, are diverse groups of chemical compounds that are highly active against larvae of mosquitoes and other insects. The IGRs in general have a good margin of safety to most non-target biota including invertebrates, fish, birds and other wildlife. They are also relatively safe to man and domestic animals. The IGR compounds do not induce quick mortality in the pre-imaginal stages treated and occur many days post treatment. This is indeed a desirable feature of a control agent because larvae of mosquitoes and other vectors are an important source of food for fish and wildlife. On account of these advantages of IGRs and the high level of activity against

Table 1. Advantages of Mosquitocidal Bacteria and Synthetic Insecticides

Mosquitocidal Bacteria

- Safe for human and animals
- Safe to handle
- Persistence in environment is less
- Resistance development is very slow process
- Attributes toxicity very specifically to target organism.
- Recycling potential is higher
- Can able to withstand polluted water as well as high
- Toxicity level id very low
- Less intensive toxicological testing is required

Synthetic Insecticides

- Rapidly killing the mosquitoes
- Wide range of mosquito species controlled by these insecticides
- Chemicals remain to be there in mosquito larval feeding zone
- Chemicals not rapidly inactivated by UV-light
- Few synthetic chemicals eventually degrade to harmless products
- Effective under varied environmental conditions

Porter et al., (1993)

target species, it is likely that IGRs could play an important role in vector control programs in the future (Mulla, 1995). They are more specific for mosquitoes than conventional insecticides. The IGRs interfere with the hormonal mechanisms of target organisms and result in various kinds of morphological, anatomical and physiological abnormalities so that the target species does not reach the final stage of development (Amalraj et al., 1988a). There is no likelihood of resistance development against these IGRs. A large number of IGRs, both juvenoids and chitin synthesis inhibitors, have been evaluated for the vector control but only very few of these are found effective and commercially feasible, e.g., diflubenzuron, methoprene, fenoxycarb (Tyagi et al., 1985, 1987; Amalraj et al., 1988b, Vasuki, 1988). The IGRs have an added advantage of being used at a relatively very low dose compared to the conventional insecticides. Ecdysone agonists are hormonally active insect growth regulators that disrupt development of larvae and are found to be active against *Ae. aegypti, An. gambiae,* and *Cx.quinquefasciatus* (Beckage et al., 2004).

2.2.4.   Genetic manipulation (SIT)

Potential applications for reducing transmission of mosquito-borne diseases by releasing genetically modified mosquitoes have undergone extensive research and mosquitoes are being created with such an application in mind in several laboratories (Boete, 2003;

Cattevuccia et al., 2003). The use of the sterile insect technique (SIT) provides a safe programme in which production, release and mating competitiveness questions related to mass-reared genetically modified mosquitoes could be answered. It also provides a reversible effect that would be difficult to accomplish with gene introgression approaches. SIT using transgenic material would provide an essentially safe and efficacious foundation for other possible approaches that are more ambitious (Benedict and Robinson 2003). Apart from the use of SIT, other approaches of developing genetically modified mosquitoes that are refractory to malaria parasite development include mapping of QTLs associated with refractoriness, use of RNAi technology to identify genes and gene pathways that can be perturbed to reduce vector's potential in pathogen transmission etc etc.

3.0   Biological control agents

Since the past few decades, there is growing realization for alternate biocontrol agents that are currently being used in controlling vector larval populations in different breeding habitats. The advantages of the biological control agents when compared with synthetic chemicals are summarized in Table 1.

3.1   Bacterial agents

In the last decade, the bacilli based mosquito larvicides popularly known as biocides or biolarvicides are becoming popular in vector control (Porter et al., 1993). Many commercial formulations are available and can be used in large-scale mosquito control operations. The major advantages of biolarvicides are: reduced application costs and safety to environment, human beings, animals and other non-target organisms. *Bacillus thuringiensis* subsp. *israelensis* (*Bti*) and *Bacillus sphaericus* (*Bs*) are entomopathogenic bacteria. Both *Bs* and *Bti* are spore forming bacteria producing upon sporulation a parasporal crystal toxic to some invertebrates, mostly insects and nematodes (Feitelson et al., 1992). The parasporal inclusion body is composed of proteins or ∂-endotoxins, varying in quantity and type depending on the strain. Each type of crystal protein is characterized by a specific host range, and based upon differences in sequence and specificity, insecticidal crystal ∂-endotoxins have classified into several groups of proteins designated *Cry* (Crickmore et al., 1995; Hofte and Whiteley 1989). These entomopathogenic bacteria are the most important biopesticides sold worldwide (Bernhard and Utz 1993) and its share of the world market of pesticides is expected to rise in the coming years. *Bt* based products are, however, limited with respect to the diversity of strains used in commercial products, and more toxins are needed to target other insect pests and to manage the emerging problem of insect resistance (Ferre et al., 1995). Large screening programs, leading to important collections of isolates, have been conducted. The need for novel crystal proteins has prompted the development of molecular approaches to quickly and easily characterize toxin genes present in *Bt* isolates. In the last few years, several PCR based methodologies, mostly multiplex PCR, which allowed the accurate determination of families of *Cry* genes (Carozzi et al., 1991) or specific ∂-endotoxin have been proposed (Bourque et al., 1993; Ceron et al., 1994). Although powerful PCR approaches are limited to the detection of already known genes and fail to detect and identify novel *Cry* genes even though various strategies have been proposed to increase their efficiency (Kuo and Chak 1996). *Bacillus sphaericus* strains have been classified into five groups, based on DNA homology, with group 11 further subdivided into 11A and 11B, and all the toxic strains are in group 11A (Krych et al., 1980). The toxic strains have been further subdivided into low toxicity strains and high toxicity strains. *Bs* SS11 is a low toxicity strain (Singer 1973) in which toxicity is produced initially in the vegetative phase of growth before the onset of sporulation (Myers et al., 1979), but the toxicity was found to be markedly unstable (Myer and Yousten 1978). In contrast to strain SS11, the high toxicity strains of *Bs*, for example, strains 1593, 2362 (Weiser 1984), and 2297 (Wickremesinghe and Mendis 1980), develop relatively stable, high toxicity at the onset of sporulation (Broadwell and Baumann 1986). The high toxicity strains of *Bs* have been shown to produce binary toxins with protein components of 41.9 and 51.4 kDa, and both these proteins are required to kill

mosquito larvae (Broadwell et al., 1990). A gene encoding a 100 kDa toxin designated *Mtx* from the low toxicity strain *Bs* SS11-1 has been cloned and sequenced (Thanabalu et al., 1992). Genes encoding the binary toxin are distributed among the high-toxicity strains, while the gene encoding the 100 kDa toxin is widely distributed among both the low and high toxicity strains (Thanabalu et al., 1991). Thus the low toxicity of *Bs* SS11-1 is not due to a low specific activity of the 100 kda toxin but could be due to poor expression or a low of stability of the toxin during sporulation in this strain. As the *Mtx* promotor resembles vegetative promoters, the present study was undertaken to test the idea that *Mtx* synthesized during vegetative growth of *Bs* is degraded by proteases during vegetative growth of *Bs* is degraded by proteases during sporulation. Several lines of evidence confirmed this notion, and unexpectedly, strains in two groups of *Bs* which do not produce protease activity against *Mtx* have been identified. Efficient expression of the cloned *Mtx* gene was obtained in both vegetative cells and sporulated cultures of *Bs* 1693, suggesting that this strain is worth investigating as a vehicle for delivering *Mtx* to mosquito larvae.

Bti is toxic to both mosquito and blackfly whereas *Bs* only to mosquitoes. Extensive field trials have been conducted with *Bti* using powder/ liquid formulation and comparatively fewer number of field-tests with *Bs* (WHO 1979). During sporulation, both produce crystals that are toxic to dipteran larvae after ingestion. Since the discovery of *Bti* in 1977 (Goldberg and Margalit, 1977), other bacteria have also been found to be highly toxic to mosquito larvae, including *Bs* (Myers and Yousten, 1980; Wickremesinghe and Mendis, 1980) and more recently, *Clostridium bifermentans* serovar *malaysia* (de Marjac et al., 1987). Among the various species of *Bt*, few strains are as toxic as *Bti* (Tabashnik et al., 1997).

Moreover, cloning and expression of the toxic genes of these organisms in other environment friendly bacteria have also made them important for further investigation. Recent advances with novel types of recombinant microorganisms with new cloning strategies and cloning the toxin genes under strong promotor for expression together with *in vitro* gene manipulation and site directed mutagenesis of the active sites for increased toxicity, have the potential to provide more effective control of mosquitoes by exploiting these two bacteria. But the toxins of *Bs* and *Bti* in particular, do not persist long in nature and require frequent application, which is a limiting factor for these organisms to be most successful and potent biolarvicide. Nevertheless, they are by far the best choice for controlling mosquitoes (Riehle and Lorena 2005).

3.2 Synergistic activity of bacteria–a new approach to control mosquitoes

B. thuringiensis strains pathogenic to insects produce two distinct types of toxin proteins, *Cry* and *Cyt* proteins (Crickmore et al., 1995). Generally, the genes encoding these proteins are located on large plasmids, and the proteins sporulation. More than 100 different Cry genes have sporulation. More than 100 different Cry genes have been identified and sequenced, and significant homologies among the amino acid sequences of this group, in combination with experimental studies, suggest they have a common mode of action, colloid-osmotic lysis (Crickmore et al., 1998). Cyt toxin, however have only been found in Bt strains that are mosquitocidal, have amino acid sequences that are unrelated to those of the Cry toxins, and can lyse a variety of cell types in vitro (Guerchicoff et al., 1997). The mode of action of Cry toxins has not been fully elucidated. However, research suggests that Cyt toxins are also involved in colloid-osmotic lysis (Hofte and Whitely 1989), but may differ in the mechanism by which lesions are formed in the cell membrane (Butko et al., 1996, 1997). One of the most interesting features of the Cry and Cyt toxin combination that is found in Bti, the susceptible upon which many commercial mosquito larvicides are based, is the effect of this combination on toxicity. This subspecies produces a crystalline parasporal inclusion containing four major toxic proteins, Cry4A (134 kDa), Cry4B (128 kDa), Cry11A (66 kDa), and CytA (27 kDa). These four proteins are assembled into separate inclusions that are enveloped together to form parasporal body. The proportion of each protein in the parasporal body, based on SDS-PAGE and electron microscopy (Ibarra and Federici 1986) is approximately 40 % Cyt A and 20 % for each of the three Cyt proteins. Studies of these proteins revealed that the toxicity of individual proteins in Bti was much less than that of the intact parasporal crystal. Subsequently, it was shown that the Cry toxins in Bti interact synergistically with the Cyt1A toxin, as well with each other, to produce this high level of activity (Poncet et al., 1994; Crickmore et al., 1995). This synergism has also been shown to be important in the relatively low rate of resistance development toward Bti in Culex mosquitoes (Georghiou and Wirth 1997) and can suppress high levels of resistance to Cry4 and Cyt11 toxins (Wirth et al., 1997). The mechanism of this synergism is not understood, but we have postulated that Cyt1A aids these toxins in binding to or inserting into the mosquito microvillar membrane (Wirth et al., 1998). The synergistic capacity of Cyt1A was extended by the recent observation that sublethal concentrations of Cyt1A combined with Bs, an unrelated mosquitocidal bacterium which does not have Cry type toxins, were synergistic and toxic towards highly resistant Culex quinquefasciatus (Wirth et al., 1997). Because, this combination was synergistic against resistant mosquitoes that have lost the capacity to bind Bs toxins (Nielsen-LeRoux et al., 1997). This same mixture might be synergistic toward a mosquito species that is naturally insensitive to Bs.

3.3   Laboratory and Field Evaluation

Both Bti and Bs respectively, have shown great promise, particularly those breeding in highly polluted water. The formulations of Bs effective against mosquito larvae are DP1593, 2362; Dulmage 2362, 2297; Abbot: DP1593: Strains B6, B64: Vectolex (G); Microgel 1593M; Biocide-SABG 6185; Vectolex (G); Spherix etc. The formulations of Bti effective against vector mosquitoes are Deltox (briquettes); Bactimos (WP); Teknar (4 formulations); dust preparation Bactimos (flowable concentrate), Vectobac (granule), Bti (ultra low volume), Vectobac 12 aqueous solution etc. These formulations would lose efficacy in cold climatic conditions and is useful in warm months or tropical climatic conditions for most of the year. Efficacy of various other formulations of Bti in the form of tablet, granule and wettable powder against different disease vector species in urban and rural areas were tested and found to be convincing.

While Bti is now commercially available as Deltox, Bactimos and Teknar in different formulations (e.g., liquid concentrate and water dispersible powder), Bs is still undergoing the marketing process and is expected to be effective against all larval stages of Anopheles, Culex and Aedes mosquitoes. These bacterial species although pathogenic to the target vector species, are non-hazardous to other beneficial and non-target organisms.

Microbial mosquito larvicides may also suffer the same fate in inducing resistance development as synthetic insecticides. For resistance management, Bti would be a good candidate to manage Bs resistance, because there is a complete lack of cross-resistance between Bti and Bs toxins. (Rao et al., 1995, Rodcharoen and Mulla, 1996; Wirth et al., 2000a,b). It is a well known fact that Bti and Bs toxins bind to different classes of specific receptors on the surface of midgut brush border membranes (Nielsen-LeRoux and Charles 1992) which could explain the lack of cross-resistance between Bti and Bs. Using Bti alone or in rotation or mixture with Bs would be a logical choice to restore Bs susceptibility in Bs resistant mosquitoes.

3.4   Mode of action

Bti produces a proteinaceous parasporal crystalline inclusion during sporulation. Upon ingestion by insects, this crystalline inclusion is solubilized in the midgut, releasing proteins called delta-endotoxins. These proteins (protoxins) are activated by midgut proteases and the activated toxins interact with the larval midgut epithelium causing a disruption in membrane integrity and ultimately leading to insect death (Poopathi et al., 1999). Exposure of C. quinquefasciatus to Bti over several generations will likely result in the selection of populations with lower sensitivity to the crystal mosquitocidal toxin.

Exposure of insect populations to single toxins, after prior exposure to complex mixtures of toxins can result in the rapid increase of resistance. To delay onset of

Table 2. Mosquitocidal bacterial strains, their toxins and host range.

Bacterial strains of bacterial toxins	Molecular weight range of bacterium	Mosquito host	Reference
Bacillus thuringiensis subsp. *israelensis* (Bti	134, 128, 78, 72, 27	*Aedes aegypti, Culex Anopheles*	Porter et al., (1993)
Bacillus thuringiensis subsp.*kurstaki* (Btk)	65	*Aedes aegypti*	- do -
Bacillus thuringiensis subsp. *jegathesan* (Btj)	81, 70, 65, 37, 26, 16	*Anopheles stephensi Culex pipiens Aedes aegypti*	Delecluse et al., (1995)
Bacillus sphaericus (Bs)	100, 51, 42, 32	*Culex, Anopheles, Aedes aegypti*	Baumann et al., (1991)
Clostridium bifermentans ser. *Malaysia*	66, 18, 16	*Anopheles, Aedes detritus, Aedes caspius Aedes aegypti*	Thiery et al., (1992)

resistance, *Bti* formulations must contain toxins that interact either with multiple receptor sites or that have different modes of action. Ingestion of *Bs* protein causes destruction of gut epithelial cells in the larvae of certain mosquito species, resulting in the death of insect. In general, the larvae of genus *Culex* are especially susceptible to this toxin, those of *Anopheles* spp are moderately susceptible and larvae of *Aedes* spp. are quite resistant. The larvicidal properties of the crystal have made *Bs* a useful agent for the biological control of mosquitoes. But there are some reports of resistance to *Bs* from field in France (Sinegre et al., 1994) Brazil (Regis et al., 1995) and India. (Rao et al., 1995, Poopathi et al., 2001, 2002, 2003a, 2003b).

The nature of change involved in resistance to *Bs* is unknown but a mutation affecting the synthesis of the target receptor molecule or inducing a change in its conformation, leading to non-functionality may be responsible. However, mosquito populations resistant to *Bs* remain susceptible to *Bti* toxins indicating that different receptors are involved (Georghiou et al., 1992). Some laboratory selected resistance to *Bt* to Indian meal moth and also cases of field selected resistance to diamondback moth in Philippines and Hawaii has been reported. Laboratory and field-selected resistance may be due to different factors. Laboratory development of resistance is more likely to involve polygenes since they are prone to be selected under conditions where biological and environmental stress factors are minimized (Georghiou et al., 1992). In contrast, development of resistance in the field is more likely to involve single major genes, which can be selected from a much wider genetic pool under stressful conditions (Georghiou and Wirth 1997).

The mechanisms of resistance to *Bs* are not yet defined (Rao et al., 1995; Rodcharoen and Mulla, 1996) but more than one mechanism seems to be involved. *Bs* is at high risk for selecting resistance in mosquito populations, because its binary toxin apparently binds to a single receptor type on midgut microvilli. A potential key strategy for delaying resistance to insecticidal proteins is to use mixtures of toxins that act at different targets within the insect, especially mixtures that interact synergistically (Wirth et al., 2005). Resistance to *B.ti* has resulted from reduced binding of the toxin to the brush border in the lumen of the insect gut (Escriche et al., 1995, Tabashnik et al., 1996) or by enhanced digestion of toxin by gut proteases (Keller et al., 1996). The six different toxin types in the *Bti* strain used for vector control were expected to retard or prevent development of a comprehensive resistance mechanism; however multi toxin resistance to *Bti* has already appeared.

3.5   Other mosquitocidal *Bt* strains

Highly mosquitocidal strains were found in various serotypes such as *B.t.canadensis, B.t.thompsoni, B.t.malaysiensis,* and *B.t.jegathesan* showing the diversity of mosquitocidal strains within *Bt* species. Both *B.t.medellin* and *B.t.jegathesan* appear good candidates for further characterization and investigation for use as microbial pesticides (Ragni et al., 1996). Polypeptides responsible for the toxicity of these strains are being identified through cloning and expression of corresponding genes (Delecluse et al., 1995). A summary of the mosquitocidal toxins produced from various entomopathogenic bacteria is given in Table 2.

3.6   Slow release formulations of bacterial agents

The formulation of bacterial agents has two important problems. One is that, they do not remain in the larval feeding zone for a longer period and thus are not effective for long duration. Also, the spores do not replicate to sufficient extent, thus, powder /liquid formulations have to be applied frequently. Hence slow release formulations were developed and were found to be active for a longer duration. But larger quantities of larvicidal material are required for use in operational programmes, which affects logistics and cost.

These formulations now commercially available in different forms have invariably a biodegradable base carrier, which is imbibed with a relatively high concentration of some larvicide so that when placed in the mosquito breeding habitat the base carrier spontaneously biodegrades releasing the larvicide slowly in desired low, yet toxic, concentration. Such a method is particularly very desirably useful in highly polluted media where repeated spraying of the conventional larvicide (e.g., fenthion) is highly cost prohibitive and time consuming *C.quinquefasciatus and C.tritaeniorhynchus* can be significantly brought under check by this methodology. These formulations are required to be sprayed on weekly basis to bring about good reduction in the larval density.

VectoBac, a tablet formulation of *Bti* and zeolite granules of Temephos (ZG) were found to be effective as larvicides of *Ae. aegypti* in Thailand (Mulla 1995). Ice granules containing endotoxins of microbial agents for the control of mosquito larvae was found to be a new application technique against *Aedes sp.* Solutions containing powder formulations of *Bti* or *Bs* were transformed into ice pellets using a special ice-making machine (Becker 2000). This new technique was demonstrated to have the following advantages over *Bti* sand granules: (1) the *Bti* ice pellets melted on the water surface and released the microbial crystals there, (2) the control agent remained inside the ice pellets during the application and were not lost by friction in the spraying equipment, and (3) the ice formulation resulted in increased swath widths, significantly reducing the cost of application (Becker, 2000). The efficacy of new water-dispersible granular (WDG) formulations of *Bti* (VectoBac) and *Bs* (VectoLex) was tried against malaria vectors and also found to be good. The effectiveness of larvicidal treatment by AQUABAC Biolarvicide (*Bti*), Biological Larvicide Aqueous Suspension) against *Cx. quinquefasciatus* in slow-flowing water ditches has been evaluated in USA and found to be effective. (Xue and Doyle 2005). The bioefficacy and residual activity of *Bti* H-14 (water-dispersible granules of VectoBac ABG 6511 and liquid formulations of VectoBac 12AS) and pyriproxyfen (insect growth regulator) for control of larvae of *Ae. aegypti* and *Ae. albopictus* were found to be effective in Malaysia (Lee et al., 2005).

3.7 Combination strategies

Prolonged efficacy of a combination of bacteria (Bti) and copepods (*Mesocyclops aspericornis*) in controlling immature forms of *Ae. aegypti* in peridomestic water containers was achieved in Thailand. Studies were carried out on the bioefficacy and residual activity of Bti

H-14 and pyriproxyfen as direct applications for control of larvae of *Ae aegypti* and *Ae albopictus* (Lee et al., 2005). Persistence, wash-resistance, and shelf life of mosquito nets treated with a water-dispersible tablet formulation of synthetic pyrethroid insecticide deltamethrin (K-O TAB) was evaluated against malaria vectors in India (Bhatt et al., 2005). Efficacy of neem oil towards *Anopheline* larvae was tested and found that application of chlorpyriphos-methyl/fenitrothion and neem oil resulted in dramatic reduction in larval density (Awad and Shimaila 2003). Dengue control measures conducted in China was based on integrated methods which included release of mosquito larvivorous fish in the drinking water storage facilities, application of larvicides to the water storage facilities in vegetable gardens, removal of discarded and unused containers and tires, improvement of household water storage facilities and increase of potable water supply through community participation which reduced considerably the dengue population levels (Wang et al., 2000).

Malaria and Japanese encephalitis are the two most serious human diseases transmitted by riceland mosquitoes, but they have been incriminated as vectors of dozens of arboviruses and other parasites and pathogens including the causal agents of West Nile and Rift Valley Fevers and lymphatic filariasis. The integrated pest management (IPM) strategy for mosquito control, also known as integrated vector control (IVC), is an ecologically based approach that may involve several complementary interventions used in combination or singly. Environmental management and chemical, biological and mechanical control, comprise the elements of IVC proposed for use in or near riceland habitats. Some of the elements of environmental management include the use of intermittent irrigation; flushing of fields; use of rice cultivars that require less water; shifting of planting schedules to avoid optimal mosquito breeding conditions; relocation of communities or use of dry belt farming around them and zooprophylaxis and other personal protection methods, especially use of insecticide-impregnated bed nets. The successful use of any particular method or combination of interventions for the control of riceland mosquitoes will depend on in-depth ecological studies on the target species and non-target organisms, sound geographic reconnaissance and effective routine sampling and evaluation. When biological control agents are considered, additional background on the environmental factors limiting their efficacy will also be needed. In addition to the technical components of the various interventions employed in integrated control, sustained suppression of rice land mosquitoes and the diseases they transmit will require a greater socio cultural supportive background, particularly in developing countries (Lacey and Lacey, 1991). The effect of different vector control interventions in rice fields, including environmental measures (i.e. alternate wet and dry irrigation (AWDI)), and biological control approaches (i.e.

bacteria, nematodes, invertebrate predators, larvivorous fish, fungi and other natural products) was evaluated against *Cx. Tritaeniorhynchus* mosquitoes transmitting Japanese encephalitis (JE) which is a disease caused by an arbovirus that is spread by marsh birds, amplified by pigs. In JE-endemic rural settings, where vaccination rates are often low, an integrated vector management approach with AWDI and the use of larvivorous fish as its main components can reduce vector populations and hence has the potential to reduce the transmission level and the burden of JE (Keiser et al., 2005).

3.8 Other Biological control agents

3.8.1 Cyclopoid copepods

Cyclopoid copepods like *Macrocyclops distictus, Mesocyclops pehpeiensis* and *Megacyclops viridis* are used as control agents against dengue vector *Aedes albopictus* in Japan (Dieng et al., 2002). *Mesocyclops* spp., aided by the corixid bug *Micronecta quadristrigata* was also utilised in Vietnam against *Aedes aegypti* (Nam et al., 2000). *Mesocyclops* sp., a predacious copepod, was found to be an effective larvicide against *Aedes aegypti* in Vietnam, (Vu et al., 2005). Use of predacious copepods of the genus *Mesocyclops* as a biological control agent; delivered by community activities of health volunteers, schools, and the public was found to be effective against *Ae. aegypti* in Vietnam (Kay and Vu, 2005).

3.8.2 Larvivorous fishes

Larvivorous fishes were the first biocontrol agents employed to control mosquito vectors. Fish have been used in many countries for malaria control by controlling vectors. Of these the common varieties utilized as biocontrol agents are the mosquito fish (*Gambusia affinis*), Guppies (*Poecilia reticulata), Aplocheilus blochii, Macropodus* and a variety of other local and indigenous fishes as per their availability in the local habitat. Many indigenous varieties of fishes are available and their larvivorous potential has been studied. In different countries the local fishes available have been explored to exploit their use against *Anopheline* and *Culicine* larvae. The ability of 2 freshwater fishes, eastern rainbow fish *Melanotaenia splendida* and flies pecked hardy head *Craterocephalus stercusmuscarum*, native to North Queensland to prey on immature *Ae. aegypti* was evaluated (Russel et al., 2001). Larvivorous fish *Oreochromis spilurus* was found to be effective against malaria vectors in Somalia (Mohamed 2003).

3.8.3 Mermithid nematodes

Mermithids like *Romanomermis iyengari, R. culicivorax* and *Octomyomermis muspratti* show very high specificity

to mosquito larvae. They must undergo part of their development within mosquito larvae and they recycle and infect mosquito larvae season after season in nature. However since they cannot tolerate extreme pH and pollution they are yet to be developed for use in polluted habitats. *R.iyengari* and *R.culicivorax* have a broad host range and are promising bio control agents of various species of mosquitoes (Petersen, 1982, 1985). Mermithid nematodes were found parasitizing *Cx. quinquefasciatus* as early as 1906 (Ross, 1906) and subsequent records show that larvae of several species of Anophelines were found infected (Iyengar, 1927). The mermithid nematode *R. Iyengari* found parasitizing mosquito larvae in paddy fields (Chandrahas and Rajagopalan, 1979), has been successfully mass cultured in the laboratory, found safe against non-target organisms and has been tested in the field also (Gajanana et al., 1978). But it is suitable for fresh water habitats alone as habitats with high pH and salinity is detrimental (Bheema Rao et al., 1979). The nematode *Strelkovimermis spiculatus* was also found to be a promising biological control agent against *Cx.quinquefasciatus* in Cuba (Rodriguez et al., 2003).

3.8.4 Dragonfly nymphs

Biocontrol potential of dragonfly nymph, *Brachythemis contaminata* against the larvae of *An. stephensi, Cx. quinquefasciatus* and *Ae. aegypti* was conducted and found that they had good predatory potential and can be used as a biological control agent for control of mosquito breeding (Singh et al., 2003).

3.8.5 Protozoa

Microsporidians such as *Nosema, Thelohania, Parathelohania, Amblyospora* and *Vavraia* have been studied in detail for mosquito control efficacy. Selective infection of *Anopheles* larvae with some ciliates belonging to the genus *Lamborella* was first reported from forest areas of Assam. Natural infection was found in the immatures of *An. barbirostris, An. hyrcanus* and *An. philippinensis*. However, none of these agents are yet ready for field application. Anopheline larvae are parasitised by *Thelohania sp.* (Kudo, 1929; Sen, 1941). *Nosema algerae* were infective to *Cx. quinquefasciatus, Ae aegypti, An.stephensi* and *Armigeres subalbatus* (Gajanana et al., 1979).

3.8.6 Predatory mosquitoes

Toxorhyncites splendens is a non-blood sucking predatory mosquito whose larvae were found to be effective in controlling *Anopheline* and *Culicine* larvae by feeding on them (Amalraj and Das, 1998). The predatory efficacy of this mosquito was also proved in field evaluation studies against thirteen species of mosquitoes esp. *Culex* spp. and *Aedes* spp. conducted in Japan (Miyagi et al., 1992). *Ae. aegypti* populations were suppressed by *Toxorhyncites splendens* larvae in household water storage containers in Jakarta (Annis et al., 1989).

3.8.7   Viruses

Several viruses such as *Iridescent virus, Densonucleosis virus, Cytoplasmic polyhedrosis virus, Nuclear polyhedrosis virus* etc. have been evaluated in the past for mosquito control (Jenkins, 1964; Roberts and Strand, 1977). These viruses attack a wide range of tissues and although they are highly lethal to their hosts, and are not very infectious (WHO, 1982b). While the problems of insufficient infectivity or virulence handicap the development of viruses as biocontrol agents, the most serious obstacle to their development and use is the non-availability of an efficient method for their mass production as the viruses are highly specific obligate pathogens.

3.8.8   Entomopathogenic Fungus

Many fungi such as *Coelomomyces, Lagenidium, Metarrhizium, Culinomyces* and *Tolypocladium* have been isolated and tested (Roberts and Strand, 1977). *Coelomomyces* is an obligate parasite with a complex lifecycle in which an alternate crustacean host is required to complete the life cycle. *Lagenidium* can be grown on artificial media and can maintain itself in a habitat without the presence of a host. But it is yet to reach large scale testing because its infective propagules pose certain problems such as fragility of zoospores and asynchronous and poor germination of oospores (WHO, 1979). The other fungi, *viz., Metarrhizium, Culicinomyces* and *Tolypocladium* could be cultured on artificial media and formulated easily but to achieve a substantial reduction in the larval population their spores have to be applied in very high doses which would be uneconomical. Also, they do not recycle in aquatic habitats. The entomopathogenic fungus, *Metarhizium anisopliae*, was found to be effective against *Anopheles gambiae* (malaria vector) and *Cx. quinquefasciatus* (filariasis vector) insect-pathogenic fungi of the *Hypocrella/Aschersonia* group might be useful as an agent for pest control. Fungal pathogens such as *Lagenidium, Coelomomyces* and *Culicinomyces* are known to affect mosquito populations, and have been studied extensively (Chandrahas and Rajagopalan, 1979). *Lagenidium giganteum* was highly pathogenic to immatures of *Cx. quinquefasciatus, Cx. tritaeniorhynchus, An. culicifacies, An. stephensi* and *An. subpictus.* There are, however, many other fungi that infect and kill mosquitoes at the larval and/or adult stage. Several isolates of deuteromycetes fungi such as *Metarhizium, Beauveria tenella, Fusarium oxysporum* were isolated indigenously and were found pathogenic to

Cx. fatigans and *An. stephensi.* Some water molds such as *Leptolegnia caudate* and *Aphanomyces laevis* were found to be naturally occurring parasites of mosquito larvae. The fungal mosquito pathogen *Leptolegnia chapmanii* (ARSEF 5499) was tested against 12 species of mosquito larvae and on species of non-target aquatic invertebrates and vertebrates was found to be effective against *Anopheline* and *Culicine* mosquitoes (Lopez et al., 2004). Efficacy of fungal metabolites of *Chrysosporium tropicum* was evaluated against *Cx. Quinquefasciatus* larvae in the laboratory was conducted and found to have promising effects (Priyanka and Prakash, 2003). *Trichophyton ajelloi*, a fungus isolated from soil, caused high larval mortality in *An. stephensi* and *Cx. quinquefasciatus* (Mohanty and Prakash, 2003). Residual sprays of fungal bio pesticides might replace or supplement chemical insecticides for malaria control, particularly in areas of high insecticide resistance (Blanford et al., 2005).

4.0   INSECTICIDE RESISTANCE AND ITS MANAGEMENT

Insecticide resistance development is an ability of a strain of an insect population to tolerate doses of toxicants, which could have otherwise, be lethal to the majority of individuals in a normal population of the same species. Insecticide resistance is expected to directly and profoundly affect the re-emergence of vector-borne diseases (Krogstad 1996) and where resistance has not contributed to disease emergence; it is expected to threaten disease control. However, careful scrutiny of current information about vector resistance (e.g., the WHO resistance database and records of control programs) shows that the full effect of resistance on control efforts is not known. Resistance surveillance is an essential step in resistance management. It aims at providing baseline data for program planning and pesticide selection before the start of control operations, detecting resistance at an early stage so that timely management can be implemented and the effect of control strategies on resistance can be continuously monitored. The challenge will be to maximize the exposure of vector control personnel and entrepreneurs to management principles and to make widely available the surveillance tools required.

5.0   PRIORITIES FOR FUTURE RESEARCH

Further inputs for developing microbial control agents should therefore be diverted to look for new agents, which have not been encountered so far to improve *Bti* and *Bs* through bioengineering and rDNA techniques on a priority basis. The immediate challenges are (i) to obtain /develop highly toxic strains so as to reduce the bulk of the product and the manufacturing cost, (ii) to develop a stable formulation capable of releasing the toxin in the larval feeding zone for prolonged periods which would obviate the high cost involved in frequent applications and also increase the operational efficiency and (iii) to engineer the toxin coding genes of Bti and Bs in alternative prokaryotic and/or eukaryotic micro organisms which can proliferate well in aquatic habitats and be readily available in the larval feeding zone. Finally, before declaring an agent as efficient, its activity should be thoroughly evaluated in proper field tests and not in simulated field tests or field conditions.

The ultimate toxin gene cloning host should have simple nutritional requirements, capable of multiplying well in mosquito breeding habitats and available in high numbers in the larval zone. The gene cloning experiments have, nevertheless, led to the understanding of structural and genetic and molecular mechanism in the genesis of toxins. With this background it is now apparent that the cloning of the toxin genes into a desired host can be realised more precisely and easily. Identification of novel mosquitocidal toxins that differ in structure and mode of action from those produced by Bs and Bti may be of great value. Especially, this could tend to improve in the mosquitocidal properties of these bacteria. New pesticides could be developed, combing in the same organism several genes encoding toxins with different specificities or levels of activities. Combining various toxins in one recombinant cell may prevent or delay the onset of resistance. Indeed, these is a report which suggests that the appearance of mosquito resistance to B. thuringiensis toxins is inversely correlated to the number of toxins used for selection, at least in the laboratory (Tabashnik, 1994). Therefore, several programs have been established to isolate and characterize mosquitocidal active Bt strains.

5.1   GENETIC ENGINEERED BACTERIA

Genetic improvement of the insecticidal activity has been approached in two different ways: Transfer of the toxin genes into non-homologous isolates to combine expression of toxic proteins to produce additive or synergistic effects and the transfer the toxin genes into baculo virus in order to improve Bti activity on host and age (Orduz et al., 1995). Vector-borne diseases impose enormous health and economical burdens throughout the world. Unfortunately, as insecticide and drug resistance spread, these burdens will increase unless new control measures are developed. Genetically modifying vectors to be incapable of transmitting parasites is one possible control strategy and much progress has been made towards this goal. Numerous effector molecules have been identified that interfere with parasite development in its insect vectors and techniques for transforming the vectors with genes encoding these molecules have been established. While the ability to generate refractory vectors is close at hand, a mechanism for replacing a

wild vector population with a refractory one remains elusive. Current research has focussed on the feasibility of using bacteria to deliver the anti-parasitic effector molecules to wild vector populations.

The cloning of mosquitocidal toxin genes from Bti and different Bs strains has allowed the re-expression of a combination of toxins from both species in one recombinant cell in an attempt to broaden the host range of the toxins and obtain beneficial synergistic effects. (Bourgouin et al., 1990, Bar et al., 1991, Trisrisook et al., 1990). Two sequences 100 % and 65 % identical to the left inverted repeat of the insertion sequence IS 240A of Bti were found starting at nucleotides 2756 and 3104, respectively. These insertion sequences are located in the same transcription direction as that of gene Cry 11 Bb1 downstream from the second inverted repeat begins an open reading frame starting at nucleotide 3194 and encoding the first 27 amino acids of a protein which shares 62.9% identity with the protein encoded by the IS240 of Bti. These types of sequences were reported initially in Bti flanking the Cry4Aal gene (Delecluse et al., 1991), later the presence of IS like sequences were described in several mosquito and non-mosquito active strains (Rosso and Delecluse, 1997). It is believed that these elements could be involved in crystal protein gene dispersion among the Bt (Declecluse et al., 1995). Homologous recombination can be applied to Bti or Bs (or other bacteria) to integrate novel combinations of mosquitocidal toxin genes into the bacterial chromosome and obtain their expression during the vegetative or sporulation phase of cell growth.

Bacterial spores are used to control mosquito larvae but spores generally sediment more rapidly from the larval feeding zone than vegetative cells. A novel approach to controlling pests utilizes heat killed, encapsulated vegetative cells of Pseudomonas fluorescens that are genetically engineered to express high levels of Bt toxins. Much effort has been put into evaluating engineered vegetative bacteria such as cyanobacteria, Caulobacter and Ancylobacter aquaticus as delivery vehicles form mosquitocidal toxins (Thanabalu, 1992a,b). Being surface dwellers, cyanobacteria, caulobacters and A.aquaticus are able to resist rapid inactivation by UV light. These bacteria are non-pathogenic to animals and grow in simple, inexpensive media. Moreover, in the reproduction of these bacteria occurs naturally in environments that are low in nutrient levels, unlike the spores of bacilli, which depend on protein rich larval cadavers for their germination. Various species of cyanobacteria have been investigated as vehicles for prolonged delivery of insecticidal toxins to the larval feeding zone. (de Marjac et al., 1987, Angsuthanasombat and Panyim, 1989, Chungjatopornachai, 1990, Murphy and Stephens, 1992). Various toxin genes from Bti and Bs have been cloned into different cyanobacteria, Caulobacter crescentus or A. aquaticus, and the recombinant cells have been shown to be significantly toxic to *Culex* and *Aedes* larvae. Although toxin expression levels remain disappointingly low, the characteristic buoyancy of the cells may prolong their larvicidal action. By 1993, approximately 27 recombinant micro-organisms had been approved for release into the environment in several countries but so far no genetically modified mosquitocidal bacteria have been released. Before seeking approval to release engineered bacteria, a number of issues must be raised and experiments performed in order to determine, as far as possible, that the modified organism would kill only mosquito larvae and not disturb the ecosystem. There are powerful arguments both for and against the release of genetically manipulated microorganisms. Protagonists point out that there is an urgent need for a new approach to control vectors of parasitic and viral diseases. The decision whether or not to release a genetically modified organism must be taken on a case by case basis and made after very careful consideration.

Currently 19 distinct mosquitocidal toxin genes are known, and novel genes will undoubtedly be found and cloned. The toxins appear to vary in their species specificity and mode of action, making it likely that particular combinations cloned in recombinant microorganisms can be chosen so as to enlarge insect host range and delay or prevent the development of resistance. The mosquito host range of the toxins is rather narrow and difficult to predict. The spores do not multiply significantly outside the larval cadaver; the spore-crystal complex is sensitive to UV light and the spores settle rapidly, limiting the duration of control. Insecticidal bacteria are costlier to produce than chemical insecticides. However, a combination of novel genetic manipulation approaches (high level expression of toxin combinations; chromosomal integration of toxin genes), coupled with existing formulation technology or the use of engineered vegetative bacteria which can exist in the upper layer of water, may well overcome these problems. The phasing out of chemical insecticides, together with the emergence of pesticide resistant mosquitoes and drug –resistant parasites, will act as a strong incentive to develop effective mosquitocidal bacteria.

Biological control of mosquito larvae mainly relies on the use of two entomopathogenic bacteria, *Bacillus sphaericus* (*Bs*) and *B.thuringiensis* serovar *israelensis* (*Bti*). Both bacteria synthesize during sporulation, proteins that assemble into crystals which are toxic for the larvae upon ingestion. *Bti* crystals are composed of four major polypeptides with molecular weights of 125, 135, 68 and 28 kDa, now referred to as Cry IVA, Cry IVB, Cry IVD and Cyt A, respectively (Priest, 1997). Genes encoding these polypeptides are located on a 72 kDa resident plasmid (Delecluse et al., 1995) and have all been cloned and expressed in various hosts. Sufficient means exist to detect and manage resistance at a higher level and with greater effectiveness. While initial detection and field surveillance for resistance will likely continue to be based upon simple bioassay, biochemical and molecular tools, the deeper understanding of how resistance arises and maintains itself in populations requires molecular genetics studies. Resistance detection should be made an integral part of all control programmes. The resources for vector control, even under emergency situations, are limited and must be used as effectively as possible. Any attempt to devise management strategies for delaying or forestalling the evolution of pesticide resistance requires a thorough understanding of the parameters influencing the selection process. More than 447 species of arthropods have now developed resistance to insecticides (Georghiou and Wirth 1997). The main weapon for countering this resistance has been the use of alternative chemicals with structures that are unaffected by cross-resistance. As resistance is now developed against many chemicals, there is need to maximize the useful life of new chemicals through their application under conditions that delay or prevent the development of resistance. Hence, an understanding of the parameters influencing the selection process is essential.

The development of new field of metabolic engineering involves the improvement of cellular activities by manipulation of enzymatic, transport and regulatory functions of the cell using recombinant DNA technology. Metabolic engineering has emerged in the past decade as an interdisciplinary field aiming to improve cellular properties by using modern genetic tools to modify pathways (Nielsen, 2001). With rapid developments in new analytical techniques and cloning procedures, it is now possible to introduce directed genetic changes in microbes and subsequently analyse the consequences of the introduced changes at the cellular level. Advances in the field of genetic engineering, sequencing of whole genomes of several organisms and developments in bioinformatics have speeded up the process of gene cloning and transformation. Metabolic engineering is therefore concerned with modifying pathways and assessing the physiological outcome of such genetic modifications in an effort to improve the degradative abilities of microorganisms (Jain et al., 2005).

6.0   CONCLUSION

Nature represents a formidable pool of bio-active compounds and is a strategic source for new and successful pesticidal products. Recent advances made in genomics, proteomics and combinatorial chemistry show that newer strategies for vector control can be successfully devised. In view of the foregoing review on mosquito control strategies, it may be concluded that vector control operations for the prevention and control of vector borne-diseases must be carried out at a cost not exceeding what the communities concerned can afford to allocate for such a purpose. Vector control operations must also be based on the ecological and population dynamics characteristics of vectors concerned. Integrated Vector Management strategy (IVM) is defined as a vector management practice which utilizes all suitable techniques and methods in an orderly and compatible manner and which aims at maintaining the vector population at a level below which disease transmission ceases to occur. The prime requisites for designing IVM strategy are careful appraisal of ecological situation, application of an appropriate method at the appropriate time with primary emphasis on environmental management utilizing highly skilled and disciplined man power, continuous monitoring and feed back, enlisting community participation and co-ordination of the functioning of different departments involved directly or indirectly in preventing mosquitogenic conditions. If designed and implemented properly the IVM strategy is cost-effective in the long run. More emphasis must be laid on field experiments with proper designs. A newer and novel bacterial toxin with different structures and modes of action has to be identified so as to minimize the risk of developing insect resistance. Studies should be pursued towards developing effective and environment friendly "green-technologies".

REFERENCES

Amalraj D, Vasuki V, Kalyanasundaram M, Tyagi BK Das PK (1988a) Laboratory and field evaluation of three insect growth regulators against mosquito vectors. Indian J. Med.Res, 87 : 24-31.

Amalraj D, Vasuki V, Sadanandane C, Kalyanasundaram M, Tyagi BK, Das PK (1988b) Evaluation of two new juvenile hormone compounds against mosquito vectors. Indian J. Med. Res, 87:19-23.

Amalraj D, Das PK (1998) Estimation of predation by the larvae of Toxorhynchites splendens on the aquatic stages of Aedes aegypti. Southeast Asian J. Trop. Med. Public Health, 29: 177-83.

Annis B, Krisnowardojo S, Atmosoedjono S, Supardi P (1989) Suppression of larval Aedes aegypti populations in household water storage containers in Jakarta, Indonesia, through releases of first-instar Toxorhynchites splendens larvae. Amer. Mosq. Control Assoc. 5 : 235-8.

Awad OM, Shimaila A (2003) Operational use of neem oil as an alternative Anopheline larvicide. Part A: Laboratory and field efficacy. East Mediterr Health J. 2003 Jul; 9(4): 637-45. East Mediterr Health J. 9(4): 637-45.

Angsuthanasombat C, Panyim S (1989) Bioasynthesis of 130-kd mosquito larvicide in the cyanobacterium Agmenellum quadruplicatum PR-6. Appl. Environ. Microbiol. 55: 2428-2430.

Bar E, Lieman-Hurwitz J, Rahamim E, Keynan A, Sandler N (1991) Cloning and expression of Bacillus thuringiensis d– endotoxin DNA in B.sphaericus. J. Invertebr. Pathol. 57, 149-158.

Baumann P, Clark MA, Baumann L, Broadwell AH (1991) Bacillus sphaericus as a mosquito pathogen: properties of the organism and its toxin. Microbiol. Rev. 55: 425-436.

Beckage NE, Marion KM, Walton WE, Wirth MC, Tan FF (2004) Comparative larvicidal toxicities of three ecdysone against on the mosquitoes Aedes aegypti, Culex quinquefasciatus, and Anopheles gambiae. Arch. Insect Biochem. Physiol. 57 : 111-22.

Becker N (2000) Bacterial control of vector-mosquitoes and black flies. In: Charles, J.-F, Delecluse A, Nielsen-LeRoux C (Eds), Entomopathogenic Bacteria: From laboratory to field application. Kluwer Academic Publishers, Dordrecht, The Netherlands.

Benedict MQ, Robinson AS (2003) The first releases of transgenic mosquitoes: an argument for the sterile insect technique. Trends Parasitol. 19 : 349-55.

Bernhard K, Utz R (1993) production of Bacillus thuringiensis insecticides for experimental and commercial uses, p 256-267. In

Entwistle, JS, Cory M, Bailey, MJ, Higgs, S (eds), Bacillus thuringiensis, an environmental biopesticides: theory and practice. John Wiley & Sons, Inc. Chichester, England.

Bhatt RM, Yadav RS, Adak T, Babu CJ (2005) Persistence and wash-resistance of insecticidal efficacy of nettings treated with deltamethrin tablet formulation (K-O TAB) against malaria vectors. J. Am. Mosq. Control Assoc. 21: 54-8.

Bheema Rao US, Gajanana A, Rajagopalan PK (1979) A note on the tolerance of the mermithid nematode, Romanomermis sp, to different pH and salinity. Indian J. Med. Res. 69: 423- 427.

Blanford S, Chan BH, Jenkins N, Sim D, Turner RJ, Read AF, Thomas MB (2005) Fungal pathogen reduces potential for malaria transmission. Science 308 : 1531-1533.

Boete A, Koella JC (2003) Evolutionary ideas about genetically manipulated mosquitoes and malaria control. Trends in Parasitol. 19: 32 – 38.

Bourgouin C, Delecluse A, de la Rorre F, Szulmajster J (1990) Transfer of the toxin protein genes of Bacillus sphaericus into Bacillus thuringiensis subsp. israelensis and their expression. Appl. Environ. Microbiol. 56, 340 - 344.

Bourque S, Valero NJ, Mercier M, Lavoie C, Levesque RC (1993) Multiplex polymerase chain reaction for detection and differentiation of the microbial insecticide Bacillus thuringiensis. Appl. Environ. Microbial. 59: 523-527.

Broadwell, AH, Baumann, L, Baumann, P. (1990) Larvicidal properties of the 42 and 51 kilodalton Bacillus sphaericus proteins expressed in different bacterial hosts: evidence for a binary toxin. Curr. Microbiol. 21: 361-366.

Broadwell AH, Baumann P (1986) Sporulation associated activation of Bacillus sphaericus larvicide. Appl. Environ. Microbiol. 52: 758-764.

Butko P, Huang F, Pusztal-Carey, Surewicz WKK (1996) Membrane permeabilization induced by cytolytic delta-endotoxin CytA from Bacillus thuringiensis var. israelensis. Biochem. 35: 11355-11360.

Butko P, Huang F, Pusztal-Carey, Surewicz WKK (1997) Interaction of the ∂-endotoxin CytA from Bacillus thuringiensis var. israelensis with lipid membranes. Biochemistry 36: 12862-12868.

Carozzi NB, Kramer VC, Warren GW, Evola S, Koziel M (1991) Prediction of insecticidal activity of Bacillus thuringiensis strains by polymerase chain reaction product profiles. Appl. Environ. Microbial. 57: 3057-3061.

Cattevuccia F, Godfray HCJ, Crisanti A (2003) Impact of genetic manipulation on the fitness of Anopheles stephensi mosquitoes. Science 299: 1225 – 1227.

Ceron J, Covarrubias L, Quintero R, Ortiz A, Ortiz M, Aranda E, Lina L, Bravo A (1994) PCR analysis of the Cry I insecticidal crystal family genes from Bacillus thuringiensis . Appl. Environ. Microbial. 60: 353-356.

Chandrahas RK, Rajagopalan PK (1979) Observations on mosquito breeding and the natural parasitism of larvae by a fungus Coelomomyces and a mermithid nematode Romanomermis in paddy fields in Pondicherry. Indian J. Med. Res. 69:63 - 67.

Chungjiatupornchai W (1990) Expression of the mosquitocidal protein of Bacillus thuringiensis subsp israelensis and the herbicide resistance gene bar in Synechocystis PCC6803. Curr. Microbiol. 21:283-288.

Curtis CF, Minjas J (1985) Expanded polystyrene for mosquito control. Parasitol. Today, 1:36.

Crickmore N, Bone EJ, Williams JA, Ellar DJ (1995) Contribution of the individual components of the ∂-endotoxin crystal to the mosquitocidal activity of Bacillus thuringiensis subsp. israelensis. FEMS Microbiol. Lett. 131: 249-254.

Crickmore N, Zeigler DR, Feitelson J, Schnepf E, Van Rie J, Lerecleus D, Baum J, Dean DH (1998) Revision of the nomenclature for the Bacillus thuringiensis pesticidal crystal proteins. Microbiol. Mol. Biol. Rev. 62: 807-813.

Das PK, Tyagi BK, Kalyanasundaram M (1986a) Vernacidae, a new insecticide–impregnated paint for controlling mosquito vector Culex quinquefasciatus and cockroach Periplanata americana. Indian J. Med. Res, 83:268-270.

Das PK, Tyagi BK, Somachari NV(1986b) Efficacy of Arosurf, a monomolecular surface film, in controlling Culex quinquefasciatus Say, Anopheles stephensi Liston and Aedes aegypti (L.). Indian J. Med. Res. 83:271-276.

Delecluse A, Bourgouin, A, Klier G, (1991) Rapoport, Nucleotide sequence and characterization of a new insertion element, IS240, from *Bacillus thuringiensis* subsp. *israelensis* Plasmid. 21: 71-78.

Delecluse A, Rosso ML, Ragni A (1995) Cloning and expression of a novel toxin gene from *Bacillus thuringiensis* susp. *jegathesan* encoding a highly mosquitocidal protein. Appl. Environ. Microbiol. 61: 4230-4235.

de Marsac NT, de la Torre F, Szulmajster J (1987) Expression of the larvicidal gene of *Bacillus sphaericus* 1593 Min the cyanobacterium *Anacystis nidulans* R2. Mol. Gene Genet. 209:396-398.

Dieng H, Boot M, Tuno N, Tsuda Y, Takagi M (2002) A pehpeiensis as laboratory and field evaluation of *Macrocyclops distinctus*, *Megacyclops viridis* and Mesocyclops control agents of the dengue vector *Aedes albopictus* in a peridomestic area in Nagasaki, Japan. Med. Vet. Entomol.16 : 285-91.

Escriche B, Tabashnik B, Finson N, Ferre J (1995) Immuno histochemical detection of binding of CryIA crystal proteins of *Bacillus thuringiensis* in highly resistant strains of *Plutella xylostella* (L.) From Hawaii. Biochem. Biophys. Res. Commun. 212:388-95.

Feitelson JS, Payne J, Kim L (1992) *Bacillus thuringiensis*: insects and beyond. Bio/Technol. 10: 271-275.

Ferre J, Escriche B, Bel Y, Van Rie J (1995) Biochemistry and genetics of insect resistance to *Bacillus thuringiensis* insecticidal crystal proteins. FEMS Microbiol. Lett. 132: 1-7.

Gajanana A, Kazmi SJ, Bheema Rao US, Suguna SG, Chandrahas RK (1978) Studies on a nematode parasite (Romanomermis sp: Mermithidae) of mosquito, larvae isolated in Pondicherry. Indian J. Med. Res. 68:242.

Gajanana A, Tewari SC, Reuben R, Rajagopalan PK (1979) Partial suppression of malaria parasites in *Aedes aegypti* and *Anopheles stephensi* doubly infected with Nosema algerae and *Plasmodium*. Indian J. Med. Res. 70:417.

Georghiou GP, Mallik JI, Wirth M, Sainato K (1992) Characterisation of resistance of *Culex quinquefasciatus* to the insecticidal toxins of *Bacillus sphaericus* (strain 2362). University of California, Mosquito Control Research, Annual Report. pp. 34-35.

Georghiou GP, Wirth MC (1997) Influence of exposure to single versus multiple toxins of *Bacillus thuringiensis* var. *israelensis* on development of resistance in the mosquito *Culex quinquefasciatus* (Diptera: Culicidae) Appl. Environ. Microbiol. 63: 1095-1101.

Goldberg LJ, Margalit J (1977) A bacterial spore demonstrating rapid larvicidal activity against *Anopheles sergentii*, *Uranotaenia unguiculata*, *Culex univitattus*, *Aedes aegypti* and *Culex pipiens*. Mosq. News 37: 355-358.

Guerchicoff A, Ugalde RA, Rubenstein CP (1997) Identification and characterization of a previously undescribed Cyr gene in *Bacillus thuringiensis* subsp. *israelensis*. Appl. Environ. Microbial. Rev. 53: 242-255.

Hofte, H, Whitely, HR (1989) Insecticidal crystal proteins of *Bacillus thuringiensis*. Microbial. Rev. 53: 242-255.

Ibarra J, Federici BA (1986) Isolation of a relatively non-toxic 65 kilodalton protein inclusion from the parasporal body of *Bacillus thuringiensis* var. *israelensis*. J. Bacterial. 165: 527-533.

Iyengar MOT (1927) Parasitic nematodes of *Anopheles* in Bengal. Trans 7th Cong Far East Assoc. Trop. Med. Calcutta 3:128.

Jambulingam P, Gunasekharan K, Sahoo SS, Hoti PK, Tyagi BK, Kalayanasundaram M (1989) Effect of permethrin impregnated bednets in reducing population of malaria vector, *Anopheles culicifacies*, in a tribal village of Orissa state. (India). Indian J. Med. Res. 89: 48-51.

Jenkins DW (1964) Pathogens, parasites and predators of medically important arthropods. Bull WHO. 30 suppl. 28-37.

Jain RK, Kapur M, Labana S, Lal B, Sharma PM, Bhattacharya D, Thakur I(2005) Microbial diversity: application of microorganism for the biodegradation of xenobiotics. Current Science 89 :101-112.

Kay B, Vu SN (2005) New strategy against *Aedes aegypti* in Vietnam. Lancet 365: 551-552.

Keiser J, Maltese MF, Erlanger TE, Bos R, Tanner M, Singer BH, Utzinger J (2005) Effect of irrigated rice agriculture on Japanese encephalitis, including challenges and opportunities for integrated vector management. Acta Trop. 95: 4057.

Keller M, Sneh B, Strizhov N, Prudovsky E, Regev A, Konez C (1996)

Digestion of delta-endotoxin by gut proteases may explain reduced sensitivity of advanced instar larvae of Spodoptera littoralis to Cry IC. Insect Biochem. Mol. Biol. 26:365-73.

Krogstad D (1996) Malaria as a re-emerging disease. Epidemiol. Rev. 18: 77-89.

Krych, VK, Johnson, JL, Yousten, AA (1980) Deoxyribonucleic acid homologies among strains of *Bacillus sphaericus*. Int. J. Syst. Bacteriol. 30: 476-484.

Kudo R(1929) Studies on microsporidia parasitic in mosquitoes. VII. Notes on microsporidia of some Indian mosquitoes. Arch. Protistenkd. 67:1.

Kuo, WS, Chak, KF (1996) Identification of novel Cry-type genes from *Bacillus thuringiensis* strains on the basis of restriction fragment length polymorphism of the PCR-amplified DNA. Appl. Environ. Microbiol. 62: 1369-1377.

Lacey LA, Lacey CM (1991) The medical importance of riceland mosquitoes and their control using alternatives to chemical insecticides J. Am. Mosq. Control Assoc. 7:132.

Lee YW, Zairi J, Yap HH, Adanan CR (2005) Integration of *Bacillus thuringiensis* H-14 formulations and pyriproxyfen for the control of larvae of *Aedes aegypti* and *Aedes albopictus*. J. Am. Mosq. Control Assoc. 21 : 84-9.

Lopez CC, Scorsetti AC, Marti GA, Garcia JJ (2004) Host range and specificity of an Argentinean isolate of the aquatic fungus *Leptolegnia chapmanii* (Oomycetes: Saprolegniales), a pathogen of mosquito larvae (Diptera: Culicidae). Mycopathologia.158 : 311-5.

Miyagi I, Toma T, Mogi M (1992) Biological control of container-breeding mosquitoes, *Aedes albopictus* and *Culex quinquefasciatus*, in a Japanese island by release of *Toxorhynchites splendens* adults. Med. Vet. Entomol. 6: 290-300.

Mohanty SS, Prakash S (2003) Laboratory evaluation of Trichophyton ajelloi, a fungal pathogen of *Anopheles stephensi* and *Culex quinquefasciatus*. J. Am. Mosq. Control Assoc.16 : 254-7.

Mohamed AA (2003) Study of larvivorous fish for malaria vector control in Somalia, 2002. East Mediterr. Health J. 9 : 618-26.

Mulla MS (1995) The future of insect growth regulators in vector control. J. Am. Mosq. Control Assoc. 11: 269-73.

Murphy RC, Stephensen Jr SE (1992) Cloning and expression of the cryIV gene of *Bacillus thuringiensis* subsp. *israelensis* in the cyanobacterium *Agmenellum quadruplicatum* PR-6 and its resulting larvicidal activity. Appl. Environ. Microbiol. 58:1650-1655.

Myers P, Yousten A (1978) Toxic activity of *Bacillus sphaericus* SSII-I for mosquito larvae. Insect. Immunol. 19: 1047-1053.

Myers P, Yousten A, Davidson, EW (1979) Comparative studies of the mosquito-larval toxin of *Bacillus sphaericus* SSII-I and 1593. Can. J. Microbiol. 25: 1227-1231.

Myers P, Yousten A (1980) Localization of mosquito larval toxin of *Bacillus sphaericus* 1593. Appl. Environ. Microbiol. 39:1205-1211.

Nam VS, Yen NT, Holynska M, Reid JW, Kay BH (2000) National progress in dengue vector control in Vietnam: survey for Mesocyclops (Copepoda), Micronecta (Corixidae), and fish as biological control agents. Am. J. Trop. Med. Hyg. 62 : 5-10.

Nielsen-LeRoux, C, Charles J-F (1992) Binding of *Bacillus sphaericus* binary toxin to a specific receptor on midgut brush border membranes from mosquito larvae. Eur. J. Biochem. 210: 585-590.

Nielsen-LeRoux C, Pasquier F, Charles J-F, Sinegre G, Gaven B, Pasteur N (1997) Resistance to *Bacillus sphaericus* involves different mechanisms in *Culex pipiens* (Diptera : Culicidae) larvae. J. Med. Entomol. 34:321-7.

Nielsen J (2001) Metabolic engineering. Appl. Microbiol. Biotechnol., 55, 263-283.

Orduz S, Restrepo N, Patino MM, Rojas W (1995) Transfer of toxin genes to alternate bacterial hosts for mosquitocidal control. Mem Inst Oswaldo Cruz, Rio de Janeiro. 90: 97-107.

Petersen JJ (1982). Current status of nematodes for biological control of insects. Parasitol. 84: 177-204.

Petersen JJ (1985) Nematodes as biological control agents: Pt I, Mermithidae. Adv. Parasitol. 14: 307-34.

Poncet S, Delecluse A, Anello G, Klier A, Rapoport G (1994) Transfer and expression of the CryIVB and CryIVD genes of *Bacillus thuringiensis* subsp. *israelensis* in *Bacillus sphaericus* 2297. FEMS Microbiol. Lett. 117: 91-96.

Poopathi S, Rao, DR (1995) Pyrethroid impregnated hessian curtains for protection against mosquitoes indoors in South India. Med. Vet. Entomol. (England)) 9:169-175.

Poopathi S, Rao DR, Mani TR, Baskaran G, Kabilan L (1999). Effect of *Bacillus sphaericus* and *Bacillus thuringiensis* var *israelensis* on the ultrastructural changes in the midgut of *Culex quinquefasciatus* Say. (Diptera: Culicidae). J. Ent. Res. 23: 347-357.

Poopathi S, Baskaran G (2001) Efficacy of *Bacillus thuringiensis* serovar *israelensis* against *Bacillus sphaericus* resistant and susceptible larvae of *Culex quinquefasciatus* Say. J. Biol. Control 15: 85 – 92.

Poopathi S, Nielsen-LeRoux C, Charles J.-F (2002) Alternative methods for preservation of mosquito larvae to study binding mechanism of *Bacillus sphaericus* toxin. J. Invert Pathol. (England) 79: 132 – 134.

Poopathi S, Baskaran. G (2003a). *Bacillus sphaericus* resistance in mosquitoes : An approach for resistance management. Proc. Natl. Acad. Sci. (Animal Sci) (India) 73, B (I) 29 – 36.

Poopathi S, Anupkumar K, Arunachalam N, Sekar V, Tyagi BK (2003b) A small scale mosquito control field trial with the biopesticides *Bacillus sphaericus* and *Bacillus thuringiensis* serovar *iaraelensis* produced from a new culture medium. Biocontrol Sci. and Technol. (USA). 13(8): 743 – 748.

Porter AG, Davidson E, Liu JW (1993) Mosquitocidal toxins of bacilli and their genetic manipulation for effective biological control of mosquitoes. Microbiol. Rev. 57, 838-861.

Priest, F, Ebdrup, L, Zahner, V, Carter, P (1997) Distribution and characterization of mosquitocidal toxin genes in some strains of *Bacillus sphaericus*. App. Environ. Microbiol. 63: 1195-1198.

Priyanka Prakash S. (2003) Laboratory efficacy tests for fungal metabolites of *Chrysosporium tropicum* against *Culex quinquefasciatus*. J. Am. Mosq. Control Assoc.19 : 404-7.

Ragni A, Thiery I, Delecluse A (1996) Characterization of six highly mosquitocidal *Bacillus thuringiensis* strains that do not belong to H-14 serotype. Curr. Microbiol. 32: 48-54.

Rao DR, Mani TR, Rajendran R, Joseph AS, Gajanana A, Reuben R. (1995) Development of a high level of resistance to *Bacillus sphaericus* in a field line of *Culex quinquefasciatus* from Kochi, India. J. Am. Mosq. Control Assoc. 11: 1-5.

Regis L, Silva-Filha MH, Claudia MF, de-Oliveira R, Eugenia MR, Sinara B, de Silva A, Furtado F (1995) Integrated control measures against *Culex quinquefasciatus*, the vector of filariasis in Recife. Mem Inst Oswaldo Cruz, Rio de Janeiro, 90 :115-119.

Roberts DW, Strand A. (1977) Pathogens of medically important arthropods. Bull. WHO, 55 (Suppl.I): 419.

Rodcharoen J, Mulla MS (1996) Cross-resistance to *Bacillus sphaericus* strains in *Culex quinquefasciatus* . J. Am. Mosq. Control Assoc. 12:247-50.

Rodriguez J, Garcia IG, Diaz M, Avila IG, Sanchez JE (2003) Pathogenic effect of the parasite nematode *Strelkovimermmis spiculatus* (Nematoda Mermithidae) in larvae of mosquito *Culex quinquefasciatus* (Diptera: Culicidae) under laboratory conditions in Cuba. J. Med. Ent. 55 : 124-5.

Ross R. (1906) Notes on parasites of mosquitoes found in India between 1895-1899. J. Hyg 6:101.

Rosso ML, Delecluse A (1997) Distribution of the insertion element IS240 among *Bacillus thuringiensis* strains. Curr. Microbiol. 34: 348-353.

Russell BM, Wang J, Williams Y, Hearnden MN, Kay BH (2001) Laboratory evaluation of two native fishes from tropical North Queensland as biological control agents of subterranean *Aedes aegypti*. J. Am. Mosq. Control Assoc. 17: 124-6.

Sen P. (1941) On the microsporidian infesting some Anophelines of India. J. Malaria Inst. India 4:257.

Sharma RC, Yadav RS Sharma VP (1985) Field trials on the application of Expanded Polystyrene (EPS) beads in mosquito control. Indian J. Malariol, 22; 107-109.

Sharma VP, Ansari MA, Mittal, PK, Razdan, RK (1989) Insecticide impregnated ropes as mosquito repellent. Indian J. Malariol. 26:179-185.

Singh RK, Dhiman RC, Singh SP (2003) Laboratory studies on the predatory potential of dragonfly nymphs on mosquito larvae. Com-mun. Dis. 35 : 96-101.

Singer S (1973) Insecticidal activity of recent bacterial isolates and their toxins against mosquito larvae. Nature (London) 244: 110-111.

Sinegre G, Babinot M, Quermel JM, Gaven B (1994) First field occurrence of *Culex pipiens* resistance to *Bacillus sphaericus* in Southern France , p.17. In Proceedings of the VII European Meeting of the Society for Vector Ecology, Barcelona, Spain, pp. 5 - 8 September, 1994.

Tabashnik E (1994) Evolution of resistance to *Bacillus thuringiensis*. Annu. Rev. Entomol. 39: 47-49.

Tabashnik E, Malvar T, Liu YB, Finson N, Borthakur D, Shin B S (1996). Cross-resistance of the diamond black moth indicates altered interactions with domain of *Bacillus thuringiensis* toxins. Appl. Environ. Microbiol. 62:2839-2844.

Tabashnik BE, Liu YB, Finson N, Masson L, Heckel DG (1997) One gene in diamondback moth confers resistance to four *Bacillus thuringiensis* toxins. Proc. Natl. Acad. Sci USA. 94: 1640-4.

Tami A, Mubyazi G, Talbert A, Mshinda H, Duchon S, Lengeler C (2004) Evaluation of Olyset insecticide-treated nets distributed seven years previously in Tanzania. Malaria J. 29: 3:19.

Thanabalu T, Hindley J, Berry C (1992a) Proteolytic processing of the mosquitocidal toxin from *Bacillus sphaericus* SSII-1. J. Bacterial. 174: 5051-5056.

Thanabalu T, Hindley J, Brenner S, Oei C, Berry C (1992b) Expression of the mosquitocidal toxins of *Bacillus sphaericus* and *Bacillus thuringiensis* subsp. *israelensis* by recombinant *Caulobacter crescentus* , a vehicle for biological control of aquatic insect larvae. Appl. Environ. Microbiol. 58, 905-910.

Thanabalu T, Hindley J, Jackson-Yap, Berry C (1991) Cloning, sequencing and expression of a gene encoding a 100-kilodalton mosquitocidal toxin from *Bacillus sphaericus* SSII-I. J. Bacteriol. 173: 2776-2785.

Thiery I, Hamon S, Gaven B, de Barjac H (1992) Host range of *Clostridum bifermentans* serovar *Malaysia,* a mosquitocidal anaerobic bacterium. J. Amer. Mosquito Control Assoc. 8: 272 – 277.

Thiery I, Hamon S, Gaven B, de Barjac H (1992) Host range of *Clostridum bifermentans* serovar *Malaysia,* a mosquitocidal anaerobic bacterium. J. Amer. Mosquito Control Assoc. 8: 272 – 277.

Trisrisook M, Pantuwatana S, Bhumiratana A, Panbangred W (1990) Molecular cloning of the 130-kilodalton mosquitocidal dendo toxin gene of *Bacillus thuringiensis* subsp. *israelensis* in *Bacillus sphaericus*. Appl. Environ. Microbiol. 56: 1710-1716.

Tyagi BK, Kalyanasundaram M, Das PK, Somachary N. (1985) Evaluation of a new compound (VCRC/INS/A-23) with juvenile hormone activity against mosquito vectors. Indian J. Med. Res. 82:9-13.

Tyagi BK, Somachari N, Vasuki, V, Das PK (1987). Evaluation of three formulations of a chitin synthesis inhibitor (fenoxycarb) against mosquito vectors. Indian J. Med. Res. 85:161-167.

Vasuki V. (1988) Role of insect growth regulators in vector control. Proc. 2nd Symp. Vector –borne Diseases, Trivandrum. pp 204-218.

Vu SN, Nguyen TY, Tran VP, Truong UN, Le QM, Le VL, Le TN, Bektas A, Briscombe A, Aaskov JG, Ryan PA, Kay BH. (2005). Elimination of dengue by community programs using Mesocyclops (Copepoda) against *Aedes aegypti* in central Vietnam. Am. J. Trop. Med. Hyg. 72 : 67-73.

Wang CH, Chang NT, Wu HH, Ho CM (2000) Integrated control of the dengue vector *Aedes aegypti* in Liu-Chiu village, Ping-Tung County, Taiwan. J. Am. Mosq. Control Assoc. 16 : 93-9.

Weiser, J (1984) A mosquito-virulent *Bacillus sphaericus* in adult *Simulium damnosum* from northern Nigeria. Zentralbl. Mikrobiol. 139: 57-60.

WHO (1979) Data sheet on the biological control agent, *Bacillus thuringiensis* serotype H-14 (de Barjac 1978). WHO Mimeographed Document. WHO/VBC/79.750 Rev. 1 VBC/BCDS/79/79.01:46.

WHO (1982a). The role of biological agents in integrated vector control and the formulation of protocols for field-testing of biological agents. Report of the sixth meeting of the scientific working group on biological control of vectors. WHO/TDR/VEC-SWG (6)/82(3): 46.

WHO (1982b) Biological control of vectors of disease. Sixth report of WHO expert committee, Vector Biology and Control. WHO Tech Rep Ser No. 679:39.

WHO (1986) Resistance of vectors and reservoirs of disease to pesticides. Tenth report of the Expert Committee on Vector Biology and Control. Technical Report Series, p.737, 87.

WHO (1999) Weekly epidemiological records. WHO Tech Rep Ser No 74, 265 – 272.

Wickremesinghe RSB, Mendis CL. (1980) *Bacillus sphaericus* spore from Sri Lanka demonstrating rapid larvicidal activity on *Culex quinquefasciatus*. Mosq. News 40 : 387-389.

Wirth MC, Georghiou GP, Federici BA (1997) *Cyt*A enables *Cry*IV endotoxins of *Bacillus thuringiensis* to overcome high levels of *Cry*IV resistance in the mosquito *Culex quinquefasciatus*. Proc. Natl. Acad. Sci. USA 94: 10536-10540.

Wirth MC, Delecluse CA, Federici BA, Walton WE (1998) Variable cross-resistance to *Cry* 11B from *Bacillus thuringiensis* subsp. *jegathesan* in *Culex quinquefasciatus* (Diptera: Culicidae) resistant to single or multiple toxins of *Bacillus thuringiensis* subsp. *israelensis*. Appl. Environ. Microbiol. 64: 4174-4179.

Wirth MC, Georghiou GP, Malik J I, Abro GH. (2000a) Laboratory selection for resistance to *Bacillus sphaericus* in *Culex quinquefasciatus* (Diptera: Culicidae) from California, USA. J. Med. Entomol. 37, 534-540.

Wirth MC, Walton WE, Federici BA (2000b) Cyt I A from *Bacillus thuringiensis* restores toxicity of *Bacillus sphaericus* against resistant *Culex quinquefasciatus* (Diptera: Culicidae). J. Med. Entomol. 37, 401-407.

Wirth MC, Park HW, Walton WE, Federici BA (2005) Cyt 1 A of *Bacillus thuringiensis* delays the evolution of resistance to Cry 11 A in the mosquito *Culex quinquefasciatus*. Appl. Environ. Microbiol. 71, 185-189.

Xue RD, Doyle MA (2005) Evaluation of upstream point treatment in flowing water ditches by Aquabac (*Bacillus thuringiensis* var. *israelensis*) against *Culex quinquefasciatus*. J. Am. Mosq. Control Assoc. 21 : 234-5.

11

Cancer investigation: A genome perspective

Varsale A. R., Wadnerkar A. S. and Mandage R. H.

Centre for Advanced Life Sciences, Deogiri College, Aurangabad, Maharashtra, India.

The completion of human genome project has evolved many techniques used to locate the human genes. The focus is mainly on the genome, transcriptome or proteome to recognise distinctive characteristics that may explain the basis of human disease and potentially envisage prospect outcomes. Cancer is one of the recent deadliest diseases. Various cancer types root problems in the generalised dealing. The objective of these investigative pursuits is to ultimately individualize treatment for each patient based on their exclusive gene expression prototypes.

Keywords: Human genome, cancer, functional genomics, transcripts, microarray, proteomics.

INTRODUCTION

Cancer is a series of biological consequences starting with growth and survival of neoplastic cells at the primary site, invasion, angiogenesis, intravasation, extravasation and finally growth at the secondary site. The disease is deadly and cure is thought to be impossible due to lack of recognition of specific biomarkers indicating detection and diagnosis of the disease at early stages (Riccardo, 2005). Newer scientific research methodologies have come into prime focus with the accomplishment of human genome project. It has opened the perception of big science in the field of biology, where huge data is generated and advanced computational technologies are required to analyze the data. For example, (a) the necessity to know a phenotype in order to define a new gene is not required, instead of starting with a phenotype and tracing it back to a sequence of DNA in order to discover the genes, it is now possible to start with the genome sequence and look for signals indicative of promoters, exons, splice junctions, and similarity to known sequences in order to discover new genes. In fact, most genes are now identified based on their DNA sequence. (b)Comparative genome analysis is used to search and identify homologous genes in an evolutionarily-related organism and the differences amongst closely related organisms also can be resolved using this technique. (c)Identification of virulence factors in various bacterial strains was done by comparing pathogenic and nonpathogenic strains (Dobrindt, 2005;

Dobrindt et al., 2003; Schmidt et al., 2004; Wick et al., 2005). Similar concepts have been used for the study of cancer. For example, Hematological malignancies are diagnosed by karyotyping (Bayani and Squire, 2002; Jotterand and Parlier, 1996).

THE GENOME

The genome is supposed to be the whole set of DNA sequence of the germ line cells of an organism. Knowledge of this DNA sequence will lead to valuable information about gene functioning. Some organisms have segmented genomes as in case of humans. Microscopic observations can visualize such units. Such units are studied previously to check their role in the cell. (Jackson, 1978; Martin and Hoehn, 1974; Pogosianz and Prigogina, 1972). Genome can be studied on nucleotide level. Whole gene analysis uniquely identifies the region of the genome that plays important role in the diseases. Cancer development is generally due to single point mutations in the gene. (Claus, 1995; Den Otter et al., 1990; Weinberg, 1983). Hence, whole genomic and focused approaches should be combined and used to find out the genes interfering cancer development. For years it is known that oncogenic mutations and suppression can manipulate the development of cancer. Specific mutations can be identified using large-scale sequencing of the genome in many cancers (Capella et al., 1991; Casey et al., 2005; Frank et al., 1999; Li et al., 1998). The type of mutation in a tumor can be a better source to have a choice of treatment. Large genomic

*Corresponding author. E-mail: rajendra.mandage@gmail.com.

rearrangements are associated with cancer in many cases. These are observed in the microscope but functional genomic techniques and can be employed to detect smaller changes. (Beheshti et al., 2003; Hoque et al., 2003; Mundle and Sokolova, 2004; Squire et al., 2003; Weiss et al., 2003).

THE TRANSCRIPTOME

The huge number of mRNA transcripts that are formed by copying or splicing of genome segments can be termed as transcriptomes (Velculescu et al., 1997). This includes all the mass of RNA that encode the different proteins which determines the gene functionality. The term comprises all the transcripts that are formed in the specific biological and all other possible conditions. A gene can provide more than one transcript. Hence, the transcriptome study will provide a big scenario of the cellular functions. Variation in transcriptome is related to the variation in cell, tissue and even in the organism. Microarray and large scale nucleotide sequencing enables the complete study of transcriptome to get the clear view of inner cellular complexity. In case of cancers, transcriptomes are better source to categorise tumors into its subclasses, which can be helpful in treatment. (Golub et al., 1999; Perou et al., 2000; Ramaswamy et al., 2001). Diagnosis of the patient can be carried out using a single or smaller or even larger group of transcripts. For example, expression level of the estrogen receptor (ER) detects the subsequent consequence in some breast tumors (Leal et al., 1995; Perin et al., 1996). Such biological markers must be associated with other tumors. Transcriptome analysis thus has the advantage of identification of decisive transcriptional marker through the screening of transcripts. Introduction of novel designed drugs requires sophisticated transcriptome analysis (Rhodes and Chinnaiyan, 2005).

THE PROTEOME

The proteome is the complete set of proteins expressed by the entire genome. As some genes code for multiple proteins; the size of proteome is greater than the total number of genes. Thus proteome is a potential target for the researchers to get clear insights into the biological phenomenon through their investigative approaches. (Kahn, 1995). Large scale and high throughput techniques are utilized to investigate proteome. Mass spectroscopy and antibody-based techniques are focused in localizing, quantitating, and structurally characterizing individual proteins or small groups of proteins. This can be achieved only if the proper structure and function of the proteins is elucidated. Proteomics is referred as the large scale analysis of the total protein content of the cell, or fluid. (Petricoin and Liotta, 2004;

Stults and Arnott, 2005; Wulfkuhle et al., 2003). The proteome can be studied by evaluating many proteins from a cell through High-throughput mechanisms or microarray techniques to study a single protein in a multitude of tissue samples. The transcriptional and post transcriptional modifications cause variation in the proteome. This is an interesting feature to study the genome functioning and cell interactions. However, it is fascinating that some cellular responses occur in the proteome and does not necessarily involve genomic or transcriptome variations. This can only happen if protein undergoes modification or as a result of protein-protein interactions or with proteins and other macromolecules. The change in expression levels of a protein or its modifications are studied in proteomic experiments in response to a stimulus. One can study a single protein or group of proteins that determine a cellular response to the treatment. In case of cancer differences in the proteome due to the mutation of a single gene, following drug treatment, or between groups of patients separated by their clinical characteristics can be studied. (e.g., histology or survival outcome) (Dephoure et al., 2005; Soreghan et al., 2003) Early detection of diseases using selective markers or the possible mechanism of drug resistance at molecular level can be achieved by evaluating proteome. (Alexander et al., 2004; Bhattacharyya et al., 2004; Hondermarck et al., 2001; Petricoin et al., 2005; Zhang et al., 2004).

Techniques for cancer investigation

As the cancerous cell advances, cellular metabolism is altered dramatically as a consequence of genetic changes in the genome. Genes may lose their functionality or can promote cellular growth. Tumor formation varies in steps and hence genetic variation amongst similar tumors is important. These variations also change the way of response to eradication. To understand a tumor in a better way, it is necessary to know its genomics. Tools used for human genomics could be routinely used for diagnosis and suggest treatment of cancer (Mount and Pandey, 2005; Yeatman, 2003). Recently, microscopic observations are made to identify tumors based on their pathology. This also involves genome analysis in a crude manner. Chromosomal anomalies are associated with other forms of cancer (Gronwald et al., 2005; Kimura et al., 2004; Meijer et al., 1998; Micci et al., 2004). High resolution screening of genomes is done nowadays using various tools (Jones et al., 2005; Nakao et al., 2004).

Expressed sequence tags

As an outcome of the human genome project it is now possible to go for high throughput sequencing. Numbers of sequence based techniques have evolved to extend biological understanding, especially diseases. Now, priority is given to transcriptome than genome. Complementary DNA libraries and sequencing is used to generate expressed sequence tags (ESTs) libraries (Kawamoto et al., 2000; Okazaki et al., 2002; Stapleton et al., 2002). These ESTs are single sequencing reads from cDNA. Variations in the human genome could be studied using these ESTs as demonstrated in 1991 (Adams et al., 1991). Not only the categorization of sequences from a cell is possible but the comparison can be made with other cells (Carulli et al., 1998; Kawamoto et al., 2000; Lindlof, 2003).

SAGE

SAGE is another useful technique used to evaluate transcriptomes (Velculescu et al., 1995). It involves short sequence generation from cDNA using enzymes and then concentrating them in a large string for sequencing. These generated short sequences are the specific markers for transcripts. Abundance of the transcript within the transcriptome can be measured quantitatively by the frequency of short sequence tags. Variation in the expression of gene can be studied under experimental conditions using this technique (Sengoelge et al., 2005; Zucchi et al., 2004). SAGE is also employed to evaluate expression of undefined genes. Yet SAGE has its own limitations. It is mainly based on complexity of cloning and hence expensive. Secondly, it is also difficult to identify sequences with short tag size; below 21 base pairs significant cDNA databases or sequenced genomes are required by some SAGE applications (Liu, 2005). SAGE technique was used to recognise transcripts that are having enhanced metastatic potential like keratin K5, cystatin S, the human homologue of yeast ribosomal S28, and the p32 subunit of human pre-mRNA splicing factor SF2(Parle-McDermott et al., 2000.). Over expression of PGP9.5, a neurospecific peptide that functions to remove ubiquitin from ubiquitinated cellular proteins, in turn protects their degradation by the proteasome-dependent pathway was demonstrated using the SAGE in more than 50% of primary lung cancers (Bittencourt et al., 2001).

Proteomics

Proteomic analysis is widely applicable to investigate clinical biomarkers (Alaiya et al., 2005; Bhattacharyya et al., 2004; Chen et al., 2005; Hondermarck et al., 2001; Srinivas et al., 2001; Steel et al., 2003; Zhang et al., 2004) . High through put proteomics was utilized to detect pancreatic cancer using serum of the patients (Bhattacharyya et al., 2004). Proteomics was applied in a multi-institutional study of women with ovarian cancer,

benign pelvic masses or pathology (Zhang et al., 2004). They exposed three distinct protein markers exclusive to ovarian cancer that could possible be used for early detection tumor markers. Proteomics has been employed for a histological diagnosis to subtype tumors whether directly from tumor samples or in attempts at early detection (Borczuk et al., 2004; Seike et al., 2005; Steel et al., 2003). So one can easily detect the tumor and the nearby tissue for changes associated to tumor growth or to gaze for microscopic tumors in an or else normal looking tissue section. These kinds of investigation will perhaps offer important information about the appearance of tumors from microscopic disease sites that cannot be obtained by any other method.

Two-dimensional polyacrylamide gel electrophoresis (2D-PAGE)

An immobilized pH gel gradient is used to separate proteins by their isoelectric points, followed by SDS PAGE to promote separation of the proteins on the basis of their molecular mass. A number of spots are resolved in a single gel representing different proteins, different isoforms of the same protein, or its post-translational modifications (Sivakumar, 2002.) Identification of biomarkers in case of breast cancer in nipple aspiration fluid has been carried out recently by 2D-PAGE. Breast carcinomas can be easily evaluated by investigating breast ductal fluids (Kuerer et al., 2002).

Mass spectrometry

Gel based techniques like isoelectric focusing, SDS-PAGE polypeptide sequencing, and liquid chromate-graphy methods like affinity, ion exchange, or reverse phase separations can be utilized to separate protein components within a complex protein mixture like serum, tissue or cellular extract. Each separated fraction is then coupled to tandem mass spectrometry peptide sequencing (LC-MS/MS) and database search to resolve its protein constituents. With mass spectrometry profiling or differential fluorescence techniques or isotopic labeling, evaluation of fractions from multiple samples can be done. Whole plasma or tissue has been analyzed by mass spectrometry (Caprioli, 2005; Steel et al., 2003). Mass spectrometry is practicable for direct tissue analysis at the single cell level (Danna and Nolan, 2006) and implicated on microdissected samples for purer tumor analysis (Greengauz-Roberts et al., 2005; Jain, 2002). Multiple locations across microscopic tissue slices can even be identified using this technique (Chaurand et al., 2004). Matrix-assisted laser desorption ionization time of flight (MALDI-TOF) utilizes peptide mass fingerprinting (PMF). Gel separated proteins digested by trypsin are measured at high accuracy (100 ppm or better).

Molecular ions from the peptide samples are produced using a laser source and then introduced in an analyzer that resolves ionized fragments on the basis of their mass-to charge (m/z) ratio(Rowley A, et al., 2000.). The Surface-enhanced laser desorption and ionization time-of-flight (SELDI-TOF) is not only able to find single protein biomarkers but is also able to identify biomarker expression patterns (Seibert et al., 2004). Mass spectroscopy has the potential to be used for systematic identification and characterization of proteins that are helpful in diagnosis as prognostic markers. (Volker et al., 2005)

Single nucleotide polymorphisms

Approximately more than 1 million common variations in the human genome have been documented in the public database generated by sequencing diverse human genomes. A search for genetic difference in the form of single nucleotide polymorphism, which indicates human diversity, is continued in this HapMap project (Thorisson and Stein, 2003). Now more than 10 million genomes have been sequenced to find out more than 2 million differences counting rare SNPs (Altshuler et al., 2005; Botstein and Risch, 2003; Gu et al., 1998; McVean et al., 2005; Sherry et al., 2001). Many techniques have evolved to utilize this information with respect to cancer. Linking of specific genotype databases with diseases is now possible with individual SNPs as markers in the genome. Cancer involves large genomic insertions or deletions that can be recognized by SNP data (Hoque et al., 2003). Polymorphism found within the regulatory genes and variation in the protein function due to difference in the coding regions of some genes can affect the expression level of the gene (Marsh, 2005). Many cases have been found where expression level or specific polymorphism affected chemotherapeutic response in the patients (Landi et al., 2003). The correlation between an individual's reaction to chemicals and the polymorphisms in his genome is studied in pharmacogenomics (Bomgaars and McLeod, 2005; Marsh, 2005; Turesky, 2004). Microarray techniques are better than traditional sequencing to locate individual single point mutations and SNPs, if specific variants are known.

Microarray techniques

Microarray is the most promising technique for functional genomics. Screening with high density DNA microarrays allows the pattern of gene expression to be compared between tumor cells and normal cells. It involves complementary joining of two nucleic acid strands to generate duplexes. With the aid of this specific complementary binding, it is easily possible to locate the

specific sequences in billions of different sequences. Entire transcriptomes or single nucleotides within the genome can be screened using microarray. Microarray analysis provides valuable information on disease pathology, progression, resistance to treatment, and response to cellular microenvironments and ultimately may lead to improved early diagnosis and innovative therapeutic approaches for cancer (Pascale and Jeremy, 2002). Alternative splicing provides variation in the transcripts by changing the exon joining patterns and difference in start and stop points. Almost 40% of human genome is supposed to show alternative splicing. (Brett et al., 2000; Mironov et al., 1999; Modrek and Lee, 2003). The alterations in the type of transcripts through alternative splicing can influence the host susceptibility towards the cancer (Mercatante and Kole, 2000; Milani et al., 2006). Such type of splicing also affect on the individual response to the therapy for same type of tumor (Mercatante and Kole, 2000). These variations in tumor types and also in the response towards tumor can be correlated using microarray technologies (Bracco and Kearsey, 2003; Veuger et al., 2002).

Cancers can be identified using transcriptomes. Microarrays are one of the important tools to provide evidences for cancer. Gene expression profiles of various histologically similar tumor types allow altering the treatment of choice. Different types of tumors based on gene expression profiles have been identified. (Bucca et al., 2004; Cao et al., 2004; Elek et al., 2000; Halvorsen et al., 2005; Hu et al., 2005; Khan et al., 1999; Lee et al., 2004; Smith,2002; Sorlie et al., 2001; Wrobel et al., 2005; Zhang and Ji, 2005) . With microarray, 78% accuracy in prediction of unknown tumor type identification has been achieved (Ramaswamy et al. 2001). Using microarray technique and cDNA, 84% accuracy was found to identify unknown tumor type. Significant gene expression differences between patients suggesting that several subtypes might exist were noted that can explain the response in therapeutic variations (Perou et al., 2000). A positive or negative therapeutic response prediction through identified gene markers has been done (Cheok et al., 2003; Kakiuchi et al., 2004; McLean et al., 2004; Staunton et al., 2001). A study of advanced non-small cell lung cancer revealed 51 genes that predicted a response to Gefitinib (Kakiuchi et al., 2004). Almost 9000 genes expressed in AML have been evaluated to check the response before and after therapy with methotrexate and mercaptopurine, given alone or in combination (Cheok et al., 2003). In a study of 8000 genes and 60 cell lines from central nervous system, renal, ovarian, leukemia, colon, and melanoma neoplasms established the strength of the genomic approach to differentiate among tumor subtypes, using cDNA microarray (Ross et al., 2000). Microarray techniques are being utilized to locate genes that may result in tumor progression and metastatic potential (Agrawal et al., 2003; Henshall et al., 2003; Ramaswamy et al., 2003; Sanchez-Carbayo et al.,

2003; van 'T Veer et al., 2002; Vasselli et al., 2003). 295 patients have been evaluated using 70 gene classifier with stage I or stage II breast Cancer (van 'T Veer et al., 2002). Hence, prediction of clinical responses based on gene expression patterns in tumors is achievable. A reverse-phase microarray approach (RPA) was utilized to compare expression of several pro-survival proteins in micro-dissected normal and prostate cancer samples. Protein expression was studied using antibodies. It was revealed that early step in the development of cancer is phosphorylation and activation of AKT/PKB (Paweletz et al., 2001). National Cancer Institute studied 60 human cancer cell lines (NCI-60) to screen compounds for anticancer activity, using RP Microarray and recognition of two promising pathological markers to distinguish colon from ovarian adenocarcinomas in the abdomen was achieved (Nishizuka et al., 2003).

Chip-on-chip technology

Identification of DNA-binding sites of transcription factors is necessary to study regulation of transcriptomes. ChIP-on-Chip technology also known as Location Analysis (LA) is used for the same, it comprises microarray chips with chromatin immunoprecipitation. In situ cross linking of specific transcription factor to its DNA-binding site is carried out. The DNA is then fragmented and immunoprecipitated using a transcription factor specific antibody. These fragments of DNA are amplified by PCR, labeled and then hybridized to array. The DNA-binding sites for specific transcription factors within the genome can be evaluated using this technique (Horak and Snyder, 2002). Verification of insilico predictions of target genes regulated by ER alpha was also done (Jin et al., 2004). Identification of new targets of the p53 gene responding to ionizing radiations was recognized using the same technique. (Jen and Cheung, 2005) Within a whole genome, the patterns of DNA methylation related to disease status can be investigated (Wilson et al., 2006). This technology provides insight into key mechanisms of methylation, histone modification, as well as DNA replication, modification, and repair. It has been used to understand not only cancer but also diseases such as diabetes and leukemia. It has also provided important insight to vital processes like cell proliferation, cell fate determination, oncogenesis, cell cycle, apoptosis and neurogenesis.

Conclusion

Completion of human genome project and advancement in whole genome analysis is providing ample opportunities for comprehensive analysis and interpretation of cancer genomes, exomes, transcriptomes, and proteomes as well as epigenomic components. The integration of these data sets with well-annotated phenotypic

and clinical data will expedite improved interventions based on the individual genomics of the patient and the specific disease. Human diseases are no more a threat to scientists. This has been possible because of the advancement in technology and the fine approach. The recognition of DNA as origin of phenotypic expression, its role in gene expression and its manipulations has lead to many discoveries one of them is to know details about some diseases like cancer. With the human genome project completion and the detailing aspects of proteomics and genomics reinforced scientists to think in the vicinity of genes responsible for the development and viability of cancer. The recent era in approach would be significant in suggesting the treatment for cancer and also be able to predict the outcomes. It would be possible to opt for an individualized treatment in near future, with the help of microarray technology and proteomics. This requires generation, storage and processing of huge data from patients. This is feasible only with the help of computer scientists. Bioinformatics will have to play a major role in analyzing the data in order to provide a correct treatment plan for an individual.

REFERENCES

Adams MD, Kelley JM, Gocayne JD, Dubnick M, Polymeropoulos MH (1991). Complementary DNA sequencing: expressed sequence tags and human genome project. Science, 252: 1651-1656.

Agrawal D, Chen T, Irby R, Quackenbush J, Chambers AF, Szabo M, Cantor A (2003). Osteopontin identified as colon cancer tumor progression marker. CR. Biol., 326: 1041-1043.

Alaiya A, Al-Mohanna M, Linder S (2005). Clinical cancer proteomics: promises and pitfalls. J. Proteome Res., 4: 1213-1222.

Alexander H, Stegner AL, Wagner-Mann C, Du Bois GC, Alexander S, Sauter ER (2004). Proteomic analysis to identify breast cancer biomarkers in nipple aspirate fluid. Clin. Cancer Res., 10: 7500-7510.

Altshuler D, Brooks LD, Chakravarti A, Collins FS, Daly MJ, Donnelly P (2005). A haplotype map of the human genome. Nature, 437: 1299-1320.

Bayani JM, Squire JA (2002). Applications of SKY in cancer cytogenetics. Cancer Invest, 20: 373-386.

Beheshti B, Braude I, Marrano P, Thorner P, Zielenska M, Squire JA (2003). Chromosomal localization of DNA amplifications in neuroblastoma tumors using cDNA microarray comparative genomic hybridization. Neoplasia, 5: 53-62.

Bhattacharyya S, Siegel ER, Petersen GM, Chari ST, Suva LJ, Haun RS (2004). Diagnosis of pancreatic cancer using serum proteomic profiling. Neoplasia, 6: 674-686.

Bittencourt Rosas SL, Caballero OL, Dong SM (2001). Methylation status in the promoter region of the human PGP9.5 gene in cancer and normal tissues. Cancer Lett., 170(1):73-79.

Bomgaars L, McLeod HL (2005). Pharmacogenetics and pediatric cancer. Cancer J., 11: 314-323.

Borczuk AC, Shah L, Pearson GD, Walter KL, Wang L, Austin JH, Friedman RA, Powell CA (2004). Molecular signatures in biopsy specimens of lung cancer. Am. J. Respir. Crit. Care Med., 170: 167-174.

Botstein D, Risch N (2003). Discovering genotypes underlying human phenotypes: past successes for mendelian disease, future approaches for complex disease. Nat. Genet., 33: 228-237.

Bracco L, Kearsey J (2003). The relevance of alternative RNA splicing to pharmacogenomics. Trends Biotechnol., 21: 346-353.

Brett D, Hanke J, Lehmann G, Haase S, Delbruck S, Krueger S, Reich J, Bork P (2000). EST comparison indicates 38% of human mRNAs contain possible alternative splice forms. FEBS Lett., 474: 83-86.

Bucca G, Carruba G, Saetta A, Muti P, Castagnetta L, Smith CP (2004).

Gene expression profiling of human cancers. Ann. NY. Acad. Sci., 1028: 28-37.

Cao QJ, Belbin T, Socci N, Balan R, Prystowsky MB, Childs G, Jones JG (2004). Distinctive gene expression profiles by cDNA microarrays in endometrioid and serous carcinomas of the endometrium. Int. J. Gynecol. Pathol., 23: 321-329.

Capella G, Cronauer-Mitra S, Pienado MA, Perucho M (1991). Frequency and spectrum of mutations at codons 12 and 13 of the c-K-ras gene in human tumors. Environ. Health Perspect , 93: 125-131.

Caprioli RM (2005). Deciphering protein molecular signatures in cancer tissues to aid in diagnosis, prognosis, and therapy. Cancer Res., 65: 10642-10645.

Carulli JP, Artinger M, Swain PM, Root CD, Chee L, Tulig C, Guerin J (1998). High throughput analysis of differential gene expression. J. Cell Biochem. Suppl., 30-31: 286-296.

Casey G, Lindor NM, Papadopoulos N, Thibodeau SN, Moskow J, Steelman S, Buzin CH (2005). Conversion analysis for mutation detection in MLH1 and MSH2 in patients with colorectal cancer. JAMA, 293: 799-809.

Chen R, Yi EC, Donohoe S, Pan S, Eng J, Cooke K, Crispin DA, Lane Z, Goodlett DR, (2005). Pancreatic cancer proteome: the proteins that underlie invasion, metastasis, and immunologic escape. Gastroenterol., 129: 1187-1197.

Chaurand P, Sanders ME, Jensen RA, Caprioli RM (2004). Proteomics in diagnostic pathology: profiling and imaging proteins directly in tissue sections. Am. J. Pathol., 165: 1057-1068.

Cheok MH, Yang W, Pui CH, Downing JR, Cheng C, Naeve CW, Relling MV, Evans WE (2003). Treatment-specific changes in gene expression discriminate in vivo drug response in human leukemia cells. Nat. Genet., 34: 85-90.

Claus EB (1995). The genetic epidemiology of cancer. Cancer Surv., 25: 13-26.

Danna EA, Nolan GP (2006). Transcending the biomarker mindset: deciphering disease mechanisms at the single cell level. Curr. Opin. Chem. Biol., 10: 20-27.

Den Otter W, Koten JW, Van der Vegt BJ, Beemer FA, Boxma OJ (1990). Oncogenesis by mutations in anti-oncogenes: a view. Anticancer Res., 10: 475-487.

Dephoure N, Howson RW, Blethrow JD, Shokat KM (2005). Combining chemical genetics and proteomics to identify protein kinase substrates. Proc. Natl. Acad. Sci. USA, 102: 17940-17945.

Dobrindt U (2005). (Patho-)Genomics of Escherichia coli. Int. J. Med. Microbiol., 295: 357-371.

Dobrindt U, Agerer F, Michaelis K, Janka A, Buchrieser C, (2003). Analysis of genome plasticity in pathogenic and commensal Escherichia coli isolates by use of DNA arrays. J. Bacteriol., 185: 1831-1840.

Elek J, Park KH, Narayanan R (2000). Microarray-based expression profiling in prostate tumors. In vivo, 14: 173-182.

Frank TS, Deffenbaugh AM, Hulick M, Gumpper K (1999). Hereditary susceptibility to breast cancer: significance of age of onset in family history and contribution of BRCA1 and BRCA2. Dis. Markers, 15: 89-92.

Greengauz-Roberts O, Stoppler H, Nomura S, Yamaguchi H, Goldenring JR (2005). Saturation labeling with cysteine-reactive cyanine fluorescent dyes provides increased sensitivity for protein expression profiling of laser-microdissected clinical specimens. Proteomics, 5: 1746-1757.

Gronwald J, Jauch A, Cybulski C, Schoell B, Bohm-Steuer B (2005). Comparison of genomic abnormalities between BRCAX and sporadic breast cancers studied by comparative genomic hybridization. Int. J. Cancer, 114: 230-236.

Gu Z, Hillier L, Kwok PY (1998). Single nucleotide polymorphism hunting in cyberspace. Hum. Mutat., 12: 221-225.

Halvorsen OJ, Oyan AM, Bo TH, Olsen S (2005). Gene expression profiles in prostate cancer: association with patient subgroups and tumour differentiation. Int. J. Oncol., 26: 329-336.

Henshall SM, Afar DE, Hiller J, Horvath LG, Quinn DI, Rasiah KK (2003). Survival analysis of genome-wide gene expression profiles of prostate cancers identifies new prognostic targets of disease relapse. Cancer Res., 63: 4196-4203.

Hondermarck H, Vercoutter-Edouart AS, Revillion F, Lemoine J (2001).

Proteomics of breast cancer for marker discovery and signal pathway profiling. Proteomics, 1: 1216-1232.

Hoque MO, Lee CC, Cairns P, Schoenberg M, Sidransky D (2003). Genome-wide genetic characterization of bladder cancer: a comparison of high-density single-nucleotide polymorphism arrays and PCR-based microsatellite analysis. Cancer Res., 63: 2216-2222.

Horak CE, Snyder M (2002). ChIP-chip: a genomic approach for identifying transcription factor binding sites. Methods Enzymol., 350: 469-483.

Hu J, Bianchi F, Ferguson M, Cesario A, Margaritora S, Granone P, Goldstraw P (2005). Gene expression signature for angiogenic and nonangiogenic non-small-cell lung cancer. Oncogene 24: 1212-1219.

Jackson LG (1978). Chromosomes and cancer: current aspects. Semin. Oncol., 5: 3-10.

Jain KK (2002). Recent advances in oncoproteomics. Curr. Opin. Mol. Ther., 4: 203-209.

Jen KY, Cheung VG (2005). Identification of novel p53 target genes in ionizing radiation response. Cancer Res., 65: 7666-7673.

Jin VX, Leu YW, Liyanarachchi S, Sun H, Fan M, Nephew KP, Huang TH, Davuluri RV (2004). Identifying estrogen receptor alpha target genes using integrated computational genomics and chromatin immunoprecipitation microarray. Nucleic Acids Res., 32: 6627-6635.

Jones AM, Douglas EJ, Halford SE, Fiegler H, Gorman PA, Roylance RR, Carter NP, Tomlinson IP (2005). Array-CGH analysis of microsatellite-stable, near-diploid bowel cancers and comparison with other types of colorectal carcinoma. Oncogene., 24: 118-129.

Jotterand M, Parlier V (1996). Diagnostic and prognostic significance of cytogenetics in adult primary myelodysplastic syndromes. Leuk Lymphoma, 23: 253-266.

Kahn P (1995). From genome to proteome: looking at a cell’s proteins. Sci., 270: 369-370.

Kakiuchi S, Daigo Y, Ishikawa N, Furukawa C, Tsunoda T (2004). Prediction of sensitivity of advanced non-small cell lung cancers to gefitinib (Iressa, ZD1839). Hum. Mol. Genet., 13: 3029-3043

Kawamoto S, Yoshii J, Mizuno K, Ito K, Miyamoto Y (2000). BodyMap: a collection of 3′ ESTs for analysis of human gene expression information. Genome Res., 10: 1817-1827.

Khan J, Saal LH, Bittner ML, Chen Y, Trent JM, Meltzer PS (1999). Expression profiling in cancer using cDNA microarrays. Electrophoresis, 20: 223-229.

Kimura Y, Noguchi T, Kawahara K, Kashima K, Daa T, Yokoyama S (2004). Genetic alterations in 102 primary gastric cancers by comparative genomic hybridization: gain of 20q and loss of 18q are associated with tumor progression. Mod. Pathol., 17: 1328-1337.

Kuerer HM, Goldknopf IL, Fritsche H, Krishnamurthy S, Sheta EA, Hunt KK (2002). Identification of distinct protein expression patterns in bilateral matched pair breast ductal fluid specimens from women with unilateral invasive breast carcinoma. High-throughput biomarker discovery. Cancer, 95: 2276-2282.

Leal CB, Schmitt FC, Bento MJ, Maia NC, Lopes CS (1995). Ductal carcinoma in situ of the breast. Histologic categorization and its relationship to ploidy and immunohistochemical expression of hormone receptors, p53, and c-erbB-2 protein. Cancer, 75: 2123-2131.

Lee YF, John M, Falconer A, Edwards S, Clark J, Flohr P, Roe T, Wang R, Shipley J, Grimer RJ, Mangham DC, Thomas JM, Fisher C, Judson I, Cooper CS (2004). A gene expression signature associated with metastatic outcome in human leiomyosarcomas. Cancer Res., 64: 7201-7204.

Li YJ, Hoang-Xuan K, Zhou XP, Sanson M, Mokhtari K, Faillot T, Cornu P, Poisson M, Thomas G, Hamelin R (1998). Analysis of the p21 gene in gliomas. J. Neurooncol., 40: 107-111.

Lindlof A (2003). Gene identification through large-scale EST sequence processing. Appl. Bioinformatics, 2: 123-129.

Marsh S (2005). Thymidylate synthase pharmacogenetics. Invest New Drugs, 23: 533-537.

Martin GM, Hoehn H (1974). Genetics and human disease. Hum. Pathol., 5: 387-405.

McLean LA, Gathmann I, Capdeville R, Polymeropoulos MH, Dressman M (2004). Pharmacogenomic analysis of cytogenetic response in chronic myeloid leukemia patients treated with imatinib. Clin. Cancer Res., 10: 155-165.

McVean G, Spencer CC, Chaix R (2005). Perspectives on Human Genetic Variation from the HapMap Project. PLoS Genet., 1: 54.

Mercatante D, Kole R (2000). Modification of alternative splicing pathways as a potential approach to chemotherapy. Pharmacol. Ther., 85: 237-243.

Micci F, Teixeira MR, Haugom L, Kristensen G, Abeler VM, Heim S (2004). Genomic aberrations in carcinomas of the uterine corpus. Genes. Chromosomes Cancer 40: 229-246.

Milani L, Fredriksson M, Syvanen AC (2006). Detection of alternatively spliced transcripts in leukemia cell lines by minisequencing on microarrays. Clin. Chem., 52: 202-211.

Mironov AA, Fickett JW, Gelfand MS (1999). Frequent alternative splicing of human genes. Genome Res., 9: 1288-1293.

Modrek B, Lee CJ (2003). Alternative splicing in the human, mouse and rat genomes is associated with an increased frequency of exon creation and/or loss. Nat. Genet., 34: 177-180.

Mount DW, Pandey R (2005). Using bioinformatics and genome analysis for new therapeutic interventions. Mol. Cancer Ther, 4: 1636-1643.

Mundle SD, Sokolova I (2004). Clinical implications of advanced molecular cytogenetics in cancer. Expert Rev. Mol. Diagn., 4: 71-81.

Nakao K, Mehta KR, Fridlyand J, Moore DH (2004). High-resolution analysis of DNA copy number alterations in colorectal cancer by array-based comparative genomic hybridization. Carcinogenesis, 25: 1345-1357.

Nishizuka S, Charboneau L, Young L, Major S, Reinhold WC (2003). Proteomic profiling of the NCI-60 cancer cell lines using new high-density reverse-phase lysate microarrays. Proc .Natl. Acad. Sci., USA, 100: 14229-14234.

Okazaki Y, Furuno M, Kasukawa T, Adachi J (2002). Analysis of the mouse transcriptome based on functional annotation of 60,770 full-length cDNAs. Nature, 420: 563-573.

Parle-McDermott A, McWilliam P, Tighe O, Dunican D, Croke, DT (2000). Serial analysis of gene expression identifies putative metastasis-associated transcripts in colon tumor cell lines. Br. J. Cancer, 83: 725-728.

Pascale F. Macgregor , Jeremy A Squire (2002).Application of Microarrays to the Analysis of Gene Expression in Cancer, Clin. Chem., 48:1170-1177.

Paweletz CP, Charboneau L, Bichsel VE, Simone NL, Chen T, Gillespie JW (2001). Reverse phase protein microarrays which capture disease progression show activation of pro-survival pathways at the cancer invasion front. Oncogene, 20: 1981-1989.

Perin T, Canzonieri V, Massarut S, Bidoli E, Rossi C, Roncadin M, Carbone A (1996).Immunohistochemical evaluation of multiple biological markers in ductal carcinoma in situ of the breast. Eur. J. Cancer, 32A: 1148–1155.

Perou CM, Sorlie T, Eisen MB, van de Rijn M, Jeffrey SS, Rees CA, Pollack JR, Ross DT (2000). Molecular portraits of human breast tumours. Nature, 406: 747-752.

Petricoin EF, Liotta LA (2004). Proteomic approaches in cancer risk and response assessment. Trends Mol. Med., 10: 59-64.

Petricoin EF, Bichsel VE, Calvert VS, Espina V, Winters M, Young L, Belluco C (2005). Mapping molecular networks using proteomics: a vision for patient-tailored combination therapy. J. Clin. Oncol., 23: 3614-3621.

Pogosianz HE, Prigogina EL (1972). Chromosome abnormalities and carcinogenesis. Neoplasma, 19:319-325.

Rhodes DR, Chinnaiyan AM (2005). Integrative analysis of the cancer transcriptome. Nat Genet 37(Suppl): S31-S37.

Riccardo Alessandro, Simona Fontana, Elise Kohn, Giacomo De Leo (2005). Proteomic strageies and their application in cancer research. Tumori, 91: 447-455.

Ross, DT, Scherf U, Eisen MB, Perou CM, Rees C (2000). Systematic variation in gene expression patterns in human cancer cell lines. Nat. Genet., 24: 227-235.

Rowley A, Choudhary JS, Marzioch M, Ward MA, Weir M, Solari RCE, Blackstock WP, (2000). Applications of protein mass spectrometry in cell biology. Methods, 20: 383-397.

Sanchez-Carbayo M, Socci N.D, Lozano J.J, (2003). Gene discovery in bladder cancer progression using cDNA microarrays. Am. J. Pathol., 163: 505-516.

Schmidt H, Hensel M, Dobrindt U, Agerer F, Michaelis K, Janka A, Buchrieser C (2004). Pathogenicity islands in bacterial pathogenesis. Clin. Microbiol. Rev., 17: 14-56.

Seibert V, Wiesner A, Buschmann T, Meuer J (2004). Surface-enhanced laser desorption ionization time-of-flight mass spectrometry (SELDI ToF-MS) and Protein Chip technology in proteomics research. Pathol. Res. Pract, Vol., 200: 83-94.

Seike M, Kondo T, Fujii K, Okano T, Yamada T, Matsuno Y, Gemma A, Kudoh S, Hirohashi S (2005). Proteomic signatures for histological types of lung cancer. Proteomics, 5: 2939-2948.

Sengoelge G, Luo W, Fine D, Perschl AM, Fierlbeck W, Haririan A, Sorensson J, Rehman TU, (2005). A SAGE-based comparison between glomerular and aortic endothelial cells. Am. J. Physiol. Renal. Physiol., 288: F1290-F1300.

Sherry ST, Ward MH, Kholodov M, Baker J, Phan L, Smigielski EM, Sirotkin K.(2001). dbSNP: the NCBI database of genetic variation. Nucleic Acids Res., 29: 308-311.

Sivakumar A (2002). 2D gels and bioinformatics--an eye to the future. In Silico Biol., 2: 507-510.

Smith DI (2002). Transcriptional profiling develops molecular signatures for ovarian tumors. Cytometry, 47: 60-62.

Soreghan BA, Yang F, Thomas SN, Hsu J, Yang AJ (2003). High-throughput proteomic-based identification of oxidatively induced protein carbonylation in mouse brain. Pharm. Res., 20: 1713-1720

Sorlie T, Perou CM, Tibshirani R, Aas T, Geisler (2001). Gene expression patterns of breast carcinomas distinguish tumor subclasses with clinical implications. Proc. Natl. Acad. Sci. USA, 98: 10869-10874.

Squire JA, Pei J, Marrano P, Beheshti B, Bayani J, Lim G, Moldovan L, Zielenska M (2003). Highresolution mapping of amplifications and deletions in pediatric osteosarcoma by use of CGH analysis of cDNA microarrays. Genes Chromosomes Cancer, 38: 215-225.

Srinivas PR, Kramer BS, Srivastava S (2001). Trends in biomarker research for cancer detection. Lancet Oncol., 2: 698-704.

Stapleton M, Liao G, Brokstein P, Hong L, Carninci P, Shiraki T, Hayashizaki Y, Champe M (2002). The Drosophila gene collection: identification of putative full-length cDNAs for 70% of D. melanogaster genes. Genome Res., 12: 1294-1300.

Staunton JE, Slonim DK, Coller HA (2001). Chemosensitivity prediction by transcriptional profiling. Proc. Natl. Acad. Sci. USA, 98: 10787-10792.

Steel LF, Shumpert D, Trotter M, Seeholzer SH, Evans AA, London WT, Dwek R, Block TM (2003). A strategy for the comparative analysis of serum proteomes for the discovery of biomarkers for hepatocellular carcinoma. Proteomics, 3: 601-609.

Stults JT, Arnott D (2005). Proteomics. Methods Enzymol., 402: 245-289.

Thorisson GA, Stein LD (2003). The SNP Consortium website: past, present and future. Nucleic Acids Res., 31: 124-127.

Turesky RJ (2004). The role of genetic polymorphisms in metabolism of carcinogenic heterocyclic aromatic amines. Curr. Drug Metab., 5: 169-180.

Vasselli JR, Shih JH, Iyengar SR, Maranchie J, Riss J, Worrell R, Torres-Cabala C, Tabios R (2003). Predicting survival in patients with metastatic kidney cancer by gene-expression profiling in the primary tumor. Proc. Natl. Acad. Sci. USA, 100: 6958-6963.

Van' T Veer LJ, Dai H, van de Vijver MJ, He YD, Hart AA, Mao M, Peterse HL, (2002). Gene expression profiling predicts clinical outcome of breast cancer. Nature, 415: 530-536.

Velculescu VE, Zhang L, Vogelstein B, Kinzler KW (1995). Serial analysis of gene expression. Sci., 270: 484-487.

Velculescu VE, Zhang L, Zhou W, Vogelstein J, Basrai MA, Bassett DE, Jr Hieter P, Vogelstein B, Kinzler KW (1997). Characterization of the yeast transcriptome. Cell, 88: 243-251.

Veuger MJ, Heemskerk MH, Honders MW, Willemze R, Barge RM (2002). Functional role of alternatively spliced deoxycytidine kinase in sensitivity to cytarabine of acute myeloid leukemic cells. Blood, 99: 1373-1380.

Volker Seibert, Matthias PA Ebert, Thomas Buschmann (2005). Advances in clinical cancer proteomics: SELDI-ToF-mass spectrometry and biomarker discovery. Briefings in Functional Genomics and Proteomics., 4: 16-26.

Weiss MM, Snijders AM, Kuipers EJ, Ylstra B, Pinkel D, Meuwissen SG, van Diest PJ, Albertson DG, Meijer GA (2003). Determination of amplicon boundaries at 20q13.2 in tissue samples of human gastric adenocarcinomas by high-resolution microarray comparative genomic hybridization. J. Pathol., 200: 320-326.

Weinberg RA (1983). Alteration of the genomes of tumor cells. Cancer, 51: 1971-1975.

Wick LM, Qi W, Lacher DW, Whittam TS (2005). Evolution of genomic content in the stepwise emergence of Escherichia coli O157:H7. J. Bacteriol., 187: 1783-1791.

Wilson IM, Davies JJ, Weber M, Brown CJ, Alvarez CE, MacAulay C, Schubeler D, Lam WL (2006). Epigenomics: mapping the methylome. Cell Cycle, 5: 155–158.

Wrobel G, Roerig P, Kokocinski F, Neben K, Hahn M, Reifenberger G, Lichter P (2005). Microarraybased gene expression profiling of benign, atypical and anaplastic meningiomas identifies novel genes associated with meningioma progression. Int. J. Cancer, 114: 249-256.

Wulfkuhle JD, Paweletz CP, Steeg PS, Petricoin EF, Liotta L (2003). Proteomic approaches to the diagnosis, treatment, and monitoring of cancer. Adv. Exp. Med. Biol., 532: 59-68

Yeatman TJ (2003). The future of clinical cancer management: one tumor, one chip. Am. Surg., 69: 41-44.

Zhang LH, Ji JF (2005). Molecular profiling of hepatocellular carcinomas by cDNA microarray (2000). WorldJ Gastroenterol., 11: 463-468.

Zhang Z, Bast RC, Yu Y, Li J, Sokoll LJ, Rai AJ (2004). Three biomarkers identified from serum proteomic analysis for the detection of early stage ovarian cancer. Cancer Res., 64: 5882-5890.

Zucchi I, Mento E, Kuznetsov VA, Scotti M, Valsecchi V (2004). Gene expression profiles of epithelial cells microscopically isolated from a breast-invasive ductal carcinoma and a nodal metastasis. Proc. Natl. Acad. Sci. USA, 101: 18147-18152.

12

Genomic imprinting: A general overview

Muniswamy K.[1] and Thamodaran P. [2]

[1]Division of Veterinary Biotechnology, Indian Veterinary Research Institute, Izatnagar- 243 122, India.
[2]Division of Veterinary Microbiology, TANUVAS, Chennai-600 007, India.

Usually, most of the genes are biallelically expressed but imprinted gene exhibit monoallelic expression based on their parental origin. Genomic imprinting exhibit differences in control between flowering plants and mammals, for instance, imprinted gene are specifically activated by demethylation, rather than targeted for silencing in plants and imprinted gene expression in plant which occur in endosperm. It also displays sexual dimorphism like differential timing in imprint establishment and RNA based silencing mechanism in paternally repressed imprinted gene. Within imprinted regions, the unusual occurrence and distribution of various types of repetitive elements may act as genomic imprinting signatures. Imprinting regulation probably at many loci involves insulator protein dependent and higher-order chromatin interaction, and/or non-coding RNAs mediated mechanisms. However, placenta-specific imprinting involves repressive histone modifications and non-coding RNAs. The higher-order chromatin interaction involves differentially methylated domains (DMDs) exhibiting sex-specific methylation that act as scaffold for imprinting, regulate allelic-specific imprinted gene expression. The paternally methylated differentially methylated regions (DMRs) contain less CpGs than the maternally methylated DMRs. The non-coding RNAs mediated mechanisms include C/D RNA and microRNA, which are invovled in RNA-guided post-transcriptional RNA modifications and RNA-mediated gene silencing, respectively. The maintenance and reprogramming of imprinting are not significantly affected by reduced expression of *Dicer1* and the evolution of imprinting might be related to acquisition of DNMT3L (*de novo* methyltransferase 3L) by a common ancestor of eutherians and marsupials. The common feature among diverse imprinting control elements and evolutionary significance of imprinting need to be identified.

Key words: Genomic imprinting, differentially methylated regions (DMRs), non-coding RNA, imprinting evolution.

INTRODUCTION

Genomic imprinting is a germline-specific epigenetic modification of the genome that results in parent-of-origin-specific expression of a small subset of genes in offspring. The concept of genomic imprinting introduced by Metz (1938) and Crouse (1960), who coined the term in the context of the unique inheritance of sex chromosomes in the dipteran insect, *Sciara coprophila*. Zygote consisting of two maternal genomes is called gynogenones or parthenogenones and zygote, which contained two paternal genomes is called androgenones. Neither of these two types of reconstituted zygote could develop to term but the former had better embryos, and the later, better development of placental tissues, which suggested that the parental genomes are functionally non-equivalent despite the fact that they have equivalent genetic information. This observation led to discovery of genomic imprinting, which indicate functional difference that is dictated by the parental origin of the genome (McGrath and

Solter, 1984; Surani et al., 1984). The first imprinted gene to be identified was the insulin-like growth factor 2 (*Igf2*), which is expressed exclusively from chromosome of paternal origin (Dechiara et al., 1991). Imprinted genes are involved in many versatile functions including development, growth, complex regulation of the physiology, metabolism of both embryonic and adult stages, and behavior of adult stage. The current list of imprinted genes and information about them can be obtained at http://www.mgu.har.mrc.ac.uk/research/imprinting/(Beechey et al., 2005). Imprinting operating at various level like genes and segments of chromosomes in mammals, whole chromosome (inactivation of paternal X-chromosome in extra embryonic tissue of mice), are the unique example of the whole haploid genome being subjected to genomic imprinting by inactivating the paternal set of chromosomes in male mealybugs (Schrader and Hughes-Schrader, 1931; Brown and Chandra, 1977). This phenomenon has been observed in flowering plant, in eutherian mammals (such as mice and humans) (Tilghman, 1999; Ferguson-Smith and Surani, 2001; Reik and Walter, 2001a) and marsupials (such as opossums and kangaroos) (O'Neill et al., 2000; Suzuki et al., 2005), but not in monotremes (egg-laying mammals such as platyplus and echidnas) (Killian et al., 2000) or birds (such as chicken) (O'Neill et al., 2000; Nolan et al., 2001; Yokomine et al., 2001, 2005).

Fate of imprints during mammalian life cycle

Expression of imprinted genes requires an epigenetic marking process that allows the transcription machinery to distinguish between expressed and repressed parental alleles of imprinted genes. Parent-of-origin specific epigenetic markers are combined at fertilization in the zygote. Just after fertilization, the paternal pronucleus undergoes active DNA demethylation. It must be an active demethylation process, because it occurs before DNA replication. However, maternal pronucleus undergoes demethylation over the first several divisions, suggesting a passive demethylation process (Rougier et al., 1998). Parent specific methylation patterns in imprinted genes are retained even in the face of genome-wide demethylation by an unknown mechanism (Reik and Walter, 2001b). These parental imprints are propagated in somatic tissue during embryogenesis. Some parental imprints in the placenta are maintained by histone modifications independent of DNA methylation. In case of primordial germ cells (PGCs), PGCs migration into the genital ridge occurs at 10.5-11.5 days post-coitus (dpc). Whereas the erasure of imprint occurs identically in PGCs of developing male and female embryo between 11.5 and 12.5 dpc (Hajkova et al., 2002; Lee et al., 2002), parental imprints are re-established later in gametogenesis in a strictly sex dependent manner. In males, this process occurs between 14.5 and 18.5 dpc, while in females it occurs in growing oocytes postnally (Ferguson-Smith and Surani, 2001; Reik and Walter,

2001b; Surani, 2001) (Figure 1). In germ cells, it is not clear whether both imprints are first erased and then the sex-appropriate imprint is established or whether the sex-appropriate imprint is retained and only the sex-inappropriate erased, though the first possibility is more likely. In PGCs, the timing of this erasure and deposition of new marks is distinct for each sex and for each gene. In male demethylation of maternal and paternal allele of *H19* is complete around 13.5 dpc, but the paternal allele is remethylated first. However, by the time the male enters meiosis, both alleles are equally methylated (Davis et al., 1999; Ueda et al., 2000). This phenomenon of asynchronous methylation of maternal and paternal alleles has also been demonstrated for the maternally methylated *Snrpn* gene in oocyte, with the maternal allele methylated first (Lucifero et al., 2004).

Expression profiles of Megs and Pegs in mammalian life cycle

Before erasure of genomic imprinting memories, both somatic cells and PGCs exhibit monoallelic expression profiles of paternally and maternally imprinted regions, which are represented as 1:1 and 1:1. After erasure (non-marked state), expression profiles change to 0:2 and 2:0; because in paternally imprinted region, paternally expressed genes (pegs) become silent and maternally expressed genes (Megs) become biallelically expressed, but in maternally imprinted region, Pegs and Megs become biallelically expressed and silent, respectively. Paternal imprinting only affects the expression profiles of paternally imprinted region producing a 2:0 and 2:0 pattern and maternal imprinting only affects the expression profiles of maternally imprinted region leading to a 0:2 and 0:2 pattern (Obata et al., 1998; Obata and Kono, 2002). It is because paternal imprint is necessary for induction of the paternally imprinted Pegs, simultaneously repressing paternally imprinted Megs, and that the maternal imprint is necessary to induce maternally imprinted Megs with simultaneous repression of maternally imprinted Pegs. The monoallelic expression profiles of Pegs and Megs are reproduced in individuals in the next generation by combining each of the parental imprinted alleles (Figure 2). The resulting reciprocal (complementary) monoallelic expression is necessary for normal mammalian development (Kaneko-Ishino et al., 2003).

DIFFERENTIAL CHARACTERISTICS OF GENOMIC IMPRINTING

Genomic imprinting not only exhibit inter and intra species difference but also between embryonic and extraembryonic tissues of an individual.

Differential characteristics between plants and animals

One obvious explanation for the divergent aspects of im-

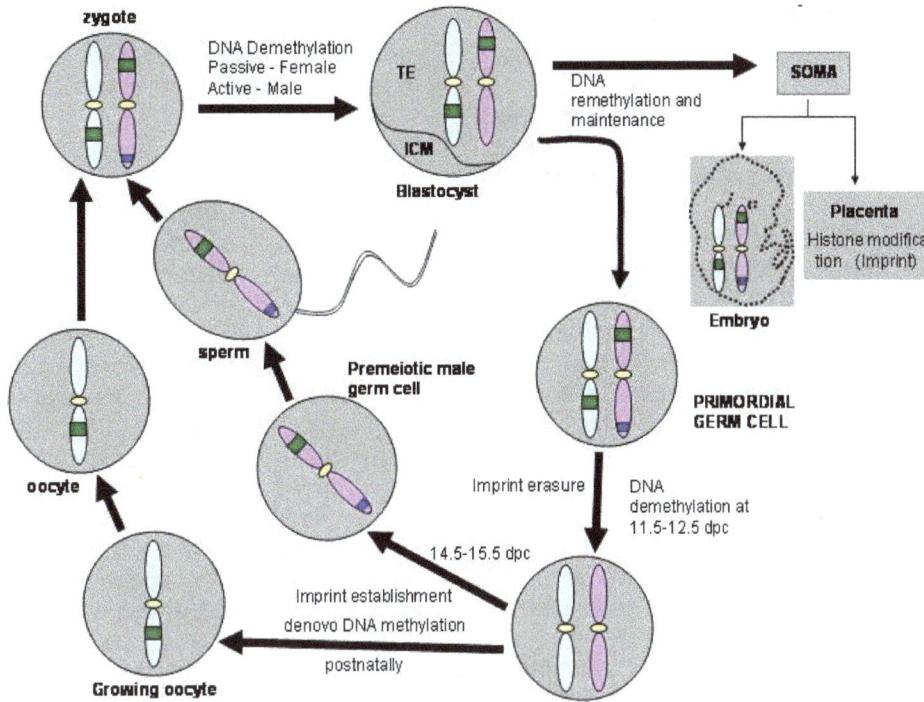

Figure 1. Ontogeny of imprint in germ cell. Gametic methylation imprint represented as green bar as well as other methylated DNA that is not retained in the early embryo as blue bar.

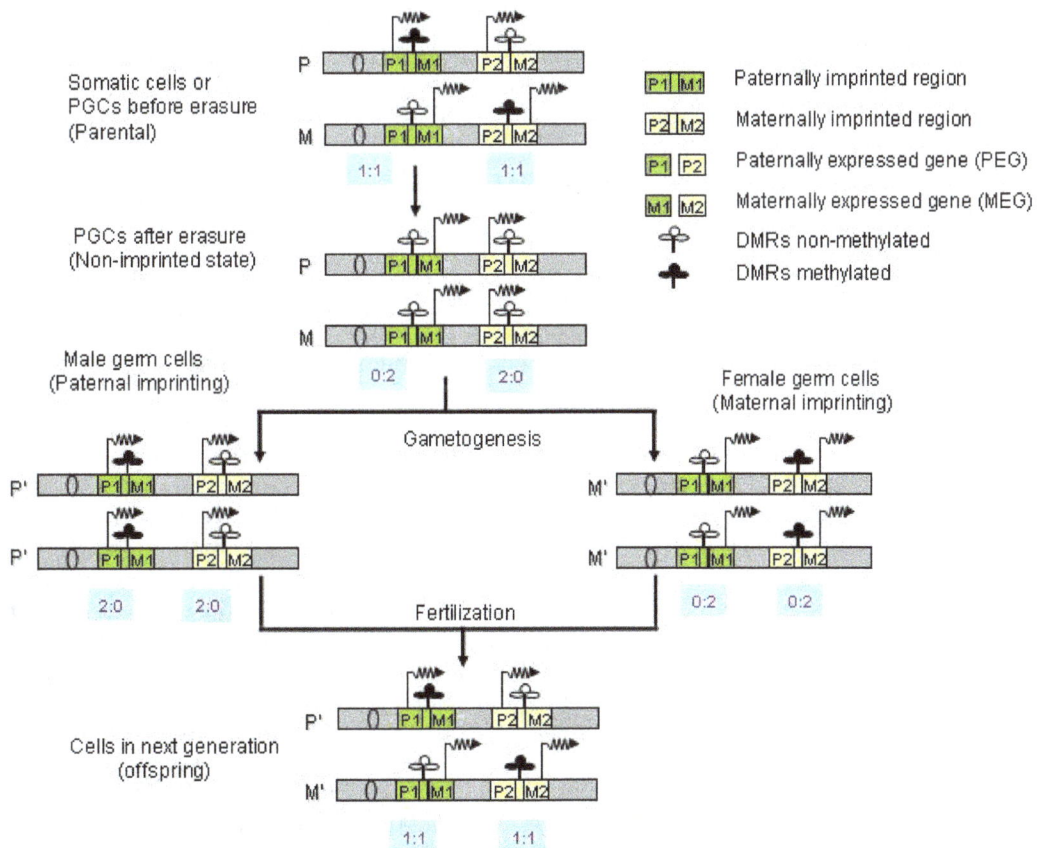

Figure 2. Pegs and Megs expression profiles during mammalian life cycle. M: Maternal chromosome, M': female germ cell chromosome, P: paternal chromosome, P': male germ cell chromosome.

printing is the difference in life history of the mammals and flowering plants (Walbot and Evans, 2003).

1. Both plants and mammals silence imprinted genes using DNA methylation and associated chromatin remodeling, but in mammals imprinted alleles are targeted for silencing, while in plants the selective activation of imprinted genes is achieved by specifically removing pre-existing methylation from the allele destined to be active (Gerhing et al., 2004; Scott and Spielman, 2004).
2. In plants imprinted gene expression is largely (or) wholly restricted to the endosperm, functional homologue of the mammalian placenta (Gehring et al., 2004), while in mammals, imprinted loci show allele specific expression across many regions of the placenta and embryo (Hu et al., 1998).
3. The need for erasing and resetting imprint mark is by passed because the cells whose descendants will eventually form gamete are participating in development throughout the life of the plant, however in mammals the germline is set aside early in embryogenesis.
4. The lack of global demethylation and resetting of imprint in plants requires maintenance but not *de novo* methyltransferase activity.

Sexual dimorphism of imprinting

Although total number of paternally expressed imprinted genes is similar to that of maternally expressed genes, the following are striking differences between maternal and paternal imprinting.

1. *De novo* methylation occur fairly late in the maturation process in growing oocyte, from midsize to metaphase II oocytes, just prior to ovulation (Lucifero et al., 2004). Methylation imprints are therefore of short duration and occur after meiotic recombination in female germline. In male, germ cell methylation mark established prior to meiosis and exists for the entire reproductive life span of the individual.
2. Most paternally repressed imprinted genes are indeed not associated with DMRs but are controlled in cis by paternally expressed nontranslated RNAs because very long term existence of germline methylation mark and large number of cell division as many as 100 times occuring in male germ cells after methylation led to high rate DMRs erosion via accumulation of C→T mutation (Crow, 2000), and the evolution of methylation independent, RNA or transcription based silencing mechanism. Whereas transient existence of germline methylation in female lead to low rate of loss of CpG sites by mutation, hence retention of methylation impriting at DMR.
3. *Dnmt3L* is required for imprint establishment but not for transposon methylation in female germ cells, but *Dnmt3L* is required for transposon methylation and has only a minor role in *de novo* methylation at imprinted loci in male germcells.

4. Many CpG island-associated promoters are subject to maternal methylation but no known promoters are subject to paternal-specific germline methylation.

Imprinting in placenta

The genes that are specifically imprinted in trophoblast are expressed on the maternal and repressed on the paternal allele revealed by the expression studies in mice and humans. In embryos, reading of the imprinting marks relies mainly on DNA methylation (Li et al., 1993; Reik and Walter, 2001b), however histone modification seem to regulate imprinting in the placental tissues (Lewis et al., 2004; Umlauf et al., 2004). DNA methylation is probably the more stable epigenetic mark in the embryo, which may be appropriate in this case for the imprint to survive in adulthood, while histone modifications may serve purpose in the short-lived placenta that is not required after birth. Imprinting maintenance in the absence of DNA methylation was first reported for *Ascl2* on distal chromosome 7, its placenta-specific imprinting persists in *Dnmt1*-deficient embryos (Caspary et al., 1998; Tanaka et al., 1999). Silencing of *Kcnq1* paternal allele in placenta involve trimethylation of lysine 27 (K27) and dimethylation of K9 on histone H3 and histone H3 are deacetylated and show complete absence of K4 methylation.

REPETITIVE ELEMENT ROLE IN IMPRINTING

Several years ago, Mary Lyon put forward an attractive model that suggests repeated elements may function as genetic waystations that support the spreading of epigenetic inactivation mechanisms along the X-chromosome (Lyon, 1998).

Retrotransposable element

The human and mouse X chromosome possess a high content of long interspersed nuclear elements (LINE1) and Alu elements which is a member of (SINE) short interspersed nuclear elements are less frequent on the human X chromosome as compared to autosomes (Bailey et al., 2000; Waterston et al., 2002; Ross et al ., 2005). It is well documented that G+C poor isochores are relatively poor in SINE and enriched in LINE elements and vice versa (Zhang et al., 2004), but it was found that the LINE and SINE content was not correlated to the relative G+C content of the analyzed regions. The overall content and also the relative distribution of retrotransposable elements is characteristic features of imprinting clusters and X chromosome. It is based on observations that the SINE depletion in imprinted regions is most evident in intergenic regions (Ke et al., 2002; Allen et al., 2003; Luedi et al., 2005) and pronounced depletion of SINEs is observed near imprinting control centers in the Beckwith-wiedemann syndrome (BWS) region, that is, between genes that exhibit consistent tissue independent imprinting (Greally, 2002).

Subfamilies of primate Alu elements were shown to be hypomethylated in sperm as compared to oocytes and somatic tissues (Hellmann-Blumberg et al., 1993; Rubin et al., 1994). In contrast, LINE1 elements are hypermethylated in sperm and somatic tissues (Howlett and Reik, 1991; Lane et al., 2003). LTRs are not significantly over-or under-represented in imprinted regions hence it less likely that they are important elements for the germline specific marking of imprinted regions. Since the imprinted genes and retrotransposed elements became *de novo* methylated during similar time in female and male germ cells respectively, they might involve similar mechanistic machineries for both types of DNA elements. Since Dnmt3a and Dnmt3l act in the male as well as in the female germlines, germline-specific protein might bind specifically to DMRs and/or repetitive elements, thereby either inducing or inhibiting epigenetic modifications. In a nutshell, retrotransposable element might create an environment for allele specific marking in the germline by recruiting epigenetic machineries to specific genomic locations.

Tandem repeats

It has been hypothesized that tandemly repeated DNA elements in or close to DMRs are involved in regulation of imprinting (Neumann et al., 1995). Tandem repeat arrays in imprinted regions in animals can be divided into two large subclasses long repeat motifs that encode small RNAs (snoRNA and miRNA) and direct tandem repeats of variegating length.

Long repeat motifs

The long repeat motifs of more than 70 bp length have sequence encoding snoRNA and miRNA. These repeat motifs are separated by unique sequence or retrotransposed elements so they may cover more than 100 kb long genomic segments. The major examples of such repeats are the snoRNA clusters in the Prader-Willi/Angelman syndrome region, and miRNAs and snoRNA in the *Dlk1/Gtl2* region.

Direct tandem repeats

The motifs range between 5 and 400 bp in length are arranged in direct head to tail order. At least 23 imprinted genes contain direct tandem repeats in or close to DMRs. Almost all the most prominent DMRs, the so called imprinting centers (ICs) that regulate mono-allelic expression of numerous neighboring genes, possess direct tandem repeats.

In contrast to the positional conservation individual sequence motifs and the surrounding DMR sequence are poorly conserved, prominent examples are the IG-DMR upstream of *Gtl2* and the imprinting centers IC1 and IC2

in the BWS region (Paulsen et al., 2001, 2005). The IC1 upstream of *H19* regulates reciprocal imprinting of the maternally expressed *H19* gene and the neighboring paternally expressed *Igf2* gene. The function of the IC1 depends on a high concentration of bound CTCF protein that can be achieved by the array of repeated binding motifs. The repeated nature of direct tandem repeat motifs may be required directly or indirectly to control the establishment and function of an allele-specific chromatin structure.

Tandem repeat motifs are among the hallmarks of imprinted genes. It is postulated that both, the local effects caused by direct tandem repeats and the large genomic effects promoted by the retrotransposon signatures may be needed to create a suitable environment for the establishment and maintenance of imprinting. In addition to these, rather general features of imprinted genes, other locus and gene specific structures like non-coding RNAs are very likely to also have a strong influence. It seems very likely that the mix of such general and locus-specific signatures promotes imprinting of genes.

Paoloni-Giacobino et al. (2006) hypothesized a model in which the primary role of tandem repeats of primary DMDs is most likely to act as highly effective epigenetic maintenance signal (to maintain the gametic methylation pattern indefinitely following fertilization). According to this model, during the fourth embryonic S phase (8 cell embryo) hemimethylated tandem repeats of the DMD attract Dnmt1o maintenance methyl transferase, this lead to its catalytic activity and inheritance of DMD sequence methylation. Dnmt1o could be attracted to the hemimethylated DMD due to the arrangement of CpG dinucleotides, the imperfect repetitiveness, and a specific chromatin structure induced by the hemimethylated DMD.

EPIGENETIC MODIFICATIONS

In this section, we discussed epigenetic modification, which will help in better understanding of genomic imprinting regulation. It is defined as heritable and reversible instructions superimposed over the DNA (Jenuwein and Allis, 2001). Epigenetic mechanisms can operate either at the level of DNA or at the level of chromatin. DNA in the form of chromatin has greatly enhanced range of possible epigenetic modification.

DNA methylation

A key attribute of genomic imprinting is the inheritance of expression potential from the gamete through all of prenatal development, and cytosine methylation is the only mark known to be transmitted by mitotic inheritance; there is no evidence of a mechanism that would mediate the inheritance of states of histone modification (Goll and Bestor, 2002). Palindromic or symmetric cytosine that lie directly 5' of guanine residue (CPG's) are the predominant target of DNA methylation in mammals and plants.

Cytosine within CNG and CNN are also commonly methylated in plants and filamentous fungi (Selker and Stevens, 1985). The methylation reaction is catalysed by DNMTs using S-adenosyl metheonine as the methyl donor. Methylation associated gene silencing is achieved either by blocking access of transcription factors to DNA, or through the recruitment of methyl-CpG binding domain (MBD) protein, which form complexes with histone de-acetylases, histone methyltransferase or chromatin remodeling proteins to generate transcriptionally refractory chromatin (Li, 2002). DNA methylation is not exclusively associated with silencing, for example the paternal allele of insulin-like growth factor type 2 (*Igf2*) is expressed by virtue of methylation within an adjacent imprinting control region which prevent binding of the enhancer-blocking zinc-finger protein CCCTC binding factor (CTCF) (Bell and Felsenfeld, 2000; Hark et al., 2000). Genomic imprint should be inherited after they are established in the gamete. Consistent with this, the methylation patterns at several DMDs have been shown to be perpetuated in pre implantation embryos and later during fetal development.

Factors regulating methylation of imprinted gene

Cis acting DNA signal for methylation

The cis-acting signals for DNA methylation have been identified at four imprinted loci discussed below.

Igf2r: *Igf2r*, a maternally expressed gene located on mouse distal chromosome 17 (Barlow et al., 1991) contain a DMD within intron 2 (region 2) that acquires maternal allele methylation. A 6-bp sequence required to protect the paternal allele from being methylated, at the 5' end of 113-bp fragment of region 2 identified by mutation is a region referred to as the allele discriminating signal (ADS). Similarly, mutations identified an 8-bp sequence at 3' end of 113-bp fragment referred to as the *de novo* methylation signal (DNS), sufficient for acquiring DNA methylation in either parental pronucleus. ADS and DNS function in methylation establishment but not for methylation maintenance.

H19: The *H19* DMD contains sequence needed for methylation maintenance but not for methylation establishment. The CTCF binding sites within the *H19* DMD are needed to prevent maternal allele methylation.

Snrpn: It is paternally expressed imprinted gene on chromosome 7. Two DNS signal and one ADS signal identified using a human transgene in mouse may function to establish methylation (Kantor et al., 2004). The 10 and 7-bp MPI sequences required for maintenance of paternal imprint is termed maintenance of paternal imprint (MPI).

Rasgrf1: The DMD and repeats together constitute a binary switch that regulates imprinting at *Rasgrf1*. The repeated sequence element is required for establishment of DMD methylation in the male germline at the endogenous

locus (Yoon et al., 2002). Repeats are also required for maintenance of methylation at *Rasgrf* in the pre-implantation embryo but are dispensable for maintenance of methylation after embryonic day 5.

The protein factors, other than DNMTs, which collaborate with the DNMTs to direct them to target DNA sequence that acquire methylation are discussed below.

Polycomb group proteins

EED is a member of polycomb group, that has been shown to be part of multimeric protein complex that has histone methyl transferase (HMT) and histone deacetylase (HDAC) activity (van der Vlag and Otte, 1999; Czermin et al., 2001). Different mechanisms for imprinted methylation and expression operate at different loci because *Eed* mutant affect only a subset of imprinting.

Chromatin modifying factor

The trans acting factors important for histone modifications, especially histone H3 lysine 9 methylation (H3mK9) and HP1, which binds H3mK9, are critical for normal DNA methylation in Neurospora (Tamaru and Selker, 2001), Arabidopsis (Jackson et al., 2002) and mice (Lehnertz et al., 2003). Though the connection between several forms of DNA methylation and histone methylation is clear, it is ambiguous how imprinted DNA methyltion is affected.

Chromatin remodeling factor

DDM1 (decrease in DNA methylation 1) encodes a member of SWI/SNF2 family, which reposition nucleosome in an ATP dependent manner affect DNA methylation (Jeddeloh et al., 1999). It is also required to maintain histone H3 methylation pattern (Gendrel et al., 2002), providing evidence for mutual interactions among histone modifications, DNA methylation and chromatin remodeling in human mutation in *ATRX*, which encode an SWI/SNF-like protein exhibit decreased DNA methylation at repetitive DNA (Gibbons et al., 2000).

Regulation of CTCF

The switching from CCCTC binding factor (CTCF) to BORIS (testis specific paralog of CTCF) expression and down regulation of CTCF (present in male and female germline) may be critical for allele specific methylation of H19. Another explanation for how CTCF may regulate allele-specific methylation is poly ADP-ribosylation of CTCF by PARP1 seems to regulate CTCF in its role as an insulator. Reale et al. (2005) showed that PARP1 can form a complex with *Dnmt1 in vivo* and possibly protect the unmethylated states of CpG islands.

Putative ADS/DNS binding protein

A specific protein present only in androgenetic ES cells bind to ADS of *Igf2*. Similarly, specific DNS binding factor is present in both androgenetic and gynogenetic ES cells extracts. These specific proteins involved in the binding to the ADS and DNS is yet to be shown to play a role in regulating methylation.

Histone modifications

The n-termini of histones, the histone-tails, protrude out of the nucleosome and the aminoacid residues at the tails are sites for various post-translation modifications, like phosphorylation, sumolylation, ubiquitination, methylation and acetylation (Li, 2002). The importance of chromatin remodelling and covalent modification of histones resulting in unique 'histone code' in maintaining the development fate of cellular lineages is being increasingly recognized. The histone code hypothesis proposes that transcriptionally active and silent regions have characteristics patterns of histone-tail modifications (Jenuwein and Allis, 2001). Transcriptionally active genes are characterized by methylation of histone H3 lysine 4 and acetylation of H3 and H4. Common hallmarks of silent genes are H3K9me, H3K27me and H4K20me (Jenuwein and Allis, 2001; Sarma and Reinberg, 2005). The complexity is further increased by mono, di and tri-methylation of lysine residues. MacroH2A, a variant of H2A, is associated with transcriptionally inactive X chromosome (Heard, 2004; Sarma and Reinberg, 2005).

The nuclear periphery is thought to contain chromosomal domain that are less transcriptionally active than domains located in the nuclear center (Kosak and Groudine, 2004).

Interactions of epigenetic modifications

Lewis and Reik (2006) predicted a model for how maternal germline DNA methylation can lead to a paternal postzygotic methylation mark. The IC2 and *Igf2r/Air* clusters contain a maternally methylated imprinting centre (IC), non-coding RNAs and postzygotic paternally methylated DMRs. The germline DMR leads to allele-specific expression of a non-coding RNA. This RNA may be responsible for recruiting polycomb protein and H3 lysine 27 methylation at pre-implantation stages. After implantation, the lysine 27 methylation recruits lysine 9 methylation. The histone methylation then target DNA methylation of a postzygotic DMR (Figure 3). In the *Igf2/H19* locus, the maternal germline methylation is necessary to create other paternal, postzygotic DMRs. On the maternal chromosome, CTCF binds to the unmethylated IC at an unknown time in early development. This leads to the establishment of allele-specific higher-order chromatin structures which then protect the IC and other DMRs from *de novo* DNA methylation. The epigenetic stages involved in the establishment of the postzygotic DMRs are unknown but a possible pathway involves histone methylation as proposed by Lewis and Reik (2006)

(Figure 3). Heterochromatin spread occurs due to recruitment of the Suvar methylases by Hp1, methylation of adjacent nucleosome followed by Hp1 binding and so on (Hall et al., 2002).

REGULATORY MECHANISM

Most of **the** imprinted gene clusters are under control of discrete DNA element called imprinting centres (ICs). The molecular and cellular mechanism by which ICs control other gene and regulatory regions in the cluster involves insulation of gene and non-coding RNAs including micro-RNAs (miRNAs) (O'Neill, 2005). The miRNAs may repress translation, induce degradation of mRNA, or be involved in chromatin remodeling. Although, the whole complex nature of the imprint itself remains to be fully elucidated, the key player of the imprint involve one or more differentially methylated regions displaying an allele-specific DNA methylation pattern, which determine the expression status of the imprinted genes (Murrell et al., 2004). In addition, the differential histone modifica-tions associated with some imprinted genes are thought to regulate chromatin organization and gene expression (Grewal and Moazed, 2003).

Imprinting centre and its characteristics

An imprinting centre is defined as one or more discrete DNA elements, which regulate imprinted gene expression and epigenotype throughout an imprinting cluster. It regulates in cis imprint resetting and imprint maintenance in the whole domain. Genes within imprinting clusters share many regulatory elements and often have similar developmental and tissue specific patterns of expression.

The first IC to be genetically characterized was the differentially methylated region (DMR) upstream of H19, which is also called differentially methylated domain (DMD) or ICR (Imprint control region). DMRs are the region in many imprinted genes, which show parent-of-origin-dependent DNA methylation patterns on one parental allele but not on the opposite allele. There are two classes of DMR, primary DMRs acquire gamete-specific methylation in either spermatogenesis or oogenesis and maintain the allelic methylation difference throughout development. Secondary DMRs establish differential methylation pattern after fertilization, most probably through the influence of primary DMRs (Lopes et al., 2003). So far, 15 primary DMRs have been identified, among which 12 are maternally methylated and 3 are paternally methylated. Loss of the primary DMRs often result in aberrant expression of associated imprinted gene and such DMRs are called imprint control regions (ICRs) (Wutz et al., 1997; Thorvaldsen et al., 1998; Fitzpatrick et al., 2002; Yoon et al., 2002; Lin et al., 2003; Williamson et al., 2004; Liu et al., 2005).

Since genomic imprinting is a cis-directed process, cer-

IC2/Air cluster   *Igf2/H19* cluster

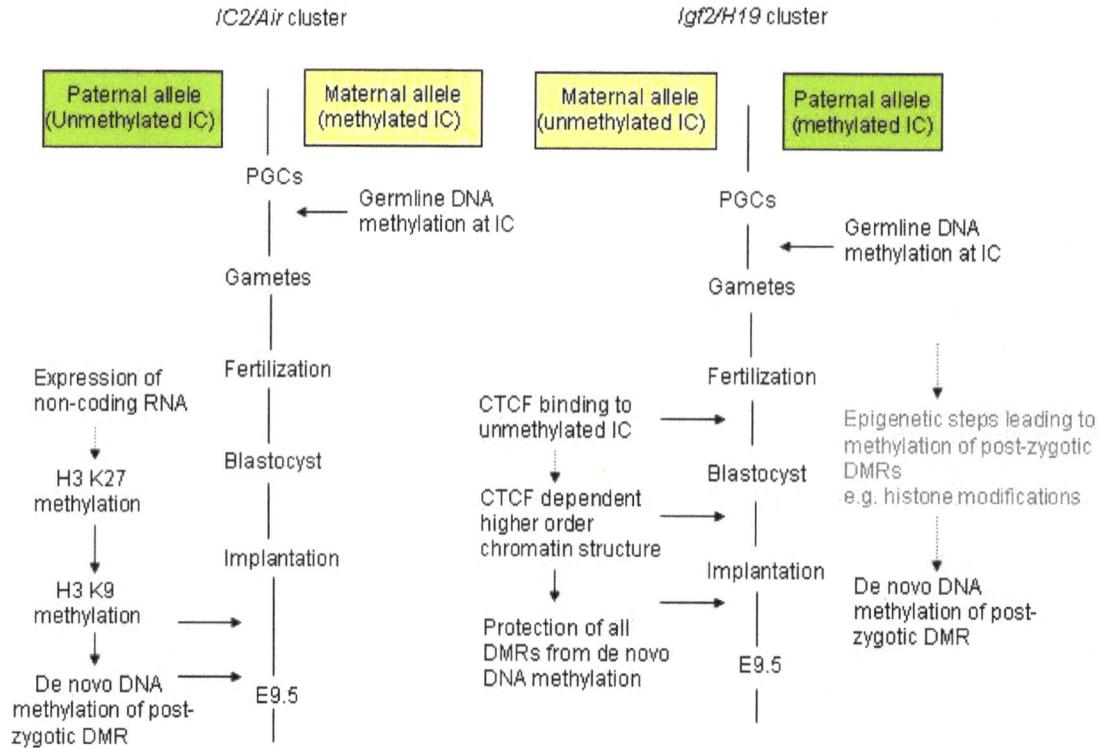

Figure 3. The predicted order of establishment of epigenetic modifications at imprinting clusters in early embryonic development. Hatched arrows indicate progression for which there is no experimental evidence. Light grey text indicate predicted steps which have not been demonstrated.

tain structural features of DMDs might indeed be the cis-acting signals directing correct establishment and maintenance of genomic imprints (Rand and Cedar, 2003). Consistent transgene imprinting was only obtained when 130 kb of genomic sequence surrounding the H19 DMD or when 300 kb of contiguous *Igf2r* sequence including DMD2 where used to generate transgenic lines (Wutz et al., 1997). It was hypothesized that large amount of contiguous genomic sequence of an imprinted locus are essential for its imprinting at an ectopic genomic location. This implies that there are two likely general genomic arrangements of the cis-acting imprinting signals. Either the signals are very large or there are multiple small signals scattered throughout the genomic region defined by the size of consistently imprinted mouse transgenes. DMDs play crucial role in both the differential expression and differential epigenetic marking of the parental alleles of imprinted genes.

Kobayashi et al. (2006) determined the extents of 15 primary DMRs in 12.5-dpc mouse embryo by sulfite sequencing. They found that DMRs have more CpGs than the whole mouse genome, but in general, CpGs are less than the non imprinted CpG islands. However, three (*Peg3, Snrpn* and *H19*) out of 15 DMRs has no CpG islands. They found that some DMRs had sharp boundaries and others had transition zones. One possible explanation for less CpG content in DMRs than non

imprinted CpG islands was due to mutation susceptibility of primary methylated DMRs CpG dinucleotides, so they gradually lost CpGs and inevitably accumulate C/T transition in successive generation during evolution. It is well known that methylated cytosine is mutable to thymine (C/T transition) by spontaneous deamination (Holliday and Grigg, 1993).

Kobayashi et al. (2006) found that the average G+C content of the paternally methylated DMRs was significantly smaller than that of the maternally methylated DMRs and paternally methylated DMRs contain less CpGs than the maternally methylated DMRs. The difference in CpG content between paternally and maternally methylated DMRs was due to longer persistence of methylation imprints in male than female germline based on observation that in male paternal methylation imprints are established in gonocyte in the fetal testis and persist in the germline throughout the reproductive life of the male (Davis et al., 1999; Ueda et al., 2000; Li et al., 2004). By contrast, the maternal methylation imprints are imposed in growing oocytes after birth (Li et al., 2004; Lucifero et al., 2004). In addition, male germline divides many times after methylation imprints are established but female germcell do not. It is also possible, however that the CpG content is one of the features recognized by the *de novo* methylation machinery. CpG islands are generally free of methylation but weak or small CpG islands may lose protection from methylation in one of the germlines and could behave as DMRs. DMRs are present in imprinted gene promoters, including antisense RNA genes, in chromatin boundaries, in silencer and activator sequences, suggesting their role in regulating monoallelic expression. The following are characteristics of imprinting centre:

1. Some large imprinting clusters can be further subdivided into domains containing separate ICs whose functions are limited to the domains.
2. All ICs contain germline (primary) DMRs. In addition to germline DMRs there are also post-zygotic (secondary) DMRs that are established after fertilization (Kierszenbaum, 2002).
3. In primordial germ cells, DNA methylation in both types of DMR is erased before the parental specific methylation is established.
4. With respect to DNA methylation, the parental alleles of imprinted genes have different levels of CPG methylation in DMD that are located at specific sites within or surrounding the gene, one parental allele is methylated on the majority of CPG dinucleotides within a DMD and the opposite parental allele is methylated on a small percentage of CPG dinucleotides or not methylated at all.
5. In addition to differential DNA methylation, ICs also show allelic differences in chromatin structure, namely Dnasel hypersensitivity and covalent modifications of histone tails.
6. In general, repressive histone modifications such as methylation at histone H3 lysine 9 and lysine 27 are found at ICs and other DMRs on the methylated allele, whereas activating histone modifications such as H3 and H4 acetylation and H3 lysine 4 methylation are found on the unmethylated allele.
7. The allele-specific difference in histone modifications mark promoter, exonic and intergenic regions (whole of the ICs cluster) in extraembryonic tissue, whereas in embryonic tissue only the DMRs are marked (Lewis et al., 2004; Umlauf et al., 2004).
8. Imprinting centre act as a chromatin insulator. An insulator is defined as a sequence of DNA that blocks enhancers from interacting with gene promoter when positioned between the two and/or acts as a barrier to the spread of transcriptionally repressive condensed chromatin.
9. The CCCTC binding factor (CTCF) is known to bind to insulators and mediate their enhancer blocking activity. There are also non-CTCF dependent insulators at imprinting loci. It is possible that YY1 or unknown proteins may bind to other ICs to give them insulator activity in a similar manner to CTCF.
10. Noncoding RNA transcribed in antisense to that of target gene, have promoters within or near ICs.
11. Methylation of DMRs are exceptional due to fact that they escape the genome-wide demethylation that takes place during the first cleavages in embryo development. They also avoid the global *de novo* methylation that takes

place post implantation (Kafri et al., 1992; Razin and Shemer, 1995).

Hence, currently the defining properties of ICs appear to be that they cis regulate other imprinted genes within the same clusters. Differential DNA methylation is established in germlines and maintained in somatic tissues of the offspring. Differential histone modifications also mark allelic differences. They act as chromatin insulator and contain promoter for non-coding RNAs.

Different mechanisms involved in regulation

Imprinting regulation probably at many loci involves insulator protein dependent and higher-order chromatin interaction (looping); non-coding RNAs mediated mechanisms; and mechanisms involving both (Dual role of ICs).

For co-ordination of epigenotype across an imprinting cluster, the first model proposed was linear spreading of DNA methylation (Turker and Bestor, 1997). However, after examining regional control of DNA methylation in the *Igf2/H19* region, Lopes et al. (2003) proposed an alternative model: long range chromatin interactions (or) looping.

Looping model

The two possible mechanisms by which chromatin looping could be involved in the epigenetic regulation of imprinting clusters was revealed by the following examples.

The first example involves *Igf2/H19* locus, the ICs of the unmethylated maternal chromosome of this locus interacts with the DMR1 region of *Igf2* and a downstream matrix attachment region (MAR3) in a CTCF dependent manner. The binding of CTCF to the IC creates an inactive chromatin domain surrounding *Igf2* which is specific to the unmethylated maternal allele, which prevents downstream enhancers directly interacting with *Igf2* promoter region. In the case of imprinted *Dlx5* locus, the methylation sensitive binding of MeCP2 (methyl binding protein) to a DMR form an inactive chromatin structure on the paternal chromosome and an active structure on the maternal allele. There is redundancy between methyl binding domain proteins (MBDs) at imprinted loci. Another method of forming discrete domain of active or silent chromatin, in large scale could involve tethering of imprinted regions to fixed structures in the nucleus. In the case of paternal allele of *Igf2/H19* locus, the chromatin structure in absence of CTCF binding due to methylation of IC seems to depend on matrix attachment regions (MARs), which attach the paternal allele to the nuclear structure. MARs on both parental alleles have also been found in large number between the two imprinting clusters on the mouse distal chromosome 7 (Purbowasito et al., 2004). Thus, MARs may also act as boundary elements to block the spreading of chromatin modifications or to form separate chromatin or loops. Labrador and Corces (2002) proposed a model in which boundary or insulator elements throughout the genome establish chromatin domains, organizing the chromatin fibres into local compartment of condensed (inactive) chromatin and decondensed (active) chromatin. This model can be extended to imprinted genes, thus an unmethylated insulator may bind to a nuclear structure dividing the imprinting cluster into two and dividing promoters and enhancers into separate compartments. If the same insulator is inactivated by methylation, they cannot form two separate domains and therefore enhancers and promoters will interact.

Mechanism involving non-coding RNAs

The best characterized mammalian non-coding RNA is *Xist*, which coats the future inactive 'X' and triggers the events which lead to gene silencing along the length of the X chromosome (Heard, 2004). In two well characterized clusters (IC2 and *Igf2r/Air*), these RNAs are expressed from the paternal allele, which silence the surrounding genes paternally. In the *Igf2r/Air* IC, the critical element is the *Air* non-coding RNA transcript. It is not known, how the *Air* RNA mediates silencing of the surrounding genes. However, possible models include an RNAi based mechanism, transcription through specific DNA elements or a chromosome coating mechanism as in X chromosome inactivation. The *Kcnqlotl* non-coding RNA from the IC2 cluster has many similar features of *Air* and *Xist* RNA. Its trancripition is initiated from the unmethylated paternal IC in the IC2 cluster and all the flanking genes are paternally silenced.

General features of Non-coding RNAs: The presence of non-coding RNA genes, often (but not always) transcribed in the opposite orientation to protein-coding genes, is a recurrent theme in imprinted domains.

a. C/D RNA: C/D type small RNAs (C/D RNA) are metabolically stable, 60-to 300-nucleotide-long, and present in eukaryotes and archaebacteria. In eukaryotic cells, they reside in nucleus, either in the nucleolus (they are called C/D snoRNAs for C/D small nucleolar RNAs) or in the cajal bodies (they are called scaRNAs for small cajal body-specific RNAs). In vertebrate, vast majority of C/D RNA genes are located within the intron of protein coding gene, more often and also within the introns of non-coding RNA genes. Most of the C/D RNAs appear to be processed from the host gene introns through exonucleolytic degradations of the debranched lariat. The human C/D RNA sequences are available at (http://www-snorna.biotoul.fr/). C/D RNAs owe their names to canonical structural motifs, the C-box (consensus 5'-PuUGAUGA-3') and the D-box (consensus 5'-CUGA-3') present close to their 5' and 3' termini, respectively. They also contain more degenerate C' and D' boxes that occupy an internal position within the RNA sequence. Many C/D RNAs contain conserved antisense elements (8 to 21 nucleotide

long) positioned upstream from the D- and/or the D'-motifs. C/D RNAs act by pairing with RNA targets on which they guide ribose methylation at specific ribonucleotide (the modified nucleotide is always paired to the fifth nucleotide upstream from the D or D' box) (Figure 4a). C/D snoRNAs modify the Pol I transcribed pre-rRNAs or the Pol III-transcribed U6 spliceosomal snRNA while scaRNAs modify the Pol II-transcribed U1, U2, U4 and U5 spliceosomal snRNAs (Figure 4a). A broad proportion of C/D RNAs-including most of the imprinted C/D RNAs is devoid of any specific RNA target, so it might target other cellular transcript (including mRNA?) or might play a different role in RNA modification guiding in which no pairing with a target RNA is needed.

b. microRNA (miRNA): It constitute the largest eukaryotic small RNA family discovered so far, they are 21-to 23-nucleotide-long single-stranded RNA molecules that are processed from one arm of an irregular 60-to 70-nucleotide-long hairpin structure called pre-miRNA. The pre-miRNA genes exhibit different genomic organization. They can be transcribed from their own promoters either as independent entities or as polycistrons, or they can be included in larger transcript units of coding or non-coding genes (Kim, 2005). A pre-miRNA gene is first transcribed by RNA pol II as a several kb long pri-miRNA (the miRNA primary transcript). Then, the RNAse-III type enzyme Drosha assisted by DGCR8 make pair of cuts on the large pri-miRNA precursor to give rise to the pre-miRNA, this cut establishes one end of the miRNA. The pre-miRNA is then translocated to the cytoplasm by exporting 5-mediated export, where another multidomain RNAse-III Dicer enzyme create other extremity of the miRNA, to yield a short RNA duplex. This duplex is unwound upon loading on the RISC (the mature miRNA-containing ribonucleoparticle), giving rise to the single-stranded mature miRNA. The mouse and human miRNA sequences are available at http://www.sanger.ac.uk/Software/Rfam/mirna). Two pathways can be adopted for miRNAs mediated gene silencing at the post-transcriptional level, by basepairing with a target RNA (Figure 4b): (i) if perfect (or almost perfect) complementarity is shared, the target RNA is directed for cleavage; (ii) If the miRNA presents a partial complementarity with an mRNA (generally in its 3' UTR part), then the translation of the target mRNA is made non-productive by still poorly characterized mechanisms. The most 5' part of the miRNA (2nd to 8th nucleotide, also called the 'seed') plays a critical role for target recognition (Doench and Sharp, 2004). Plant miRNA with the full complementarity to its target can also act at the translation level (Aukerman and Sakai, 2003). In plant systems, miRNA has been shown to direct asymmetric DNA methylation within the gene it targets (Bao et al., 2004). Whether such a mechanism operating in the nucleus at the DNA level can also take place in mammals is unknown. By computational predictions of miRNA target, about 10 to 30% of mRNA populations are potentially

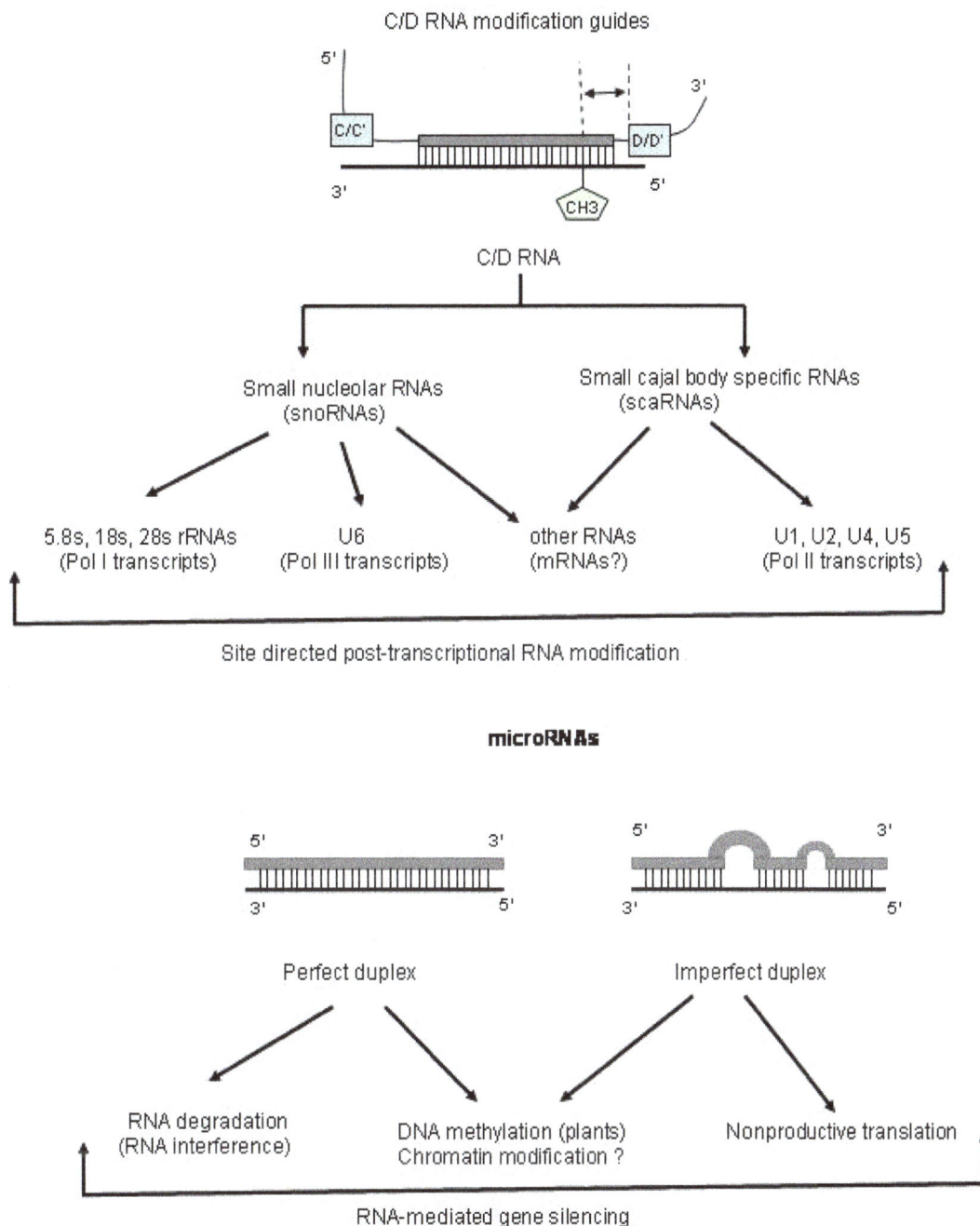

Figure 4. Function of (a). C/D RNA modification guides and (b). miRNAs. The sequence complementary to cognate target RNA are depicted as grey region of the strand.

targeted by miRNAs. But concretely, few genetic data about the biological role of vertebrate miRNA are available.

Imprinted small RNA genes: More than 100 C/D RNA and miRNA genes were predicted by use of computational methods and systematic DNA cloning strategies in two imprinted loci of mammals: human 14q32 (so called callipyge domain) and 15q11-q13 (so-called prader-willi domain). They seem to share several structural and functional characteristics: (i) they are grouped into clusters of homologous repeated gene copies, with most embedded within introns of large non-coding genes subject to alternative splicing; (ii) They are not found in non-eutherian mammals; (iii) they display a tissue specific expression pattern with prepondent expression in adult brain; (iv) they lack an obvious functional antisense element against a cellular transcript except few and their molecular and biological functions remain highly elusive.

The target identification for small RNA is difficult task because imprinted small RNAs are specific to eutherian species (they were not detected in chicken or even opossum brain). Which considerably limits the use of comparative sequence analysis for target identification, and functional parts of the small RNAs involved in target specificity (C/D snoRNA sequence upstream from the D/D' boxes or the 5' seed for miRNAs) can vary notably from one copy to the other, suggesting that they may target different RNA species. The miRNAs and RNAi related processes play a key role in neuronal function, brain morphogenesis as well as stem cell self-renewal and cell fate decisions and behavioural phenotypes. A miRNA (miR-32) is at the basis of an antiviral defense mechanism in human cells (Le Cellier et al., 2005). In *Schizosaccharomyces pombe*, Dicer produced siRNAs derived from centromeric repeats are incorporated into RITS (RNA-induced transcriptional silencing complexes), a RISC-related complex, to direct heterochromatin formation and gene silencing at the transcriptional level (Matzke and Birchler, 2005). Several lines of evidence in mammalian system support the existence of nuclear RNA-mediated gene silencing mechanisms: (i) small RNA-mediated RNA degradation is active in the nucleus (Robb et al., 2005); (ii) DNA methylation and histone H3K9 methylation can be triggered by small interfering RNAs (Kawasaki and Taira, 2004; Morris et al., 2004), although this may not occur in all experimental systems; (iii) Dicer-deficient mouse ES cells show reduced levels of DNA and histone H3K9 methylation of centromeric DNA, high levels of centromeric repeats (Kanellopoulou et al., 2005). A conditional loss of functional mutant of Dicer in hybrid human-chicken cells also causes accumulation of transcript from alpha-satellite sequences and leads to abnormal mitotic cells, presumably due to a defect in the formation of centromeric heterochromatin (Fukagawa et al., 2004; Kanellopoulou et al., 2005). Thus, Dicer-related RNAi machinery is required for the silencing of centromeric heterochromatin in vertebrates. Repeated sequence are known to attract silencing and considering the involvement of small RNAs and the RNAi machinery in the construction of repressive chromatin in mammals, it is legitimate to propose that imprinted small RNAs-and more especially the large miRNA gene at 14q32 could account for certain epigenetic regulation at the basis of genomic imprinting. In many RNA-mediated gene silencing mechanism, like X chromosome inactivation (mammals), RNA-directed DNA methylation (plants) or heterochromatinization at the sexual maternal locus (*S. pombe*), the RNA and/or the RNAi machinery is mainly required to initiate the epigenetic state. Hence, miRNAs might guide epigenetic marking early during development (in the germline), but once the imprints are installed on gamete haplogenomes, they might be dispensable for the subsequent maintenance and reading of the imprints. DNA methylation plays a central role in genomic imprinting. First, there is interest to investigate

their potential involvement in DNA methylation, including non-canonical contexts as observed in plants (on cytosine in CpNpG and asymmetric sequence contexts). Additionally, miRNAs could help to install an allele-specific chromatin state through recruitment of histone modification complexes.

RNA induced methylation: RNA directed DNA methylation (RdDM) was first demonstrated by Wassenegger et al. (1994) and has been extensively studied in plants (Chan et al., 2005). RdDM is carried out by double stranded RNA (dsRNA), which may be produced from transcription through inverted repeats. Double stranded RNA is then cleaved by the RNAase III family enzyme Dicer into small interfering RNAs (siRNAs) of 21-26 length which are able to direct methylation of homologous sequences. The involvement of histone modifiers and DNMTs in RdDM indicate a complex system of regulation of DNA in plants. siRNAs have since been shown to direct DNA methylation in human cells (Kawasaki and Taira, 2004; Morris et al., 2004). These finding suggest that, a similar mechanism may act at imprinted loci, in which dsRNA could be formed by transcription through inverted repeats often present in it or simultaneous transcription of sense and antisense transcript. Martienssen (2003) proposed a mechanism by which tandem repeat can continuously produce siRNAs via the use of RNA dependent RNA polymerase (RdRP). RdRP is identified in RNA viruses, yeast and plants, but it is yet to be identified in mammalian systems. From the common appearance of tandem repeats at imprinted loci and at heterochromatized regions elsewhere in the genome, it is tempting to speculate that such repeats regulate imprinted methylation by an RNA dependent mechanism. At first, miRNAs were identified as translational repressors by binding to the 3' UTRs of target genes but recent evidence also supports a role for miRNAs in DNA methylation (Bao et al., 2004).

Dual role of ICs

The IC2/*Kcnq1* and the *Igf2r/Air* clusterS contain placental-specific imprinted genes on one side of the IC and genes imprinted in both embryonic and extra embryonic lineages on the other. In the mouse IC2/*Kcnq1* cluster these two sets of genes are clearly regulated by IC in two different ways. One involves CTCF binding to the insulator on the unmethylated paternal IC and using higher-order chromatin structures to restrict enhancer access. On the maternal chromosome CTCF binding is blocked by DNA methylation. This mechanism is active in all lineages and results in stable imprinting of genes on the telomeric side of the locus. The other mechanism occurs only in the extraembryonic lineage but may begin in the pre-implantation embryo. This involves the *Kcnq1otI* RNA, possibly coating the chromosome and recruiting repressive histone modifications to the placen-tally imprinted genes on the centromeric side of paternal allele. The histone modifications are recruited to genes on both sides of the IC but are probably irrelevant on the telomeric side since the stronger, looping-based mecha-nism takes precedence. The RNA/histone methylation mechanism is much less stable, is not maintained through the later stages of gestation. The *Air* RNA is important for imprinting in both embryonic and extraembryonic line-ages in *Igf2r/Air* cluster (Sleutels et al., 2002) but there are clear differences between the placental specific imprinted genes which have no associated DMRs and *Igf2r* which exhibits methylation dependent imprinting in many tissues.

The PWS/AS cluster IC is described as bipartite. Establishment of imprinting on the paternal chromosome requires the PWS-IC in cis whereas establishment of imprinting on the maternal chromosome requires the AS-IC. In addition, the tissue-specific maternal expression of *Ube3a* requires the *Ube3a* antisense RNA. Thus there may be as many as three mechanism regulating imprinting in this cluster.

Role of dicer in imprinting

In fission yeast, Dicer and Argonaute are required not only for RNAi but also for transcriptional silencing of centromeric repeats by chromatin modifications (Hall et al., 2002; Volpe et al., 2002). Dicer generated siRNA form the RNA induced initiation of the transcriptional gene silencing (RITS) complex, which contains Argonaute protein and is required for heterochromatin silencing (Noma et al., 2004). Dicer is required for centromeric heterochromatin silencing also in chicken and mouse cell (Fukagawa et al., 2004; Kanellopoulou et al., 2005). In mammals, although recent papers have reported that the introduction of siRNAs complementary to the promoter regions of some genes induces DNA methylation (Kawasaki and Taira, 2004; Morris et al., 2004), the native target of transcriptional gene silencing induced by siRNA have been unclear. In addition, an RNAi-related phenomenon called cosuppression was observed for an imprinted gene, *U2af1-rs1* (Hatada et al., 1997). The mechanism underlying cosuppression is very similar to those of RNAi because there are common RNA intermediates and also similar genes are required in the silencing pathway (Hamilton and Baulcombe, 1999; Hannon, 2002).

Fukasawa et al. (2006) suggested that the reduced expression of *Dicer1* did not affect the genomic imprinting and a very low level of Dicer enzyme would be sufficient for silencing of functional genes. With respect to this, maintenance of imprinting requires a very low level of DNA methylase activity because imprinting continues during the early embryonic stages when the DNA methylase activity is reduced to a minimum. *Dicer1* was not required for the maintenance of transcriptional silencing at pericen-

centromeric satellite sequence, the maintenance of DNA methylation and X chromosome inactivation in female cells, and the stable shutdown of developmentally regulated gene (Cobb et al., 2005). The similarity between X chromosome inactivation and genomic imprinting indicate that, *Dicer1* may not be required for maintenance of genomic imprinting.

HYPOTHESIS FOR EVOLUTION OF GENOMIC IMPRINTING

Conflict hypothesis or kinship theory

According to this theory, the situation of direct codependence results in competition for resources between foetus and mother. The paternal expressed gene like *Igf2* will favour growth of each individual foetus at the expense of all other foetuses and mother, resulting in larger progeny that have a competitive advantage over those of other males. But the maternal genome will favour an equal distribution of resources among all foetuses and preservation of itself for future pregnancies (Haig and Graham, 1991; Moore and Haig, 1991). Though it is plausible and intellectually pleasing, all imprinting phenomena cannot be explained (Hurst and Mc Vean, 1997; Iwasa, 1998). This theory has been challenged by the lack of phenotype observed in knockout mice of the maternally imprinted small nuclear ribonucleoprotein N (Snrpn), which was expected to show an overgrowth phenotype and paternal expressed Zn finger and C2 domain protein (Zac) has growth inhibitory effect. A possible prediction by Haig (2006) is that inhibitors of demands located on the X-chromosome and enhancers of demand located on autosomes, because X-linked loci are maternally derived two-thirds of the time by contrast to autosomal loci which are maternally derived half of the time. The prediction that the X- chromosome should express a bias towards inhibiting the demands that offspring impose on their mothers is called the hypothesis of X-Linked Inhibitory Bias (XLIB).

Complementation hypothesis

It is proposed from the mechanistic point of view, imprintng regulation in somatic and germ cells by Lee et al. (2002) and Kaneko-Ishino et al. (2003). It argues that genomic imprinting is essential for mammalian development as a mechanism regulating complementary or reciprocal expression profiles of paternal and maternal genomes, because Pegs and Megs cannot be expressed from the same chromosome simultaneously, even when the parental imprints are completely erased. In addition, it also rescues Pegs and Megs involved in development and growth from catastrophic situation, in which the expression of either half of the imprinted gene was lost.

Barrier to parthenogenesis in mammals

Genomic imprinting prevents accidental or unexpected parthenogenesis, which is life threatening and undesirable in females, because food and environmental factors like temperature and climatic conditions suitable for breeding is seasonal. It requires genetic contribution from both parents, and is evolutionarily advantageous in producing variation by mixing genetic information. The "paternal dual barrier theory" state that two sets of coordinately imprinted genes, Igf2-H19 and Dlk-Gtl2, function as a critical barrier to parthenogenetic development in order to render paternal contribution obligatory for the descendants of mammals (Kono, 2006).

Host defense mechanism

According to this proposal, genomic imprinting arose as an accessory system by which mammalian genome represses exogenous DNA sequences using DNA methylation. However, it fails to explain why imprinting occurs exclusively in mammals because DNA methylation and retrotransposons are not unique to mammals. It also does not explain why all imprinted genes are not methylated.

Ovarian time bomb theory

Genomic imprinting by placing control of placental development on the paternal genome would have a protective effect from trophoblastic tumorigenesis in females, which could become malignant in the absence of genomic imprinting. But it does not explain imprinting of neither the paternal genomes nor why genes which are not involved in placental development are still imprinted.

Novel placental hypothesis

The significant relationship between placental formation and genomic imprinting in mammals and observation that most imprinted gene are expressed in placenta lead to proposal of this hypothesis by Kaneko-Ishino et al. (2003). It is assumed that imprinted gene is regulated to ensure appropriate expression in placental tissue, which enabled the ancestral mammal to form placental structures.

EVOLUTIONARY LINK BETWEEN DNMT3L AND IMPRINTING

One of the important functions of DNA methylation in mammals is to regulate genomic imprinting. In mammals, DNMT3A and DNMT3B are so-called de novo methyltransferase, which create new methylation patterns on non-methylated DNA (Bestor, 2000). DNMT1 (DNA Methyltransferase1) is a maintenance methyltransferase, which methylates the newly synthesized hemimethylated DNA strand after DNA replication. de novo methyltransferase 3A (DNMT3A) and a related protein with no methyltransferase activity, DNMT3L, have been shown to be essential for the establishment of germline specific methylation imprints associated with imprinted genes (Bourc'his et al., 2001; Hata et al., 2002; Kaneda et al., 2004). After fertilization, DNMT1 propagates the methyltion patterns and regulates parent-of-origin-specific gene expression in somatic tissues of embryo and adults (Li et al., 1993).

A possible link is obtained between the existence of this protein and the evolution of genomic imprinting from the observation that DNMT3L is present in eutherians and marsupials but likely to be absent in birds and fish. Thus, DNMT3L is a key regulator of genomic imprinting, and acquisition of this gene via gene duplication in a common ancestor of eutherians and marsupials may have been a critical event in the evolution of imprinting. The original function of DNMT3A is probably essential to the survival and/or development of vertebrates but a mechanism evolved only in placental mammals utilized this enzyme to establish the parent-of-origin-specific methylation imprints.

CONCLUSION

The possession of DNMT3L and placenta during mammalian evolution are probable key events for evolution of imprinting in placental mammals. In essence, the presence and consequently also the absence of distinct types of repetitive elements may influence the accessibility of DMRs to the specific epigenetic modification machineries in the parental germlines. Although, the whole nature of imprinting mechanisms needs to be unraveled, the higher order chromatin (looping) and non-coding RNAs may play significant role in imprinting. It has been proven that low level of Dicer did not affect imprinting. Even though several theory has been proposed, there is need for unified hypothesis for imprinting evolution that encompasses all imprinted gene.

Future research

We are at initial stages of this phenomenon, despite being the tremendous amount of knowledge we have amassed just over two decades. Ingenious and insightful experiment will be required to show the whys and hows of imprinting. In germ cell, the query of whether all parental imprint are erased and established freshly or sex specific imprint are retained, while other nonspecific imprint are erased need to be solved. Elucidation of various mechanisms involved in regulation of imprinting, enlighten other scientific area like stem cell research, cloning and imprinting disorder. There are exciting question for future research on the evolutionary origins of autosomal imprinting in the placenta and its link to imprinted X inactivation.

Although the role of cis acting DNA signal like ADS and DNS is attributed to DNA methylation, the *in vivo* function need to be established. There is need for unified hypothesis for imprinting evolution that encompasses all imprinted gene.

ACKNOWLEDGEMENTS

We sincerely apologize to all the authors whose work, due to space limitation is not cited. The authors thank Dr. Annabelle Lewis and Dr. W. Reik for permission to reproduce flowchart (Figure 3) from their published papers.

REFERENCES

Allen E, Horvath S, Tong F, Kraft P, Spiteri E, et al. (2003). High concentrations of long interspersed nuclear element sequence distinguish monoallelically expressed genes. Proc. Natl. Acad Sci. USA 100: 9940-9945.

Aukerman MJ, Sakai H (2003). Regulation of flowering time and floral organ identity by a Micro RNA and its APETALA2-like target genes. Plant Cell 15: 2730-2741.

Bailey JA, Carrel L, Chakravarti A, Eichler EE (2000). Molecular evidence for a relationship between LINE-1 elements and X chromosome inactivation: the Lyon repeat hypothesis. Proc. Natl. Acad. Sci. USA 97: 6634-6639.

Bao N, Lye KW, Barton MK (2004). Micro RNA binding sites in *Arabidopsis* class III HD-ZIP mRNA are required for methylation of the template chromosome. Dev. Cell 7: 653-662.

Barlow DP, Stoger R, Herrmann BG, Saito K, Schweifer N (1991).The mouse insulin-like growth factor type-2 receptor is imprinted and closely linked to the *Tme* locus. Nature 349: 84-87.

Beechey CV, Cattanach BM, Blake A, Peters J (2005). Mouse imprinting data and references. http://www.mgu.har.mrc.ac.uk/research/imprinting/ .

Bell AC, Felsenfeld G (2000). Methylation of a CTCF-dependent boundary controls imprinted expression of the Igf2 gene. Nature 405: 482-485.

Bestor TH (2000). The DNA methyltransferases of mammals. Hum. Mol. Genet. 9: 2395-2402.

Bourc'his D, Xu GL, Lin CS, Bollman B, Bestor TH (2001). Dnmt3L and the establishment of maternal genomic imprints. Science 294: 2536-2539.

Brown SW, Chandra HS (1977). Chromosome imprinting and the differential regulation of homologous chromosomes. In: Goldstein, L., Prescott, D.M.'s Cell Biology: A Comprehensive Treatise, Vol. I. Academic Press, New York.

Caspary T, Cleary MA, Baker CC, Guan XJ, Tilghman SM (1998). Multiple mechanisms regulate imprinting of the mouse distal chromosome 7 gene cluster. Mol. Cell Biol. 18: 3466-3477.

Chan SW, Henderson IR, Jacobsen SE (2005). Gardening the genome: DNA methylation in *Arabidopsis thaliana*. Nat. Rev. Genet. 6: 351-360.

Cobb BS, Nesterova TB, Thompson E, Hertweck A, O'Connor E, Godwin J et al., (2005). T cell lineage choice and differentiation in the absence of the RNase III enzyme Dicer. J. Exp. Med. 201: 1367-1373.

Crouse HV (1960). The controlling element in the sex chromosome behaviour in Sciara. Genetics 45: 1429-1443.

Crow FC (2000). The origins, patterns and implications of human spontaneous mutation. Nat. Rev. Genet. 1: 40-47.

Czermin B, Schotta G, Hulsmann BB, Brehm A, Becker PB, Reuter G, Imhof A (2001). Physical and functional association of SU(VAR)3-9 and HDAC1 in Drosophila. EMBO Rep. 2: 915-919.

Davis TL, Trasler JM, Moss SB, Yang GJ, Bartolomei MS (1999). Acquisition of the H19 methylation imprint occurs differentially on the parental alleles during spermatogenesis. Genomics 58: 18-28.

DeChiara TM, Robertson EJ, Efstratiadis A (1991). Parental imprinting of the mouse insulin-like growth factor II gene. Cell 64: 849-859.

Doench JG, Sharp PA (2004). Specificity of microRNA target selection in translational repression. Genes Dev. 18: 504-511.

Ferguson-Smith AC, Surani MA (2001). Imprinting and the epigenetic asymmetry between parental genomes. Science 293: 1086-1089.

Fitzpatrick GV, Soloway PD, Higgins MJ (2002). Regional loss of imprinting and growth deficiency in mice with a targeted deletion of KvDMR1. Nat. Genet. 32: 426-431.

Fukagawa T, Nogami M, Yoshikawa M, Ikeno M, Okazaki T, Takami Y, Nakayama T, Oshimura M (2004). Dicer is essential for formation of the heterochromatin structure in vertebrate cells. Nat. Cell Biol. 6: 784-791.

Fukasawa M, Mortia S, Kimura M, Horii T, Ochiya T, Hatada I (2006). Genomic imprinting in Dicer 1-hypomorphic mice. Cytogenet. Genome Res. 113: 138-143.

Gehring M, Choi Y, Fischer RL (2004). Imprinting and seed development. Plant Cell 16: S203-S213.

Gendrel AV, Lippman Z, Yordan C, Colot V, Martienssen RA, (2002). Dependence of heterochromatic histone H3 methylation patterns on the Arabidopsis gene DDM1. Science 297: 1871-1873.

Gibbons RJ, McDowell TL, Raman S, O'Rourke DM, Garrick D, Ayyub H, Higgs DR (2000). Mutations in ATRX, encoding a SWI/SNF-like protein, cause diverse changes in the pattern of DNA methylation. Nat. Genet. 24: 368-371.

Goll MG, TH Bestor (2002). Histone modification and replacement in chromatin activation. Genes Dev. 16: 1739-1742.

Greally JM (2002). Short interspersed transposable elements (SINEs) are excluded from imprinted regions in the human genome. Proc. Natl. Acad. Sci USA 99: 327-332.

Grewal SI, Moazed D (2003). Heterochromatin and epigenetic control of gene expression. Science 301: 798-802.

Haig D (2006). Intragenomic politics. Cytogenet. Genome Res. 113: 68-74.

Haig D, Graham C (1991). Genomic imprinting and the strange case of the insulin-like growth factor II receptor. Cell 64: 1045-1046.

Hajkova P, Erhardt S, Lane N, Haaf T, El-Maarri O, Reik W, Walter J, Surani MA (2002). Epigenetic reprogramming in mouse primordial germ cells. Mech. Dev. 117: 15-23.

Hall IM, Shankaranarayana GD, Noma K, Ayoub N, Cohen A, Grewal SI (2002). Establishment and maintenance of a heterochromatin domain. Science 297: 2232-2237.

Hamilton AJ, Baulcombe DC, (1999). A species of small antisense RNA in posttranscritpional gene silencing in plants. Science 286: 950-952.

Hannon GJ (2002). RNA interference. Nature 418: 244-251.

Hark AT, Schoenherr CJ, Katz DJ, Ingram RS, Levorse JM, Tilghman SM (2000). CTCF mediates methylation sensitive enhancer-blocking activity at the H19/Igf2 locus. Nature 405: 486-489.

Hata K, Okano M, Lei H, Li E (2002). Dnmt3L cooperates within the Dnmt3 family of de novo DNA methyl-transferases to establish maternal imprints in mice. Development 129: 1983-1993.

Hatada I, Nabetani A, Arai Y, Ohishi S, Suzuki M, Miyabara S, Nishimune Y, Mukai T (1997). Aberrant methylation of an imprinted gene U2af1-rs1(SP2) caused by its own transgene. J. Biol. Chem. 272: 9120-9122.

Heard E (2004). Recent advances in X-chromosome inactivation. Curr. Opin. Cell. Biol. 16: 247-255.

Hellmann-Blumberg U, Hintz MF, Gatewood JM, Schmid CW (1993). Developmental differences in methylation of human Alu repeats. Mol. Cell Biol. 13: 4523-4530.

Holliday R, Grigg GW (1993). DNA methylation and mutation. Mutat Res. 285: 61-67.

Howlett SK, Reik W (1991). Methylation levels of maternal and paternal genomes during preimplantation development. Development 113: 119-127.

Hu JF, Orugantu H, Vu TH, Hoffman AR (1998). Tissue-specific imprinting of the mouse insulin-like growth factor II receptor gene correlates with differential allele-specific DNA methylation. Mol. Endocrinol. 12: 220-232.

Hurst LD, Mc Vean GT (1997). Growth effects of uniparental disomies and the conflict theory of genomic imprinting. Trends Genet. 13: 436-433.

Iwasa Y (1998). The conflict theory of genomic imprinting: how much can be explained? Curr. Top. Dev. Biol. 40: 255-293.

Jackson JP, Lindroth AM, Cao X, Jacobsen SE (2002). Control of CpNpG DNA methylation by the KRYPTONITE histone H3 methyltransferase. Nature 416: 556-560.

Jeddeloh JA, Stokes TL, Richards EJ (1999). Maintenance of genomic methylation requires a SW12/SNF2-like protein. Nat Genet. 22: 94-97.

Jenuwein T, Allis CD (2001). Translating the histone code. Science 293: 1074-1080.

Kafri T, Ariel M, Brandeis M, Shemer R, Urven L, McCarrey J, Cedar H, Razin A (1992). Developmental pattern of gene-specific DNA methylation in the mouse embryo and germline. Genes Dev. 6: 705-714.

Kaneda M, Okano M, Hata K, Sado T, Tsujimoto N, Li E, Sasaki H (2004). Essential role for de novo DNA methyltransferase Dnmt3a in paternal and maternal imprinting. Nature 429: 900-903.

Kaneko-Ishino T, Kohda T, Ishino F (2003). The regulation and biological signifciance of genomic imprinting in mammals. J. Biochem. (Tokyo) 133: 699-711.

Kanellopoulou C, Muljo SA, Kung AL, Ganesan S, Drapkin R, Jenuwein T, Livingston DM, Rajewsky K (2005). Dicer-deficient mouse embryonic stem cells are defective in differentiation and centromeric silencing. Genes Dev. 19: 489-501.

Kantor B, Makedonski K, Green-Finberg Y, Shemer R, Razin A, (2004). Control elements within the PWS/AS imprinting box and their function in the imprinting processes. Hum Mol. Genet. 13: 751-762.

Kawasaki H, Taira K (2004). Induction of DNA methylation and gene silencing by short interfering RNAs in human cells. Nature 431: 211-217.

Ke X, Thomas NS, Robinson DO, Collins A (2002). The distinguishing sequence characteristics of mouse imprinted genes. Mamm. Genome 13: 639-645.

Kierszenbaum AL (2002). Genomic imprinting and epigenetic reprogramming: unearthing the garden of forking paths. Mol. Reprod. Dev. 63: 269-272.

Killian JK, Byrd JC, Jirtle JV, Munday BL, Stoskopf MK, MacDonald RG, Jirtle RL (2000). *M6P/IGF2R* imprinting evolution in mammals. Mol. Cell. 5: 707-716.

Kim VN (2005). MicroRNA biogenesis : coordinated cropping and dicing. Nat Rev. Mol. Cell Biol. 6: 376-385.

Kobayashi H, Suda C, Abe T, Kohara Y, Ikemura T, Sasaki H (2006). Bisulfite sequencing and dinucleotide content analysis of 15 imprinted mouse differentially methylated regions (DMRs): paternally methylated DMRs contain less CpGs than maternally methylated DMRs. Cytogenet. Genome Res. 113: 130-137.

Kono T (2006). Genomic imprinting is a barrier to parthenogenesis in mammals. Cytogenet. Genome Res. 113: 31-35.

Kosak ST Groudine M (2004). Form follows function. The genomic organization of cellular differentiation. Genes Dev. 18: 1371-1384.

Labrador M, Corces VG (2002). Setting the boundaries of chromatin domains and nuclear organization. Cell 111: 151-154.

Lane N, Dean W, Erhardt S, Hajkova P, Surani A et al. (2003). Resistance of IAPs to methylation reprograming may provide a mechanism for epigenetic inheritance in the mouse. Genesis 35: 88-93.

Le Cellier CH, Dunoyer P, Arar K, Lehmann-Che J, Eyquem S, Himber C, Saib A, Voinnet O (2005). A cellular microRNA mediates antiviral defense in human cells. Science 308: 557-560.

Lee J, Inoue K, Ono R, Ogonuki N, Kohda T, Kaneko-Ishino T, Ogura A, Ishino F (2002). Erasing genomic imprinting memory in mouse clone embryos produced from day 11.5 primordial germ cells. Development 129: 1807-1817.

Lehnertz B, Ueda Y, Derijck AA, Braunschweig U, Perez-Burgos L, Kubicek S, Chen T, Li E, Jenuwein T, Peters AH (2003). Suv39h mediated histone H3 lysine 9 methylation directs DNA methylation to major satellite repeats at pericentric heterochromatin. Curr. Biol. 13: 1192-1200.

Lewis A, Mitsuya K, Umlauf D, Smith P, Dean W, Walter J, Higgins M, Feil R, Reik W (2004). Imprinting on distal chromosome 7 in the placenta involves repressive histone methylation independent of DNA methylation. Nat. Genet. 36: 1291-1295.

Lewis A, Reik W (2006). How imprinting centres work. Cytogenet Genome Res. 113: 81-89.

Li E (2002). Chromatin modification and epigenetic reprogramming in mammalian development. Nat. Rev. Genet. 3: 662-673.

Li E, Beard C, Jaenisch R (1993). Role for DNA methylation in genomic imprinting. Nature 366: 362-365.

Li JY, Less-Murdock DJ, Xu GL, Walsh CP (2004). Timing of establishment of paternal methylation imprints in the mouse. Genomics 84: 952-960.

Lin SP, Youngson N, Takada S, Seitz H, Reik W, Paulsen M, Cavaille J, Ferguson-Smith AC (2003). Asymmetric regulation of imprinting on the maternal and paternal chromosomes at the *Dlk1/Gtl2* imprinted cluster on mouse chromosome 12. Nat. Genet. 35: 97-102.

Liu J, Chen M, Deng C, Bourc'his D, Nealon JG, Erlichman B, Bestor TH, Weinstein LS (2005). Identification of the control region for tissue-specific imprinting of the stimulatory G protein alpha-subunit. Proc. Natl. Acad. Sci. USA 102: 5513-5518.

Lopes S, Lewis A, Hajkova P, Dean W, Oswald J, Forne T, Murrell A, Constancia M, Bartolomei M, Walter J, Reik W (2003). Epigenetic modifications in an imprinting cluster are controlled by a hierarchy of DMRs suggesting long-range chromatin interaction. Hum. Mol. Genet. 12: 295-305.

Lucifero D, Mann MR, Bartolomei MS, Trasler JM (2004). Gene-specific timing and epigenetic memory in oocyte imprinting. Hum. Mol. Genet. 13: 839-849.

Luedi PP, Hartemink AJ, Jirtle RL (2005). Genome-wide prediction of imprinted murine genes. Genome Res. 15: 875-884.

Lyon MF (1998). X-chromosome inactivation: a repeat hypothesis. Cytogenet. Cell Genet. 80: 133-137.

Martienssen RA (2003). Maintenance of heterochromatin by RNA interference of tandem repeats. Nat. Genet. 35: 213-214.

Matzke MA, Birchler JA (2005). RNAi-mediated pathways in the nucleus. Nat. Rev. Genet. 6:24-35.

McGrath J, Solter D (1984). Completion of mouse embryogenesis requires both the maternal and paternal genomes. Cell 37: 179-183.

Metz CW (1938). Chromosome behaviour, inheritance and sex determination in *Sciara*. Am. Nat. 72: 485-520.

Moore T, Haig D (1991). Genomic imprinting in mammalian development : a parental tug-of-war. Trends Genet. 7: 45-49.

Morris KV, Chan SW, Jacobsen SE, Looney DJ (2004). Small interfering RNA-induced transcriptional gene silencing in human cells. Science 305: 1289-1292.

Murrell A, Heeson S, Reik W (2004). Interaction between differentially methylated regions partitions the imprinted genes *Igf2* and *H19* into parent-specific chromatin loops. Nat. Genet. 36: 889-893.

Neumann B, Kubicka P, Barlow DP (1995). Characteristics of imprinted genes. Nat. Genet. 9: 12-13.

Nolan CM, Killian JK, Petitte JN, Jirtle RL (2001). Imprint status of *M6P/IGF2R* and *IGF2* in chickens. Dev. Genes Evol. 211: 179-183.

Noma K, Sugiyama T, Cam H, Verdel A, Zofall M, Jia S, Moazed D, Grewal SI (2004). RITS acts in *cis* to promote RNA inteference-mediated transcriptional and post-transcriptional silencing. Nat. Genet. 36: 1174-1180.

Obata Y, Kaneko-Ishino T, Koide T, Takai Y, Ueda T, Domeki I, Shiroishi T, Ishino F, Kono T (1998). Disruption of primary imprinting during oocyte growth leads to the modified expression of imprinted genes during embryogenesis. Development 125: 1553-1560.

Obata Y, Kono T (2002). Maternal primary imprinting is established at a specific time for each gene throughout oocyte growth. J. Biol. Chem. 277: 5285-5289.

O'Neill MJ (2005). The influence of non-coding RNAs on allele-specific gene expression in mammals. Hum. Mol. Genet. 14 Spec. No. 1: R113-120.

O'Neill MJ, Ingram RS, Vrana PB, Tilghman SM (2000). Allelic expression of *IGF2* in marsupials and birds. Dev. Genes. Evol. 210: 18-20.

Paoloni-Giacobino, Chaillet JR, (2006). The role of DMDs in the maintenance of epigenetic states. Cytogenet. Genome Res. 113: 116-121.

Paulsen M, Khare T, Burgard C, Tierling S, Walter J (2005). Evolution of the Beckwith-Wiedemann syndrome region in vertebrates. Genome Res. Commun. 15: 146-153.

Paulsen M, Takada S, Youngson NA, Benchaib M, Charlier C et al. (2001). Comparative sequence analysis of the imprinted *Dlk-Gtl2* locus in three mammalian species reveals highly conserved genomic elements and refines comparison with the *Igf2-H19* region. Genome Res. 11: 2085-2094.

Purbowasito W, Suda C, Yokomine T, Zubair M, Sado T, Tsutsui K, Sasaki H (2004). Large-scale identification and mapping of nuclear matrix-attachment regions in the distal imprinted domain of mouse chormosome 7. DNA Res. 11: 391-407.

Rand E, Cedar H (2003). Regulation of imprinting : A multitiered process. J. Cell Biochem. 88: 400-407.

Razin A, Shemer R (1995). DNA methylation in early development. Hum. Mol. Genet. 4: 1751-1755.

Reale A, Matteis GD, Galleazzi G, Zampieri M, Caiafa P (2005). Modulation of DNMT1 activity by ADP-ribose polymers. Oncogene 24: 13-19.

Reik W, Walter J (2001a). Evolution of imprinting mechanisms : the battle of the sexes begins in the zygote. Nat. Genet. 27: 255-256.

Reik W. Walter J (2001b). Genomic imprinting : parental influence on the genome. Nat. Rev. Genet. 2: 21-32.

Robb GB, Brown KM, Khurana J, Rana TM (2005). Specific and potent RNAi in the nucleus of human cells. Nat. Struct. Mol. Biol. 12:133-137.

Ross MT, Grafham DV, Coffey AJ, Scherer S, McLay K et al. (2005). The DNA sequence of the human X chromosome. Nature 434: 325-337.

Rougier N, Bourc'his D, Gomes DM, Niveleau A, Plachot M, Paldi V, Viegas-Pequignot E (1998). Chromosome methylation patterns during mammalian preimplantation development. Genes Dev. 12: 2108-2113.

Rubin CM, VandeVoort CA, Teplitz RL, Schmid CW (1994). Alu repeated DNAs are differentially methylated in primate germ cells. Nucleic Acids Res. 22: 5121-5127.

Sarma K, Reinberg D (2005). Histone variants meet their match. Nat. Rev. Mol. Cell Biol. 6: 139-149.

Schrader F, Hughes-Schrader S (1931). Haploidy in metazoa. Q. Rev. Biol. 6: 411-438.

Scott RJ, Spielman M (2004). Imprinting in plants and mammals – the same but different? Curr. Biol. 14: R201-R203.

Selker, EU, Stevens JN (1985). DNA methylation at asymmetric sites is associated with numerous transition mutations. Proc. Natl. Acad. Sci. USA 82: 8114-8118.

Sleutels F, Zwart R, Barlwo DP (2002). The non-coding Air RNA is required for silencing autosomal impriinted genes. Nature 415: 810-813.

Surani MA (2001). Reprogramming of genome function through epigenetic inheritance. Nature 414: 122-128.

Surani MA, Barton SC, Norris ML (1984). Development of reconstituted mouse eggs suggests imprinting of the genome during gameto-genesis. Nature 308: 548-550.

Suzuki S, Renfree MB, Pask AJ, Shaw G, Kobayashi S, Kohda T, Kaneko-Ishino T, Ishino F (2005). Genomic imprinting of *IGF2*, *p57^{KIP2}* and *PEG1/MEST* in a marsupial, the tammar wallaby. Mech. Dev. 122: 213-222.

Tamaru H, Selker EU (2001). A histone H3 methyltransferase controls DNA methylation in *Neurospora crassa*. Nature 414: 277-283.

Tanaka M, Puchyr M, Gerstenstein M, Harpal K, Jaenisch R, Rossant J, Nagy A (1999). Parental origin-specific expression of *Mash2* is established at the time of implantation with its imprinting mechanism highly resistant to genomic-wide demethylation. Mech. Dev. 87: 129-142.

Thorvaldsen JL, Duran KL, Bartolomei MS (1998). Deletion of the *H19* differentially methylated domain results in loss of imprinted expression of *H19* and *Igf2*. Genes Dev. 12: 3693-3702.

Tilghman SM (1999). The sins of the fathers and mothers: genomic imprinting in mammalian development. Cell 96: 185-193.

Turker MS, Bestor TH (1997). Formation of methylation patterns in the mammalian genome. Mutat. Res. 386: 119-139.

Ueda T, Abe K, Miura A, Yuzuriha M, Zubair M, Noguchi M, Niwa K, Kawase Y, Kono T, Matsuda Y, Fujimoto H, Shibata H, Hayashizaki Y, Sasaki H (2000). The paternal methylation imprint of the mouse *H19* locus is acquired in the gonocyte stage during foetal testis development. Genes Cells 5: 649-659.

Umlauf D, Goto Y, Cao R, Cerqueira F, Wagschal A, Zhang Y, Feil R (2004). Imprinting along the *Kcnq1* domain on mouse chromosome 7 involves respessive histone methylation and recruitment of Polycomb group complexes. Nat. Genet. 36: 1296-1300.

van der Vlag J, Otte AP (1999). Transcriptional repression mediated by the human polycomb-group protein EED involves histone deacetylation. Nat. Genet. 23: 474-478.

Volpe TA, Kidner C, Hall IM, Teng G, Grewal SI, Martienssen RA (2002). Regulation of heterochromatic silencing and histone H3 lysine-9 methylation by RNAi. Science 297: 1833-1837.

Walbot V, Evans MM (2003). Unique features of the plant life cycle and their consequences. Nat. Rev. Genet. 4: 369-379.

Wassenegger M, Heimes S, Riedel L, Sanger HL (1994). RNA-directed de novo methylation of genomic sequences in plants. Cell 76: 567-576.

Waterston RH, Lindblad-Toh K, Birney E, Rogers J, Abril JF et al. (2002). Initial sequencing and comparative analysis of the mouse genome. Nature 420: 520-562.

Williamson CM, Ball ST, Nottingham WT, Skinner JA, Plagge A, Turner MD, Powles N, Hough T, Papworth D, Fraser WD, Maconochie M, Peters J (2004). A *cis*-acting control region is required exclusively for the tissue-specific imprinting of *Gnas*. Nat. Genet. 36: 894-899.

Wutz A, Smrzka OW, Schweifer N, Schellander K, Wagner EF, Barlow DP (1997). Imprinted expression of the *Igf2r* gene depends on an intronic CpG island. Nature 389: 745-749.

Yokomine T, Kuroiwa A, Tanaka K, Tsudzuki M, Matsuda Y, Sasaki H (2001). Sequence polymorphisms, allelic expression status and chromosome locations of the chicken *IGF2* and *MPR1* gene. Cytogenet Cell Genet. 93: 109-113.

Yokomine T, Shirohzu H, Purbowasito W, Toyoda A, Iwama H, Ikeo K et al. (2005). Structural and functional analysis of a 0.5-Mb chicken region orthologous to the imprinted mammalian *Ascl2/Mash2-Igf2-H19* region. Genome Res. 15: 154-165.

Yoon BJ, Herman H, Sikora A, Smith LT, Plass C, Soloway PD (2002). Regulation of DNA methylation of *Rasgrf1*. Nat. Genet. 30: 92-96.

Zhang Z, Carriero N, Gerstein M (2004). Comparative analysis of processed pseudogenes in the mouse and human genomes. Trends Genet. 20: 62-67.

13

Mechanisms and molecular genetic bases of rapid speciation in African cichlids

Kazhila C. Chinsembu

University of Namibia, Faculty of Science, Department of Biological Sciences, Private Bag13301, Windhoek, Namibia.

African cichlid fishes are a textbook model of evolution in motion but the molecular genetic bases and mechanisms involved in their rapid speciation largely remain elusive. Emerging experimental evidence now suggests that African cichlids have undergone rapid speciation due to a combination of their molecular genetic potential and the influences of the environment on this potential. The genetic potential of the cichlids lies mainly in the ecomorphological plasticity of their feeding apparatus and their strong sexual selection. Putative genes that underlie the phenotypic variations in African cichlids are beginning to be unravelled but their coverage in the literature remains modest and scattered. This review forms one of the first comprehensive attempts to consolidate emerging data that explain various genes and mechanisms underlying explosive speciation in this family of fishes. The review analyzes the modes of African cichlid speciation, radiation-in-stages model, molecular genetic bases of plastic pharyngeal jaws and teeth and signature genes for sexual selection premised mainly on nuptial colour patterns, egg dummies and maternal mouth brooding, opsins and the sensory drive hypothesis. Explaining sexual selection mechanisms based on colour patterns and sensitivity to light is crucial to understanding African cichlid species biodiversity and conservation in polluted lakes.

Key words: African cichlids, rapid speciation, sexual selection, genes.

TABLE OF CONTENTS

INTRODUCTION

In the last 10 million years, African cichlids (Family Cichlidae; Order Perciformes) have undergone such unexpected rates of speciation that they are now a model species for show-casing evolution in motion. Cichlid fishes are tropical freshwater fish that are unsurpassed by any other vertebrate group in terms of the sheer number of species (> 3,000), variety of body shapes, assortment of colouration, behavioural diversity and degree of trophic and ecological specialisation (Fryer and Iles, 1972; Meyer, 1993; Stiassny and Meyer, 1999; Kornfield and Smith, 2000; Kocher, 2004).

Since the publication of the first reports on the East African fish fauna at the end of the 19th century (Boulenger, 1898a; 1898b), the exceptional diversity of fishes in the family Cichlidae has attracted the attention of evolutionary biologists. Cichlids were considered an aberrantly species-rich group, yet the variation endowed within the cichlids has made them a textbook model for the study of rapid speciation and diversification, and animal behaviour. The multiple invasion model suggests that each of the East African Great Lakes was colonized by multiple lineages which had evolved independently in time (Fryer, 1977) and space (Mayr, 1942). However, the model cannot adequately explain the explosive speciation and enormous diversity of African cichlids (Danley and Kocher, 2001).

Molecular analyses have revealed that the phylogenetic relationships among the major lineages of cichlids are consistent with an initially Gondwanaland distribution, with the Indian and Madagascar representatives forming the most basal lineages and the reciprocally mono-phyletic African and American lineages being sister groups. This distributional pattern is congruent with models of vicariance biogeography rather than overseas dispersal (Zardoya et al., 1996; Streelman et al., 1998; Farias et al., 2000, 2001; Sparks, 2004). Only about a dozen extant species represent the most basal para-phyletic lineages of cichlids from India/Sri Lanka and Madagascar, which are the two landmasses that split off first from the supercontinent of Gondwanaland between 165 and 130 million years ago (MYA). The Americas are inhabited by an estimated 400 - 500 cichlid species.

In Africa, the cichlid centre of biodiversity is in East Africa where they inhabit several large and small lakes and have formed the so-called "species flocks" with sometimes hundreds of endemic species in each of these lakes. The greatest diversity of cichlids is in Lakes Victoria, Malawi and Tanganyika where a total of > 2000 species occur (Salzburger and Meyer, 2004). With an age of 9 - 12 million years, Lake Tanganyika is the oldest of East Africa's Great Lakes (Cohen et al., 1993, 1997) followed by Lake Malawi with a probable age of 2 - 5 million years (Johnston and Ng'ang'a, 1990; Delvaux, 1995). Lake Victoria is the youngest, with an estimated age between 250,000 and 750,000 years (Johnston et al., 1996). These Lakes contain varying numbers of cichlid species: Lake Tanganyika (200-250 species), Lake Malawi (500 - 700, could even be 1000 species) and Lake

Victoria (500 or more) (Seehausen, 1996; Seehausen et al., 2003; Verheyen et al., 2003; Snoeks et al., 1994).

Although Lakes Victoria and Malawi have very high species richness and morphological diversity of cichlids and have been studied as model systems for explosive speciation and ecomorphological diversification, the rivers associated with Lake Victoria and Lake Malawi have low cichlid species richness and morphological diversity. This difference has been attributed to the observation that rivers lack the wealth of ecological opportunity that drives cichlid adaptive radiation in lakes (Joyce et al., 2005). In their study, Joyce et al. (2005) showed that the situation is different in Southern Africa. Haplochromine cichlids in 5 Southern African rivers (Upper Congo, middle/upper Zambezi, Okavango, Cunene and Limpopo) show species richness and ecomorphological diversity similar to that in Lake Victoria and Lake Malawi. Joyce et al. (2005) found numerous sympatric haplochromines different in shape and size within each of the rivers surveyed. Within this Southern African radiation, mitochondrial DNA haplotypes comprise 6 clades within which a large number of closely related haplotypes have arisen from a small number of more divergent haplotypes.

On the other hand, several morphologically different and geographically distant (allopatric) species have very similar haplotypes. These phylogenetic relationships suggest a species flock that emerged rapidly and simul-taneously in many geographically distant rivers, in which very similar haplotypes are found several thousand kilometres apart, yet for which strongly differentiated haplotypes are found within single populations. Through the use of geological evidence, Joyce et al. (2005) contend that the high diversity among Southern African riverine cichlids arose from Lake palaeo-Makgadikgadi, a lake that disappeared about 2000 years Before Present. The centre of this extinct lake is now a saltpan north of the Kalahari Desert. They showed that Lake palaeo-Makgadikgadi hosted a rapidly evolving cichlid species radiation comparable in morphological diversity to that in the extant African Great lakes. This Lake palaeo-Makgadikgadi stock of cichlids later seeded all major river systems of Southern Africa with ecologically diverse cichlids.

Modes of African cichlid speciation

Although Charles Darwin proposed the mechanism of speciation in his book On the origin of species by means of natural selection (Darwin, 1859), the mechanisms through which new species are generated remain a major problem and the evolutionary forces and genes that drive speciation are still not well understood (Kocher, 2004). In the case of the African cichlids, biologists worldwide are still intrigued by their explosive speciation in the Great Lakes of East Africa. Mayr (1963, 1984) promoted allopatic

*Corresponding author. E-mail: kchinsembu@unam.na.

speciation, a gradual divergence of populations with completely separate geographic ranges.

Evidence against allopatric speciation begun to mount when Meyer et al. (1990) established through mitochondrial DNA evidence that the enormous haplochromine cichlid species in Lake Victoria evolved from a single ancestral stock. Indeed after the publication of geological data that Lake Victoria was completely dry during the recent Ice Age about 12, 400 years before present (Johnston et al., 1996), it became even more untenable that over 500 species of cichlids could have evolved within Lake Victoria in such an extremely short evolutionary time frame. Thus the old hypothesis of allopatric speciation in Lake Victorian cichlids (Greenwood, 1964) was discarded because it could not account for the explosive speciation of cichlids in the lake (Seehausen, 1996).

It has become apparent that the great diversity of East African Great lakes cichlid species (the cichlid problem) contradicts traditional models of speciation (Salzburger and Meyer, 2004). There is now a wealth of evidence that parapatric (divergence of populations with adjacent geographic ranges) and sympatric speciation (divergence of populations in the same geographic area) also take place (Kocher, 2004). Despite abandoning the hypothesis of allopatric speciation, the question of how such large numbers of cichlid species (and not the other families of fishes found in these lakes) evolved rapidly into a large number of closely related but morphologically diverse species in such a short period of time continues to bother biologists. Hence recent studies in evolution and ecology have focused on selective forces that are responsible for the differentiation of populations regardless of the gene flow among incipient species (Kocher, 2004).

It is now becoming clear that sexual selection is an important force in the origination of new species (Higashi et al., 1999). Disruptive sexual selection on colour polymorphisms has caused sympatric speciation and accounts for the rapid evolution of cichlid diversity in Lake Victoria (Seehausen and Van Alphen, 1999). Intrinsic properties that cause disruptive selection elevate sympatric speciation rates (Seehausen and Van Alphen, 1999). But rapid radiations only occur in parts of the lake with relatively good visual conditions (Seehausen et al., 1997). When colour vision is impaired by water pollution and eutrophication, disruptive sexual selection becomes an ineffective driver of sympatric speciation (Seehausen and Van Alphen, 1999).

Male-male competition is also an important agent of diversification in African cichlids (Seehausen and Schluter, 2004). Closely related species of cichlids with very different colours are often sympatric or parapatric (Seehausen and Van Alphen, 1999; Danley et al., 2000). Competition between males for breeding territories promotes colour diversification thereby setting the stage for speciation (Seehausen and Schluter, 2004). Colourful males dominate breeding territories that they defend very

fiercely against other males. Thus males of the same colour compete more heavily than males of different colours. This leads to different colour distribution patterns and can explain the origin of colour diversity and colour morphs (Seehausen and Schluter, 2004). Rare male colour morphs that increase in populations under negative frequency-dependent selection can lead to stable colour polymorphisms or sympatric speciation. Salzburger et al. (2009) discussed the possibility of acoustic, olfactory and behavioural cues in mate recognition and choice, but the genetic bases of these traits are still anectodal.

Radiation-in-stages model

Ecological and behavioural factors have had the largest effect on the diversification of African cichlids. The wealth of ecomorphological and behavioural traits that have led to the rapid speciation and diversity of African cichlids has now become known as the radiation-in-stages model. This 3-stage radiation model (Streelman and Danley, 2003) explains the extraordinary species richness of cichlids. The primary stage consists of divergence into different habitats. In the secondary radiation, ecomorphological differentiations occur and the tertiary stage consists of sensory behavioural diversification.

The first radiation resulted in the divergence of the rock-dwelling species from the sand-dwelling species. In Lake Malawi, this split resulted into about 200 species of rock-dwelling and sand-dwelling cichlids. This mechanism of cichlids species formation is consistent with allopatric speciation (Jonsson and Jonsson, 2001). Adaptation to the rock and sand macro-habitats resulted in the divergence of many morphological and behavioural characteristics such as body shape, trophic morphology, habitat preference, colour patterning and reproductive behaviours (Danley and Kocher, 2001).

In the second radiation, ecomorphological selection of the feeding apparatus led rock-dwellers to an adaptive innovative switch from the mandibular oral jaw to the pharyngeal jaw. Cichlids therefore acquired 2 sets of jaws: true mandibular oral jaws which shape the normal mouth to suck scrape and bite off bits off food and internal pharyngeal jaws derived from the fifth gill arch, found in the throat and used to mash, macerate, slice or pierce the morsel before it is ingested (Salzburger and Meyer, 2004). The innovations in the cichlid pharyngeal jaw led to the utilization of novel prey and ultimately to the trophic diversification of these fishes (Hulsey, 2006). Pharyngeal jaw modification is therefore a key innovation that determines what prey fishes exploit, hence some cichlids are molariforms that specialize in crushing molluscs, while others are papilliforms that utilize prey requiring less force to process (Hulsey, 2006). The specialized organisation of the pharyngeal jaw apparatus is a shared derived characteristic (synapomorphy) among all cichlids (Liem, 1973).

The jaws are exceedingly versatile and adaptable and can change in form even within the lifetime of a single individual (Meyer, 1993). The oral jaw can grow specialized teeth which allow the cichlids to gather different types of foods. The division of labour between the oral jaws (mandibles) and the pharyngeal jaws has made cichlids become very efficient feeders, allowing them to capture and process a very wide variety of food. This specialized anatomy of the jaws is 1 morphological explanation given for the rapid speciation of cichlids. With the additional pharyngeal jaw, the mandibular oral jaw was afforded the evolutionary opportunity to diversify into specialized mouthparts for collecting different foods. Oral jaw specializations enabled the cichlids to exploit new food resources thereby allowing trophic diversification (Bootsma et al., 1996). Cichlid genera diversified in response to competition for trophic resources.

Feeding habits, dietary preferences and trophic morphological differences now account for the 10 - 12 genera of Lake Malawi cichlids (Danley and Kocher, 2001). The oral jaw has allowed a variety of foraging strategies. As a result, cichlids exploit a range of trophic niches usually occupied by several families, if not orders of fishes (Greenwood, 1964). Thus progression from ecological diversification to the refinement of trophic oral jaw structures is the major event in the secondary radiation of cichlids (Seehausen, 1996; Sturmbauer, 1998). This is consistent with the Darwinian adaptation of beaks to different food and ecological habitats in finches on the Galapagos Islands (Grant, 1981).

The third radiation, which resulted into the diversification of extant species, is mainly based on sexual selection (Danley and Kocher, 2001). While morphological features that evolved during the primary and secondary radiations have remained conserved, sexual selection now accounts for the incredible species richness of African cichlids. After elaborate experimentation, it was shown that cichlid fish species of Lake Victoria are sexually isolated by mate choice (Seehausen and Van Alphen, 1998, 1999). Male secondary sexual characteristics and female mating preferences determine mating behaviour. Sexual selection on male nuptial colours was a central driving force in the diversification of haplochromine cichlids in Lake Victoria (Seehausen and Schluter, 2004). It has been shown that direct mate choice of females for differently coloured males maintains reproductive isolation and diversity among sympatric cichlids (Seehausen, 1996). Females have a strong preference for males of a particular colour when light conditions are sufficiently good.

Competition for mates and assortative mating drives the diversification process. Mate choice by the females also allows full sympatric speciation as in the models of Lande (1982), Turner and Burrows (1995), and Payne and Krakaeur (1997). The abundance of sympatrically occuring colour morphs as well as the common absence of mating barriers other than behavioural ones suggests that sympatric speciation has played an important role in the explosive speciation and adaptive radiation of cichlids. However, the increased turbidity of the water due to pollution has broken down reproductive barriers leading to hybridization with other species, and a decline in cichlid species in certain parts of Lake Victoria (Seehausen et al., 1997).

With the increasing use of molecular biology techniques, the interest has now shifted towards the identification of genetic variation and isolation of putative genes responsible for phenotypic differences within and between closely related species. Thus genetic markers are being used to characterize population structure and trace the phylogenetic relationships among species.

Genetic bases of jaw and teeth morphology

Fish species that exploit hard or attached prey and those that feed on highly mobile prey, have evolved prescribed mandibular morphologies that correspond to the mechanical requirements of the feeding apparatus. Fish that prey on hard food evolve short and stout jaws for efficient biting, and those that eat mobile prey have developed gracile jaws for suction feeding (Albertson et al., 2005). Within this ecological prism, cichlids have used the biting-suction feeding instruments as a key point of evolutionary departure. The first step in elucidating the genetic architecture of the cichlid oral jaw apparatus was reported by Albertson et al. (2003a). They showed that skeletal differences in the head and oral jaw apparatus were inherited together, suggesting a degree of pleiotropy in the genetic architecture of this character complex. Alleles implicated in the evolution of jaw morphology were differentially distributed in cichlids that were algal scrapers and suction feeders. Hybrids showed that 4 to 11 genes were involved in the evolution of different jaw morphologies. The genes affect the shape of skeletal elements, although some of them were known to influence several other traits (Albertson et al., 2003b). Terai et al. (2002) found that a chromosomal region marked by polymorphism of the bone morphogenetic protein 4 gene (bmp4) contributes to variations in the shape of the upper and lower jaw elements in an explosively speciated lineage of East African cichlids. The same gene would later be shown to affect jaw morphology in mice (Wilson and Tucker, 2004). Consistent with the role of bmp4 as a putative genetic signature for the divergence of jaw morphology, Terai et al. (2002) found a high rate of amino acid substitutions in the pro-domain of the bmp4 protein.

The bmp4 gene is an attractive candidate for the evolution of craniofacial diversity in vertebrates and its regulation underlies some aspects of cichlid evolution. Albertson et al. (2005) demonstrated that bmp4 has the potential to alter mandibular morphology in a way that mimics adaptive variation among cichlid species. Increased levels of *bmp4* were associated with biting /crushing morphologies; hence *bmp4* is a major player in the evolutionary drive from gracile to robust jaws. The *bmp4* gene has also been found to be important in determining beak shape diversity among avian species, and interestingly, *bmp4* signalling is thought to have contributed to the evolution of beak shape in Darwin's finches (Albertson et al., 2005).

The range of dental diversity in Lake Malawi cichlids is amazingly high: some species possess about 10 teeth in a single row, or as many as 700 teeth in up to 20 rows (Fraser et al., 2008). Cichlid teeth are as diverse as their jaws and are essential components of the trophic machinery. Cichlid species also differ in tooth size, spacing and shape. In terms of tooth shape, some cichlids are unicuspid, bicuspid or tricuspid species. Tooth shape is highly correlated with the feeding ecology of cichlid fishes (Kocher, 2004). A unicuspid dentition is suited to generalist feeding; hence cichlids with unicuspid teeth are piscivorous, zooplanktivorous and insectivorous species, while those with tricuspid teeth are specialized algal scrapers (Fraser et al., 2008). Tooth development is a classic example of tissue and genic interactions. In vertebrates, there are genes that are essential for teeth development. For example, gene s*hh* is crucial for correct establishment of the global dental programme (Fraser et al., 2008). It is expressed during tooth morphogenesis and marks the bell-shaped dental epithelium. Fraser et al. (2008) showed that a combination of *shh* and *pitx2* were necessary for a competent field of tooth initiation.

Tooth shape differences between bicuspids and tricuspids are choreographed by a small number of genes. Cuspid number and morphology are regulated by antagonistic actions of extracellular signalling ligands such as fibroblast growth factors (fgfs) and bone morphogenetic proteins (bmps) secreted from transitory enamel knots (EKs) (Streelman et al., 2003a; Fraser et al., 2008). These proteins are active during tooth initiation and morphogenesis and affect the sites in the jaw where teeth develop, as well as the shape of individual teeth. During tooth initiation, the expression of genes *bmp4* and *fgf8* in the epithelium control the mesenchymal expression of genes *pax9* and *msx1*, which direct tooth formation and position. The expression of gene *bmp4* in the mesenchyme promotes formation of the primary EK. Then bpms and fgfs secreted from the EK modulate cusp development (Fraser et al., 2008); b*mp4* specifies unicuspid dentition and *fgf8* codes for multicuspid dentition (Streelman and Albertson, 2006). Both unicuspid and multicuspid teeth develop under the control of homeobox genes expressed in the mesenchyme; *msx1/2* for unicuspids and *barx1*, *dlx1/2* and *lhx6/7* for multicuspids.

It has been suggested that divergent dentitions in cichlids were driven by the genes *pitx2*, *eda* and *wnt7b* and and their interactions with *shh* and *edar* (Fraser et al.,

2008). The genes *eda* and *wnt7b* regulate initial tooth germ size and position within rows. These genes are responsible for putting tooth rows in jaws and teeth in tooth rows. Fraser et al. (2008) suggested that initiation of new tooth rows follows a copy and paste mechanism where the dental expression network is redeployed for each new tooth row. Later, it was demonstrated that a core gene network involving *bmp2*, *bmp4*, *dlx2*, *eda*, *dlx2*, *edar*, *pax9*, *pitx2*, *runx2*, *shh* and *wnt7b* is expressed commonly on cichlid oral and pharyngeal jaws (Fraser et al., 2009). These genes are core markers of dental epithelial initiation and are associated with variations in oral jaw tooth row number, tooth number within rows and the spacing of teeth. It was postulated that this core dental network represents a conserved set of molecules for tooth development, adding that it was likely that nature had never made a tooth without this core genetic network (Fraser et al., 2009).

Signature genes for sexual selection

Sexual selection is a special case of natural selection. Sexual selection acts on an organism's ability to successfully copulate with a mate. Selection makes many organisms go to extreme lengths for sex, often evolving elaborate body features to lure mates. In Lakes Malawi and Victoria cichlids, females select their mates based on male nuptial colouration (Seehausen and Van Alphen, 1998). Thus there is a possibility that female choice of male nuptial colours is a special driving force for speciation (Seehausen et al., 1999). It has been postulated that female cichlids have preferences for different male colours and that these female preferences cause reproductive isolation between incipient species.

Several genes carry the signatures of sexual selection and speciation in African cichlids. Signature genes that underlie phenotypic differences in cichlids can be located using genetic mapping techniques. It is now possible to locate chromosomal regions responsible for any quantitative trait in cichlids. Putative genes that confer phenotypic differences among various species are identified by positional cloning. Here, we review the putative genes that elucidate the molecular genetic raw materials that fuel phenotypic variations, explosive speciation and evolutionary success of African cichlids.

Colour patterns

Closely related species of cichlids exhibit different colour patterns that in turn direct the choice of mating partners (Seehausen and Van Alphen, 1998). The orange-blotch (OB) pattern is found in many species of Lakes Victoria and Malawi cichlids. After conducting many genetic crosses, Seehausen et al. (1999) proposed that OB is produced by an X-linked gene that is modified by an autosomal locus in Lakes Victoria and Malawi cichlids. A region of conserved synteny has been found around the *OB* locus with a strong association to the *cski* marker located 2 cM from *OB*. Streelman et al. (2003b) also found that the *OB* locus was tightly linked to the *c-ski1* gene. Terai et al. (2002b) found a high rate of splicing events in the *hagoromo* gene, a putative gene involved in the development of pigment patterns in *zebra fish*. This gene has a complex pattern of alternative splicing which produces numerous splice variants in each species (Terai et al., 2003). These splicing patterns seem to be species-specific but the differences in splicing are yet to be linked to specific colour patterns.

Egg-dummy and maternal mouth brooding genes

One characteristic feature of haplochromines is their possession of egg-spots on the anal fins of males (Goldschmidt, 1991). These egg-spots mimic real eggs and are therefore called egg-dummies. The egg-dummies play an important role in the mating behaviour of these maternal mouthbrooding fish (Hert, 1989) and are now known to have mediated the speciation of African cichlids (Goldschmidt and de Visser, 1990). Haplochromine egg-dummies form in juvenile male fish and begin to brighten when the young males reach sexual maturity. Anal fin egg-dummies also form in females but they are less colourful than those in males. In most riverine and rock-dwelling haplochromines, the egg-dummies are made up of a conspicuous yellow-red central area and a less transparent outer ring.

A female with mature eggs approaches the territory of courting males and lays eggs on the lake bed. Later the female swallows the eggs in their mouth. The female then pursues a male with bright anal fin egg-spots or egg dummies. The female tries to ingest the egg dummies and in the process she brings her mouth closer to the genital opening of the male. The male then releases sperms into the female's mouth thereby fertilizing the eggs. The molecular genetic basis of this elaborate inheritable mating behaviour has now been traced to the formation of pigment cells.

The egg-spots consist of pigment cells called xanthophores (Salzburger et al., 2007). Using fluore-scent-based detection methods, reverse transcriptase PCR and *in situ* hybridization experiments, it has recently been shown that colony-stimulating factor I receptor a (*Csfl ra*) is the gene that mediates the production of yellow xanthophores in male haplochromine egg-dummies (Salzburger et al., 2007). *Csfl ra* is a type III receptor tyrosine kinase gene which is 2.9 kb long and has 21 exons. It has a cysteine-rich extracellular ligand-binding domain composed of five immunoglobulin-like chains, a transmembrane domain and an intracellular domain with two separate tyrosine kinase domains (Salz-burger et al., 2007).

Receptor tyrosine kinases are key components of cell signalling networks and play crucial roles in physiological processes. These signal transduction cascades detect, amplify, filter and process a variety of environmental and intercellular cues. Their N-terminal extracellular domain binds ligands such as growth factors and hormones. Their C-terminal domain has kinase activity whereby transfer of a phosphate to tyrosine residues activates a signal trans-duction cascade that leads to xanthophore pigment gene expression.

2 important molecular genetic architectures and mechanisms could have accelerated the evolution of egg-dummies in haplochromines:

(a) Adaptive sequence evolution: The *Csfl ra* gene has more non-synonymous substitutions than synonymous ones (Salzburger et al., 2007). By definition, non-synonymous mutations easily change the translated amino acid sequence of proteins because they are under constant evolutionary pressure (Schattner and Diekhans, 2006). Sequence comparisons of haplochromine and non-haplochromine species revealed that several regions of the *Csfl ra* gene give a dN/dS ratio of greater than one, demonstrating that positive selection pressure has greatly changed the *Csfl ra* protein in haplochromines (Salzburger et al., 2007). The many variant forms of the *Csfl ra* protein suggest that novel modifications of existing signal transduction mechanisms evolved in haplochromines. This provided the genetic raw material which produced many routes in the expression of xanthophore pigment genes and resulted in an enormous variety of egg-dummies that characterise the explosive speciation of haplochromines.

(b) Duplication of receptor tyrosine kinase paralogons: Gene and genome duplications are important mechanisms for the evolution of phenotypic complexity, diversity, innovation, and origin of novel functions in the development of organisms (Ding et al., 2008). In many teleosts, several genes (*mitf, sox10, tyrosinase*) involved in pigment cell development are retained in duplicate copies (Hoegg et al., 2004) and this explains why teleosts possess greater diversity in colouration than tetrapods (Bagnara, 1998). In the cichlids of the East African Great Lakes, the *csfl ra* gene has also undergone several duplications (Braasch et al., 2006). It is now accepted that genome duplications and the expansion of the receptor tyrosine kinase signalling cascade proteins increased the repertoire of pigment cells that led to the enormous diversity pigment cell innovations found in cichlid egg-dummy patterns. The *pdgfr β-csfl r* locus was also thought to mediate new gene functions that could further drive the evolution of cichlid colouration (Braasch et al., 2006).

Opsin genes

Differences in opsin visual pigments alter visual sensitivity and dictate mate preferences. In Lake Malawi cichlids, an ultraviolet (UV)-sensitive cone pigment was found to detect the UV reflectance common to many blue cichlids (Carleton et al., 2000). Terai et al. (2002) and Sugawara et al. (2002) found the signatures of positive selection for rhodopsin and long wavelength sensitive opsin genes. These genes allowed cichlid visual pigments to evolve rapidly and adapt to changing water quality parameters in different habitats of the lake. Consistent with this theory, Seehausen et al. (1997) showed that differences in visual sensitivity have influenced the evolution of colour patterns and sexual selection in cichlids adapted to living in turbid and clear parts of Lake Victoria.

Sensory drive hypothesis

Sensory drive is a hypothesis about how communication signals are designed to work effectively (Boughman, 2002). In the case of mating signals, the hypothesis explains how signals are best designed to attract mates. Thus the sensory drive hypothesis predicts that females mate more often with male phenotypes that they detect more easily and evolve to prefer signals that are conspicuous and easy to detect in the environment (Kawata et al., 2007). In guppies, orange spots are a visual cue in female mate choice (Endler, 1991; 1992). In sticklebacks, females prefer males with a larger area of red (Boughman, 2001; 2002). Both of these visual mate choice scenarios are influenced by environmental light (Gamble et al., 2003).

According to the sensory drive hypothesis, easy-to-detect signals are likely favoured by the choosy mates (Boughman, 2002). Thus females (usually the choosy ones) often prefer signals that are conspicuous, for example, long feathers, bright colours, complex vocalizations, or bizarre extensions of male morphology such as horns and eye stalks (Boughman, 2002). Inherent properties of signals such as their colour, intensity, or size, affect signal conspicuousness and detection by females (Boughman, 2002), including the following 3 processes described by Endler (1993):

(1) Habitat transmission (passage of signals through the habitat).
(2) Perceptual tuning (perceptual adaptation to local habitat).
(3) Signal matching (matching of male signals to female perception).

The sensory drive hypothesis explains how these three processes shape the evolution of inherent signal properties.

In cichlids, Gray and Mckinnon (2007) and Chunco et al. (2007) postulated that sensory drive could cause the appearance of colour polymorphisms and eventually lead to speciation (Kawata et al., 2007) even when the population is not geographically isolated (Seehausen et al.,

2008). Such evolution of sympatric species was earlier shown to occur in the sticklebacks of British Columbia (Boughman, 2001). According to Kawata et al. (2007), sensory drive hypothesis argues that divergent sensory adaptation in different habitats may lead to pre-mating isolation upon secondary contact of populations.

Speciation by sensory drive has traditionally been treated as a special case of speciation as a by-product of adaptation to divergent environments in geographically isolated populations (Kawata et al., 2007). However, if habitats are heterogeneous, local adaptation in the sensory systems may cause the emergence of reproductively isolated species from a single unstructured population. In Lake Victoria, habitat heterogeneity has been attributed to differences in ambient light regimes caused by water pollution and eutrophication (Seehausen et al., 1997).

In Lake Victoria cichlids, females prefer males with colours which the females perceive as intense or conspicuous (Seehausen and Van Alphen, 1998). Colour perception is determined by several different components (Kelber et al., 2003), one such component is sensitivity at a given wavelength of light (Kawata et al., 2007). Retinas of cichlid fish inhabiting relatively blue-shifted environments are more sensitive to blue light than the retinas of those from red-shifted environments (Carleton et al., 2005). Closely related species of cichlids in Lake Victoria differ in their retinal absorption spectra (Meer and Bowmaker, 1995). Cichlids with different retinal absorption spectra also have different male mating and breeding colouration (Kawata et al., 2007). In fact, the major peaks in the retinal absorption spectra match the most common breeding colours (Kawata et al., 2007). These observations had led earlier workers to the suggestion that evolution of the visual system could drive the speciation process in cichlids (Seehausen et al., 1997). Recent evidence now shows that in single unstructured populations of cichlids, sensitivity to light of different wavelengths is determined by heritable variation in the absorption spectra of opsins (Carleton et al., 2005; Mann et al., 2006). This can be demonstrated via three mechanisms of colour tuning.

First of all, changing the amino acid sequence of opsin genes, the so-called spectra-tuning, causes changes in peak absorption spectra of the visual pigments (Yokoyama and Radlwimmer, 2001). Fixed genetic differences were found in the LWS opsin locus between closely related populations of Lake Victoria cichlid fish (Terai et al., 2002). Variation in the amino acid sequence of opsin proteins may lead to differences in mate choice signals (Terai et al., 2002).

Secondly, colour vision can be changed by varying the transcription and translation of different opsin genes. When a larger amount of LWS opsin protein is expressed in the retina than the medium and short wavelength-sensitive opsins, the individual could be more sensitive to light of longer wavelengths (Kawata et al., 2007). In Lake

Malawi cichlids, Carleton et al. (2008) also found that differential expression of unique subsets of cone opsin genes produces drastic differences in the visual pigments of these fishes. Cichlids have 5 - 6 cone opsins but express only 3 of them in adults (Kawata et al., 2007). Some species of Lake Malawi cichlids that live in different light environments express complimentary subsets of opsin genes (Carleton and Kocher, 2001).

Thirdly, Carleton et al. (2008) suggested that cichlids have 4 cone opsin genes but additional duplications have produced 7 distinct cone opsin genes (sws1, sws2b, sws2a, RH2b, RH2aβ, RH2aα and LWS) which produce visual pigments that are spectrally distinct from each other.

Accordingly, 3 molecular genetic mechanisms produce differences in the visual pigment genes and spectral tuning; amino acid substitutions, differential expression and gene duplications of cone opsins. Given these 3 mechanisms, evolutionary adaptation of the visual system could lead to the divergence between populations in female preference for male nuptial colour (Kawata et al., 2007). Upon secondary contact, such divergence can cause pre-mating isolation and sympatric speciation by sexual selection (Herder et al., 2006; Schliewen et al., 1994; Seehausen and Van Alphen, 1999; Shaw et al., 2000). This happens as follows (Kawata et al., 2007):

(1) Spectral sensitivity evolves as an adaptation to environmental (ambient) light regimes.
(2) A female prefers to mate with a male whose nuptial colour reflects at the wavelength that she most intensely perceives.
(3) The female's sensitivity to light of a given wavelength depends on the absorption spectra of her visual pigments.

Recently, Seehausen et al. (2008) have demonstrated the sensory drive hypothesis in island populations of cichlid fish. They determined ecological, population genetic and molecular bases of divergent evolution in the cichlid visual system and linked it to divergence in male colouration and female mate preferences and finally demonstrated that this plethora of ecological and molecular forces can cause differentiation of neutral loci and reproductive isolation in sympatric cichlids inhabiting different light gradients in Lake Victoria (Seehausen et al., 2008). Male fish living at different water depths develop different colours and females prefer conspicuously coloured males due to variations in the Long-wavelength sensitive (LWS) opsin gene. Females have mating preferences for conspicuously coloured males. The LWS opsin genotype determines female mating preferences. Divergence in LWS systems is associated with divergence in male colours and female mating preferences. This causes differentiation of neutral loci, leading to reproductive isolation and speciation.

A 4-step sensory drive speciation model has been pro-

posed by Seehausen et al (2008):

(1) Divergent natural selection between light regimes at different water depths acts on LWS.
(2) Sexual selection for conspicuous colouration also becomes divergent because perceptual biases differ between light regimes.
(3) Their interaction generates initial deviation from linkage equilibrium between LWS and nuptial colour alleles.
(4) Subsequent disruptive selection due to reduced fitness of genotypes with a mismatch between LWS and colour alleles causes speciation, possibly involving reinforcement-like selection for mating preferences where male nuptial colour serves as a marker trait for opsin genotype. See Box 1 for a simpler synthesis of the sensory drive hypothesis in African lake cichlids.

Box 1: A model on sensory drive in African lake cichlids

The Seehausen et al. (2008) model explains how the physics of light predisposes African lake cichlids to find mating partners and then over time, drives the cichlid population towards sexual selection and sympatric speciation. Water tends to absorb red light and leaves blue light to travel to the bottom. This explains why deep water bodies appear blue. But in a slightly polluted and cloudy lake, blue light dominates the visual environment near the surface (because the blue light is absorbed by debri), while red light dominates the visual environment in deeper waters.

Cichlid fish populations live along a slope, from shallow waters at the lake's shores to deeper waters at the bottom of the lake. Thus some cichlids spend more of their time in blue light near the surface and red light in deeper waters. These differences induce genetic variation, allowing cichlids in shallow water to have genes that enable them to perceive blue light and deeper-dwelling cichlids are imbued with genes that favour them to see red light. Thus shallow-dwelling cichlids have genes that confer blue-light advantage and deeper water cichlids have genes that confer red-light fitness.

Therefore in different parts of the cichlid habitats, specific colour-sensitive genes are favoured by natural selection. Over many generations and if the cichlids do not move a lot within their range, blue-sensitivity will be common in cichlids near the surface and red sensitivity will prevail in cichlids in deeper water further down the slope. This natural selection acting on light sensitivity can divide the cichlid population and together with sexual selection, the divergence is exacerbated even further. Since female cichlids choose brightly coloured males for mating, it follows therefore that blue males living in deep water have difficulties finding female mates, for two reasons; there is little blue light in deep bottom water, making blue male fish appear duller than the red males and females in deep water are less sensitive to blue light.

For these 2 reasons, bottom-dwelling blue male cichlids are doomed to a useless sex life and so are the red male cichlids living near the surface. After many cycles of selective breeding, the cichlid population will diverge into 2 different sub-populations that prefer to mate with fish of similar colouration, light sensitivity and habitat. Female cichlids will tend to prefer males of their own habitat because their colours are congruent to their colour sensitivity. In the longer term, the 2 sub-populations of cichlids will no longer mate with each other and will evolve into separate species. (Adapted from: http://evolution.berkeley.edu/evolibrary/news/090310_cic hlidsspeciation)

Concluding Remarks

Studies in African cichlids are just beginning to unravel the interactions among genotype, phenotype and the environment. Molecular genetic studies, in particular have focussed on determining the ecological and evolutionary mechanisms that drive the speciation of cichlids in African lakes and rivers. These studies have illustrated the species richness and ecomorphological diversity of cichlids and documented various selective pressures involved in the ecomorphological differentiation and species divergence of African cichlids. The rapid speciation of African lake cichlids is now firmly rooted in their sexual selection and choice of mating partners based on nuptial colour patterns. Putative genes that underlie phenotypic variation and speciation of African cichlids are being elucidated. This research presents the potential to discover the full range of the cichlids species in African lakes and rivers.

Many cichlids, particularly the tilapias, are important food fishes, while others are valued game fish for anglers. Many cichlid species are also highly valued in the aquarium trade (Loiselle, 1994; Chapman, 1992). Cichlids are also the family of vertebrates with the largest number of endangered species, most of these found in the haplochromine group. According to the 2007 International Union for Conservation of Nature and Natural Resources red list, 156 cichlid species are currently listed as vulnerable, 40 species are listed as endangered, while 69 species are listed as critically endangered. 6 species, *Haplochromis ishmaeli*, *Haplochromis lividus*, *Haplochromis perrieri*, *Paretroplus menarambo*, *Platytaeniodus degeni* and *Yssichromis* sp. *nov.* 'argens' are extinct in the wild, while at least 39 species, most from the genus *Haplochromis*, have become extinct since the early 1990s (IUCN, 2006).

In the past few decades, the number of Lake Victoria cichlid species has declined and many species have collapsed into each other. Deforestation, nutrient run-off into the lake, eutrophication and pollution has conspired to make the lake's water more and cloudier and murkier.

Poor water quality makes it very difficult for female fish to choose mates based on colour, making the cichlid species that rely on visual signals for mate choice to interbreed. Ultimately, interbreeding has resulted in fewer species of cichlids. The evidence presented here points to how cichlid species biodiversity can be diminished by poor water quality emanating from human pollution. Given that cichlids are commercially important food fishes in Africa, current studies that focus on molecular genetic bases and mechanisms of speciation in cichlids will aid in correct decision-making in fisheries management, conservation, breeding and aquaculture.

Further work is still needed to map genes for phenoltypic traits associated with speciation, such as jaw morphology and adult colour patterns. The goal is to identify the genes responsible for speciation of these fishes, and to study the geographic distribution of allelic variants among populations and species in the wild. A key element to studying the diversification in cichlids is the development of genomic resources to support the identification of genes underlying the phenotypic differences among species. The idea to sequence the whole genome of an African cichlid would help decipher the full range of cichlid genes and help shed more light on the molecular genetic bases of speciation. The current work involving the sequencing of the Nile tilapia may help reveal genomic hotspots responsible for the cichlids' ability to adapt and to diversify (Salzburger, 2009). The 'cichlidomics' era is now at hand. Further work to elucidate the signal transduction pathways involved in the expression of speciation genes is also needed.

ACKNOWLEDGEMENTS

Many thanks to Dr Cyprian Katongo for reading through the first draft of the manuscript and to Professor Walter Salzburger for availing some of the literature.

REFERENCES

Albertson RC, Streelman JT, Kocher TD (2003a). Directional selection has shaped the oral jaws of Lake Malawi cichlid fishes. Proc. Nat. Acad. Sci. USA 100: 5252-5257.

Albertson RC, Streelman JT, Kocher TD (2003b). Genetic basis of adaptive shape differences in the cichlid head. J. Hered. 94: 291-301.

Albertson RC, Streelman JT, Kocher TD, Yelick PC (2005). Integration and evolution of the cichlid mandible: the molecular basis of alternate feeding strategies. Proc. Nat. Acad. Sci. USA 102: 16287-16292.

Bootsma HA, Hecky RE, Hesslein RH, Turner GF (1996). Food partitioning among Lake Malawi nearshore fishes as revealed by stable isotope analysis. Ecology 77: 1286-1290

Boughman JW (2001). Divergent sexual selection enhances reproductive isolation in sticklebacks. Nature 411: 944-948.

Boughman JW (2002). How sensory drive can promote speciation. Trends Ecol. Evol. 12: 571-577.

Boulenger GA (1898a). Catalogue of the fresh-water fishes of Africa in the British Museum (Natural History). London: British museum for Natural History.

Boulenger GA (1898b). Report on the fishes recently obtained by Mr.

J.E.S. Moore in Lake Tanganyika. Proc. Zool. Soc. Lond. 1898 (pt.3): 494-497.

Braasch I, Salzburger W, Meyer A (2006). Asymmetric evolution in two fish-specifically duplicated receptor tyrosine kinase paralogons involved in teleost colouration. Mol. Biol. Evol. 23:1192-1202.

Carleton KL, Harosi FI, Kocher TD (2000). Visual pigments of African cichlid fishes: Evidence for ultraviolet vision from microspectrophotometry and DNA sequences. Vision Res. 40: 879-890.

Carleton KL, Kocher TD (2001). Cone opsin genes of African cichlid fishes: tuning spectral sensitivity by differential gene expression. Mol. Biol. Evol. 18(8): 1540-1550.

Carleton KL, Parry JWL, Bowmaker JK, Hunt DM, Seehausen O (2005). Colour vision and speciation in Lake Victoria cichlids of the genus *Pundamilia*. Mol. Ecol. 14: 4341-4353.

Carleton KL, Spady TC, Streelman JT, Kidd MC, McFarland WN, Loew ER (2008). Visual sensitivities tuned by heterochronic shifts in opsin gene expression. BMC Biology 6: 22. Open access: http://www.biomedcentral.com/1741-7007/6/22.

Chapman FA (1992). Culture of hybrid tilapia: a reference profile. University of Florida Institute of Food and Agricultural Sciences.

Chunco AJ, Mckinnon JS, Servedio MR (2007). Microhabitat variation and sexual selection can maintain male colour polymorphisms. Evolution 61: 2504-2515.

Cohen AS, Lezzar KE, Tiercelin JJ, Soreghan M (1997). New palaeogeographic and lake-level reconsructions of Lake Tanganyika: implications for tectonic, climatic and biological evolution in a rift lake. Basin Res. 9: 107-132.

Cohen AS, Soreghan MJ, Scholz CA (1993). Estimating the age of formation of lakes, an example from Lake Tanganyika, East African Rift system. Geology 21:511-514.

Danley PD, Markert JM, Arnegard ME, Kocher TD (2000). Divergence with gene flow in rock-dwelling cichlids of Lake Malawi. Evolution 54: 1725-1737.

Danley PD, Kocher TD (2001). Speciation in rapidly diverging systems: lessons from Lake Malawi. Mol. Ecol. 10: 1075-1086

Darwin C (1859). On the origin of species by means of natural selection. London, UK: John Murray.

Delvaux D (1995). Age of Lake Malawi (Nyasa) and water level fluctuations. Mus R Afr Centr Tervuren (Belgium), Dept. Geol. Min. Rapp. Annu. 1995: 99-108.

Ding G, Sun Y, Li H, Wang Z, Fan H, Wang C, Yang D, Li Y (2008). EPGD: a comprehensive web resource for integrating and displaying eukaryotic paralo/ paralogon information. Nucleic Acids Res. 36: D255-D262; doi:10.1093/nar/gkm924

Endler JA (1991). Variation in the appearance of guppy colour patterns guppies and their predators under different visual conditions. Visual Res. 31(3): 587-608.

Endler JA (1992). Signals, signal conditions, and the direction of evolution. Am. Nat. 139: 125-153.

Endler JA (1993). Some general comments on the evolution of and design of animal communication systems. Philos. Trans. R. Soc. Lond. Ser. B340: 215-225.

Farias IP, Ort G, Sampaio I, Schneider H, Meyer A (2001). The cytochrome b gene as a phylogenetic marker: the limits of resolution for analyzing relationships among cichlid fishes. J. Mol. Evol. 53: 89-103.

Farias IP, Orti G, Meyer A (2000). Total evidence: molecules, morphology, and the phylogenetics of cichlid fishes. J. Exp. Zool. 288: 76-92.

Fraser GJ, Bloomquist RF, Streelman TD (2008). A periodic pattern generator for dental diversity. BMC Biology 6: 32. Open access: http://www.biomedcentral.com/1741-7007/6/32.

Fraser GJ, Hulsey CD, Bloomquist RF, Uyesugi K, Manley NR, Streelman JT (2009). An ancient gene network is co-opted for teeth on old and new jaws. Plos Biology 7(2): e1000031. Open access: http://biology.plosjournals.org/perlserv/?request=get-document &doi=10.1371/ journal.pbio.1000031.

Fryer G, Iles TD (1972). The cichlid fishes of the Great Lakes of Africa: their biology and evolution. Oliver & Boyd: Edinburgh.

Fryer G (1977). Evolution of species flocks of cichlid fishes in African lakes. Zeitschrift fuer Zoologische Systematik und Evolutions for schung

15: 141-163.

Gamble SA, Lindholm A, Endler JA, Brooks RC (2003). Environmental variation and the maintenance of polymorphism: the effect of ambient light. Ecol. Lett. 6: 463-472.

Goldschmidt T (1991). Egg mimics in haplochromine cichlids (Pisces, Perciformes) from Lake Victoria. Ethology 88:177-190.

Goldschmidt T, de Visser J (1990). On the possible role of egg mimics in speciation. Acta Biotheor. 38:125-134.

Grant PR (1981). Speciation and the adaptive radiation of Darwin's finches. Am. Sci. 69: 653-663.

Gray SM, Mckinnon JS (2007). Linking colour polymorphism maintenance and speciation. Trends Ecol. Evol. 22: 71-79.

Greenwood PH (1964). Explosive speciation in African lakes. Proceedings of the Royal Institute of Great Britain 40: 256-269.

Herder F, Nolte A, Pfaender J, Schwarzer J, Hadiaty RK, Schliewen UK (2006). Adaptive radiation and hybridization in Wallace's Dreamponds: evidence from sailfin silversides in Malili Lakes of Sulawesi. Proc. R Soc B: Biol. Sci. 273: 2209-2217.

Higashi M, Takimoto G, Yamamura N (1999). Sympatric speciation by sexual selection. Nature 402: 523-526.

Hulsey CD (2006). Function of a key morphological innovation : fusion of the cichlid pharyngeal jaw. Proc. R. Soc. B. 273: 669-675.

Hoegg S, Brinkmann H, Taylor JS, Meyer A (2004). Phylogenetic timing of the fish-specific genome duplication correlates with the diversification of teleost fish. J. Mol. Evol. 59: 190-203.

IUCN (2006). IUCN red list of threatened species. List of recently extinct Cichlidae. www.iucnredlist.org (downloaded on 20 April, 2009).

Johnston TC, Ng'ang'a P (1990). Reflections on a rift lake. In Katz, B.J. (Ed). Lacustrine basin exploration: case studies and modern analogs. Am. Assoc. Pet. Geol. Memoir: pp. 113-135.

Johnston TC, Scholz CA, Talbot MR, Kelts K, Ricketts RD, Ngobi G, Beuning K, Ssemmanda I, McGill JW (1996). Late Pleistocene desiccation of Lake Victoria and rapid evolution of cichlid fishes. Sci. 273: 1091-1093.

Jonsson B, Jonsson N (2001). Polymorphism and speciation in Arctic charr. J. Fish Ecol. 58(3): 605-638.

Joyce DA, Hunt DH , Bills R, Turner GF, Katongo C, Duftner N, Sturmbauer C, Seehausen O (2005). An extant cichlid fish radiation emerged in an extinct Pleistocene lake. Nature 435: 90-95.

Kawata M, Shoji A, Kawamura S, Seehausen O (2007). A genetically explicit model of speciation by sensory drive within a continuous population in aquatic environments. BMC Evol. Biol. 7: 99. doi: 10.1186/1471-2148-7-99. Available from: http://www.biomedcentral.com/1471-2148/7/99.

Kelber A, Vorobyev M, Osorio D (2003). Animal colour vision-behavioural tests and physiological concepts. Biol. Rev. 79: 81-118.

Kocher TD (2004). Adaptive evolution and explosive speciation: the cichlid fish model. Nature Rev. Genet. 5: 288-298.

Kornfield I, Smith PF (2000). African cichlid fishes: model systems for evolutionary biology. Annu. Rev. Ecol. and Syst. 31: 163-196.

Lande R (1982). Rapid origin of sexual isolation and character divergence in a cline. Evolution 36(2): 213-223.

Liem KF (1973). Evolutionary strategies and morphological innovations: cichlid pharyngeal jaws. Syst. Zool. 22: 425-441.

Loiselle PV (1994). The cichlid aquarium. Tetra Press: Melle, Germany.

Mayr E (1984). Evolution of fish species flocks: a commentary. In: Echelle AA, Kornfield I (Eds). Evolution of fish species flocks. University of Maine: Orono Press. pp 3-11.

Mayr E (1942). Systematics and the origin of species. Columbia University Press: New York.

Mayr E (1963). Animal species and evolution. Harvard University Press: Cambridge, Mass.

Meer HJ, Bowmaker JK (1995). Interspecific variation of photoreceptors in four co-existsing haplochromine cichlid fishes. Brain. Behav. Evol. 45(4): 232-240.

Meyer A, Kocher TD, Basasibwaki P, Wilson AC (1990). Monophyletic origin of Lake Victoria cichlid fishes suggested by mitochondrial DNA sequences. Nature 347: 550-553.

Meyer A (1993). Phylogenetic relationships and evolutionary processes in East African cichlid fishes. Trends Ecol. Evol. 8: 279-284.

Payne RJH, Krakauer DC (1997). Sexual selection, space and speciation. Evolution 51: 1-9.

Salzburger W, Braasch I, Meyer A (2007). Adaptive sequence evolution in a colour gene involved in the formation of the characteristic egg-dummies of male haplochromine cichlid fishes. BMC Biol. 5: 51; doi:10.1186/1741-7007-5-51.

Salzburger W, Meyer A (2004). The species flocks of East African cichlid fishes: recent advances in molecular phylogenetics and population genetics. Naturwissenschaften 91: 277-290.

Salzburger W (2009). The interaction of sexually and naturally selected traits in adaptive radiations of cichlid fishes. Mol. Ecol. 18: 169-185.

Schattner P, Diehhans M (2006). Regions of extreme synonymous codon selection in mammalian genes. Nucleic Acids Res. 34:1700-1710.

Schliewen UK, Tautz D, Paabo S (1994). Sympatric speciation suggested by monophyly of crater lake cichlids. Nature 368: 629-633.

Seehausen O, Schluter D (2004). Male-male competition as a diversifying force in Lake Victoria cichlid fishes. Proceedings of the Royal Society of London B.271: 1345-1353.

Seehausen O, Terai Y, Magalhaes IS, Carleton KL, Mrosso HDL, Miyagi R, Van der Sluijs I, Schneider MV. Maan ME, Tachida H, Imai H, Okada N (2008). Speciation through sensory drive in cichlid fish. Nature 455: 620-626.

Seehausen O, Van Alphen JJM (1998). The effect of male colouration on female mate choice in closely related Lake Victoria cichlids (*Haplochromis nyererei* complex). Behav. Ecol. Sociobiol. 42: 1-8.

Seehausen O, Van Alphen JJM (1999). Can sympatric speciation by disruptive sexual selection explain rapid evolution of cichlid diversity in Lake Victoria? Ecol. Lett. 2: 262-271.

Seehausen O, Van Alphen JJM, Lande R (1999). Colour polymorphism and sex ratio distortion in a cichlid fish as an incipient stage in sympatric speciation by sexual selection. Ecol. Lett. 2: 367-378.

Seehausen O, Van Alphen JJM, Witte F (1997). Cichlid fish diversity threatened by eutrophication that curbs sexual selection. Science 277: 1808-1811.

Seehausen O, Van Alphen JM (1999). Can sympatric speciation by disruptive sexual selection explain rapid evolution of cichlid diversity in Lake Victoria? Ecol. Lett. 2(4): 262-271.

Seehausen O, Schluter D (2004). Male-male competition and nuptial-colour displacement as a diversifying force in Lake Victoria cichlid fishes. Proc. Biol. Sci. 271(1546): 1345-1353.

Seehausen O, Van Alphen JJM (1998). The effect of male colouration on female mate choice in closely related Lake Victoria cichlids (Haplochromis nyererei complex). Behav. Ecol. Sociobiol. 42: 1-8.

Seehausen O (1996). Distribution of and reproductive isolation among color morphs of a rock-dwelling Lake Victoria cichlid (*Haplochromis nyererei*). Ecol. Freshw. Fish 5: 195-202.

Seehausen O, Koetsier E, Schneider MV, Chapman LJ, Chapman CA, Knight ME, Turner GF, Van Alphen JJM, Bills R (2003). Nuclear markers reveal unexpected genetic variation and a congolese-nilotic origin of the lake victoria cichlid species flock. Proc. R. Society Lond B270: 129-137.

Shaw PW, Turner GF, Idid MR, Robinson RL, Carvalho GR (2000). Genetic population structure indicates sympatric speciation of Lake Malawi pelagic cichlids. Proc R Soc London Series B: Biol. Sci. 267: 2273-2280.

Snoeks J, Ruber L, Verheyen E (1994). The Tanganyika problem: comments on the taxonomy and distribution patterns of its cichlid fauna. In: Martens K, Goddeeris G, Coulter GW (Eds). Speciation in ancient lakes, Stuttgart: Schweizerbart'sche Verlagsdruckerei, pp. 355-372.

Sparks JS (2004). Molecular phylogeny and biogeography of the Malagasy and South Asian cichlids (Teleostei: Perciformes: cichlidea) Mol. Phylogenet. Evolution 30: 599-614.

Stiassny MLJ, Meyer A (1999). Cichlids of the Rift Lakes. Sci. Am. 280: 64-69.

Streelman JT, Albertson RC (2006). Evolution of novelty in the cichlid dentition. J. Exp. Zool. (Mol. Dev. Evol.)306B: 216-226.

Streelman JT, Danley PD (2003). The stages of vertebrate evolutionary radiation. Trends Ecol. Evol. 18: 126-131.

Streelman JT, Webb JF, Albertson RC, Kocher TD (2003a). The cusp of evolution and development: a model of cichlid tooth shape diversity. Evol. Dev. 5: 600-608.

Streelman JT, Albertson RC, Kocher TD (2003b). Genome mapping of the orange blotch colour pattern in cichlid fishes. Mol. Ecol. 12: 2465-2471.

Streelman JT, Zardoya R, Meyer A, Karl SA (1998). Multi-locus phylogeny of cichlid fishes (Pisces: Perciformes): evolutionary comparison of microsatellite and single-copy nuclear loci. Mol. Biol. Evol. 15: 798-808.

Sturmbauer C (1998). Explosive speciation in cichlid fishes of the African Great Lakes: a dynamic model of adaptive radiation. J. Fish Biol. 53: 18-36.

Sugawara T, Terai Y, Okada N (2002). Natural selection of the rhodopsin gene during the adaptive radiation of East African Great Lakes cichlid fishes. Mol. Biol. Evol. 19: 1807-1811.

Terai Y, Morikawa N, Kawakami K, Okada N (2002). Accelerated evolution of the surface amino acids in the WD-repeat domain encoded by the *hagoromo* gene in an explosively speciated lineage of East African cichlid fishes. Mol. Biol. Evol. 19: 574-578.

Terai Y, Morikawa N, Kawakami K, Okada N (2003). The complexity of alternative splicing of *hagoromo* mRNAs is increased in an explosively speciated lineage in East African cichlids. Proc. Nat. Acad. Sci. USA 100: 12798-12803.

Terai Y, Morikawa N, Okada N (2002). The evolution of the pro-domain of bone morphogenetic protein (Bmp4) in an explosively speciated lineage of East African cichlid fishes. Mol. Biol. Evol. 19: 1628-1632.

Terai Y, Mayer WE, Klein J, Tichy H, Okada N (2002). The effect of selection on long wavelength-sensitive (LWS) opsin gene of Lake Victoria cichlid fishes. Proc. Nat. Acad. Sci. USA 19: 1550-1556.

Turner GF, Burrows MT (1995). A model of sympatric speciation by sexual selection. Proc. R. Soc. Biol. 260: 287-292.

Verheyen E, Salzburger W, Snoeks J, Meyer A (2003). Origin of the superflock of cichlid fishes from Lake Victoria, East Africa. Science 300: 325–329.

Wilson J, Tucker AS (2004). FGf and bmp signals repress the expression of Bapx1 in the mandibular mesenchyme and control the position of the developing jaw point. Dev. Biol. 266: 138-150.

Yokoyama S, Radwimmer FB (2001). The molecular genetics and evolution ofred and green colour vision in vertebrates. Genetics 158: 1697-1710.

Zardoya R, Vollmer DM, Craddock C, Streelman JT, Karl SA, Meyer A (1996). Evolutionary conservation of microsatellite flanking regions and their use in resolving the phylogeny of cichlid fishes (Pisces: Perciformes). Proceedings of the Royal Society of London 263: 1589-1598.

14

Emerging trends in enhancement of cotton fiber productivity and quality using functional genomics tools

N. Manikanda Boopathi[1*] and R. Ravikesavan[2]

[1]Department of Plant Molecular Biology and Biotechnology, Coimbatore- 641003, India.
[2]Department of Cotton, Tamil Nadu Agricultural University, Coimbatore- 641003, India.

Cotton, the most preferred natural fiber in the world, is the mainstay of global economy for several centuries. However, the fiber productivity has reached its plateau in the past decade which forced the research community to develop high-yielding and high quality cotton cultivars. In this genomics era, cotton researches focussed on two aspects: identification of genes for important agronomic traits and manipulation of such genes in view of developing elite cotton cultivar. Despite the complexity of the molecular mechanisms underlying its development, the study of the cotton fiber has become a trait of primary interest besides biotic and abiotic stress resistance. Albeit several strategies, functional genomics approach offers new unprecedented opportunities for identification of complex network of genes involved in fiber productivity and quality. Recent years have witnessed a better understanding of the plethora of genes affecting cotton fibre. Molecular, cellular and developmental changes related to fiber development have been identified through high-throughput EST projects and microarray analysis coupled with cotemporary biological tools. Despite impressive progress, the genomics and post-genomics revolution will be applicable in plant breeding only when they can elucidate the relationship between variation in phenotypic traits and the variation in gene sequences and/or expression. To this end, there is an immediate demand for integration of disciplines such as structural genomics, transcriptomics, proteomics and bioinformatics with plant physiology and breeding. Integration of multidisciplinary approaches is indispensable in upcoming cotton improvement programs since cotton is an important renewable resource that needs to be preserved for future generations.

Key words: Cotton fibre, expressed sequence tags, functional genomics, microarray, transcriptomics.

INTRODUCTION

Cotton, a high valued agricultural commodity for more than 8000 years, has long been recognized as a vital component of the global economy (Arpat et al., 2004). Cotton production provides income for approximately 100 million families and approximately 150 countries are involved in cotton import and export (Chen et al., 2007a). All parts of the cotton plant are useful and it has hundreds of uses. No other fiber comes close to duplicating all of the desirable characteristics combined in cotton. In addition to the fiber used in textile manufacturing, cotton seed is used to produce oil, seed meal (rich in essential amino acids which is lacking in most seed crops) and seed hulls (used for mulch and cattle feed). It has been estimated that 180 million people depend, either directly or indirectly, on the production of cotton for their livelyhood (Benedict and Altman, 2001).

Globally, cotton is grown on 32 million hectares (mha) with approximately 71% of the production in developing countries (FAO, 2000; http://apps.fao.org/page /collections?subset=agriculture). India, USA and China are the main producers of world production. India has larger sown cotton area (9 mha) than any country in the world

(Clive, 2006) and produces almost three million tonnes of all qualities and staple lengths of cotton per year. The cotton industry in India has 1,543 spinning units, more than 281 composite mills, 1.72 million registered looms and an installed capacity of 36.37 million spindles (Kambhampati et al., 2005). Cotton provides a livelihood to more than 60 million people in India by way of support in agriculture, processing and use of cotton in textiles and also contributes 30% to the Indian agricultural gross domestic product and thus cotton is a very important cash crop for Indian farmers (Barwale et al., 2004). Albeit India's cotton area repre-senting 25% of the global area of cotton, it produced only 12% of world production. Yields of cotton in India are low, with an average yield of 300 kg/ha compared to the world average of 580 kg/ha (Clive, 2006). The major limiting factors to both cotton production and quality in India are biotic and abiotic stresses. As with many cotton growing areas of the world, major damage is due to insect pests, especially the boll-worm complex, sucking pests and viruses. The productivity is still worsened by abiotic stress such as drought and heat. It is worth to mention here that most of the cotton in India is grown under rainfed conditions and about a third is grown under irrigation (Sundaram et al., 1999), which also experiences water stress during certain growth periods. Rising production costs to combat biotic and abiotic stresses and stagnant pricing are the additional factors that threaten cotton production. The low pricing of recent years due to poor quality of fiber resulted to biotic and abiotic stresses and has forced many growers to plant alternative crops, even in the face of farm subsidies. Hence, to cope with the growing demand on cotton fiber and by products, genetic enhancement of cotton is indispensable which will ensure competitiveness in the market of this natural-renewable product with petroleum-derived synthetic fibres, given the projected future decline in petroleum reserves. Moreover, modifications to expand the use of seed derivatives for food and feed could profoundly benefit the diets and livelihoods of millions of people in food-challenged economies (Chen et al., 2007b).

The genus *Gossypium* L. has long been a focus of genetic, systematic and breeding research and has a long history of improvement through breeding, with sustained long-term yield gains. *Gossypium* spp. consists of at least 45 diploid and 5 allotetraploid species. The allotetraploid cotton species, which include two commercially important cultivated species, *G. hirsutum* L., and *G. barbadense* L., were generated by A- and D-compound genomes (Fryxell, 1979). The best living models of the ancestral A- and D-genome parents are *G. herbaceum* and *G. raimondii*, respectively (Endrizzi et al., 1985). These four genome groups have received special attention, since they have been domesticated for their abundant seed trichomes. The diploid donor of the allopolyploid A_T genome [where the T subscript indicates the A genome in the tetraploid (AD) nucleus], was a species

much like the modern *G. arboreum* or *G. herbaceum*, whereas the allopolyploid D_T genome is derived from a progenitor similar to the modern *G. raimondii* species. These well-established relationships provide a phylogenetic framework to investigate the evolution of gene expression both in terms of domesticated fiber production and polyploidy (Udall et al., 2007). Understanding the contribution of the A and D subgenomes to gene expression in the allotetraploids may facilitate improvement of fiber traits (Chen et al., 2007a).

Interestingly, the A genome species produce spinnable fiber and are cultivated on a limited scale, whereas the D genome species do not (Applequist et al., 2001). More than 95% of the annual cotton crop worldwide is *G. hirsutum* (Upland or American cotton) and *G. barbadense* (the extra-long staple or Pima cotton) which accounts for less than 2% (National Cotton Council, 2006. http://www.cotton.org). Yield and fiber quality of upland cotton varieties have declined over the last decade – a downward trend that has been attributed to erosion in genetic diversity of breeding stocks and an increased vulnerability to environmental stresses (Meredith, 2000). The level of genetic diversity is low in *G. hirsutum*, especially among agriculturally elite types, as revealed by all means of assessment (Lacape et al., 2007a).

Increasing diversity is therefore essential to genetic improvement efforts. Each of the three major approaches to increasing genetic diversity that is, mutagenesis, germplasm introgression and transformation has advantages and disadvantages. Interspecific germplasm introgression is particularly attractive in that it utilizes a broad germplasm base, can be targeted to one or more specific traits/genes or modulated to include thousands of genes/even entire genomes and is readily coupled to marker-assisted genome analysis and selection (Saha et al., 2006). Though, quantitative trait loci (QTL) mapping and marker aided selection has potential application in genetic improvement of cotton for higher productivity, its application is not yet documented in cotton breeding program due to poor knowledge on physiological and genetic nature of fibre quality and productivity traits, low and complex heritability of investigated traits, genotype X environment interactions etc., (Lacape et al., 2007b). Although introgression of genes across species boundaries is difficult, it is quite desirable because the gene pools of cultivated species do not contain all of the desired alleles. Alternatively, mutagenesis and transgenic technology has been proposed. However, currently they have limited applications due to several technical reasons such as non availability of novel genes, lack of efficient method to alter/transfer large genetic element etc., (Wilkins et al., 2000). Thus, the paucity of information about genes that control important traits impedes the genetic improvement of cotton.

Fiber represents over 90% of the total value of the cotton crop and the genetic improvement of fiber properties is certainly a major target trait besides biotic and abiotic stress resistance since these stresses also ultimately affect the final productivity and quality of fiber. The cotton fiber is a complex biological system that is the net result of the intricate interplay of elaborate developmentally regulated pathways consisting of literally thousands of genes and gene products and as such has been a difficult subject to tackle using conventional approaches, especially as cotton does not lend itself easily to genetic analysis (Taliercio and Boykin, 2007). The high value per hectare of cotton and global textile market demand for increased fiber uniformity, strength, extensibility and quality clearly justify the importance of new and innovative approaches towards evaluating and under-standing genetic mechanisms of fiber qualities (Saha et al., 2006). Gene discovery and strategies to produce the desired pattern of expression and phenotype are major goals that lay ahead for biotechnology programs. In this context, the use of genomic tools and resources to facilitate breeding using molecular approaches (Wilkins and Arpat, 2005) and characterization of the cotton fiber transcriptome (Arpat et al., 2004) are considered as key strategies for the genetic improvement of cotton. Besides its economic importance, cotton fiber is an outstanding model for the study of plant cell elongation, cell wall and cellulose biosynthesis (Kim and Triplett, 2001). The fiber is composed of nearly pure cellulose, the largest component of plant biomass. Compared to lignin, cellu-lose is easily convertible to biofuels. Translational genomics of cotton fibre and cellulose may lead to the improvement of diverse biomass crops (Chen et al., 2007a). Thus, functional genomics of cotton fiber development has several folds of applications.

SCOPE OF FUNCTIONAL GENOMICS

It has been only in the last decade that researchers have begun to focus on studying the underlying developmental mechanisms that control fiber properties as the basis for manipulating biological and cellular processes to improve fiber characteristics. The genetic complexity of the cotton fiber transcriptome is currently estimated to consist of approximately 18000 genes in the genome of cultivated diploid species. The cotton fiber transcriptome in allotetraploid species is similarly estimated at approxima-tely 36000 genes and to include homoeologous loci from both the A_T and D_T genomes (Arpat et al., 2004). The high genetic complexity of the fiber transcriptome in both diploid and tetraploid species accounts for a significant proportion (45 - 50%) of all the genes in the cotton genome (Wilkins and Arpat, 2005). However, despite approximately 1.5 million years of evolution following the polyploidization event, polyploidy has not been accompanied by rapid genome change, as the genome organization and gene sequences of orthologous loci from the A and D genomes of diploid and tetraploid cotton species are highly conserved (Senchina et al., 2003; Rong et al., 2004). Moreover, spatial and temporal expression pat-

terns have been evolutionarily conserved (Cedroni et al., 2003). Thus, fiber gene function is highly conserved in the genomes of wild and cultivated species, as well as diploid and tetraploid species, despite millions of years of evolutionary history. The phenotypic variation in fiber properties therefore is more likely one of the quantitative differences in gene expression as opposed to differences in the genotype at the DNA level (Wilkins and Arpat, 2005). Further studies, hence, are required to understand the genes, their copy number and specific function in fiber development.

The direct or indirect aims of all functional genomics programs focusing on fiber development are to define the function of the genes involved and to find candidate genes to improve fiber productivity and quality. This helps to sieve a smaller subset of genes which can be employed to draw up a final list of candidate genes (or master regulators) that could be used in future innovative breeding program for superior fiber productivity and quality (Chen et al., 2007a). Several efforts have led to the creation of the first fiber model (Wilkins and Jernstedt, 1999) that serves as the current framework for cotton biotechnology programs worldwide to alter the timing, rate, and/or duration of fiber elongation.

As for methods of cloning fiber-related genes, differential screening of the fiber cDNA library (DD-RT PCR) was most popular in the earlier days. At the same time, sequencing randomly selected cDNA clones from the fiber library, PCR amplification using gene-specific probes and suppression subtractive hybridization were shown to be useful in isolating cotton-fiber-related genes. Recently, expressed sequence tag (EST) projects and microarray technology have gained recognition as means for discovering novel genes. The results were confirmed by (reverse) northern blotting and/or quantitative real time PCR techniques. Thus techniques to monitor global gene expression rely either on hybridisation (microarrays) or on PCR (real-time PCR, cDNA-AFLP and differential display) or, more recently on massively parallel signature sequencing (http://mpss.udel.edu) based methods. However, it is noteworthy to mention that a combination of different profiling methods is likely to be most informative (Jansen and Nap, 2001), since no method is perfect and each one of them has its own pros and cons. At the time of this writing, microarrays are considered as a powerful method to simultaneously measure relative expression levels for thousands of genes. Two main types of microarrays have been developed in cotton, namely cDNA-based and oligonucleotide-based arrays. Based on their technical specifications, cDNA and oligonucleotide microarrays may differ in sensitivity and dynamic range for detecting variation in mRNA abundance as well as in power to discriminate between related target sequences (Wilkins and Arpat, 2005).

Limitations and constraints of the different microarray platforms (Mah et al., 2004; Rensink and Buell, 2005) or in comparison with PCR-based techniques have been emphasized in several instances (Reijans et al., 2003; Tan et al., 2003). The principal limitation for the microarray technique in fiber transcriptome analysis, irrespective of its types, is that only a fraction of genes, those for which DNA sequence are available can be investigated.

As of 1st September, 2008, 369872 *Gossypium* sequences were in GenBank, including 39232 ESTs from *G. arboreum* (A), 63577 from *G. raimondii* (D), 265793 from *G. hirsutum* (AD tetraploid), 1023 from *G. barbadense* and 247 from *G. herbaceum* var. *africanum* (http://www.ncbi.nlm.nih.gov/dbEST/dbEST_summary.html). Non-redundant ESTs have been used both to develop sequence - specific markers for genetic mapping and to construct microarrays for the identification of candidate genes involved in fiber cell initiation and elongation (Arpat et al., 2004; Lee et al., 2006; Shi et al., 2006; Wu et al., 2006; Udall et al., 2007). For example, Udall et al. (2007) created a long oligonucleotide microarray for cotton because of its low manufacturing cost, flexibility in design, homogeneous melting temperatures (Tm) and relative ease of adding probes. A small EST assembly (~ 45,000 ESTs) was previously used to generate oligonucleotide probes for cotton fiber (Arpat et al., 2004). A larger scale EST assembly (> 150,000 ESTs) was recently produced as a community-wide effort by cotton researchers (Udall et al., 2006). Subsequent additions of cotton ESTs to Genbank (> 210,000 ESTs) have been compiled into a large EST assembly (TIGR Cotton Gene Index: http:// compbio.dfci.harvard.edu/tgi). These two assemblies constitute nearly all of the known genic sequence from cotton. An increasing number of sequence resources (bacterial artificial chromosomes (BACs) and ESTs) in *Gossypium* have been used to design fiber cDNA microarrays for functional studies by another group of scientists (Zhang et al., 2001; Wu et al., 2006). Recent efforts from the cotton community lead to the public release of 2 cotton microarrays (essentially from fiber ESTs), including a 24 K GeneChip® Cotton Genome Array from Affymetrix (http://www. Affymetrix.com /products/arrays/specific/cotton.affx); and a 23 K oligonucleotide (60-70mer) microarray by Udall et al. (2007).

Two essential considerations of microarray quality include the number of targeted genes and the broad utility of the microarray for specific tissues or treatments. Regarding the first consideration, the 22,778 genes described by Udall et al. (2007) include perhaps 46 - 60% of the total genic diversity, given that the total number of genes in the cotton genome may be approximately 40,000 – 50,000 (Hawkins, 2006). Regarding the second consideration, a detailed analysis of the probes revealed that ~ 7,300 probes represented genes expressed in specific tissues or under specific conditions (Udall et al., 2007). These two considerations suggest that the oligonucleotides selected for the cotton oligonucleotide microarray have a broad diagnostic utility while potentially targeting tissue specific transcripts expressed under a va-

riety of conditions. It is worth mentioning that the sequences and annotations of all the probes are publicly available via a web-based query (http://cottonevolution.info) or by request.

Molecular, cellular and developmental aspects of fiber development

Cotton fibers are seed hairs and they originate from the epidermal cells of the ovular surface. It is well known that fiber development is composed of four overlapping stages: fiber cell initiation and enlargement from -3 to 1 day postanthesis (DPA), fiber elongation after anthesis until 25 DPA, secondary cell wall cellulose deposition from 15 DPA to 50 DPA and fiber cell dehydration and maturation after 45 DPA (Basra and Malik, 1984). Thus, the near-synchronous growth of 500,000 terminally differentiated single-type fiber cells per ovule (Havov et al., 2008) is characterized by four major discrete developmental stages – differentiation/initiation, expansion/elongation, primary cell wall (PCW) synthesis, secondary cell wall (SCW) synthesis and maturity (Wilkins and Jernstedt, 1999). Among these developmental stages the productivity and quality of cotton depends mainly on two processes: fiber initiation, which determines the number of fibres present on each ovule and fiber elongation, which determines the final length and strength of each fiber. As all plant cells undergo cell expansion to some degree during growth and development, rapidly elongating cotton fibres offer a unique single-celled model to study;

(i) The molecular and cellular mechanisms that regulate the rate and duration of cell expansion, and hence, govern cell size and shape, and in the case of cotton, important agronomic traits as well (Wilkins and Arpat, 2005)

(ii) Cell wall development.

(iii) Cellulose biosynthesis (John and Crow, 1992).

The type of growth mechanism (diffuse versus tip) associated with the exaggerated growth rate of rapidly elongating cotton fibers has been debated for decades (Seagull, 1990; Wilkins and Jernstedt, 1999). However, structural and physiological data provide compelling evidence for diffuse growth (Tiwari and Wilkins, 1995). Interestingly, no genes known to be specific to tip-growing cell types (for example pollen and root hairs) have been identified in developing cotton fibres, further supporting a diffuse-growth mechanism during rapid polar elongation (Wilkins and Arpat, 2005). Polar elongation of developing cotton fibres via diffuse growth is controlled at the cellular level by the transverse orientation of microtubules, the sites of cell wall loosening, the cell wall deposition in the extracellular matrix and polar vesicular trafficking (Cosgrove, 2001; Tiwari and Wilkins 1995). Fiber length is dictated by the rate and duration of cell expansion, which is in turn, governed by developmental programs that co-ordinately regulate cell turgor, the driving force of cell expansion and cell wall loosening (Ruan et al., 2001). The transition from PCW to SCW synthesis, which occurs during the latter stage of expansion between ~ 16 and 21 DPA, is distinguished by the re-orientation of microtubules and cellulose micro fibrils to steeply pitched helical arrays (Seagull, 1992) in anticipation of SCW synthesis to produce a thick cell wall consisting of > 94% cellulose. Coincident with the termination of fiber elongation at ~ 21 DPA is an increase in fiber strength (Hsieh, 1999), presumably due to cross-linking of cellulosic and non-cellulosic matrices (Carpita and Gibeau, 1993), that is also accompanied by a major loss (~ 36%) of high molecular weight noncellulosic polymers in the PCW (Shimizu et al.,1997). Developmental programs regulate the temporal synthesis of fiber PCW and SCW, which differ significantly in structure and composition. While the thin PCW (0.2 - 0.4 µm) deposited during fiber elongation contains < 30% cellulose, the thick SCW (8 - 10 µm) is composed of > 94% cellulose (Meinert and Delmer, 1977). In addition, the degree of polymerization of cellulose micro fibrils also varies, being < 5,000 in PCW and ~ 14,000 in SCW (Marx-Figini, 1966). In some domesticated varieties, cotton fibers may attain a final length of 6 cm or about one-third the height of an entire *Arabidopsis* plant (Kim and Triplett, 2001).

Thus, rapid and simultaneous elongation occurs in millions of fiber cells in the cotton boll without concurrence of cell division and multicellular develop-ment. On the day of anthesis (flower opening), approximately one in four epidermal ovular cells has already been destined to become a cotton fiber, initially appearing as a spherical protrusion. Only about a third of all the epidermal cells become fibres (Berlin 1986), although the exact proportion varies between genotypes and in response to hormone levels (Gialvalis and Seagull 2001; Rahman 2006). A greater understanding of the molecular processes that regulate which cells become fibres and molecular and physiological mechanisms of fiber development could increase the ability to either breed for, or engineer, cotton plants with a higher density of fibres and hence, a higher yield.

Functional genomics and fiber development

The cotton fiber transcriptome has attracted a lot of attention in recent years (Wilkins et al., 2007). Many fiber-specific genes involved in fiber cell initiation, fiber elongation or cell wall biogenesis have been identified from the comparisons of normal (wild-type) versus fiber mutants of *G. hirsutum* species. Few reports have also investigated the mechanisms and genes underlying the important developmental differences between *G. hirsutum* and *G. barbadense* (Ruan et al., 1997; Wu et al., 2005; Wu et al., 2006).

Since the first report of John and Crow (1992), who had cloned the E6 gene through differential screening of a fiber cDNA library in 1992, several reports have shown that many genes were expressed preferentially in cotton fibers (Wilkins and Arpat, 2005). Among these cotton fiber genes, some are highly expressed during early fiber development, some are predominantly expressed during fiber SCW deposition and some show high expression during the entire fiber development. Like E6, H6 and B6, their exact functions are not clear; however, the primary structures of putatively encoded proteins, developmental regulation and tissue specificity suggest that they are likely important for fiber development. On the other hand, some genes have definite functions in cotton fiber development for example, *GhExp1*, which specifically accumulates in developing cotton fibres, encodes a cell wall protein and regulates cell wall loosening by the disruption of non-covalent bonds between wall components (Harmer et al., 2002).

Most of the cotton fiber transcriptome was identified by a single gene discovery project (Arpat et al., 2004; http://cfgc.ucdavis.edu/) and numerous studies targeted the rapidly elongating cotton fibres for a number of the following compelling reasons:

i The rate and duration of fiber elongation governs important agronomic properties, such as yield and fiber length,
ii The exaggerated growth of elongating fibres, underpinned by high levels of metabolic activity was expected to be especially gene rich from a gene discovery perspective
iii Isolation of fibres free of contamination from surrounding complex tissues permit an unambiguous look at the transcriptome of a single cell within a biologically relevant framework (Wilkins and Arpat, 2005).

Two general approaches undertaken to maximize gene discovery, included "deep" sampling of a high-quality fiber cDNA library to generate ESTs and sequential rounds of normalization to remove highly redundant gene sequences and thereby identify rare gene transcripts.
However, the judicious selection of a cultivated diploid species (*G. arboreum* L.) as a model for fiber development proved especially rewarding, as the rate of gene discovery was enhanced at least two-fold simply by avoiding redundancy due to polyploidy. To define the cotton fiber transcriptome, more than 46,000 ESTs from rapidly elongating fibers were generated from *G. arboreum* L. (Arpat et al., 2004). The transcriptome of cotton fiber across a developmental time-course, from a few days post anthesis through PCW and SCW and maturation stages were evaluated. This revealed that there were dynamic changes in gene expression between PCW and SCW biogenesis. Among them, transcripts for cell wall structure and biogenesis, the cytoskeleton and energy/carbohydrate metabolism were the three major functioning groups during the rapid elongation of fiber cells (Arpat et al., 2004; Wilkins and Arpat, 2005).

It is also important to identify the genes that are preferentially expressed in fiber tissues of *G. hirsutum* L., which is highly adapted to the present environment and the most widely cultivated species. Further, similar biochemical pathways may have diverged during evolution and thus can create a possible drawback for the identification of *G. hirsutum* fiber specific candidate genes when the knowledge from *G. arboreum* EST projects is translated (Salentijn et al., 2007). Hence, ESTs were developed from fast elongating fiber of *G. hirsutum* (Shi et al., 2006). Results from these studies were useful to co-clude that the genes are very much essential for fiber development in commercial cultivars. Detailed sequence comparisons showed that significant sequence divergence exists between the two species (Shi et al., 2006).

Transcriptomic model for fiber development: A proposal

It has been indicated elsewhere that fiber development is a highly complex metabolic process and it involves the expression of nearly half of the cotton genome for complete development of matured fiber (Havov et al., 2008). Based on the results of several studies the following transcriptomic model for fibre development is proposed. This model may be considered as preliminary since detailed functional and physiological studies yet to be performed to get a lucid understanding on this model. It should be noted that many of the genes up-regulated in the fiber initials have been identified as either unclassified or unknown (Arpat et al., 2004; Wilkins et al., 2007; Havov et al., 2008) could fuel future research and hence a clearer picture on this model with respect to fibre cell initiation, expansion and cell wall growth in cotton can be expected in the near future. More than half of all genes were up-regulated during at least one stage of fibre development. Genes implicated in vesicle coating and trafficking were found to be over expressed throughout all stages of fiber development, indicating their important role in maintaining rapid growth of this unique plant cell (Havov et al., 2008). The following sections describe the developmental stage in the specific expression of genes and their putative role in fibre development.

Initiation

A unique feature of cotton seed development is that ~ 30% of the ovule epidermal cells initiate into fibers from the outermost layer of integument at anthesis (Ruan et al., 2001). On the day of anthesis the cotton fiber initial cells swell out from the ovule surface and so are clearly distinguished from adjacent epidermal pavement cells. Since impressive progress has been made only on the later elongation stage (Ji et al., 2003; Arpat et al., 2004; Shi et al., 2006), it resulted in a rudimentary understanding of the molecular events at the early initiation and cell expansion stage. Fortunately, a small set of cotton

mutants that lack seed fibres (Li et al., 2002; Lee et al., 2006; Wu et al., 2006) and contemporary biological dissecttion methods (Wu et al., 2007) have facilitated the discovery of a few genes expressed early during fiber development.

Transcript profiling and ovule culture experiments both indicate that several phytohormones mediate cotton fiber initiation. Auxin and gibberellins appear to promote early stages of fiber initiation. Ovules cultured *in vitro* become competent to produce fiber in response to auxin and giberellic acid (Graves and Stewart, 1988). Fiber initiation also requires brassinosterol production (Sun et al., 2005). Interestingly, the phytohormone-related genes were induced prior to the activation of MYB-like genes, suggesting an important role of phytohormones in cell fate determination (Chen et al., 2007b). However, several attempts have been made to alter the expression of genes involved in auxin and cytokinin biosynthesis in the fibers but no favourable phenotypic changes were observed in the resultant transgenic plants (John, 1999). This indicates complicate interactions of genes and phytohormones during fiber initiation. Fiber differentiation is evident *in vivo* by -1 DPA when microtubules reorient in epidermal cells destined to differentiate into fibers (Ryser, 1999). By 1 DPA, fiber initials bulge from the surface of the ovule. Protein biosynthesis and nucleoli size increase in very young fibers (Van't Hof, 1998). *In vitro* cultured ovules indicated that mRNA synthesis is required for fiber initiation up to 2 DPA and the ovules remained competent to initiate fibers up to 5 DPA (Triplet, 1998). Thus, the period of fiber initiation ends at 2 DPA and may extend to 5 DPA.

Genes that peak in expression during fiber initiation then decrease in expression during elongation would be expected to play a specific role in fiber initiation. Several well annotated genes with a fiber initiation-specific pattern of expression give potentially new insight into fiber initiation. Synchronously differentiating fibers represent a valuable developmental model to determine how developmental signals are integrated to control differentiation and elongation of fiber and how these signalling pathways differ between ovular and leaf trichomes. Fiber initiation requires transcription of several new genes and therefore transcription factors are likely to play an important role in fiber initiation. The Myb109 and Myb2 transcription factors are expressed in fiber initials (Suo et al., 2003). Evaluation of fiber less cotton mutants has identified genes differentially expressed in very young fiber, including transcription factors (Lee et al., 2006). The Myb2 transcription factor is able to comple-ment *Arabidopsis thaliana* trichome mutants and activate expression of R22-like (RDL) gene expressed in fiber initials (Wang et al., 2004). Additionally, the RDL gene along with genes involved in cell structure, long chain fatty acid biosyn-thesis and sterol biosynthesis have been identified as those absent or reduced in a fiber less mutant of cotton (Li et al., 2002). Most of these genes are expressed in 1 DPA ovules. A second round of fiber initiation occurs that produces the short linters or fuzz fibers. Yang et al. (2006) deposited a large number of ESTs into the Genbank database from cDNA libraries of whole ovules spanning the period from - 3 to +3 DPA augmenting those already collected by Wu et al. (2005). They were able to identify a large number of transcription factor genes (as expected) orthologous to genes in *Arabidopsis*, some known to be involved in *Arabidopsis* trichome formation, but have yet to localise expression of any of these genes specifically to cotton fibers or fiber initials.

It has also been documented that the expression of both Myb109 and Myb2 transcription factors were abundant in 1 DPA fibers and persisted into the elongation stage of fiber development (Taliercio and Boykin, 2007). Of the approximately 624 putative tran-scription factors represented on the microarray, five were regulated similarly to Myb109 and Myb2 and therefore were candidates to play a role in controlling fiber initiation. Further, Taliercio and Boykin (2007) have also identified five transcrip-tion factors with a similar pattern of expression that could play a role in fiber development. One of the notable examples is two genes similar to CAPRICE/TRIPCHON (CPC). CPC acts as a negative regulator of trichome development in *Arabidopsis* (Schellmann et al., 2002). One of the putative CPC genes was down regulated in 1 DPA fiber compared to ovules. The inhibitors described for *Arabidopsis* are not down regulated in trichomes; therefore it is not possible to draw a conclusion based on gene expression about which putative CPC gene in cotton was more likely involved in fiber development. If CPC genes in cotton act as inhibitors of fiber initiation, reducing expression of these genes with interfering RNAs would be expected to increase the number of fibers (Taliercio and Boykin, 2007). Therefore a transgenic cotton line with reduced CPC expression could be agronomically valuable.

Wu et al. (2007) used laser capture micro-dissection coupled with cDNA microarray and found that except for a few regulatory genes, the genes that are up-regulated in the cotton fiber initials relative to epidermal cells predominantly encode proteins involved in generating the components for the extra cell membrane and PCW, carbohydrates and lipids needed for the rapid cell expansion of the initials. Overall, there were few genes that differed markedly between the fiber initials and the epidermal pavement cells- the most different being threefold to fivefold up- or down-regulated. The majority of differential genes were only about twofold more or less expressed in fiber initial cells. The absence of large differences may have been because the individual cell types were not completely pure. This has been a criticism of laser capture micro-dissection methodology, particularly with rigid plant tissues (Day et al. 2005), but does not contradict the enormous power of this technology.

Amongst the classes of genes expressed in the initials, many of the genes were involved in DNA metabolism despite cell division having ceased in these differentiated

cells (Wu et al., 2007). Various authors have suggested that cotton fiber cells may undergo endoreduplication to amplify their DNA content to support their specialised function like *Arabidopsis* leaf trichomes (Szymanski and Marks 1998). Endoreduplication in cotton fiber development, however, remains a contentious topic. Some earlier reports suggested a limited increase in DNA content in fiber nuclei (Van't Hof, 1999) while others could detect no difference in DNA content of 14 - 25 DPA fibers (Taliercio et al. 2005). Wu et al. (2006) concluded that fiber initials at 0 DPA undergo at least one round of endoreduplication, so the abundance of DNA metabolism genes is not surprising. Others have suggested that the increase in DNA content may be associated with enlargement of the nucleolus (Kim and Triplett 2001).

In another study, Taliercio and Boykin (2007) had confirmed that an increase in the endoplasmic reticulum occurred in fiber initials on the day of anthesis (Ryser, 1999) and persisted through 3 DPA. In addition to that, they have identified consistent increase in membrane associated component proteins which played a role in early fiber development. They have also noticed genes associated with novel regulation of brassinosterols, GTP mediated signal transduction (which play roles in vesicle trafficking) and cell cycle control and components of Ca^{2+} mediated signalling pathway. They argued that the presence of more Ca^{2+} in fiber initials than other ovule cells and the differential expression of calmodulin and calmodulin binding proteins indicated a role for Ca^{2+} in fiber development. It is likely that the marked and long lasting increase in endoplasmic reticulum in fiber initials was unique to the ovular trichomes, indicating an early departure between the developmental programs that give rise to ovular and leaf trichomes. This increase in endoplasmmic reticulum was consistent with the increase in golgi bodies reported in fiber initials (Taliercio and Boykin, 2007). Abundant endoplasmic reticulum may play a role in biosynthesis and transport for components of the rapidly expanding cell membrane, cell wall and cuticle. Indeed, analysis of genes differentially regulated during fiber initiation and elongation identify numerous genes associated with these developmental pathways. Hence an increase in the endoplasmic reticulum may represent the first stages of fiber elongation since increase demands for cell membrane, PCW, and cuticle production which persist through the elongation phase of fiber development.

Genes other than transcription factors can have profound affects on expression of other genes. Expression of some other types of regulatory genes increased in 1 DPA fibers and persisted in 10 DPA fibers. Examples include receptor kinases, calmodulin, calmodulin binding proteins and lumen receptors. Increased expression of a calmodulin gene unique to fibers and differential expression of calmodulin binding proteins were also observed (Taliercio and Boykin, 2007) as indicated above. It seems likely that a calmodulin mediated signaling pathway exists that either causes or responds to the redistribution of calcium into the endoplasmic reticulum. Interestingly, deesterified pectins increase in fiber initials (Turley and Vaughn, 1999). Deesterified pectins bind calcium; therefore it is likely that the cell walls may also compete for Ca^{2+}. Manipulating expression of the calmodulin or manipulating calcium levels *in vitro* may determine whether a calcium mediated pathway exists that causes or responds to the increase in endoplasmic reticulum and what role a calmodulin mediated response to Ca^{2+} plays in fiber development.

Documentation of common set of genes for fiber initials across studies has been shown only in few cases and this may be partly due to different type of experimental materials and protocols used in various studies. Fiber cell determination is believed to occur a few days before there are any visual changes in the fiber initials (Ramsey and Berlin 1976), so ideally microarray comparison should be carried out at -2 to -3 DPA, but at this stage all the cell types look identical. As more early fiber genes and promoters are continuously characterised, a combination of laser capture micro-dissection and transgenic plant technologies to tag initiating fiber cells with non-destructive visual markers, such as green fluorescent protein (GFP) will eventually allow the profiling of gene expression changes occurring up to 3 days before anthesis when there are no outwardly visible differences between cells starting to differentiate into fibres and their adjacent non-fiber epidermal cells (Wu et al., 2007).

Elongation

Many important fiber traits such as the length, shape, structure and composition of the fiber cell, are determined during this stage of fiber development. Genetic control of fiber size and shape, and hence agronomic properties is governed by the rate and duration of fiber expansion and elongation which in turn, positively correlates to fiber length (Wilkins and Arpat, 2005). Although some progresses have been made in recent decade (Xu et al., 2007), little is known on how these genes regulate fiber elongation. During the most active elongation period (5 to 20 DPA), vigorous cell expansion with peak growth rates of > 2 mm/day are observed in upland cotton (Ji et al., 2003).

As discussed earlier, fiber elongation occurs by a diffuse growth mechanism (Tiwari and Wilkins, 1995). Many genes expressed during the elongation stage of fiber differentiation relate to cell expansion, cell wall loosening and osmoregulation. Ovule culture studies confirmed a role for brassinosterols during fiber elongation in addition to fiber initiation (Sun et al., 2005). Based on genomic, genetic, molecular, biological and physiological studies it was found that ethylene plays a major role in promoting cotton fiber elongation. The role for ethylene in fiber elongation was confirmed when longer fibers were obtained with the addition of ethylene to ovule culture.

Furthermore, ethylene may promote cell elongation by increasing the expression of sucrose synthase, tubulin and expansin genes (Shi et al., 2006). Although a review on cotton fiber (Kim and Triplett, 2001) states that cytokinins, abscisic acid and ethylene inhibit fiber development, this statement was based on experiments that did not in fact include ethylene in the ovule culture studies (Shi et al., 2006). However, the role of brassinosteroid in fiber cell elongation appears to be less prominent than that of ethylene, for it was only modestly effective. Although the actions of brassinosteroid and ethylene on fiber elongation are not interdependent, it appears that they do not act completely independently either and each hormone positively modulates the synthesis of the other. Such positive interactions between the two hormones potentially contribute to the extreme elongation of fiber cells (Shi et al., 2006).

Each cotton single fiber cell elongates from 10 to 15 mm up to 2.5 to 3.0 cm by ~ 16 DPA before it switches to SCW cellulose synthesis (Tiwari and Wilkins, 1995). The rapid fiber elongation is believed to be driven by high turgor with a highly extendable PCW (Ruan et al., 2001). Cell turgor in plants is achieved largely through the influx of water driven by a relatively high concentration of osmoticum within a cell (Cosgrove, 1997). The accumulation of osmoticum into fibers may be coupled with the transmembrane proton gradient, because the plasma membrane H^+-ATPase gene is expressed strongly during the rapid phase of fiber elongation (Smart et al., 1998). This H^+ pump could also acidify the apoplast for cell wall loosening (Cosgrove, 1997). At this time, the expansin gene of major importance in mediating cell wall extension (Cosgrove, 1997), is expressed in fibers, although its temporal expression pattern over the elongation period is not clear (Orford and Timmis, 1998). The results of Ruan et al. (2001) provide a remarkable explanation of how the gating plasmodesmata and the expression of genes for solute import and cell wall loosening are developmentally coordinated to potentially control single-cell elongation. Fiber plasmadesmata are initially opened (0 - 9 DPA) but closed at ~ 10 DPA and reopened at 16 DPA and it is accompanied by a gradual switch from simple to branched forms of plasmadesmata. However, it is difficult to assign functional implications for such a structural change and mechanism responsible for the reversible gating of fiber plasmodesmata (Ruan et al., 2001). Coincident with the transient closure of the plasmadesmata, the sucrose and K^+ transporter genes are expressed maximally in fibers at 10 DPA with sucrose transporter proteins predominantly localized at the fiber base. Consequently, fiber osmotic and turgor potentials were elevated, due to increased accumulation of soluble sugars, driving the rapid phase of elongation. The higher turgor can be maintained in fibers due to the closure of plasmadesmata. The expansin gene was highly expressed at the early phase of elongation (6 to 8 DPA) and decreased rapidly afterwards. Given the increased rigidity of the fiber cell wall, indicated by the decreased expression of expansin, it is almost certain that the higher turgor in fibers at this stage plays a critical role in driving the rapid-fiber elongation (Ruan et al., 2001). This shows that fiber elongation is initially achieved largely by cell wall loosening and finally terminated by increased wall rigidity and loss of higher turgor. This has also been shown in another study that an increase in fiber K^+ concentrations, most probably leading to higher turgor, has yielded longer fibers with higher quality (Cassman et al., 1990).

In an another study, a group of genes related to turgor regulation and the cytoskeleton such as the plasma membrane proton-translocating ATPase, vacuole-ATPase, proton-translocating pyrophosphatase (PPase), phosphoenolpyruvate carboxylase, major intrinsic protein (MIP) and α-tubulin were found to be involved in cotton fiber elongation (Smart et al.,1998). Similarly, four genes viz; putative gibberellin-regulated protein, putative tonoplast intrinsic protein, putative plasma membrane intrinsic protein and putative membrane protein were identified in the 10 DPA fiber subtracted library and they were found to be expressed during early fiber development (Liu et al., 2006). The role of these genes in fiber elongation is yet to be tested albeit their biochemical activity is established in some other studies.

The translation elongation factor 1A, eEF1A, plays an important role in protein synthesis, catalyzing the binding of aminoacyl-tRNA to the A-site of the ribosome by a GTP-dependent mechanism. To investigate the role of eEF1A for protein synthesis in cotton fiber development, nine different cDNA clones encoding eukaryotic translation elongation factor 1A were isolated from G. hirsutum fiber cDNA libraries (Xu et al., 2007). The isolated genes (cDNAs) were designated as nine cotton elongation factor 1A genes viz., GhEF1A1 - GhEF1A9 and all of them share high similarity at nucleotide level (71 - 99% identity) and amino acid level (96 – 99% identity) with conserved GTP-binding domain in the deduced amino acid sequence. The different members of eEF1A gene family in the plant may originate from a series of gene duplications during evolution. Of the nine GhEF1A genes, five are expressed at relatively high levels in young fibers. Further analysis indicated that expressions of the GhEF1As in the fibre are highly developmental-regulated suggesting that protein biosynthesis is very active at early fiber elongation to meet the rapid elongation of fiber cells in cotton (Xu et al., 2007).

Determining protein sequences and their expression changes in defined growing stages will further enhance understanding on fiber development mechanisms. In a proteomics study, cytoskeleton-related proteins, enzymes involved in flavonoid biosynthetic pathway, putative S-adenosylmethionine synthases, allergin like proteins and annexin were identified besides few unidentified proteins (Yao et al., 2006). Among them, the cytoskeleton related proteins are responsible for directing polar expansion during the rapid elongation period which contributes sig-

nificantly to fiber shape (Arpat et al., 2004).

When cotton fiber cells underwent maximal expansion (12 - 15 DPA), transcripts of several genes except for PPase accumulated to the highest levels, then declined at the beginning of SCW deposition. GhGLP1 transcripts coding for a germin-like protein, accumulated to their highest levels during the period of fiber expansion, followed by a sharp decline when the rate of cell expansion decreased (Kim and Triplett, 2001). While germins and GLPs appear to be involved in defence mechanisms in some plants, both biotic and abiotic stresses down-regulated the expression of GhGLP1. Expression of two other genes associated with fiber cell elongation, expansin and α-tubulin 1, also declined when the same abiotic and biotic stresses were applied (Kim and Triplett, 2001). Numerous functions have been proposed for dicot GLPs. However, to date, there is little direct evidence for how these proteins function in vivo. The association of maximal GhGLP1 expression with stages of maximal cotton fiber elongation suggests that some GLPs may be important for cell wall expansion (Kim and Triplett, 2001).

In developing cotton fibers, there is one major expansin and several minor isoforms, suggesting that the isoforms may perform specialized roles during fiber elongation (Arpat et al., 2004). Indeed, suppression of expression of all isoforms results in a shorter, coarser fiber (Wilkins and Arpat, 2005). An interesting possibility for such specialized roles is the potential for discrete subcellular localization of the various isoforms, such that the major isoform is preferentially targeted to the tip of the cell and functions primarily in polar elongation, while the minor species are differentially targeted to regions of the fiber likely to promote lateral expansion of the fiber. It would therefore be very interesting to determine if the various expansin isoforms are localized to defined regions of elongating fibers at the subcellular level. Moreover, major isoforms for functionally important genes may well account for the bulk of genetic variability asssociated with major QTL for fiber quality and as such provide candidate genes for genetic mapping and marker-assisted selection (Wilkins and Arpat, 2005).

Expression profiles that compare transcript abundance during and after fiber elongation revealed that ~ 20% of the fibre transcriptome, including large numbers of metabolism related genes and cell wall proteins, is down-regulated coincident with termination of cell expansion (Arpat, 2004). Besides a large group of fiber genes that perform basic functions during elongation, the switch in developmental programs from PCW to SCW synthesis and the termination of cell expansion is accompanied by dynamic changes in gene expression. About 2,500 highly and moderately expressed fiber genes were down-regulated in terminal stages of fiber elongation which function selectively or preferentially during cell expansion (Arpat et al., 2004). An increase in cellulose and expression of genes encoding cellulose synthase marks the end of the rapid elongation stage of fiber development.

Primary and secondary cell wall synthesis

In addition to a large diverse group of constitutively expressed genes, expression profiling of the transcript-ome revealed two developmentally regulated stage-specific expression patterns that follow rapid cell elongation: PCW and SCW biogenesis (Wilkins and Arpat, 2005). The PCW is laid down during the elongation phase, lasting up to 25 DPA. Synthesis of the secondary wall commences prior to the cessation of the elongation phase and continues to 40 DPA, forming a wall (5 - 10 µm) of almost pure cellulose (Wilkins and Arpat, 2005) and cell wall biosynthesis is a major synthetic activity in fiber cells. The cell wall components are synthesized and transported by a functionally integrated membrane system of endoplasmic reticulum, golgi complex and plasmalemma. Newly synthesized cell wall polypeptides are released into the endoplasmic reticulum lumen before transportation and incorporation into cell walls. Glycosylation of structural proteins, as well as polymerization of hemicelluloses and pectin, takes place in the golgi complex from which the products are released into the plasmalemma through the fusion of vesicles (Basra and Malik, 1984).

In addition to cellulose, cotton fiber also contains smaller amounts of pectins, hemicellulose, waxes, proteins and inorganic salts. The mechanism by which cellulose micro fibrils are produced and assembled along with the other components is not fully understood. Microtubules that are situated in the cytoplasm directly adjacent to the developing cell wall may participate in microfibril organization (Seagull, 1992). Some of the protein constituents of cotton fiber (enzymatic, structural or regulatory) are unique to fiber cells and are likely to influence the development and properties of cotton. Evaluation of proteins from fibers of different development mental stages and other cotton plant tissues by two-dimensional gel electrophoresis has revealed fiber-specific proteins that are developmentally regulated (Yao et al., 2006).

Although it is clear that no gross changes occur in the RNA population, the protein content may change during cell wall development. Earlier measurement of protein content in the cotton cell wall (weight percent) has shown to decrease from a high of 40% in 5 DPA fibers to < 2% in 18 DPA fibers (Mienert and Delmer, 1977). Measured as weight per unit length, the fiber wall protein content peaks in ~ 16 DPA fibers before rapidly declining (Mienert and Delmer, 1977). The level of steady-state E6 RNA was high in 20 DPA fibers and persisted beyond 24 DPA fibers. However, there was a sharp decline of E6 proteins in 15 DPA and older fibers (John and Crow, 1992). Thus, the decline of E6 protein in 15 DPA fibers and the near absence of E6 protein in 24 DPA fibers as shown by John and Crow, (1992) can be interpreted to be due to a protein degradation occurring in fiber cells during this time period. This degradation must be selective since a major biosynthetic activity, cellulose synthesis/deposition, occurs in fibers 24 DPA and older and protein compo-

nents necessary for this event would therefore be expected to be preserved. As revealed by in silico expression analysis, cytoskeleton and cell wall-related genes are by far, among the most abundant gene transcripts during PCW synthesis in elongating fibers, while metabolism-related genes account for the vast majority of moderately expressed genes (Wilkins and Arpat, 2005).

PCW and SCW are markedly different in terms of structure and composition (Carpita and Gibeau 1993). In cotton fibers, a thin (0.2 - 0.4 mm) PCW deposited during fiber elongation contains < 30% cellulose, whereas the thick (8 - 10 mm) SCW is composed of > 94% cellulose, no lignin and few proteins (Meinert and Delmer, 1977). A gene encoding H^+-pyrophosphatase is constitutively expressed but enzymatic activity is temporally regulated; suggesting that this particular proton pump plays a functional role during the developmental switch from PCW to SCW synthesis (Smart et al., 1998). Temporal regulation of vacuolar H^+-ATPase gene expression, which closely parallels fiber growth rate, is accompanied by a corresponding change in protein abundance and enzymatic activity (Smart et al., 1998). In addition to that, many cell wall and carbohydrate metabolism related genes are temporally regulated during SCW synthesis (Arpat et al., 2004) in a manner typified by the developmental regulation of GhCesA1 and GhCesA2 genes encoding the catalytic subunit of cellulose synthase. Expression of GhCesA genes initially detected at low levels during the latter stages of fiber expansion, signals the onset of SCW synthesis (Wilkins and Arpat, 2005). Expression dramatically increases coincident with the rate of cellulose synthesis to reach peak levels at approximately 24 DPA (Meinert and Delmer, 1977, Pear et al., 1996).

CelA1 and CelA2, two homologs of the bacterial celA genes that encode the catalytic subunit of cellulose synthase, were cloned from cotton fiber, and they showed expression in developing cotton fibers at the onset of SCW cellulose synthesis (Pear et al., 1996). The polypeptide encoded by the CelA1 DNA fragment with four highly conserved subdomains had the ability to bind the UDP-glucose (UDP-Glc) in vitro and UDP-Glc is the substrate to synthesize cellulose (1,4-β-D-glucan) (Pear et al., 1996). Similarly, screening of 20 DPA fiber subtracted library has shown that arabinogalactan protein and fiber glycosyl hydrolase family 19 protein, were found to be highly expressed in fibers with the maximal transcription level during the developmental switch from elongation to cellulose deposition (Liu et al., 2006).

Termination of elongation

Termination of fiber elongation is accompanied by a corresponding decrease in growth rate, transcriptional activity (Kosmidou-Dimitripoulou, 1986) and protein complexity (Graves and Stewart, 1988). At this stage, crosslinking of cellulose micro fibrils and non-cellulosic matrices presumably "fix" the structure of the PCW (Wilkins and Jernstedt, 1999), resulting in the first significant increase in fiber strength (Hsieh 1999). Although it has been speculated that programmed cell death plays a role in fiber maturity, virtually nothing is known about the latter stages of fiber development as molecular studies are hindered by the inability to isolate RNA from fibers much past 25 DPA when cellulose is being deposited (Wilkins and Arpat, 2005).

Cotton fiber cells do not synthesize SCW cellulose until their elongation process stops at ~ 16 to 18 DPA (Basra and Malik, 1984). The undetectable level of expansin gene indicates that the initial highly extendable PCW of elongating fiber has become quite a rigid synthesis (Ruan et al., 2001). Consistent with this change is the suberization of fiber SCW (Ryser, 1999).

The deposition of hydrophobic suberin in the basal part of fibers excludes the apoplastic pathway for solute import (Ryser, 1999). This could be the structural basis for the reopening of fiber plasmodesmata (Ruan et al., 2001). The dramatically reduced expressions of sucrose and K$^+$ transporter genes have been documented at 16 DPA (Kühn et al., 1997).

This reduction in transporter expression is in agreement with the shift back to the symplastic pathway of solute import into fibers. Although symplastic sucrose import into fibers is sustained by the activity of sucrose synthase in the cytosol of fibers, the import of K$^+$ into fibers was greatly reduced after 15 DPA (Ruan et al., 1997). This may contribute to the decrease of osmotic and turgor potentials in fibers and slow down the elongation (Cassman et al., 1990). Furthermore, the reopening of fiber plasmodesmata at ~ 16 DPA would release higher turgor in fibers, if any, to a level similar to that present in the seed coat (Ruan et al., 2001). Together, these studies propose that fiber elongation could be terminated by the combination of increased cell wall rigidity and loss of higher turgor.

Until now, the total mechanism of fiber growth and development has been unclear, even though the expression patterns and putative function of array of genes have been analysed in details as shown above. Thus fiber cells, though devoid of any critical functions in the cotton plant except seed dispersal, contain large numbers of active genes common to other cell types along with fewer active genes unique to fiber.

A second point that became clear during transcriptome analysis was that no major changes occurred in the mRNA population during PCW and SCW synthesis stages and no subset of genes that are exclusively expressed during a given developmental stage was detected. Thus, it seems likely that most of the genes in the fiber are active throughout its development. Alternatively, gene transcription ceases early in fiber development, but differential mRNA utilization occurs during growth and thus the protein content may change during later fiber developmental stages. It is believed that advances in 'omics' studies may open up new avenues in this area in the near future.

Functional genomics of biotic and abiotic stress resistances

It is noteworthy that under field conditions, biotic and abiotic stresses are the major reasons for poor seed set which leads to both yield loss and lower fiber quality. Hence, it is important to understand the impact of these stresses on fiber development. Though there has been no such direct studies made in cotton, the molecular response of cotton with respect to external environmental stresses were documented. For example, during the process of *Verticillium dahliae* infection, the resistant and susceptible plant varieties respond differently. Molecular cloning of the transcripts related to the cotton *Verticillium* wilt response have shown that several defense genes such as chitinase, β-1,3-glucanase and PR-10, phenylalanine ammonialyase and those encoding several metabolic and signaling enzymes were activated and their expressions changed radically (Zhang and Klessing, 2001). In an another study, suppressive subtractive hybridization method was employed to identify differentially expressed ESTs in the *V. dahliae* infection process and many up-regulated and down-regulated novel ESTs, including those involved in synthesizing signal molecules, oxidative metabolism and those related to stress tolerance were isolated (Zuo et al., 2005). Cap-turing these transcripts and characterizing their roles helped in explaining the molecular resistance mechanism of cotton *Verticillium* wilt.

In a similar transcriptome study of cotton- *Xanthomonas campestris pv. malvacearum* interaction, clones from a cDNA library were used to identify host genes expressed in upland cotton leaves following inoculation (Patil et al., 2005). Microarray analysis indicated that 98% of the analysed genes (which were enriched) that are involved in response to inoculation were significantly up-regulated at one or more of the sampling times. Of these, 63% had sequence similarity to plant genes associated with defence responses, that is, to genes that function in disease resistance/defence, protein synthesis/turnover, secondary metabolism, signalling, stress induced/ programmed cell death or code for pathogenesis-related proteins or retrotransposon-like proteins. Some of the genes in this study (17%) exhibited a maximum in differential transcript abundance at an early time. Hence, Patil et al. (2005) proposed a working hypothesis that it is quickness of up-regulation after inoculation that matters rather than the final intensity of up-regulation. Therefore, identification of early up-regulated genes would have potential applications in developing host resistance to bacterial blight.

Global expression profiling during abiotic stress has been documented in rice, wheat, maize and other crops (Vij and Tyagi, 2007). However, very scarce expression profiles are available with cotton abiotic stress response. Such large scale EST projects are useful in cotton to aid in the identification of the genes involved in abiotic stress. Water stress has a very critical impact in flower development, boll formation and fiber maturity. Identification of genes involved in drought tolerance at the flowering and boll formation phases would be useful to enhance fiber productivity and quality under water limited environments. This has been exemplified by Lan et al. (2005) who found that more than one-half of the pollination/fertilization-related genes in rice were regulated by dehydration stress, indicating that water stress may be a crucial factor during pollination and fertilization.

Domestication of cotton involved selection for fiber-related traits, typically under irrigated conditions. This selection pressure may have inadvertently narrowed the genetic base and diluted the alleles that once enabled cotton to survive arid conditions in the wild (Rosenow et al., 1983). Drought tolerance may be re-introduced into elite lines with the aid of DNA markers for QTL associated with the trait. Saranga et al. (2004) identified drought-related QTLs for several physiological and crop productivity parameters in cotton. The next step to characterize osmotic adjustment and other drought-related QTLs in cotton is to identify the genes that correspond to these QTLs. The candidate gene approach is one method to achieve this goal (Pflieger et al., 2001). Generally, the main limitation hindering utilization of this method is the availability of genes associated with a particular trait. Research utilizing microarray technology offers a solution in that a list of candidate genes related to the phenotype of interest can be produced. However, in cotton, only fiber-based microarray chips are currently available. Fiber development and abiotic stress resistance in cotton has several similarity in gene expression as discussed below and so the available chips can be used to characterize abiotic stress responsive genes in cotton.

Growth is possible if turgor pressure is greater than a minimum threshold. Turgor pressure, in turn, is related, to the osmotic potential and to the transport coefficient for water uptake (Cosgrove, 1997). From the previous sections it is clear that the cotton fibre is a cell that elongates rapidly within a period of approximately 21 DPA. The cell faces unique challenges in maintaining proper balance between turgor pressure and cell wall extensibility due to the dilution of cellular osmoticum that results from water influx during rapid polar expansion. To cope, cotton fibers up-regulate a suite of genes that aide in solute accumulation and cell wall loosening (Ruan et al., 2001). Root cells face similar turgor versus cell wall extensibility challenges during water deficit stress. For growth to occur in well-watered conditions, cell turgor pressure must again be higher than threshold turgor pressure but unlike fiber cells, the main force that root cells must counter to maintain turgor pressure is the soil water potential (Hendrix et al., 2004). In a drought situation, the water potential of the soil is further reduced and becomes a greater hindrance to growth. To counter this, plants increase the accumulation of cellular solutes and cell wall loosening enzymes (Cosgrove, 1997; Ruan et al., 2001). These actions serve to increase turgor pressure and extensibility

and, thus enable continued cell growth despite the low water potential of the surrounding soils. Though the above situations are driven by different stimuli (that is, fiber cells: rapid expansion versus root cells: soil water potential), cellular physiology demands the accumulation of solute and loosening of cell walls. Accordingly, gene expression profiles may be similar.

Nevertheless, a proof of concept is needed to address the validity of utilizing fiber-based microarrays to study drought tolerance in other tissues. Both *in silico* and biological expression analyses were used to assess the genetic similarity of these two phenomena (Hendrix et al., 2004). The data from the *in silico* analysis were supplemented with direct comparisons of the root and fiber transcriptomes using fiber-based microarray chips. Of the 5,506 genes, 95.9% (5,282) were expressed in both tissues. However, 169 fiber and 55 root genes were identified as tissue specific genes. Thus, utilization of new fiber-based microarrays for gene discovery during water-deficit stress is a feasible approach. However, expanding the fiber-based chips to include genes from other tissues would further facilitate gene discovery related to water-deficit stress and may eventually lead to complete characterization of drought-related QTLs in cotton (Hendrix et al., 2004).

It must be kept in mind that the basic task of the identification of key gene(s), whose manipulation will ultimately affect crop performance in response to abiotic stress, is highly complex and difficult to decipher because of the polygenic nature of the stress response. In addition, the plant's response to each stress is unique, and thus the response to multiple stresses will also be different. For example, global expression profiling of a plant's response to abiotic stress conditions has shown that, although overlap may occur for different abiotic stresses such as cold, salt, dehydration, heat, high light and mechanical stress, a set of genes unique to each stress response is also seen (Vij and Tyagi, 2007). However, most of the studies carried out to investigate the performance of plants under abiotic stress conditions have not focused on this aspect, making it an important area of concern especially as it is known that plants are exposed to multiple environmental stresses in the field. Further, the response to abiotic stress is also developmentally regulated (Vinocur and Altman, 2005). For instance, in plant species such as rice, wheat, tomato, barley and corn, salt tolerance increases with an increase in plant age. Moreover, it has been shown that QTLs associated with salt tolerance in the germination stage in barley, tomato and *Arabidopsis* are different from QTLs associated with the early stage of growth (Vij and Tyagi, 2007). Furthermore, in transgenic studies on crop plants such as rice, the majority have not evaluated the effect of stress on grain yield. It is apparent that an understanding of the abiotic stress responsive network will require a considerable amount of time and resources, but a systematic and concerted effort will ensure that only the most suitable genes are identified for crop improvement. The task can be shortened by integrating the information already available and by avoiding the repetition of effort or branching away from the main focus. The final list of candidate genes and their alleles identified through this approach must be subjected ultimately to field trials to determine their efficacy.

Databases

Comprehensive study of any genome depends on the availability of a saturated and fully integrated genetic and physical map of cotton. Hence, all the information should be collected and made publicly available. There are several genome databases that are exclusively developed to serve the cotton research community. They are: The International Cotton Genome Initiative (ICGI; http://icgi.tamu.edu/), The Cotton Genome Database (CottonDB; http://www.cottondb.org;), The Cotton Microsatellite Database (CMD; (http://www.cottonssr.org), Comparative Evolutionary Genomics of Cotton (http://cottonevolution.info/), TropGENE Database (http://tropgenedb.cirad.fr/en/cotton.html), Cotton Genome Centre (http://cottongenomecenter. ucdavis.edu), The Cotton Diversity Database (http://cotton.agtec.nga.edu), the Cotton Portal (http://gossypium.info), National Center for Biotechnology Information (http://ncbi.nlm.nih.gov) for EST resources and BACMan resources at Plant Genome Mapping Laboratory (www.plantgenome.uga.edu). CottonDB provides genomic, genetic and taxonomic information including germplasm, markers, genetic and physical maps, trait studies, sequences and bibliographic citations. The cotton portal offers a single port of entry to participating cotton web resources. One participating resource, the Cotton Diversity Database (Gingle et al., 2006), provides for an interface relating to performance trial, phylogenetic, genetic and comparative data and is closely integrated with comparative physical, EST and genomic (BAC) sequence data, expression profiling resources and the capacity for additional integrative queries. Cotton oligo-gene microarrays consisting of approximately 23,000 70-mer oligos designed from 250,000 ESTs can be found at the web site http://cottonevolution.info/microarray. There is a great need to expand bioinformatic infrastructure for managing, curating and annotating the cotton genomic sequences that will be generated in the near future. The cotton sequence database of the future should be able to host and manage cotton information resources in cotton using community accepted genome annotation, nomenclature and gene ontology. Some existing databases may be upgraded to effectively handle a large amount of data flow and community requests, but additional resources will be sought to support key bioinformatic needs.

Future perspectives

Advances in technologies for harvesting specific cell types and in amplifying mRNA pools for expression profiling have stimulated studies of the transcriptome at the cellular level in plants (Galbraith and Bimbaum, 2006). However, these experiments have a common obstacle of sample preparation and single cell-type isolation that could impact interpretations of gene expression. Measuring gene expression in a single, abundant cell type will not have much experimental induced error since the developmental stages are overlapping. The problem is more aggravated when different developmental stages of fiber tissues are studied. Thus, the sampling time and state of the material are crucial for the experiment's results. Methods have been developed to eliminate these obstacles at each developmental time-point of fiber tissues (Taliercio and Boykin, 2007). However, simple and easy protocols which can be routinely used in laboratories need to be developed.

Furthermore, progress in the systematic survey of the genes crucial for fiber development is hampered by several factors including non availability of complete cotton genome, common method to assess microarray data quality, control measures to avoid false positives and poor performance in evolving common strategy to analyze the high throughput gene expression profiles etc. Careful experimental design and exploitation of suitable experimental samples along with the following considerations may fuel future research in cotton fiber development at molecular level.

Most of the studies in the literature were conducted in controlled green house experiments (of course, in field conditions in some cases) and several experimental controls such as internal, positive and negative controls, calibration spike-in controls, transgene and vector controls, ratio spike in controls, blank and buffer controls interspersed among the cotton oligoNTs and biological and technical replications were included to avoid experimental error. Thus, care has been taken to ensure the quality of the research (or genes that are expressed preferentially in fiber tissues). However, the impact of biotic and abiotic factors in the developmental process of the fiber, which has a demonstrated influence on fiber quality, is not discussed in the available literature and remains to be tested.

Given that the total number of genes in the cotton genome may be approximately 40,000 (Chen et al., 2007a), the available cotton DNA chips represents ~ 40% of the cotton genome. Analysing all the cotton genes in a single experiment may give a different picture. Based on the estimate of ~ 14,000 genes in the fiber transcriptome of a cultivated progenitor species and evidence of homoeologous genes from the A_T and D_T genomes in allotetraploid species, the fiber transcriptome increases to an estimated 28,000+ genes, making redundancy an issue in gene discovery projects (Arpat et al., 2004). Until now, the absolute mechanism of fiber elongation and cellulose biosynthesis has been unclear, even though the expression patterns and putative function of the reported genes have been analyzed in details. Although so many related genes were isolated from varied plants, the precise function of the above said genes is not clear. This is mainly because of the recalcitrant nature of matured fibers which do not lend themselves for mRNA isolation. Hence, a proper method has to be identified to grasp the novel genes involved in matured fiber development.

Though proteomics help to get a complete picture of proteins and their role in fiber development, it is hard to grasp fiber proteome due to several reasons such as: presence of interfering material such as cell walls, phenolic compounds and other secondary metabolites that will severely disturb protein separation and analysis (Yao et al., 2006). In addition to that, labile proteins can be lost during the preparation of experimental samples and therefore, must be extracted from tissues by non-destructive techniques such as vacuum infiltration or recovered from liquid culture media from cell suspension cultures or seedlings. As yet, there is no efficient procedure to release proteins that are strongly bound to the extracellular matrix (Jamet et al., 2006). Structural proteins, for instance, extensins or proline-rich proteins, can be cross-linked via di-isodityrosine bonds and until now, extensins have been eluted with salts before their insolubilization from cell suspension cultures. Another difficulty is the separation of polypeptides by classical two-dimensional gel electrophoresis. For instance, basic glycoproteins are poorly resolved by this technique (Jamet et al., 2006). Such kinds of methods require specific methods of isolation and separation since the results are highly dependant on reliability of the purify-cation methods. Yet another limitation is that biochemical function of only a small proportion of the identified proteins have been demonstrated and/or determined based on the assumptions that proteins sharing conserved domains have the same activity. Hence, the remaining proteins (domains of unknown function) remain as a challenge for elucidation of their biological function. In addition to that quantitative data on proteome is still in its infant stage and protein-protein interactions and protein with other metabolites remains to be revealed. All these data combined with genetics, biochemistry and molecular biology can lead to a better understanding of roles of genes/proteins in fiber development.

The fundamental challenge in transcriptomic and proteomic studies is that precise prediction of structure and function of genes. The databases used for annotation are not complete though they have robust data on metabolic pathway (Taliercio and Boykin, 2007). Pathways associated with these may have not been identified due to poor representation of the pathway on the microarray and lack of annotated genes associated with these pathways etc. Jamet, (2004) has provided some examples of misleading annotations that were owing to error in experimental design or in data interpret-tation. A careful and critical bioinformatic analysis of DNA and/or protein sequences therefore appears to be an absolute requirement

before starting a transcriptome analysis or discussing the results from such analysis and the relevance of bioinformatic predictions to biological data should be checked whenever possible to prevent mistakes. Certainly, there is a long way to go in determining how these genes work during fiber development. Kim and Triplett, (2004) also concluded that sequence similarity is insufficient evidence to predict accurately how proteins work in plant cells and thus supporting the need for biochemical assay and other related studies.

In addition to cDNA and oligonucleotide microarrays, tiling-path arrays have been used to study gene expression in plants (Vij and Tyagi, 2007). The advantage of tiling-path arrays over conventional microarrays is that they are not stuck-up with the gene structure and hence provide unbiased and more accurate information about the transcriptome. In addition, they provide information about transcriptional control at the chromosomal level. The use of tiling-path arrays could help to provide novel information about fiber transcriptome at the genome-wide level.

The quickly expanding knowledge on gene function and the availability of whole genome sequences of model plants is expected to offer new perspectives to solve the problems encountered in genetically and physiologically complex traits in commercial crops - which is referred to as "plant translational genomics" (Salentijn et al., 2007). The most promising tool for quick implementation of this knowledge is the candidate gene approach. The candidate gene approach is based on the assumption that genes with a proven or predicted function in a 'model' species (functional candidate genes) or genes that are enriched from a particular tissue or develop-mental pathway or genes that are co-localized with a trait-locus (positional candidate genes) could control a similar function or trait in an arbitrary crop of interest / target crop (Salentijn et al., 2007). In this respect, the recent progress in high-throughput profiling of the proteomics and metabolomics in *Arabidopsis* trichome development may enable the investigation of the concerted expression of thousands of genes and their possible role in fiber development. The multidisciplinary approaches are expected to contribute novel information toward a more comprehensive understanding of regula-tion of fiber development.

Jansen and Nap (2001) proposed the merger of genomics and genetics into the novel concept of genetical genomics: the expression levels of genes or cluster of genes are analysed within a segregating population. In genetical genomics, gene expression profiles are quantitatively assessed within a segregating population, and expression quantitative trait loci (eQTL) can be mapped like classical QTLs (Lacape et al., 2007b). The approach provides a novel way of discove-ring, at a genetic level, regulators of gene expression acting either in cis or trans relative to the target gene. The eQTL position may coincide with the gene itself displaying cis regulation (Kirst et al., 2005) or be different, thus revealing transacting factors controlling expression. A common feature of eQTL studies is the detection of 'hotspots' of trans-acting eQTL (Keurenjes et al., 2007), interpreted as regions rich in regulatory genes that co-regulate many downstream targets. Population-and genome-wide expression analyses also provides novel opportunities for correlating expression data to phenotypic/functional consequences. However, it should be noted that QTL regions appear often quite complex and approximate and may contain hundreds of genes. Consequently, the actual involvement of the candidate gene in most cases remains to be confirmed by genetic and physical mapping, positional cloning, expression analysis or genetic transformation experiments (Salentijn et al., 2007). Cost-saving alternatives to large genome-wide and population-wide analyses with minimal loss of informativeness have been proposed: analysing pooled samples of phenotypically extreme members of the population (Borevitz et al., 2003), or concentrating on genotypically-selected individuals (Xu et al., 2005).

Decoding cotton genomes will be a foundation for improving understanding of the functional and agronomic significance of polyploidy and genome size variation within the *Gossypium* genus. Sequenced cotton genomes will ultimately stimulate fundamental research on genome evolution, polyploidization and associated diploidization, gene expression, cell differentiation and development, cellulose synthesis, cell growth, molecular determinants of cell wall biogenesis, and epigenomics which will be useful in the sustainable production of high-yielding and high-quality fiber, seed and biomass (Chen et al., 2007a). Future characterization and utilization of cotton genome sequence information should integrate functional and structural genomic resources at the molecular and *in silico* levels, sequence full-length cDNAs for genome annotation and expression assays, perform detailed annotation of the cotton genome sequence to support gene discovery and map-based cloning in this species, implement a large-scale platform for identifying DNA sequence diversity (single nucleotide polymorphisms and genome specific polymorphisms), facilitate high-resolution whole-genome association studies, develop genomic tiling arrays to support gene expression and epigenomic analysis of biological and agronomic traits and sequence and annotate small RNAs and microRNAs and identify their targets (Chen et al., 2007a).

It should also be noted that there is no single cotton cultivar that has been grown globally. Hence, findings made in one variety/accession should not be generalized for the rest of the varieties since each cultivar differ in their character, habit and performance etc. When there is no similarity among the transcriptome and proteome of cell types, then identical functional roles for the same gene or protein among the cultivars cannot be expected. Hence, before making any final conclusion a comprehensive comparison is required.

In brief, characterization of the fiber transcriptome using genomic approaches has provided a development-tal framework to design strategies for the genetic improvement of yield and fiber quality and therefore has immediate applications in agricultural biotechnology. The next major task at hand will be the functional analysis of unannotated genes using reverse genetic approaches, which is much promising in light of recent advances in cotton transformation and regeneration technology (Wilkins et al. 2000; Wilkins and Arpat, 2005). In addition to the potential for bioengineering fiber properties in the future, significant headway should be made to exploit the fiber's ESTs to genetically map the fiber transcriptome as a step toward marker-assisted selection by molecular breeding programs (Lacape et al., 2007b).

Conclusion

Global gene expression analysis will be an important tool for unravelling genetic architecture and the connections between genotypic and phenotypic variation, but the results of such studies require careful interpretation (Holland, 2007). Despite the wealth of efforts in genetic mapping, transcriptomics and DNA sequencing to decipher the molecular determinants for the quality of the cotton fiber is difficult to predict and in what aspects these studies will eventually impact breeding processes. To directly relate gene action to expression phenotypes will require genetic mapping approaches, such as eQTL mapping, although this will remain challenging because of the need to assay expression levels in large numbers of genotypic samples. Even so, eQTL mapping is rapidly gaining recognition as a valuable approach for closing the gap between (structural) genetics and (functional) genomics (Lacape et al., 2007b). The compilation through meta-analysis of fiber QTL data from various studies (since the majority of the markers used are cross referenced in other populations) and the integration of QTL data with expression data (eQTL) would help to identify chromosomal regions important for fiber quality as well as important candidate genes influencing fiber quality and ultimately facilitate the breeding of superior genotypes (Chen et al., 2007a). A good parallel approach may be to search for candidates in commercial cultivars that are having naturally superior fiber qualities. Several studies performed to compare the structural differences in the genomes have shown that the difference is in the expression pattern, rather than in the presence or absence of particular genes (Gingle et al., 2006). Hence, the comparison of gene expression profiling between contrasting genotypes with respect to fiber quality can be extended to transcription profiling at the QTL level, and the genes identified at such QTLs may potentially be better candidates for superior fiber quality. Research is in progress in a recently sanctioned project on genetic improvement of cotton through marker aided selection using a mapping population developed from commercial cotton cultivars adapted to target the environment. Harnessing the full potential of functional genomics will require a multidisciplinary approach and integrated know-ledge of the molecular and other biological pro-cesses of fiber development. This information has immediate applications in breeding programs geared towards genetic improvement of cotton yield and fiber quality which is the ultimate aim and outreach of these efforts.

ACKNOWLEDGEMENT

This work is supported by the Department of Biotechno-logy, Ministry of Science and Technology, government of India under the Program Support for Research and Deve-lopment in Agricultural Biotechnology at TNAU.

REFERENCES

Applequist WL, Cronn R, Wendel JF (2001). Comparative development of fibrein wild and cultivated cotton. Evol. Dev. 3: 3-17.

Arpat AB, Waugh M, Sullivan JP, Gonzales M, Frisch D, Main DO, Wood T, Leslie A, Wing RA, Wilkins TA (2004). Functional genomics of cell elongation in developing cotton fibers. Plant Mol. Biol. 54: 911-929.

Barwale RB, Gadwal VR, Zehr U, Zehr B (2004). Prospects for Bt cotton technology in India. AgBioForum 7(1&2): 23-26.

Basra AS, Malik CP (1984). Development of the cotton fiber. Int. Rev. Cyt. 89: 65-113.

Benedict JH, Altman DW (2001). Commercialization of transgenic cotton expressing insecticidal crystal protein. Science Publishers Inc, Enfield, NH.

Berlin JD (1986). The Outer Epidermis of the Cottonseed, The Cotton Foundation, Memphis.

Borevitz JO, Liang D, Plouffe D., Chang H, Zhu T, Weigel D, Berry CC, Winzeler E, Chory J (2003) Large-scale identification of single-feature polymorphisms in complex genomes. Genome Res. 13: 513-523.

Carpita NC, Gibeau DM (1993). Structural models of primary cell walls in flowering plants: Consistency of molecular structure with the physical properties of the walls during growth. Plant J. 3: 1-30.

Cassman KG, Kerby TA, Roberts BA, Bryant DC, Higashi SL (1990). Potassium nutrition effects on lint yield and fibrequality of Acala cotton. Crop Sci. 30: 672-677.

Cedroni ML, Cronn RC, Adams KL, Wilkins TA, Wendel JF(2003). Evolution and expression of MYB genes in diploid and polyploid cotton. Plant Mol. Biol. 51: 313-325.

Chen ZJ, Lee, JJ, Woodward AW, Han Z, Ha M, Lackey E (2007a). Functional genomic analysis of early events in cotton fibredevelopment, World Cotton Research Conference 4, Lubbock, Texas, USA.

Chen ZJ, Scheffler BE, Dennis E, Triplett BA, Zhang T, Guo W, Chen X, Stelly DM, Rabinowicz PD, Town CD, Arioli T, Brubaker C, Cantrell R, Lacape JM, Ulloa M, Chee P, Gingle AR, Haigler CH, Percy R, Saha S, Wilkins T, Wright RJ, Deynze AV, Zhu Y, Yu S, Abdurakhmonov I, Katageri I, Kumar PA, Rahman M, Zafar Y, Yu J, Kohel RJ, Wendel JF, Paterson AH (2007b). Towards sequencing cotton (Gossypium) genome. Plant Physiol. 145: 1303-1310.

Clive J (2006). Global status of commercialized biotech/GM crops: 2006, ISAAA Brief Number, ISAAA, Ithaca, New York.

Cosgrove DJ (1997). Relaxation in a high-stress environment: The molecular bases of extensible cell walls and cell enlargement. Plant Cell 9: 1031-1041.

Cosgrove DJ (2001). Wall structure and wall loosening: a look backwards and forwards. Plant Physiol. 125: 131-134.

Day RC, Grossniklaus U, Macknight RC (2005). Be more specific! Laser-assisted microdissection of plant cells. Trends Plant Sci. 8: 397-406.

Endrizzi JE, Turcotte EL, Kohel RJ (1985). Genetics, cytology, and evolution of Gossypium. Adv. Genet. 23: 271-375.

Fryxell PA (1979). The Natural History of the Cotton Tribe, Texas A&M University Press, College Station, TX.

Galbraith DW, Birnbaum K (2006). Global studies of cell type-specific gene expression in plants. Ann. Rev. Plant Biol. 57: 451-475.

Gialvalis S, Seagull RW (2001). Plant hormones alter fibreinitiation in unfertilized, cultured ovules of Gossypium hirsutumI. J. Cotton Sci. 5: 252-258.

Gingle AR, Yang H, Chee PW, May OL, Rong J, Bowman DT, Lubbers EL, Day JL, Paterson AH (2006). An integrated web resource for cotton. Crop Sci. 46: 1998-2007.

Graves AD, Stewart MJ (1988). Chronology of the differentation of cotton (Gossypium hirsutum L.) fibrecells. Planta 175: 254-258.

Harmer SE, Orford SJ, Timmis JN (2002). Characterization of six α-expansin genes in Gossypium hirsutum (upland cotton). Mol. Genet. Genom. 268: 1-9.

Hawkins JS, Kim H, Nason JD, Wing RA, Wendel JF (2006). Differential lineage-specific amplification of transposable elements is responsible for genome size variation in Gossypium. Genome Res. 16: 252-261.

Hendrix B, Stewart JMcD, Wilkins TA (2004). Gene expression in developing cotton fibers as a model of water deficit stress in cotton roots. AAES Research Series. 533: 134-140.

Holland JB (2007). Genetic architecture of complex traits in plants. Current Opinion Plant Biol. 2: 3-15.

Havov R, Udall JA, Hovav E, Rapp R, Flagel L, Wendel JF (2008). A majority of cotton genes are expressed in single-celled fiber. Planta 227: 319-329.

Hsieh YL (1999). Structural development of cotton fibers and linkages to fibrequality, Hawthorne Press Inc., NY.

Jamet E (2004). Bioinformatics as a critical prerequisite to transcriptome and proteome studies. J. Expt. Bot. 55: 1977-1979.

Jamet E, Canut H, Boudart G, Pont-Lezica F (2006). Cell wall proteins: a new insight through protemics. Trends Plant Sci. 11: 33-39.

Jansen RC, Nap JP (2001). Genetical genomics: the added value from segregation. Trends Genet. 7: 388-391.

Ji SJ, Lu YC, Feng JX, Wei G, Li J, Shi YH, Fu Q, Liu D, Luo JC, Zhu YX (2003). Isolation and analysis of genes preferentially expressed during early cotton fibredevelopment by subtractive PCR and cDNA array. Nucleic Acids Res. 31: 2534-2543.

John ME (1999). Genetic engineering strategies for cotton fibremodification. Food Products Press, New York.

John ME, Crow LJ (1992). Gene expression in cotton (Gossypium hirsutum L.) fiber: Cloning of the mRNAs. PNAS, (USA). 89: 5769-5773.

Kambhampati U, Morse S, Bennet R, Ismael Y (2005). Perceptions of the impacts of genetically modified cotton varieties: a case study of the cotton industry in Gujarat, India. AgBioForum 8(2&3): 161-171.

Keurenjes JJB, Fu J, Terpstra IR, Garcia JM, van den Ackerveken G, Snoek LB, Peeters AJM, Vreugdenhill D, Koorneef M, Jansen RC (2007). Regulatory network construction in Arabidopsis by using genome-wide gene expression quantitative trait loci. PNAS (USA) 104: s1708-1713.

Kim HJ, Triplett BA (2001). Cotton fibregrowth in planta and in vitro. Models for plant cell elongation and cell wall biogenesis. Plant Physiol. 127: 1361-1366.

Kirst M, Basten CJ, Myburg AA, Zeng ZB, Sederoff RR (2005). Genetic architecture of transcript-level variation in differentiating xylem of a Eucalyptus hybrid. Genet. 169: 2295-2303.

Kosmidou-Dimitripoulou K (1986). Hormonal Influences on FibreDevelopment, Cotton Foundation, Memphis.

Kühn C, Franceschi V, Schulz A, Lemoine R, Frommer WB (1997). Macromolecular trafficking indicated by localization and turnover of sucrose transporters in enucleate sieve elements. Sci. 275: 1298-1300.

Lacape JM, Claverie M, Jacobs J, Llewellyn D, Arioli T, Derycker R, Chiron NF, Giband M, Jean J, Viot C (2007b). Reconciliation of genetic and genomic approaches to cotton fibrequality improvement, World Cotton Research Conference 4, Lubbock, Texas, USA.

Lacape JM, Dessauw D, Rajab M, Noyer JL, Hau B (2007a). Microsatellite diversity in tetraploid Gossypium germplasm: assembling a highly informative genotyping set of cotton SSRs. Mol. Breed. 19: 45-58.

Lan L, Li M, Lai Y, Xu W, Kong Z, Ying K, Han B, Xue Y (2005). Microarray analysis reveals similarities and variations in genetic programs controlling pollination/fertilization and stress responses in rice (Oryza sativa L.). Plant Mol. Biol. 59: 151-164.

Lee JJ, Hassan OSS, Gao W, Wei NE, Kohel RJ, Chen XY, Payton P, Sze SH, Stelly DM, Chen ZJ (2006). Developmental and gene expression analyses of a cotton naked seed mutant. Planta 223: 418-432.

Li CH, Zhu YQ, Meng YL, Wang JW, Xu KX, Zhang TZ, Chen XY (2002). Isolation of genes preferentially expressed in cotton fibers by cDNA filter arrays and RT-PCR. Plant Sci. 163: 1113-1120.

Liu D, Zhang X, Tu L, Zhu L, Guo X (2006). Isolation by suppression-subtractive hybridization of genes preferentially expressed during early and late fibre development stages in cotton. Mol. Biol. 40: 741-749.

Mah N, Thelin A, Lu T, Nikolaus S, Kühbacher T, Gurbuz Y, Eickhoff H, Klöppel G, Lehrach H, Mellgard B, Costelio CM, Schreiber S (2004). A comparison of oligonucleotide and cDNA-based microarray systems. Physiol. Genomics. 16: 361-370.

Marx-Figini M (1966). Comparison of the biosynthesis of cellulose in vitro and in vivo in cotton bolls. Nature 210: 747-755.

Meinert MC, Delmer DP (1977). Changes in biochemical composition of cell wall of cotton fibreduring development. Plant Physiol. 59: 1088-1097.

Meredith JWR (2000). Continued progress for breeding for yield in the USA? in: U. Kechagia, (Ed.), Proceedings of the World Cotton Research Conference II, Athens, Greece. pp. 97-101.

Orford SJ, Timmis JN (1998). Specific expression of an expansin gene during elongation of cotton fibers. Biochem. Biophys. Acta 1398: 342-346.

Patil BA, Pierce ML, Phillips AL, Venters BJ, Essenberg M (2005). Identification of genes up-regulated in bacterial blight resistant upland cotton in response to inoculation with Xanthomonas campestris pv. malvacearum. Physiol. Mol. Plant Patho. 67: 319-335.

Pear JR, Kawagoe Y, Schreckengost WE, Delmer DP, Stalker DM (1996). Higher plants contain homologs of the bacterial celA genes encoding the catalytic subunit of cellulose synthase. PNAS (USA) 93: 637-642.

Pflieger S, Lefebvre V, Causse M (2001). The candidate gene approach in plant genetics: A review. Mol. Breed. 7: 275-291.

Rahman H (2006). Number and weight of cotton lint fibres: variation due to high temperatures in the field. Australian J Agri. Res. 57: 583-590.

Ramsey JC, Berlin JD (1976). Ultra structure of early stages of cotton fibredifferentiation. Bot. Gaz. 137: 11-19.

Reijans M, Lascaris R, Groeneger AO, Wittenberg A, Wesselink E, van Oeveren J, de Wit E, Boorsma A, Voetdijk B, van der Spek H, Grivell LA, Simons G (2003). Quantitative comparison of cDNA-AFLP, microarrays, and GeneChip expression data in Saccharomyces cerevisiae. Genomics 6: 606-618.

Rensink WA, Buell CR (2005). Microarray expression profiling resources for plant genomics. Trends Plant Sci. 10: 603-609.

Rong J, Abbey C, Bowers JE, Brubaker CL, Chang C, Chee PW, Delmonte TA, Ding X, Garza JJ, Marler BS, Park CH, Pierce GJ, Rainey KM, Rastogi VK, Schulze SR, Trolinder NL, Wendel JF, Wilkins TA, Williams-Coplin TD, Wing RA, Wright RJ, Zhao X, Zhu L, Paterson AH (2004). A 3347-locus genetic recombination map of sequence tagged sites reveals features of genome organization, transmission and evolution of cotton (Gossypium). Genetics 166: 389-417.

Rosenow DT, Quisenberry JE, Wendt CW, Clark LE (1983). Drought tolerant sorghum and cotton germplasm. Agric. Water Manage. 7: 207-222.

Ruan YL, Chourey PS, Delmer PD, Perez-Grau L (1997). The differential expression of sucrose synthase in relation to diverse patterns of carbon partitioning in developing cotton seed. Plant Physiol. 115: 375-385.

Ruan YL, Llewellyn DJ, Furbank RT (2001). The Control of Single-Celled Cotton Fibre Elongation by Developmentally Reversible Gating of Plasmodesmata and Coordinated Expression of Sucrose and K+ Transporters and Expansin. Plant Cell, 13: 47-57.

Ryser U (1999). Cotton FibreInitiation and Histodifferentiation, Hawthorn Press.

Saha S, Jenkins JN, Wu J, McCarty JC, Gutierrez OA, Percy RG, Cantrell RG, Stelly DM (2006). Effects of chromosome-specific introgression in upland cotton on fibreand agronomic traits. Genetics 172: 1927-1938.

Salentijn EMJ, Pereira A, Angenent CG, van der Linden F, Krens MJM, Smulders M, Vosman B (2007). Plant translational genomics: from model species to crops. Mol. Breed. 20: 1-13.

Saranga Y, Jiang CX, Wright RJ, Yakir D, Paterson AH (2004). Genetic dissection of cotton physiological responses to arid conditions and their inter-relationships with productivity. Plant Cell Env. 27: 263-277.

Schellmann S, Schnittger A, Kirik V, Wada T, Okada K, Beermann A, Jürgens G, Hülskamp M (2002). TRIPTYCHON and CAPRICE mediate lateral inhibition during trichome and root hair patterning in Arabidopsis. The EMBO J. 21: 5036-5046.

Seagull RW (1990). Tip growth and transition to secondary cell wall synthesis in developing cotton hairs, Academic Press, San Diego.

Seagull RW (1992). A quantitative electron microscopic study of changes in microtubule arrays and wall microfibril orientation during in vitro cotton fibre development. J. Cell Sci. 101: 561-577.

Senchina DS, Alvarez I, Cronn RC, Liu B, Rong J, Noyes RD, Paterson AH, Wing RA, Wilkins TA, Wendel JF (2003). Rate variation among nuclear genes and the age of polyploidy in Gossypium. Mol. Biol. Evol. 20: 633-643.

Shi YH, Zhu SW, Mao XZ, Feng JX, Qin YM, Zhang L, Cheng J, Wei LP, Wang ZY, Zhu YX (2006). Transcript profiling, molecular biological and physiological studies reveal a major role for ethylene in cotton fibrecell elongation. Plant Cell 18: 651-664.

Shimizu Y, Aotsuka S, Hasegawa O, Kawada T, Sakuno TF, Hayashi T (1997). Changes in levels of mRNAs for cell wall-related enzymes in growing cotton fibrecells. Plant Cell Physiol. 38: 375-378.

Smart LB, Vojdani F, Maeshima M, Wilkins TA (1998). Genes involved in osmoregulation during turgor driven cell expansion of developing cotton fibers are differentially regulated. Plant Physiol. 116: 1539-1549.

Sun Y, Veerabomma S, bdel-Mageed HA, Fokar M, Asami T, Yoshida S, Allen RD (2005). Brassinosteroid Regulates FibreDevelopment on Cultured Cotton Ovules. Plant Cell Physiol. 46: 1384-1391.

Sundaram V, Basu AK, Krishna Iyer KR, Narayanan SS, Rajendran TP (1999). Handbook of cotton in India. Mumbai, India: Indian Society for Cotton Improvement.

Suo J, Liang X, Pu L, Zhang Y, Xue Y (2003). Identification of GhMYB109 encoding a R2R3 MYB transcription factor that expressed specifically in fibreinitials and elongating fibers of cotton (Gossypium hirsutum L.). Biochimica et Biophysica acta, pp. 25-34.

Szymanski DB, Marks MD (1998). GLABROUS1 Overexpression and TRIPTYCHON alter the cell cycle and trichome cell fate in Arabidopsis. Plant Cell, 10: 2047-2062.

Taliercio EW, Boykin D (2007). Analysis of gene expression in cotton fibreinitials. BMC Plant Biol. 7: 22.

Taliercio EW, Hendrix B, Stewart JM (2005). DNA content and expression of genes related to cell cycling in developing Gossypium hirsutum (Malvaceae) fibers. American J. Bot. 92: 1942-1947.

Tan PK, Downey TJ, Sptiznagel EL, Xu P, Fu D, Dimitrov DS, Lempicki RA, Raaka BM, Cam MC (2003). Evaluation of gene expression measurements from commercial microarray platforms. Nucleic Acids Res. 31: 5676-5684.

Tiwari SC, Wilkins TA (1995). Cotton (Gossypium hirsutum L.) seed trichomes expand via diffuse growing mechanism. Canadian J. Bot. 73: 746-757.

Triplett BA (1998). Stage-Specific Inhibition of Cotton FibreDevelopment by Adding a-Amanitin to Ovule Cultures. In vitro Cell. Dev. Biol. 34: 27-33.

Turley RB, Vaughn KC (1999). The primary walls of cotton fibers contain an escheating pectin layer. Protoplasma, 209: 237.

Udall JA, Flagel LE, Cheung F, Woodward AW, Hovav R, Rapp RA, Swanson JM, Lee JJ, Gingle AR, Nettleton D, Town CD, Chen ZJ, Wendel JF (2007). Spotted cotton oligonucleotide microarrays for gene expression analysis. BMC Genomics. 8: 81.

Udall JA, Swanson JM, Haller K, Rapp RA, Sparks ME, Hatfield J, Yu Y, Wu Y, Dowd C, Arpat AB, Sickler BA, Wilkins TA, Guo JY, Chen XY,

Scheffler J, Taliercio E, Turley R, McFadden H, Payton P, Klueva N, Allen R, Zhang D, Haigler C, Wilkerson C, Suo J, Schulze SR, Pierce ML, Essenberg M, Kim H, Llewellyn DJ, Dennis ES, Kudrna D, Wing R, Paterson AH, Soderlund C, Wendel JF (2006). A global assembly of cotton ESTs. Genome Res.16: 441-450.

Van't Hof J (1998). Production of micronucleoli and onset of cotton fibregrowth. Planta, 205: 561-566.

Van't Hof J (1999). Increased nuclear DNA content in developing cotton fiber cells. American J. Bot. 86: 776-779.

Vij S, Tyagi AK (2007). Emerging trends in the functional genomics of the abiotic stress response in crop plants. Plant Biotech. J. 5: 361-380.

Vinocur B, Altman A (2005). Recent advances in engineering plant tolerance to abiotic stress: achievements and limitations. Current Opinion Biotech. 16: 123-132.

Wang S, Wang JW, Yu N, Li CH, Luo B, Gou JY, Wang LJ, Chen XY (2004). Control of Plant Trichome Development by a Cotton Fibre MYB Gene. Plant Cell 16: 2323-2334.

Wilkins T (2007). Cotton biotechnology: current status and future prospects. World Cotton Research Conference 4, held on September 10-14, 2007 (Lubbock, Texas, USA).

Wilkins TA, Arpat AB (2005). The cotton fibretranscriptome. Physiologia Planta. 124: 295-300.

Wilkins TA, Jernstedt JA (1999). Molecular genetics of developing cotton fibers., Hawthorne Press, New York.

Wilkins TA, Rajasekaran K, Anderson DM (2000). Cotton Biotechnology. Critical Rev. Plant Sci. 15: 511-550.

Wu Y, Llewellyn DJ, White R, Ruggiero K, Al-Ghazi Y, Dennis ES (2007). Laser capture microdissection and cDNA microarrays used to generate gene expression profiles of the rapidly expanding fibreinitial cells on the surface of cotton ovules. Planta 226: 1475-1490.

Wu Y, Machado A, White RG, Llewellyn DJ, Dennis ES (2006). Identification of early genes expressed during cotton fiber initiation using cDNA microarrays. Plant Cell Physiol. 47: 107-127.

Wu Y, Rozenfeld S, Defferrard A, Ruggiero K, Udall JA, Kim HR, Llewellyn DJ, Dennis ES (2005) Cycloheximide treatment of cotton ovules alters the abundance of specific classes of mRNAs and generates novel ESTs for microarray expression profiling. Mol. Genet. Genomics 274: 477-493.

Xu WL, Wang XL, Wang H, Li XB (2007). Molecular characterization and expression analysis of nine cotton GhEF1a genes encoding translation elongation factor 1A. Gene. 389: 27-35.

Xu Z, Zou F, Vision T (2005). Improving quantitative trait loci mapping resolution in experimental crosses by the use of genotypically selected samples. Genetics. 170: 401-408.

Yang SS, Cheung F, Lee JJ, Ha M, Wei NE, Sze SH, Stelly DM, Thaxton P, Triplett B, Town CD, Chen ZJ (2006). Accumulation of genome-specific transcripts, transcription factors and phytohormonal regulators during early stages of fibrecell development in allotetraploid cotton. Plant J. 47: 761-775.

Yao Y, Yang YW, Liu JY (2006). An efficient protein preparation for proteomic analysis of developing cotton fibers by 2-DE. Electrophoresis. 27: 4559-4569.

Zhang S, Klessig DF (2001). MAPK cascades in plant defense signaling. Trends Plant Sci. 6: 520-527.

Zuo K, Wang J, Wu W, Chai Y, Sun X, Tang K (2005). Identification and characterization of differentially expressed ESTs of Gossypium barbadense infected by Verticillium dahliae with suppression subtractive hybridization. Mol. Biol. 39: 191-199.

15

Future challenges in environmental risk assessment of transgenic plants with abiotic stress tolerance

Mohammad Sayyar Khan[1,2]

[1]Gene Research Center, University of Tsukuba, 1-1-1-Tennoudai, Ibaraki, Tsukuba, Japan, 305-8572.
[2]Institute of Biotechnology and Genetic Engineering (IBGE), Khyber Pukhtunkhwa Agricultural University Peshawar, Pakistan.
E-mail: sayyar khan kahn_sagettarius@yahoo.com.

Environmental risk assessment of transgenic plants is a prerequisite to their release into the target environment for commercial use. Risk assessment of the first generation transgenic plants with simple monogenic traits has been carried out with principles and guidelines enlisted in the Cartagena Protocol on Biosafety. For more complex traits such as abiotic stress tolerance, there is a growing need to examine for additional considerations in the risk assessment process based on the different nature of this trait. The salt tolerance-inducing *codA* gene is a representative of many abiotic stress tolerance genes that confer salt stress tolerance in transgenic plants. In comparison with simple monogenic Bt trait, the future challenge to environmental release of abiotic stress tolerance genes lies in the question whether these genes such as the salt tolerance-inducing *codA* will need additional considerations in the risk assessment process?. In the present work, we discussed the nature of abiotic stress tolerance trait, environmental risk assessment issues and comparison of the risk assessment elements on *Bt* and salt tolerance-inducing *codA* genes to examine needs for additional considerations in the risk assessment process. We concluded and recommended that the use of abiotic stress tolerance genes such as the salt tolerance-inducing *codA* gene in transgenic plants does not need additional considerations in risk assessment.

Key words: Transgenic plants, abiotic stress tolerance, environmental risk assessment, salt tolerance-inducing *codA*.

INTRODUCTION

Since commercialization in 1994, there has been substantial progress in the development and uses of transgenic plants. In 2009, the number of countries that adopted commercial cultivation of transgenic plants reached to 25 (James, 2009). During the period from 1996 to 2008, the total area under transgenic crops expanded from 1.7 million to 125 million ha (Qaim and Subramanian, 2010).

Environmental risk assessment has been a key issue of concern surrounding transgenic plants and their release into the environment (Nuclear Regulatory Commission, 2002; Andow and Zwahlen, 2006; Chandler and Dunwell, 2008). Since development of the first generation of transgenic plants with insect resistance and herbicide tolerance, issues related to environmental risk assessment have been dealt with according to the principles and guidelines laid down in the Cartagena Protocol on Biosafety to the Convention on Biological Diversity, and those formulated by the Organization for Economic Cooperation and Development (OECD) (Sumida, 1996; Baum and Madkour, 2006; Nickson, 2008). These guiding principles are being used to ensure that genetically modified plants with new traits do not pose adverse effects to the environment and to human and animal health.

During the last two decades, a large number of crop plants have been engineered with genes conferring abiotic stress tolerance traits such as salt, drought and extreme temperatures (Cherian et al., 2006; Bhatnagar et al., 2008). In the near future, these plants will be deployed for use in the abiotic stressed environment. Due to the complex nature of abiotic stress tolerance, the emerging challenges confronting deployment of these transgenic plants are mainly regulatory and environmental risk assessment (De Greef, 2004; Wolt et al., 2009). Transgenic plants with abiotic stress tolerance genes are confronted with issues, such as increased potential of persistence and invasiveness, and unpredictable non-target effects (Mallory and Zapiola, 2008; Warwick et al., 2009; Wolt et al., 2009). Recently, some abiotic stress tolerant transgenic crop plants such as maize, wheat, cotton and trees such as poplar and eucalyptus are under field trials (office of the gene technology regulator, 2008a, b, c, d; Kikuchi et al., 2006). Some other plants, engineered with genes encoding regulatory proteins, osmoprotectants and ion transporters are already in the pipeline to enter field trials. Therefore a scientific debate is timely to focus on the risk assessment issues on transgenic crops with genes conferring abiotic stress tolerance.

For environmental risk assessment of transgenic plants with abiotic stress tolerance genes, there has been a wide consensus that the current risk assessment paradigms are scientifically-sound and sufficiently robust. However, due to the specific nature of abiotic stress tolerance trait, there is a growing need to investigate whether risk assessment of these traits requires additional needs. To achieve this purpose, a comparative analysis of the basic risk assessment issues on insect resistance and abiotic stress tolerant transgenic plants was conducted to find: 1) whether additional risk assessment elements to be considered in the use of abiotic stress tolerance genes such as the salt tolerance-inducing *codA* gene; 2) whether different strategies or measurements are needed in the risk assessment methodologies to assess these additional risk considerations.

THE CARTAGENA PROTOCOL ON BIOSAFETY (CPB)

The cartagena protocol on biosafety (CPB) originated from the Convention on biological diversity (CBD) in 2000 and came into force in 2003 (Kinderlerer, 2008). It is a legal framework that deals with international movement of living modified organisms (LMOs). The main characteristics of the CPB are: 1) it distinguishes between import of LMOs for planting or for food; 2) it establishes risk assessment and risk management of LMOs; 3) it establishes biosafety clearing house (BCH); 3) it provides for capacity building, public awareness and participation; 4) socio-economic considerations (Baum and Madkour, 2006). In order to evaluate the potential risks of the LMOs in the target environment, the CPB provides certain regulations, listed in Annex 3 of the document. The main objective of this risk assessment process is to evaluate the potential adverse effects of LMOs on the conservation and sustainable use of biological diversity in the likely potential receiving environment, taking also into account risks to human health. The major principles listed in

Annex 3, state that risk assessment should be carried out in a scientifically sound and transparent manner, risks associated with LMOs and/or their products should be considered in the context of the risks posed by the non-modified parental organism in the potential receiving environment, and risk assessment should be carried out on a case-by-case basis. Moreover, an important element of Annex III is the precautionary principle which states that "lack of scientific knowledge or scientific consensus should not necessarily be interpreted as indicating a particular level of risk, an absence of risk, or an acceptable risk." Environmental risk assessment of transgenic plants uses the precautionary approach, because a lack of knowledge and unavailability of appropriate data regarding a potential ecological risk makes the assessment process difficult (Kinderlerer 2008). Based on the precautionary approach, a risk that has no scientific evidence or very low possibility to cause harm is considered for assessment.

PROCESS OF ENVIRONMENTAL RISK ASSESSMENT

Environmental risk assessment is a process based on scientific principles that aimed to evaluate the potential adverse effects of transgenic plants on the environmental entities of value. Risk is often defined as the probability of direct or indirect harm that a potential hazard may cause to the environment and human health and is the combination of the probability of environmental exposure to the hazard and the probability that the adverse effect will occur (Andow and Zwahlen 2006; Galatchi 2006). The conventional risk assessment process follows four major steps which are: hazard identification and assessment that examine the potential hazard; exposure assessment that includes levels and likelihood of exposure; effects assessment; and finally risk characterization that integrates hazard, magnitude of the potential consequences and the likelihood of occurrence (Nuclear Regulatory Commission, 1983; EPA ,1998).

In case of transgenic plants, the initial step of risk assessment is termed as problem formulation (Figure 1). Problem formulation as the beginning of risk assessment, is an important step that leads the risk assessment process to successful risk characterization provided all individual components of this step are fully defined and integrated (Raybould, 2006; Wolt et al., 2009). Problem formulation begins with identifying assessment endpoints, developing a conceptual model and analysis plan (EPA, 1998). Nickson (2008) elaborated the problem formulation step with all its components in the context of environmental risk assessment of transgenic plants. The author emphasized the need for clear identification of assessment endpoints that must be some valued ecological entities to which the adverse effects are assessed. Developing a conceptual model is essential for generating risk hypothesis that leads to make assumptions

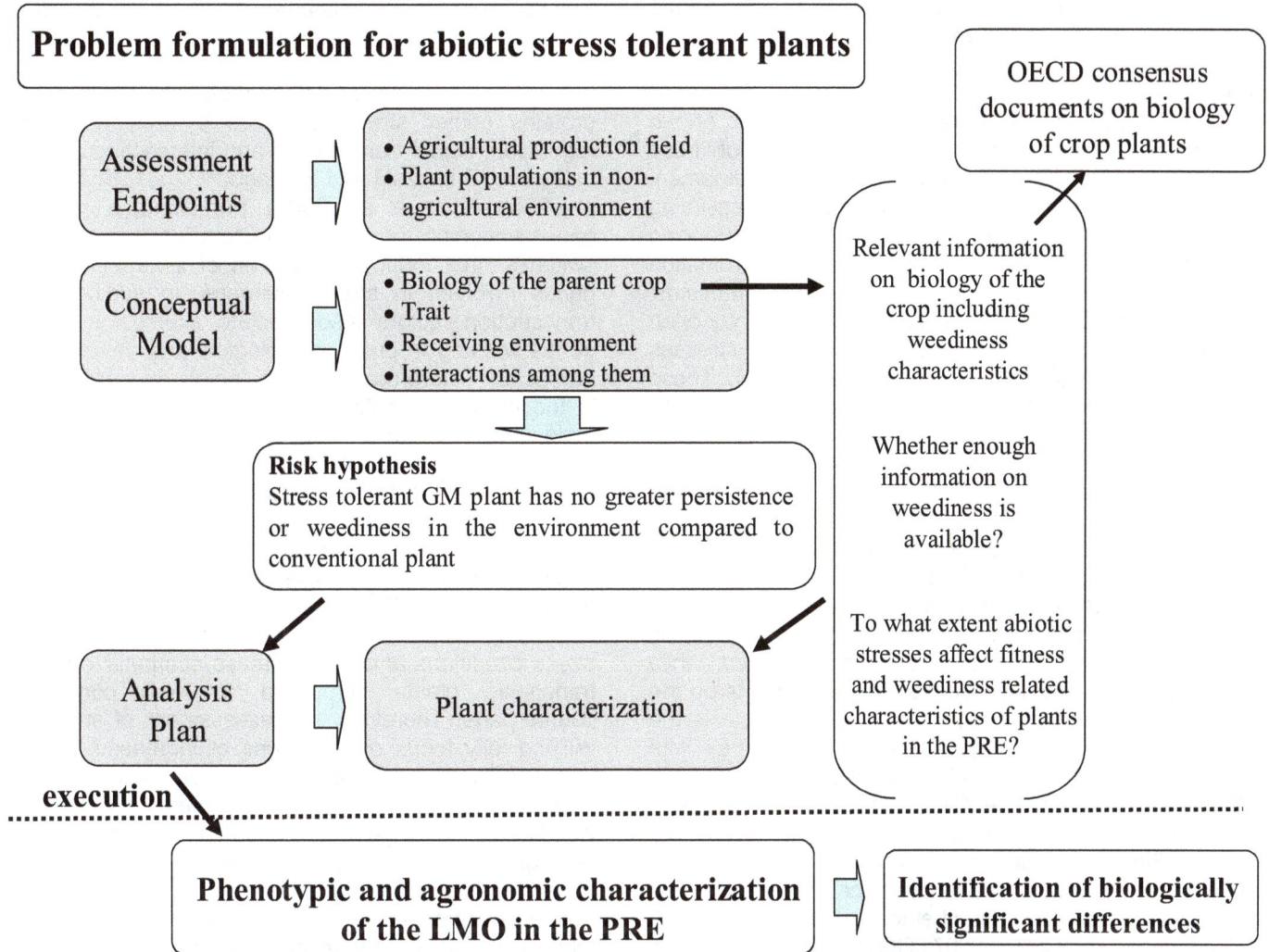

Figure 1. Schematic representation of the various steps of problem formulation in environmental risk assessment of transgenic plants. Problem formulation as the first step of risk assessment has three steps i.e. identifying assessment endpoints, developing a conceptual model and developing an analysis plan. Analysis plan is then executed for phenotypic and agronomic characterization of the LMO in the potential receiving environment and based on this, biologically significant differences are identified which are then subjected to the risk assessment process (modified from Nickson 2008).

regarding the potential effects of the stressor on the assessment endpoints. A conceptual model establishes links among the assessment endpoints, the stressor, exposure pathways and potential environmental effects. Analysis plan as the last step of problem formulation describes the nature of data needed and the kind of approach that is used for data collection.

Since commercialization of the first generation of transgenic plants with insect resistance and herbicide tolerance, risk assessment and its components have been successfully used to evaluate adverse environmental effects of these crops. Until now, no large environmental adverse effects of these crops have been documented and such risk assessment results provide limited basis for assessing risks of future transgenic plants modified with complex traits (Andow and Zwahlen, 2006; Wolt, 2009).

The risk assessment process has never been complete and has continuously evolved over the years to address the emerging environmental constraints associated with transgenic plants. The new generation transgenic plants with complex traits may raise environmental concerns with unpredictable ecological consequences, which will require careful consideration of the existing risk assessment methodologies. Abiotic stress tolerance is one of such traits under investigation for potential environmental risks. Transgenic plants modified with abiotic stress tolerance genes are confronted with questions such as whether the risk assessment process will rely on the same elements with conservative methodologies for assessing ecological impacts; and whether additional issues to be included for consideration in order to assess their environmental effects.

Risk assessment of transgenic plants with genes conferring abiotic stress tolerance

The current strategy of developing transgenic plants with increased abiotic stress tolerance is mainly based on the use of regulatory and metabolic genes. Many of these genes may affect several aspects of plant development and fitness through their important roles in regulating gene expression, signal transduction, and influencing metabolic pathways. Due to this complex coordination among various elements of stress response mechanisms, the introduced genes with one stress tolerance may often influence responses to multiple abiotic stresses (Chinnusamy et al., 2004; Chan et al., 2009). These complex changes may also have the potential to induce unintended or secondary effects, which are considered to have unpredictable ecological consequences.

A number of crops and traits are under confined field trials in Canada, Australia and the United States (Warwick et al., 2009; Beckie et al., 2010). In addition, transgenic plants with drought, salt and other abiotic stress tolerance genes have entered field trials for risk assessment studies throughout the world (Table 1). Risk assessment studies of some transgenic plants have been completed, while others are still under investigation for their effects on the environment and biodiversity. One such example is the risk assessment of transgenic wheat and barley with drought tolerance, under field trials in Australia (OGTR, 2008a, b). Environmental risk assessment of transgenic wheat with salt tolerance was also conducted (OGTR, 2005). Similarly, transgenic eucalyptus, cotton and sugarcane with abiotic stress tolerance genes are under field trials for risk assessment studies (BCH, 2005; 2007; OGTR 2008c, d; OGTR 2007; OGTR 2009). Moreover, the Monsanto led drought tolerant maize is under development that is expected to be launched in the USA in 2012 and in Sub Saharan Africa by 2017 (James, 2008). Studies on other drought tolerant crops such as soybean and cotton are in the pipeline (Monsanto, 1995).

Assessment of adverse environmental effects of these plants is based on the same basic risk assessment paradigm that has been used for Bt-transgenic crops. However, there is a growing need to search for additional needs on a case-by-case basis that may be required in the risk assessment of transgenic plants with abiotic stress tolerance genes. In the following section we will discuss the specific nature of abiotic stress tolerance trait and the environmental concerns that may arise. This will highlight the needs for any further considerations to be taken in the risk assessment process.

Nature of abiotic stress tolerance trait

The case of transgenic crops with genes conferring abiotic stress tolerance is different and more complicated than that of insect resistant or herbicide tolerant plants.

Abiotic stress tolerance is a quantitative trait controlled by many genes working in several stress response pathways (Vinocur and Altman, 2005). Transformation of crop plants with genes encoding regulatory, metabolic and membrane proteins confer stress tolerance by influencing gene regulation, signal transduction and intersecting metabolic pathways (Shinozaki and Yamaguchi, 2007; Bhatnagar et al., 2008; Khan et al., 2009). For example, regulatory genes encoding transcription factors function as "master switches" that induce expression of a large number of genes involved in stress tolerance. In addition, these transcription factors also mediate expression of other genes working in plant physiological and developmental processes. Therefore, the term "abiotic stress tolerance" is too limiting to define and encircle the magnitude of molecular, metabolic and physiological changes that occur at the whole plant level. The process involves changing the whole architecture of the plant to confer abiotic stress tolerance. In such situation, the risk assessment process will focus on the whole plant and the receiving environment, where the plant is introduced (Chaves et al., 2003). Abiotic stress tolerance is a fitness enhancing trait that confers selective advantage under stress conditions and may increase competitive ability of transgenic plants compared to their conventional counterparts. Therefore, risk assessment of such crops will logically focus on questions of increased volunteer and weediness potential in agricultural environments and invasiveness in natural environments. Other than weediness and invasiveness potential, evaluation of other ecological and non-target effects will require full understanding of the stress-associated physiological and metabolic changes that occur during abiotic stress tolerance. So far, the technology has met with limited success to fully explore the knowledge and understanding of the required metabolic changes that occur during abiotic stress tolerance (Vinocur and Altman, 2005). Due to these knowledge gaps in understanding of stress-associated metabolic profiling, the problem formulation step in the risk assessment would require an appropriate comparative approach and analysis plan to consider the consequences of metabolic changes with respect to weediness, invasiveness and other non-target effects (Wolt et al., 2010).

ABIOTIC STRESS TOLERANCE, FITNESS, WEEDINESS AND INVASIVENESS POTENTIAL

Abiotic stress tolerance is considered to be a fitness enhancing trait that increases the reproductive and vegetative growth and competitive ability of plants subjected to selection pressure. Compared to the first generation insect resistant and herbicide tolerant crop plants which have not been more invasive in natural habitats (Beckie et al., 2006; Beckie and Owen, 2007), the second generation transgenic plants with abiotic stress

Table 1. Examples of abiotic stress tolerant transgenic crop plants under field trials for risk assessment studies.

Abiotic stress category	Transgene	Crop	Implementing organization	Reference
Drought tolerance	TaDREB2/TaDREB3	Wheat/Barley	The University of Adelaide	OGTR (2008a)
Drought tolerance	CCI	Wheat	Victorian department of primary industries	OGTR (2008b)
Drought tolerance	Pyrroline-5-carboxylate reductase (P5CR)	Soybean	The agricultural research council (ARC), South Africa	De Ronde (2005)
Drought tolerance	OsDREB1A, ZmDof1	Sugarcane	BSES Limited, Australia	OGTR (2009)
Salt tolerance	Ornithine aminotransferase (oat)	Wheat	Grain Biotech Australia, Pty Ltd	OGTR (2005)
Salt tolerance	Choline oxidase (codA)	Eucalyptus camaldulensis,	University of Tsukuba, Japan	Japan Biosafety Clearing-House (2005) and Kikuchi et al. (2006)
Salt tolerance	codA	Eucalyptus globulus	University of Tsukuba, Japan	Japan Biosafety Clearing-House (2007)
Water use efficiency	Transcription factors	Cotton	Monsanto Australia Limited	OGTR (2008c)
Water logging	Adh/Pdc2	Cotton	CSIRO Australia	OGTR (2008d)
Water use efficiency/Nitrogen use efficiency	OSMAX3, OSMAX4-1, SOTB1, EcTPSP, AtMYB2, ZmDOF1	Sugarcane	BSES Limited, Australia	OGTR (2006)
Freeze tolerance	CBF transgene	Eucalyptus	ArborGen	USDA (2009)

(-) Information is not known; CCI, confidential commercial information; OGTR, Office of the Gene Technology Regulator; CSIRO, Commonwealth Scientific and Industrial Research Organisation

tolerance genes have more potential to confer increased fitness under stress conditions. Increased fitness advantage of transgenic plants, in turn, may confer persistence or volunteer potential in agricultural environments and invasiveness in natural environments (Ellstrand and Hoffman, 1990; Ellstrand, 2001; Lu, 2008). Grant (1981) indicated that abiotic stress tolerance trait may extend the range of transgenic plants beyond the area where they were previously cultivated, to areas closer to their wild relatives. This may increase the likelihood of transgenic plants to sexually hybridize with their wild relatives, to which they had never previously hybridized due to geographic isolation. In similar fashion, drought or salt tolerance trait transfer to wild and weedy relatives could confer them selective advantage under abiotic stress conditions. Compared to adjacent plant populations, the wild relatives with increased reproductive and vegetative growth may cause damage or replace the former. For example, salt tolerance in transgenic crops and their weedy hybrids may enable them to colonize, reproduce and spread to saline areas where other plant species can not easily grow (Warwick et al. 2009). This situation is quite different from that of insect resistant Bt crops, where the ecological impact assessment of transgene escape to weedy relatives is difficult to predict due to limited knowledge of the role of Bt-susceptible herbivores in regulating the density and range of crop weedy relatives. It is speculated that the abiotic stress tolerance transgenes in wild relatives will have the same fitness advantage under stress condition with unpredictable ecological consequences. The ecological consequences may appear in the form of abundance of weeds, replacing or damaging populations of other species in natural environments. In agricultural fields, increased fitness may increase the volunteer and weediness potential of transgenic plants resulting increased burden on weed management practices.

Fitness may be defined as the ability of an individual of a certain genotype to reproduce or fitness is the number of alleles that an individual contributes to the next generation (Orr, 2009). An allele that increases survival or seed production has higher fitness and it will increase and multiply in the population. Transgenes may confer significant ecological advantage, if they increase the recipient plants competitiveness and invasive ability and decline in herbivores or plant pathogens that limit their growth and reproduction (Weis, 2005). However, fitness enhancing transgenes may not necessarily increase weediness and invasiveness potential due to the many plant characteristics associated with weediness. Backer (1991) listed 13 plant characteristics that may contribute to weediness. Some of these are; 1) rapid seedling growth, discontinuous germination and prolonged seed production; 2) high seed output under favourable conditions; 3) self compatible; 4) germination and seed production under a wide range of environmental conditions; 5) special adaptations for dispersal; 6) highly competitive through allelochemicals and; 7) vigorous vegetative reproduction in case the plant is perennial. These characteristics have played an important role in the evolution of weedy species. Crop plants that are domesticated to agricultural conditions have lost many of these characteristics and have acquired domestication traits which are opposite to weediness traits.

Abiotic stresses are one of the factors that limit the ability of crop plants to develop self sustaining populations under cultivated or natural environments. Tolerance to abiotic stress may increase biomass, survivorship and fecundity in crop volunteers and wild and weedy relatives under stress conditions. However, increase in these parameters such as fecundity may not predict population expansion or invasiveness (Cummings and Alexander, 2002). Weed species may be good models to predict the ecological impact of a single or multiple stress tolerance traits in the host/target environments (Warwick et al., 2009). The consensus point is that there are limited data available which could evaluate the potential for increased weediness or invasiveness in a transgenic plant with fitness-enhancing abiotic stress tolerance trait. This poses a serious challenge for evaluation of weediness potential during phenotypic and agronomic characterization of the transgenic plant in the risk assessment process.

Although these concerns regarding abiotic stress tolerant transgenic plants have been raised in several research articles, evidences rather establish that there is almost no or negligible risk of transgenic abiotic stress tolerant plants to become weeds or attain increased persistence or volunteer potential in agricultural environment and invasiveness in natural environment. It is well established that the present day modern abiotic stress tolerant crop varieties, developed through conventional approaches also involved the same alteration of stress-related physiological and metabolic profiles. However, none of these conventionally bred crop varieties was found with an increased potential of persistence in agricultural habitats and invasiveness in natural habitats. In addition, pollen mediated gene flow is a natural process that occurs between crops and their weedy and wild relatives. From conventionally developed drought and salt tolerant plants, no reports have been documented so far that could claim transfer of these

fitness enhancing traits to their wild relatives resulting adverse ecological consequences. Moreover, biological regulation of pollen-mediated gene flow through chloroplast transformation and production of apomictic seeds could be attractive options for mitigating environmental constraints that could arise from these plants. However, these technologies are not fully developed to be implemented for practical application (Watanabe et al., 2005).

The escape of transgenic plants to non-agricultural environment and establish there as weeds is not logical in the sense that crop plants have adapted to agricultural environments through a long process of domestication. During this process, these crop plants have lost most of the typical weed like characteristics such as seed shedding, bare seeds production, and rapid vegetative growth. The OECD consensus documents on the biology of crop plants reveal that these crops have almost no weediness characteristics, can not compete with grasses, trees and shrubs and are unable to establish in non-agricultural environments[1].

Moreover the existing management practices can be easily used to control any volunteers of these abiotic stress tolerant transgenic plants in agricultural environment. Therefore, these plants do not pose any weediness or invasiveness concern for the environment. As the plants lack weediness characteristics, therefore modification with fitness enhancing abiotic stress tolerance genes can not make them potential weeds or to cause them invasive in non-cultivated areas. In natural environments, the spread and reproduction of weedy plants are regulated by many factors including many biotic and abiotic stresses, and soil nutrient conditions. Therefore, enhancing fitness of these plants by abiotic stress tolerance genes may not increase their spread and reproduction, as these are controlled by many factors.

Selective advantage under abiotic stress conditions

The signals of abiotic stress stimuli are mostly overlapping, there is a possibility that the inserted genes conferring one type of abiotic stress tolerance might also affect molecular response mechanisms to other abiotic stresses (Taylor and McAinsh, 2004; Yoshioka and Shinozaki, 2009). If a transgenic plant with genes conferring drought or salt tolerance, also acquires increased survival under cold stress might persist in agricultural habitats for longer durations compared to non transgenic plants. Under cold stress, seeds of transgenic plants may attain cold hardiness and remain dormant in soil compared to that of non-transgenic plants (SBC, 2007). In next generations, the germinated seeds in the form of volunteer plants may potentially compete with the crop under cultivation for space, CO_2, light, moisture and

[1]http://www.oecd.org/document/51/0,3343,en_2649_34387_1889395_1_1_1_1,00.html

soil nutrients (White et al., 2004). In case, the transgenic plant escapes to natural habitats may become invasive resulting damage to plant community structure. If the transgene escapes through pollen to non-transgenic plants of the same crop or weedy relatives may also produce the same consequences. In other words, this may expand the geographic range of the species beyond the cultivation area, inflicting damage on surrounding natural plant community structure.

On the other hand, changes in the ABA metabolism may also alter plant responses to cold stress. Many of the genes involved in abiotic stress tolerance work in an ABA-mediated signaling pathway (Tuteja, 2007). Other than its function in stress signaling, ABA also regulates key processes in seeds such as dormancy and storage of lipids (Kermode, 2005). Any genetic modification that targets ABA metabolism may alter seed characteristics so that the seeds survive under cold winter for longer times resulting increased potential for persistence in agricultural environments and invasiveness in natural environments (SBC, 2007).

Despite these concerns, the potential of salt and drought tolerant transgenic plants to become persistent, weedy or invasive is extremely low. The growth and spread of plants in agricultural and natural environment is regulated by many environmental factors including biotic and abiotic factors such as pests and diseases, salt, drought, low and high temperatures, UV irradiation, anoxia, soil nutrient conditions and other environmental factors. In addition, the cross-protection involves physiological and metabolic burdens, due to which transgenic plants may not have enhanced fitness and potential to become weeds.

Selective (dis) advantage of transgenic plants under biotic stress

Transgenic plants with salt and drought tolerance genes may also show a slight selective advantage or disadvantage to biotic stress conditions. There exists a cross-talk between molecular response mechanisms to abiotic and biotic stresses. Plant hormones such as abscisic acid (ABA), jasmonic acid (JA), ethylene (ET) and salicylic acid (SA) regulate plant responses to environmental stresses at the molecular level involving signal recognition, signal transduction, signal response and a multidimensional network of gene expression and regulation (Vinocur and Altman, 2005; Fujita et al., 2006). The ABA-dependent signaling pathway regulates stress-inducible gene expression through several positive and negative regulators (Shinozaki et al., 2003), while ET, JA and SA regulate biotic stress signaling upon pathogen infection. The two different hormonal pathways are not totally independent of each other as there exist some level of synergistic and antagonistic actions during response generation to biotic and abiotic stress stimuli. For example, one study found that ABA and JA antagonistically

regulated the expression of salt stress-inducible transcripts in rice (Moons et al., 1997). Apart from that, several studies have also demonstrated that the ABA and ethylene signaling pathways also interact antagonistically to modulate plant development (Beaudoin et al., 2000; Ghassemian et al., 2000). The antagonistic action of ABA and JA-ET signaling pathways to modulate responses to biotic and abiotic stresses has been demonstrated in ABA and ethylene signaling mutants of Arabidopsis (Anderson et al., 2004). Authors of this study found that exogenous ABA application suppressed both basal and JA-ethylene-activated transcription of defense genes. By contrast, the mutation of ABA synthesizing genes resulted in up-regulation of transcription of JA-ethylene responsive defense genes. In addition to the above, they also demonstrated that by disruption of AtMYC2 encoding a basic helix-loop-helix transcription factor and a positive regulator of ABA signaling resulted in transcription activation of JA-ethylene responsive defense genes. These results showed that this antagonism in ABA and JA-ethylene signaling pathways may be a strategy adopted by plants to avoid simultaneous expression of biotic and abiotic stress related genes.

Despite the phenomenon of cross-talk between biotic and abiotic stress responses, it is highly unlikely that the abiotic stress tolerant transgenic plants may show a changed response to populations of herbivores, predators, parasitoids and pathogens.

Other unintended effects

Genetic engineering of plants with genes conferring various traits may result into unintended effects. These unintended effects may include; 1) insertional effect, changed expression of a gene at the site of insertion; 2) pleiotrophic effect, altered expression of an unrelated gene at an other loci through changing its chromatin structure, methylation pattern and regulation of signal transduction or transcription; 3) generation of new products through interaction of the introduced protein with endogenous molecules; 4) high level transgene expression and the resultant metabolic burden and; 5) secondary effects due to changed substrate or product levels (OGTR, 2008b). In addition to other adverse outcomes, unintended effects may also result into weediness potential. However, these unintended effects are not restricted only to plants developed through transgenic technology. Other non-transgenic approaches of plant development may also have the potential to generate unintended effects. For example a potato variety developed through conventional biotechnology accumulated high levels of toxic glycoalkaloids (Haslberger, 2003).

Interactions with target and non-target organisms

There are no reasons to assume that genes conferring

salt and drought stress tolerance (for example, codA, DREBs and antiporters) will have effects on target organisms (herbivores, parasites and pathogens) or non-target organisms such as pest predators, beneficial insects, pollinators and populations of other organisms. In fact for abiotic stress tolerance trait, no target species to which adverse effects are evaluated may be defined. Neither the abiotic stress tolerance gene such as the codA, DREBs and antiporters, their encoded proteins, and their endproducts have no known effects on these mentioned organisms, nor is the engineered abiotic stress tolerance trait in transgenic plants aimed to confer resistance to biotic stresses. On the other hand, there has been a slight cross-talk between biotic and abiotic stress response pathways, due to which transgenic plants with salt and drought tolerance genes may show slightly changed responses to biotic factors. However, the likelihood of this to happen is extremely low.

COMPARISON OF RISK ASSESSMENT ON INSECT RESISTANCE BT AND SALT TOLERANCE TRAIT IN TRANSGENIC PLANTS (SPECIFIC CASE OF SALT TOLERANCE-INDUCING-CODA GENE)

One of the fundamental principles listed in Annex 3 of the Cartagena Protocol on Biosafety is to conduct risk assessment on a case-by-case basis. Based on this principle, the risk assessment process deals transgenes on individual basis, taking also into account nature of the host plant, the trait, the potential receiving environment and the likely interactions among them. So far, the current risk assessment procedures have been used to evaluate and identify the adverse environmental effects of first generation transgenic plants engineered with simple monogenic traits such as insect resistance and herbicide tolerance. And there has been a wide consensus in the scientific community that the current risk assessment procedures are equally applicable to new generation transgenic plants engineered with genes conferring abiotic stress tolerance. However, risk assessment may consider additional measures based on the nature of the transgene and the trait in question.

For the purpose of searching for additional considerations for the salt tolerance-inducing codA gene, a comparison of the risk assessment elements on Bt and codA is summarized in Table 2. The codA gene is isolated from the soil bacterium, Arthrobacter globiformis. The codA gene encodes an enzyme called choline oxidase (COD) that works in the biosynthetic pathway of glycine betaine, an osmoprotectant. Glycine betaine, in turn, protects vital cellular organelles, enzymes and membranes from the damaging effects of abiotic stresses including salt stress (Gorham, 1995). During the current decade, a large number of transgenic plants have been developed that harbored the bacterial codA gene. These transgenic plants exhibited multiple abiotic stress tolerance. However, in many plants, the codA gene

conferred enhanced salt tolerance and transgenic plants showed improved vegetative and reproductive growth (Chen and Murata, 2008; Khan et al., 2009). For commercial use of these transgenic plants under the target saline soils, risk assessment studies are important to evaluate the adverse environmental effects.

The assessment of potential environmental effects of salt tolerance-inducing codA gene is based on the current risk assessment procedures. However, depending on the nature of salt tolerance-inducing codA gene, there is a need to look for additional considerations in the risk assessment. In case of insect resistance Bt transgenic plants, the main issues which are considered for risk assessment include 1) competitiveness, weediness and volunteer potential; 2) gene flow to wild and weedy relatives and increased weedines and invasiveness potential in agricultural and natural environments; 3) adverse effects on non-target organisms such as predators, parasitoids, beneficial insects, and endangered and charismatic insects, and soil microbial activities; 4) production of harmful compounds that may affect biodiversity; 5) and assessment of resistance evolution in the target insects against the Bt proteins. The risk assessment practice on insect resistance transgenic plants is not new and has been carried out for the last several years with environmental release for commercial use. On the other hand, the risk assessment of abiotic stress tolerance genes such as salt tolerance-inducing codA is still under development. Based on the nature of salt tolerance-inducing codA transgene in transgenic plants, the current risk assessment is considered to evaluate 1) competitiveness, persistence and volunteer potential; 2) gene flow to wild and weedy relatives and weedines and invasiveness potential in agronomic and natural environments; 3) and production of harmful substances (allelopathic influence) that could affect biodiversity including plant communities, interacting insects, and other organisms; 4) and soil microbe analysis. These issues are already under consideration for risk assessment of insect resistance Bt transgenic plants. For salt tolerance-inducing codA transgenic plants, these risks will be evaluated in the same way as for insect resistance Bt plants. In the following section, these issues are discussed in a more detailed way.

Both insect resistance and salt tolerance are fitness enhancing traits that may increase the competitive ability of transgenic plants compared to the surrounding plant communities. Transgene flow to wild and weedy relatives may confer selective advantage, which may increase their weedines and invasiveness potential in both agricultural and natural environments. In case of insect resistance Bt transgenic plants, the ecological consequences resulting from Bt gene escape to weedy relatives in natural environment are unpredictable and less defined. The knowledge is still limited regarding whether the same Lepidopteran insects feed on wild relatives and to what extent these insect pests regulate the survival and spread

Table 2. Comparison of risk assessment elements on insect resistant-Bt and salt tolerance-inducing *codA* gene.

Environmental concerns/Hazards	Insect resistance *Bt* (cry) genes	Salt tolerance-inducing *codA*
Competitiveness and weediness potential	Bt derived proteins may confer selective advantage under *Lepidopteran* attack, which may increase competitiveness and weediness potential	The *codA* gene confers selective advantage under salt stress that may increase fitness of plant The fitness advantage may affect competitiveness and weediness potential in agricultural habitat Generally crop plants are non-competitive and have very low or negligible weedy characteristics
Gene flow to wild and weedy relatives	Bt-derived proteins may confer increased weediness and invasiveness potential. Invasiveness in natural habitat depends on the degree at which the spread of weedy relatives are regulated by *Lepidopteran* insects	The *codA* gene escape may increase invasiveness potential of wild relatives in natural environment However, the selective advantage is limited and unable to affect the spread and survival of wild relatives Other environmental factors may still regulate and limit the spread of such wild plants
Gene flow frequency	No significant effect on pollen viability and dispersal characteristics	There is unlikely that the *codA* gene may affect pollen metabolism, viability and dispersal characteristics
Production of harmful compounds	Risk assessment considers evaluation of harmful compounds that may negatively affect biological diversity	Although *codA* is not likely to produce harmful compounds, allelopathic assessment is considered in risk assessment
Toxicity to non-target and beneficial insects	The Bt-derived proteins are specifically toxic to *Lepidopteran* insects. Risk assessment considers evaluation of adverse effects on non-target insects and other organisms	The *codA* gene, choline oxidase, GB and the salt tolerance trait have no adverse effects on insect populations and other organisms This concern is not considered in risk assessment
Effects on soil micro-organisms	Assessment of soil micro-organisms and their activities is an essential element in the environmental risk assessment of Bt trait	The *codA*, choline oxidase and GB have no known effects on soil microbes and their activities However, salt tolerance through changed water and nutrient uptake may affect microbial diversity and their enzymatic activities
Development of insect resistance to Bt toxins	The risk of resistance development in insect pests to Bt toxins is well established	No such risks

of these weedy relatives. As discussed in the previous section, abiotic stress tolerance is the product of gene action working in multiple stress response pathways. These responses are mostly overlapping and there exists a cross-talk between them. Due to this phenomenon, transgenic plants with stress tolerance genes often exhibit tolerance to multiple abiotic stresses. In similar fashion, the *codA* gene manipulation in several transgenic plant species conferred salt tolerance and also drought and extreme temperature tolerance in some cases (Chen and Murata, 2008). The selective advantage of salt tolerant transgenic plants under cold stress, for example may increase the persistence or volunteer potential in the following cropping seasons. Transgene escape to wild and weedy relatives may confer selective advantage under salt stress compared to the surrounding plant populations. The consequences could be predicted in the form of improved vegetative and reproductive growth of wild relatives under salt stress resulting damage and loss to other plant communities.

The third environmental concern that is considered during risk assessment for both insect resistance and salt tolerance-inducing codA transgenic plants is the risk of production of harmful compounds due to transgene presence, position or pleiotrophic effects. For this purpose, various allelopathic tests are conducted. The purpose of allelopathic assessment is to determine whether the presence of transgene, its encoded protein and the end product in dried parts of transgenic plants or in the root exudates in the soil have allelopathic influence on the growth and germination of the surrounding plant communities and soil microbe activities. In case of salt tolerance-inducing codA, there is no reason to assume that the encoded protein and the final product, glycine betaine will produce harmful compounds. However, based on the precautionary principle, risk assessment on salt tolerance-inducing codA considers evaluation of allelopathic effects on the surrounding plant communities and valued soil microbe activities.

Moreover, evaluation of adverse effects on non-target organisms is an essential element of the risk assessment of insect resistant Bt transgenic plants. The non-target effects of insect resistance Bt trait are well studied. Knowledge of the mechanism of Bt toxicity to insects is well established that provides a basis for evaluation of adverse effects on non-target organisms. In some instances, adverse effects have been documented on some non-target and beneficial insects such as green lacewing and the monarch butterfly (Hilbeck et al., 1998; Losey et al., 1999). On the other hand, there are no reasons to assume that the salt tolerance-inducing codA gene will have any direct, indirect, immediate or delayed effects on non-target organisms. There have been no reports documented to date that the codA gene, its product or the salt tolerance trait have posed any adverse effects on non-target organisms including valued soil microbe activities. In contrast, there exists a slight cross-talk between abiotic and biotic stress responses and due to this phenomenon, transgenic plants with abiotic stress tolerance genes may show slightly altered responses to biotic factors. However, the chances of this phenomenon to happen are extremely rare.

Based on the above discussion, it is clear that the environmental risk assessment issues confronted with salt tolerance-inducing codA gene are already under consideration for the current generation transgenic traits such as insect resistance Bt trait. Moreover, the current risk assessment procedures will be used in the same way as used for insect resistance Bt trait. In comparison with insect resistance Bt trait, the salt tolerance-inducing codA does not need additional considerations or extra measurements in the risk assessment methodology. This is also evident from the environmental risk assessment of transgenic Eucalyptus plants transformed with the salt tolerance-inducing codA gene. Environmental risk assessment of transgenic Eucalyptus provides the only practical example of evaluation of environmental risks of the salt tolerance-inducing codA gene.

BIOSAFETY ASSESSMENT OF TRANSGENIC EUCALYPTUS IN JAPAN: A CASE STUDY OF BIOSAFETY ASSESSMENT OF SALT TOLERANCE-INDUCING CODA GENE

In Japan, two transgenic eucalyptus tree species, Eucalyptus camaldulensis and Eucalyptus globulus were transformed with the salt tolerance-inducing codA gene. Transgenic plants of both species exhibited salt tolerance under semi-confined conditions (Kikuchi et al., 2006; Yu et al., 2009). These transgenic eucalyptus plants were evaluated for environmental biosafety in a net-house (type II use) under the Japanese law on environmental biosafety for future field application (type 1 use).

In Japan, the biosafety of transgenic plants considers four items for assessment (Taniguchi et al., 2008). These include assessment of 1) competitiveness and weediness potential; 2) production of harmful compounds, allelopathic effects on biodiversity; 3) cross-ability, gene flow to wild relatives and increased invasiveness potential; 4) and other properties. Under net-house conditions (Type II use), environmental biosafety of three Eucalyptus camaldulensis genotypes was evaluated for the above mentioned risk assessment items. In environmental risk assessment, allelopathic studies on transgenic and their non-transgenic genotypes are important to investigate production of any harmful compounds. Some plants have allelopathic influence in nature and the transgene presence may enhance their negative effects on the surrounding plant communities and non-target organisms interacting with them. During environmental risk assessment, the direct negative effects of these transgenic plants on the biodiversity are evaluated and compared with the non-transgenic control plants.

In case of codA transgenic plants, the allelopathic effects are unlikely. The codA product, choline oxidase catalyzes the glycine betaine biosynthetic pathway. Glycine betaine is naturally produced in several plant species where its increasing content is directly correlated with increased stress tolerance. Also in transgenic plants, the increased glycine betaine accumulation conferred increased stress tolerance without any negative effects on plant growth. Therefore, the codA transgene, its encoded choline oxidase enzyme and the end product glycine betaine should not produce harmful substances or allelopathic effects. This is evident from the allelopathic studies conducted as part of environmental risk assessment of salt tolerant *codA* transgenic *Eucalyptus camaldulensis* and *Eucalyptus globulus* plants.

Transgenic *E. camaldulensis* plants harbored the *codA* gene under the constitutive *CaMV35S* promoter. These transgenic plants were found salt tolerant compared to their non-transgenic counterparts under controlled conditions. For environmental safety of these transgenic plants, allelopathic tests were conducted to evaluate the negative effects that could arise due to the *codA* transgene, its encoded protein and the glycine betaine product (Kikuchi et al., 2009). The authors of this study used four different tests for allelopathic assessment. They used sandwitch method to determine differences in root and hypocotyl growth of lettuce seeds grown on media that contained dried leaves from the three transgenic and non-transgenic genotypes. Soil germination test was conducted to evaluate differences in germination of lettuce seeds grown on soil taken from transgenic and non-transgenic genotypes. Volatile and phenolic compounds in transgenic and non-transgenic genotypes were analyzed through Gas chromatography and HPLC respectively. The authors reported no significant differences between the three *codA*-transgenic *E. camaldulensis* genotypes and their non-transgenic counterparts for the studied parameters. In addition, the salt tolerant transgenic *E. camaldulensis* plants harboring the *codA* transgene were evaluated for adverse effects on soil microbe activities and compared with those of non-transgenic control plants (Japan Biosafety Clearing-House 2005). No differences were found between microbial activities in soil samples taken from both transgenic and non-transgenic Eucalyptus plants grown in a special-netted house. Similarly, for environmental biosafety assessment, impact on allelopathic effect and soil microbes was investigated on salt tolerant *codA*-transgenic *E. globulus* and the non-transgenic plants grown under special-netted house conditions (Yu et al. 2008; Lelmen et al. 2009). No significant differences were found between transgenic and non-transgenic plants for both allelopathic effects and soil microbe communities.

These tests provided useful informations regarding the environmental safety of *codA* gene. From the environmental biosafety studies on the salt tolerance-inducing *codA* transgenic Eucalyptus plants, it was also pointed out that; 1) the risk assessment approaches and procedures used for environmental biosafety of *codA* transgenic eucalyptus were the same as currently under use for the first generation transgenic plants; 2) and risk assessment of *codA* did not need additional considerations under the Japanese government. In conclusion, the use of salt tolerance-inducing *codA* in transgenic plants are using the same risk assessment procedures as used for other transgenic crops modified with for example, insect resistance trait. Therefore, the *codA* gene does not need additional considerations in environmental risk assessment.

Examples of other field trials and risk assessment studies conducted on salt and drought tolerant transgenic plants by various regulatory authorities

Many transgenic plants with abiotic stress tolerance genes are under field trials for environmental risk assessment studies in Australia, Canada, USA, New Zealand, Philippines and the European countries. These are listed in Table 1.

The Monsanto developed drought tolerant maize will be ready for commercial use in the United States by 2012, and it has already applied for commercial release of transformation event MON 87460 in USA, Canada, Australia/New Zealand, Philippines and the European Union.. Much of the information regarding field trials and risk assessment, however has been undisclosed. Some of the details on risk assessment of MON 87460 are now available from the summary notification information format (SNIF), submitted by Monsanto. MON 87460 expresses a cold shock protein B (CspB) isolated from *Bacillus subtilis*. The transgenic maize is developed to perform better and show reduced yield loss under limited water conditions compared to the conventional counterpart. The information revealed that the MON 87460 is equivalent to conventional maize under well watered conditions. However under limited water, transgenic maize performs better in terms of reduced grain loss compared to the conventional maize. Under severe water limitation, transgenic maize like the conventional maize will be unable to grow. During comparative assessment, MON 87460 revealed no significant differences with the conventional maize except the intended trait that conferred a selective advantage only under limited water level that negatively affect plant yield. However, this selective advantage under limited water conditions is highly insufficient to alter the transgenic maize as a volunteer plant or its escape to non-agricultural fields. There may be some other important environmental factors that would limit the survival of maize in the potential receiving environment. The SNIF did not mention about cross-talk between different abiotic stress responses, or whether transgenic maize with the CspB protein also shows tolerance to other abiotic stresses other than the intended drought tolerance.

Transgenic sugarcane lines expressing the *OsDREB1A* and *ZmDof1* genes were developed that showed drought tolerance and improved nitrogen use efficiency (NUE). These transgenic lines were assessed for environmental risks. Regarding risk assessment, OGTR concluded that the limited and controlled release of these transgenic sugarcane lines pose negligible risks to the environment (OGTR, 2009). Regarding weediness potential of these transgenic sugarcane lines, it was mentioned that it is highly unlikely that the introduction of these genes could change all of the characteristics that regulate or limit the persistence of sugarcane. Moreover, the occurrence of any unintended pleiotrophic effects could have been detected during the pre-trial stage. However, further uncertainties, if still existed could be judged through containment measures and monitoring.

Transgenic wheat lines transformed with the Ornithine aminotransferase gene showed salt tolerance under greenhouse and field conditions (OGTR, 2005). The OGTR risk assessment and risk management plan concluded that modification of wheat with the Ornithine aminotransferase gene and the resultant proline accumulation confers selective advantage in some environmental conditions compared to conventional counterpart. However, this modification is unlikely to alter other characteristics associated with weediness potential. Transgenic wheat and barley (DIR077) contained drought responsive transcription factors from wheat, *TaDREB2* and *TaDREB3* (OGTR 2008a). About risk assessment and risk management, the OGTR concluded that the use of these genes pose negligible risks to human health and the environment and these negligible risks do not require specific risk treatment measures.

Moreover, several genes have been transferred into cotton to increase its water use efficiency (OGTR, 2006; 2008). The OGTR in the RARMP concluded that these genes are not likely to alter all of the characteristics that limit the spread and persistence of cotton such as dormancy, seed survival in soil for long time, length of life cycle, large amount of seed dispersal. Moreover, transgenic cotton showing multiple tolerance with these genes is not likely to show enhanced fitness; rather the multiple stress tolerance will make the plant less fit due to metabolic and physiological burdens.

In these field trials of transgenic plants with various abiotic stress tolerance genes, it is mentioned in their RARMP that 1) these genes may confer selective advantage to transgenic plants only under stress condition that enables the plant to perform better than the conventional counterpart. Under non-stress condition, the transgenic and non-transgenic plants are equivalent; 2) Due to cross-talk in stress response mechanisms, transgenic plants with these genes may show stress tolerance other than the intended one. However, this selective advantage under other stress conditions may not alter all the characteristics associated with weediness or invasiveness traits, or in other words may not make transgenic plants weedy or persistent in agricultural or natural environments: 3) these genes are mostly taken from plants, and their introduction or over expression in the host plants modify an already existing stress response mechanism. Therefore, it is concluded that these genes, their encoded proteins and the end products will pose no harm to the environment and living organisms including humans. For future large scale release, the RARMP concluded that additional information will be required regarding characteristics indicative of weediness including measurement of altered reproductive capacity, tolerance to environmental stresses, and disease susceptibility.

CONCLUSION AND RECOMMENDATIONS

Transgenic plants with abiotic stress tolerance genes are under development. Future environmental release of these transgenic plants requires assessment of their adverse effects. The abiotic stress tolerance genes are using the same risk assessment procedures as used for the current first generation transgenic plants with insect resistance Bt and herbicide tolerance genes. However, depending on the different nature of abiotic stress tolerance trait, needs for additional considerations should be examined in the risk assessment process. In comparison with insect resistance Bt genes, the abiotic stress tolerance genes are mostly taken from plants and their encoded proteins or at least their end products are not new to plants. The encoded proteins of these genes modify the already existing stress response mechanism and do not introduce novel pathways or functions in transgenic plants, which may pose adverse effects to the environment and human health. The use of these genes such as the salt tolerance-inducing *codA* gene in transgenic plants causes no risks to biodiversity including plant communities and non-target organisms. Therefore, the use of abiotic stress tolerance genes (for example, salt tolerance-inducing *codA* gene) does not need additional considerations in the already established and rigorous risk assessment procedures. In addition, as indicated in the field trials and risk assessment studies on several transgenic plants, the use of abiotic stress tolerance genes confer tolerance only under stress condition that enable transgenic plants to reduce yield losses compared to the conventional plants. Moreover, the salt and drought tolerance genes do not confer weediness, persistence and invasiveness potential to transgenic plants compared to the conventional non-transgenic plants. In agricultural and natural environments, salt and drought stresses are not the only factors that limit the growth and spread of plants. Regarding invasiveness, it is generally observed that the salt and drought tolerance of transgenic plants is considerably lower than the levels required sustaining and thriving in extreme saline and drought affected natural environments. Therefore, there is no risk of these plants to invade natural environments. Despite these observations, the abiotic stress tolerance trait in transgenic plants involves metabolic and physiological changes and the potential of selective advantages under other abiotic stresses. The uncertainties due to these specific aspects and their potential effects on weediness, persistence and invasiveness must be addressed through further information on weediness of the host plant, implementing an appropriate risk management strategy to control potential weeds and volunteers and continuous monitoring of the LMO in the potential receiving environment as mentioned in the protocol. In addition, the following points should be focused in order to overcome any uncertainties.

1) A well organised problem formulation: Increased emphasis is needed on elements of problem formulation (Figure 1). Defining assessment end points and developing a comprehensive conceptual model and analysis plan are essential to address the uncertainties that may arise in the risk assessment of salt and drought tolerant transgenic plants. Familiarity with biology of the crop, characteristics of the trait and the potential receiving environment is important for developing conceptual model and generating risk hypothesis. For generation of conceptual model of an abiotic stress tolerant LMO, informations on the biology of the crop will particularly include information on weediness characteristics of the conventional plant. In addition, questions will be raised such as whether enough information on weediness is available? And to what extent abiotic stresses affect fitness and weediness related characteristics of plants in the potential receiving environment? The OECD consensus documents provide a valuable source of information on the biology of the crop plants including information on weediness, volunteer and invasiveness potential. These all information will be used to develop an analysis plan for plant characterization in the potential receiving environment. Identification of meaningful differences between the LMO and the conventional plant will be subjected for further assessment.

2) Weediness characteristics: For weediness and invasiveness potential, the relevant information on weediness should be collected during the problem formulation step. However, there is limited data available that could evaluate the potential of increased weediness or invasiveness of a transgenic crop plant with fitness-enhancing abiotic stress tolerance trait and its comparison with the conventional counterpart. This poses a challenge in risk assessment. To meet this challenge, emphasis should be placed on; 1) phenotypic and agronomic traits associated with fitness and weediness characteristics in conventional crop plants and; 2) the effects of salt and drought stress tolerance on these traits. Baseline informations are needed on the characteristics of weeds in general and on the factors that limit the spread and persistence of conventional crop plants in particular (OGTR 2008). Further efforts are needed to understand; 1) the factors that control the population size and range of both the crop volunteers and their wild relatives and; 2) the degree by which abiotic stresses such as salt and drought regulate the survival and reproduction of crop plants in the field

3) Comprehensive phenotypic and agronomic characterization in the potential receiving environment: As stress tolerance genes may involve physiological and metabolic changes that confer selective advantage under abiotic stresses due to cross talk among various stress response pathways, careful phenotypic and agronomic characterization is needed to identify meaningful differences between the LMO and the conventional counterpart. The unintended changes may be considered for further evaluation in the risk assessment process. In relation to phenotypic characterization of the LMO in the potential receiving environment, the following points should be focused on:

1) Consideration of other environmental conditions prevalent in the potential receiving environment, while planning comparative analysis. Potential receiving environment for salt and drought stress tolerance transgenic plants will be the same as that for non-transgenic conventional plants. However, in some cases genes with one stress tolerance may also confer tolerance to other stresses. Therefore, these stresses should also be considered in that environment.

2) Choice of the conventional comparator (may be a commercial salt, drought tolerant variety) for comparative phenotypic and agronomic assessment in the potential receiving environment

3) Knowledge of and data availability on the response of conventional plant to the stress condition and to the potential receiving environment. If the conventional comparator has not been under cultivation in that area, comparative analysis will be a challenge.

ACKNOWLEDGEMENT

The author is grateful to all members of Watanabe Lab, Gene Research Centre, University of Tsukuba, Japan for their valuable advices and improvement of the manuscript.

REFERENCES

Anderson PL, Hellmich RL, Sears MK, Sumerford DV, Lewis LC (2004). Effects of Cry1Ab-expressing corn anthers on monarch butterfly larvae. Environ. Entomol., 33: 1109-1115.

Andow DA, Zwahlen C (2006). Assessing environmental risks of transgenic plants. Eco. Lett., 9: 196-214.

Backer HG (1991). The continuing evolution of weeds. Econ. Bot., 45: 445-449.

Baum M, Madkour M (2006). Development of transgenic plants and their risk assessment. Arab J. Plant Prot., 24: 178-181.

Beaudoin N, Serizet C, Gosti F, Giraudat J (2000). Interactions between Abscisic Acid and Ethylene Signaling Cascades. Plant Cell, 12: 1103-1116.

Beckie HJ, Hall LM, Simard MJ, Leeson JY, Willenborg CJ (2010). A framework for postrelease environmental monitoring of second-generation crops with novel traits. Crop Sci., 50: 1587-1604.

Beckie HJ, Harker KN, Hall LM, Warwick SI, Legere A, Sikkema PH, Clayton GW, Thomas AG, Leeson GY, Seguin-Swartz G, Simard MJ (2006). A decade of herbicide-resistant crops in Canada. Can. J. Plant Sci., 86: 1243-1264.

Beckie HJ, Owen MDK (2007). Herbicide resistant crops as weeds in North America: CAB Reviews. Perspect. Agric. Vet. Sci. Nutr. Nat. Resour., 2: 044. doi: 10.1079/PAVSNNR20072044.

Bhatnagar-Mathur P, Vadez V, Sharma KK (2008). Transgenic approaches for abiotic stress tolerance in plants: Retrospect and prospects. Plant Cell Rep., 27: 411-424.

Chan Z, Grumet R, Loescher W (2009). Salt tolerance and global gene expression analysis of transgenic mannitol-producing *Arabidopsis thaliana*. Plant and Animal Genome Conference XVII, 2009. Available from internet: Http://www.intl-pag.org/17/abstracts/P07a_PAGXVII_663.html (abstr.).

Chandler S, Dunwell JM (2008). Gene flow, risk assessment and the environmental release of transgenic plants. Crit. Rev. Plant Sci., 27: 25-49.

Chaves MM, Maroco JP, Pereira JS (2003). Understanding plant responses to drought—from genes to the whole plant. Funct. Plant Biol., 30: 239-264.

Chen THH, Murata N (2008). Glycinebetaine: An effective protectant against abiotic stress in plants. Trends Plant Sci., 13: 499-505.

Cherian S, Reddy MP, Ferreira RB (2006). Transgenic plants with improved dehydration-stress tolerance: Progress and future prospects. Biol. Plant, 50: 481-495.

Chinnusamy V, Schumaker K, Zhu JK (2004). Molecular genetic perspectives on cross-talk and specificity in abiotic stress signalling in plants. J. Exp. Bot., 55: 225-236.

Cummings CL, Alexander HM (2002). Population ecology of wild sunflower: effects of seed density and post-dispersal vertebrate seed predation on numbers of plants per patch and seed production. Oecologia, 130: 272-280.

De Greef W (2004). The Cartagena Protocol and the future of agbiotech. Nat. Biotechnol., 22: 811-812.

Ellstrand NC (2001). When transgenes wander, should we worry? Plant Physiol., 125: 1543-1545.

Ellstrand NC, Hoffman CA (1990). Hybridization as an avenue for escape of engineered genes. Bioscience, 40: 438-442.

EPA (1998). Guidelines for Ecological Risk Assessment. U.S. Environmental Protection Agency, Risk Assessment Forum, Washington, DC, EPA/630/R095/002F, 1998. Available from internet: http://cfpub.epa.gov/ncea/cfm/recordisplay.cfm?deid=12460.

Galatchi LD (2006). Environmental Risk Assessment. In Simeonov L, Chirila E (eds). Chemical as Intentional and Accidental Global Environmental Threats. Springer. DOI: 10.1007/978-1-4020-5098-5_1.

Ghassemian M, Nambara E, Cutler S, Kawaide H, Kamiya Y, McCourt P (2000). Regulation of abscisic acid signaling by the ethylene response pathway in *Arabidopsis*. Plant Cell, 12: 1117-1126.

Gorham J (1995). Betaines in higher plants: Biosynthesis and role in stress metabolism. In. Wallsgrove RM (ed). "Amino Acids and Their Derivatives in Higher Plants." Society for Experimental Biology Seminar Series. Cambridge University Press, Cambridge, 56: 171-203.

Haslberger AG (2003). Codex guidelines for GM foods include the analysis of unintended effects. Nat. Biotechnol., 21: 739-741.

Hilbeck A, Baumgartner M, Fried PM, Bigler F (1998). Effects of transgenic *Bacillus thuringiensis* corn-fed prey on mortality and development time of immature *Chrysoperla carnea* (Neuroptera: Chrysopidae). Environ. Entomol., 27: 480-487.

James C (2008). Global Status of Commercialized Biotech/GM Crops: 2008. ISAAA Brief No. 39. ISAAA: Ithaca, NY. Available from internet: http://croplife.intraspin.com/Biotech/global-status-of-commercialized-biotech-gm-crops-2008/.

James C (2009). Global Status of Commercialized Biotech/GM Crops: 2009 The first fourteen years, 1996-2009. SAAA Brief No. 41. ISAAA: Ithaca, NY. Available from internet:http://www.isaaa.org/resources/publications/briefs/41/executi vesummary/default.asp.

Japan Biosafety Clearing-House (2005). Eucalyptus tree containing salt tolerance inducing gene *codA* derived from *Arthrobacter globformis* (*codA*, *Eucalyptus camaldulensis* Dehnh.) (12-5B, 12-5C, and 20-C) 05-26P-0001, -0002, and -0003. 2005-10-12). 2005. Available from internet: http://www.bch.biodic.go.jp/english/lmo_2005.html.

Japan Biosafety Clearing-House (2007). Cartagena protocol on biosafety for the convention on biological diversity, 2007. Available from internet: http://www.bch.biodic.go.jp/english/lmo_2007.html.

Kermode AR (2005). Role of abscisic acid in seed dormancy. J. Plant Growth Reg., 24: 319-344.

Khan MS, Yu X, Kikuchi A, Asahina M, Watanabe KN (2009). Genetic engineering of glycine betaine biosynthesis to enhance abiotic stress tolerance in plants. Plant Biotechnol., 26: 125-134.

Kikuchi A, Kawaoka A, Shimazaki T, Yu X, Ebinuma H, Watanabe KN (2006). Trait stability and environmental biosafety assessments on three transgenic *Eucalyptus* lines (*Eucalyptus camaldulensis* Dehnh. *codA* 12-5B, *codA* 12-5C, *codA* 20-C) conferring salt tolerance. Breed. Res., 8: 17-26 (in Japanese).

Kikuchi A, Yu X, Shimazaki T, Kawaoka A, Ebinuma H, watanabe KN (2009). Allelopathy assessments for the environmental biosafety of the salt-tolerant transgenic *Eucalyptus camaldulensis*, genotypes *codA* 12-5B, *codA* 12-5C, and *codA* 20C. J. Wood Sci., 55: 49-153.

Kinderlerer J (2008). The Cartagena Protocol on Biosafety. Collect. Biol. Rev., 4: 12-65.

Lelmen K, Yu X, Kikuchi A, Watanabe KN (2009). Impact of transgenic Eucalyptus conferring salt tolerance on soil microbial communities. 115th meeting of the Japanese Society of Breeding, 27th-28th March, 2009, Epochal Hall, Tsukuba, Japan.

Losey JE, Rayor LS, Carter ME (1999). Transgenic pollen harms monarch larvae. Nature, 399: 214.

Lu BR (2008). Transgene escape from GM crops and potential biosafety consequences: An environmental perspective. Collect. Biol. Rev., 4: 41-66.

Mallory-Smith C, Zapiola M (2008). Gene flow from glyphosate-resistant crops. Pest. Manage. Sci., 64: 428-440.

Nickson TE (2008). Planning environmental risk assessment for genetically modified crops: Problem formulation for stress-tolerant crops. Plant Physiol., 147: 494-502.

NRC-Nuclear Regulatory Commission (1983). Risk Assessment in the Federal Government: Managing the Process. National Academy Press, Washington, DC, P 191.

NRC-Nuclear Regulatory Commission (2002). Environmental Effects of Transgenic Plants: The Scope and Adequacy of Regulation. National Academy Press, Washington, DC, p. 342.

Office of the Gene Technology Regulator (OGTR) (2005). Risk Assessment and Risk Management Plan for DIR 053/2004: Field trial of genetically modified salt tolerant wheat on saline land. Available from internet: http://www.health.gov.au/internet/ogtr/publishing.nsf/Content/dir053-2 004.

Office of the Gene Technology Regulator (OGTR) (2006). Risk Assessment and Risk Management Plan for DIR 070/2006: Limited and controlled release of GM sugarcane with altered plant architecture, enhanced water or improved nitrogen use efficiency. Available from internet: http://www.health.gov.au/internet/ogtr/publishing.nsf/Content/dir070-2 006.

Office of the Gene Technology Regulator (OGTR) (2008). Risk Assessment and Risk Management Plan for DIR 081/2007: Limited and controlled release of cotton genetically modified for water use efficiency. Available from internet: http://www.health.gov.au/internet/ogtr/publishing.nsf/Content/dir081-2 007.

Office of the Gene Technology Regulator (OGTR) (2008a). Risk Assessment and Risk Management Plan for DIR 077/2007: Limited and controlled release of wheat and barley genetically modified for enhanced tolerance to abiotic stresses or increased beta glucan.. Available from internet: http://www.health.gov.au/internet/ogtr/publishing.nsf/Content/dir077-2 007.

Office of the Gene Technology Regulator (OGTR) (2008b). Risk Assessment and Risk Management Plan for DIR 080/2007: Limited and controlled release of genetically modified wheat for drought tolerance. Available from internet: http://www.health.gov.au/internet/ogtr/publishing.nsf/Content/dir080-2 007.

Office of the Gene Technology Regulator (OGTR) (2008d). Risk Assessment and Risk Management Plan for DIR 083/2007: Limited and controlled release of cotton genetically modified for enhanced waterlogging tolerance. Available from internet:_

http://www.health.gov.au/internet/ogtr/publishing.nsf/Content/dir083-2 007.

Office of the Gene Technology Regulator (OGTR) (2009). Risk Assessment and Risk Management Plan for DIR 095: Limited and controlled release of sugarcane genetically modified for altered plant growth, enhanced drought tolerance, enhanced nitrogen use efficiency, altered sucrose accumulation, and improved cellulosic ethanol production from sugarcane biomass. Available from internet: http://www.health.gov.au/internet/ogtr/publishing.nsf/Content/dir095.

Orr HA (2009). Fitness and its role in evolutionary genetics. Nat. Rev. Gen., 10: 531-541.

Qaim M, Subramanian A (2010). Benefits of transgenic plants: A socioeconomic perspective. In. Kempken F, Jung C (eds). Genetic Modification of Plants, Biotechnology in Agriculture and Forestry. Springer-Verlag, Berlin Heidelberg., pp. 615-629.

Raybould A (2006). Problem formulation and hypothesis testing for environmental risk assessments of genetically modified crops. Environ. Biol. Res., 5: 119-125.

SBC-Schenkelaars Biotechnology Consultancy (2007). Novel aspects of the environmental risk assessment of drought-tolerant genetically modified maize and omega-3 fatty acid genetically modified soybean (Commissioned by the GMO Office of the National Institute for Public Health and the Environment, the Netherlands, 2007). Available from internet:http://www.sbcbiotech.nl/news/item/Novel_aspects_risk_asse ssment_novel_generation_GM_crops/97?mid=100050.

Shinozaki K, Yamaguchi-Shinozaki K (2007). Gene networks involved in drought stress response and tolerance. J. Exp. Bot., 58: 221-227.

Sumida S (1996). OECD's biosafety work on "large-scale" releases of transgenic plants. Field Crops Res., 45: 187-194.

Taniguchi T, Ohmiya Y, Kurita M, Tsuboyama SM, Kondo T, Park YW, Baba K, Hayashi T (2008). Biosafety assessment of transgenic poplars overexpressing xyloglucanase (AaXEG2) prior to field trials. J. Wood Sci., 54: 408-413.

Taylor JE, Mcainsh MR (2004). Signalling crosstalk in plants: Emerging issues. J. Exp. Bot., 55: 147-149.

Tuteja N (2007). Abscisic acid and abiotic stress signaling. Plant Sig. Behav., 2: 135-138.

Vinocur B, Altman A (2005). Recent advances in engineering plant tolerance to abiotic stress: Achievements and limitations. Curr. Opin. Biotechnol., 16: 123-132.

Warwick SI, Beckie HJ, Hall LM (2009). Gene flow, invasiveness, and ecological impact of genetically modified crops. Ann. New York Acad. Sci., 1168: 72-99.

Watanabe KN, Sassa Y, Suda E, Chen CH, Inaba M, Kikuchi A (2005). Global political, economic, social and technological issues on trasngenic crops. Plant Biotechnol., 22: 515-522.

Weis AE (2005). Assessing the ecological fitness of recipients. In. Poppy GM, Wilkinson MJ (eds). Gene flow from GM plants. Oxford, UK, Blackwell Publishing. doi: 10.1002/9780470988497.ch6.

White JA, Harmon JP, Andow DA (2004). Ecological Context for Examining the Effects of Transgenic Crops in Production Systems. J. Crop. Improv., 12: 457-489.

Wolt JD (2009). Advancing environmental risk assessment for transgenic biofeedstock crops. Biotechnol. Biofuels, 2: 27.

Wolt JD, Keese P, Raybould A, Fitzpatrick JW, Burachik M, Gray A, Olin SS, Schiemann J, Sears M, Wu F (2010). Problem formulation in the environmental risk assessment for genetically modified plants. Transgenic Res., 19: 425-436.

Yoshioka K, Shinozaki K (2009). Signal crosstalk in plant stress responses. Wiley-Blackwell. ISBN: 978-0-8138-1963-1.

Yu X, Kikuchi A, Lelmen E, Ahmad D, Matsunaga E, Shimada T, Watanabe K (2008). Environmental biosafety assessments of transgenic Eucalyptus conferring salt tolerance in Japan. 10th International Symposium on the Biosafety of Genetically Modified Organisms, November 16th-21st, 2008, Wellington, New Zealand.

Yu X, Kikuchi A, Matsunaga E, Morishita Y, Nanto K, Sakurai N, Suzuki H, Shibata D, Shimada T, Watanabe KN (2009). Establishment of the evaluation system of salt tolerance on transgenic woody plants in the special netted-house. Plant Biotechnol., 26: 135-141.

16

Genetic engineering, ecosystem change, and agriculture: an update

Lawrence C. Davis

Department of Biochemistry,Kansas State University,141 Chalmers Hall,Manhattan, KS 66506

Genetically modified organisms (GMOs), alternatively called biotech crops, dominate soybean and cotton production and are rapidly increasing their fraction of market share for maize and rice in the U.S. Engineered canola is important in Canada, soybeans are dominant in Argentina and Brazil, and cotton is prominent in China and India. Adoption is much slower elsewhere, in large part due to concerns for potential ecosystem effects that may occur through development of weedy plants, by selection of herbicide resistant weeds and by effects of insecticidal proteins on nontarget insects. The precautionary principle is invoked by critics concerned that one must know in advance the effects of GMOs before releasing them. Alteration of weed species composition of agricultural fields is well documented to occur under herbicide selection pressure. Gene flow to wild relatives of crop plants can be shown under herbicide selection, and one instance (sunflower) is provided for insect resistance transfer leading to increased seed production by a weedy relative. Detailed stewardship programs have been developed by seed producers to minimize risks of gene flow. Although herbicides and insecticides are known to have major effects on agroecosystems, the ecosystem impacts of GMOs per se, thus far appear to be small.

Key words: gene-flow, herbicide-resistant weeds, genetically engineered crops, Bt maize, Roundup Ready soybeans

1. Introduction

Recent advances in genetic technology and molecular biology have allowed greater molecular level understanding of many biochemical pathways, particularly for several model organisms and agricultural crops including rice and maize. Modification of pathways and products holds great potential for enhancing agriculture. Even prior to the recent sequencing successes, there was an effort to enhance the capabilities of crop plants through introduction or alteration of genes. In this review a number of examples are considered, mostly dealing with herbicide resistance and natural pesticide proteins. There is no discussion of animals or microbes, and wherever possible peer-reviewed literature from the past four years used. Earlier works may be found cited therein. Some information is only available on internet sites, and a significant fraction of articles are open-access. Tested URLs as of July 2006 are given.

Considering herbicide resistance as an example, there are both instances where gene transfer has been effected by recombinant DNA techniques (Laurent et al., 2000; Dill, 2005; Duke, 2005), and instances where selective breeding has been used to produce comparable results (Sebastian et al., 1989; Tan et al., 2005). There are thus far few indications that recombinant DNA per se will lead to outcomes that are qualitatively different from those that are available with conventional advanced breeding strategies.

When examining genetic engineering and ecosystem change in relation to agriculture, there are both events for which probabilities can be defined and socio-political considerations related to perceptions of risk. A probability of 1 in a million is small, but during a billion events, such as seed pollination, there will be about 1000 occurrences. Some individuals, organizations and governments have expressed concern that through recombinant DNA techniques it may be possible to produce genetically engineered plants that may have a qualitatively different impact on ecosystems and people than conventionally bred and selected plants. Some of the perceived risks are discussed by Madsen and Sandoe (2005) and Devine (2005). Perceptions of risk, and frequencies of actual events are not linearly correlated but the topics are inextricably entwined. This review focuses mainly on what is being done through genetic engineering and advanced plant breeding, for which probabilities of occurrence (such as gene flow frequency) may be established, but one cannot avoid some discussion of perceived risks. Major perceived risks include potentially catastrophic ecosystem alterations such as have occurred with invasive weeds. For more extensive discussion of both perceived risks and quantitatively measured events see for instance the published proceedings of the 8th International Symposium on Biosafety of Genetically Modified Organisms (GMOs) (ISBR, 2004).

Although thousands of transgenic crop plants have been approved for field testing in the U.S., the rate of approval and commercial release of engineered plants has decreased greatly in the past five years. Duke (2005) indicates that there were only seven new approvals for herbicide resistance world-wide from 2000-2003, compared to 37 in the six years prior to that. High costs for research and development are considerations but costs of regulatory approval and trade restrictions may be larger factors (Devine. 2005). Few species of modified crop plants have been submitted for approval in Canada since 2000. They include alfalfa, cotton, lentil, maize, potato, rice, soybean, sunflower, sugar beet, and wheat (Canada, 2006).

In Table 1, I have attempted to summarize some of the perceived risks of genetically modified organisms (GMOs), specifically agricultural crop plants. Recent research that addresses these risks is discussed throughout the body of this review. Where it is known, a quantitative assessment is given.

2. Historical perspective on ecosystem change

Humans have always had effects on ecosystems. The use of fire, and effective hunting weapons, has produced profound changes in the flora and fauna across whole continents. As human populations increased, their impact increased disproportionately. While settlement and the development of agriculture allowed greater populations to survive on smaller areas, it also increased the ecological impact in those settled areas. Domestication of plant species and establishment of large areas of uniform cultivars of only a few species greatly decreased biodiversity in cultivated areas. Even a century ago, there were few large landscape areas that were not affected by the human presence.

Very recently in an evolutionary time scale, systematic plant breeding and highly mechanized agriculture which makes use of fossil fuels, together have resulted in huge areas being converted to production of relatively few species over large areas. During the mid to late 20th century, herbicides and pesticides further narrowed the abundance of species living in cultivated areas. This occurred prior to the advent of genetic engineering. Now it is not uncommon to find areas up to1000 ha with >90% of the planted consisting of one cultivar of one species, such as wheat or maize. Few borders, hedgerows, woodlots or pastures remain in large portions of the U.S. and other highly mechanized agricultural production areas. This naturally supports fewer kinds of microbes, insects and animals than would a more diverse plant population. In the 21st century, genetic engineering may be used in a number of ways to once more alter the modes of production being used, possibly over even larger areas. In this review I will look at a few examples that may help us gain an appreciation of how genetic engineering might affect ecosystems, in comparison to current practices.

Table 1. A Guide to Comments on Perceived Risks of GMOs

GMO Category	Perceived risk	Section discussing the perceived risk
Herbicide resistance genes	weediness of crop	Crops as weeds
Herbicide resistance genes	gene and trait transfer to weedy relatives	Gene flow from desired plants to others; selecting natural herbicide resistance vs genetically engineered resistance
Herbicide resistance genes	overuse of herbicides	Fire, tillage and herbicides as management tools; ecosystem impacts of herbicides
Herbicide resistance genes	unanticipated emergent traits of GMOs	Producing herbicide resistance in crop plants without genetic engineering; risk of novel traits vs risk of genetic engineering; perceived risk vs quantifiable effects
Insect resistance genes	toxicity to nontarget insects	Potential insect population shifts
Insect resistance genes	gene transfer to weedy relatives, upsetting predator control	Gene flow from desired plants to others; demonstrable ecosystem impact of a transgene migration
Insect resistance genes	human sensitivity to introduced protein	not addressed here
Insect resistance genes	unanticipated emergent traits of GMOs; altered plant composition	Risk of novel traits vs risk of genetic engineering; perceived risk vs quantifiable effects
Phytoremediation genes	unanticipated emergent traits of GMOs	Enhanced phytoremediation

3. Fire, tillage, and herbicides as management tools

Fire has long been used as a tool to manage ecosystems for human benefit (O'Neill 2006). Fire suppression rapidly leads to huge changes in a landscape such as the prairies and forests of the U.S. Within prairies, lack of fire results in invasion by woody species and suppression of the grassland species (Konza, 2002). Few herbicides have ever been used to produce such extensive changes in species composition as occurs with repeated fires, with the possible exception of mangrove extermination during the Vietnam war where close to 5 million ha of land were treated up to three times (Buckingham, 1983). In forests, fire alters species composition, yet fire suppression ultimately results in more extreme wildfires, yielding major upsets in seemingly stable ecosystems (O'Neill, 2006). Slash and burn agriculture obviously has some comparable effects. Both natural and controlled burns release above-ground mineral nutrients quickly, which herbicides do not. Fire lowers N content of the remaining material, selecting for plants that demand low N (Konza, 2002). Herbicide application results in nutrients remaining tied up in biomass, until natural decomposition releases them to become available for new plant growth. Changes of N content are less. Tillage is intermediate between fire and herbicide in rate of nutrient release. Compared to the impacts of presence or absence of fire, or application of herbicides, the impact of genetically modified organisms per se is likely to be relatively small, though different, depending on how the modified organisms are applied. To the extent that it allows substitution of herbicides for tillage, it will reduce the rate of nutrient turnover.

4. Some present applications of genetically modified plants

Among the earliest transgenic plants that were introduced to commercial markets there were potatoes resistant to the Colorado potato beetle, herbicide resistant flax, a tomato with delayed ripening, and squash with multiple virus resistance. Only the last of these is still on the market (Byrne et al., 2006). Cost/profit considerations and consumer acceptance issues have focused the market in a few countries and on large acreage crops where there are not large numbers of different cultivars that must be tested one by one. Prominent examples are discussed below.

Lentils, alfalfa, sugar beet and wheat have been submitted for approval in Canada but commercial application is very limited or non-existent thus far.

4.1 Engineered herbicide resistance

Weedy plants are characterized by their abilities to grow on disturbed areas, rapidly, with low inputs, displacing more desirable species. When agriculturally desirable species are not able to out-compete weeds, herbicides and cultivation have typically been used to reduce weed populations. Some weeds, such as nutsedge (*Cyperus esculentus*) (CWMA, 2006) or Tropical Spiderwort (*Commelina benghalensis*) (Ferrell et al. 2005), have growth habits that make them relatively resistant to cultivation as a means of control. Selective herbicides are often more effective and less expensive than mechanical means. For about half a century broad spectrum herbicides such as 2,4-dichlorophenoxyacetic acid (2,4-D) have proven useful to control whole genera of weeds. But with 2,4-D, control of many grassy weeds is not possible, only broad-leafed species and sedges are reasonably susceptible. Typically, the weed population shifts in response to herbicide treatments, giving rise to a different set of troublesome weeds, when some are removed (Itoh, 1994; Owen and Zelaya, 2005; Purecelli and Tuesca, 2005).

An alternative strategy for weed control is to protect the desired crop plant against a herbicide, either through selection of resistance, or engineering of resistance genes, and then to apply a broad-spectrum herbicide. In principle this can produce a clean crop with no weeds. The best example of this strategy is use of "Roundup Ready" soybeans and other crops within the U.S. Use of resistance to the broad spectrum herbicide glyphosate (trade name Roundup) presupposed acceptance of genetically engineered plants, because producing resistance to that herbicide required introduction of a bacterial gene for aromatic amino acid synthesis, resistant to the herbicide. The advantage to farmers was so obvious from their standpoint that this herbicide strategy was rapidly adopted to the large majority of U.S. soybean production within a few years.

Five years after introduction, 70% of all U.S. soybean production was glyphosate resistant. In Argentina adoption was more rapid and extensive, reaching 98% (Dupont, 2006). In Brazil, the number two exporter of soybeans, GM crops were illegal prior to 2004 (but were grown). It is estimated that in Brazil over 9 million ha were planted with glyphosate resistant soybeans in 2005 (James, 2006). For the U.S. the area exceeded 30 million ha (Monsanto, 2006). Soybeans are the majority of all biotech crops planted world-wide (>50 million ha), with maize having about 40 % as much area (James, 2006). For maize, over half of all planted area in the U.S. contains introduced genes, about 1/4 contains herbicide resistance (USDA, 2006). Insect resistance is a more important in maize (see below). Triazine herbicides are generally effective in maize plantings as it is inherently tolerant to that herbicide class, so no engineering is required.

Recently a patent application shows that glyphosate resistance can be obtained by direct selection of soybeans using glyphosate as a seed selection (Davis, 2005a). Now we can no longer distinguish genetically engineered vs no-engineered glyphosate-resistant plants at the phenotypic level, although there will be an obvious difference at the DNA level, because the engineered plants contains DNA from other, bacterial species.

4.2 Engineered herbicide detoxification

An alternative strategy of detoxifying glyphosate by acetylation is now being announced, in combination with resistance to another herbicide, targeting acetolactate

synthase (ALS) under the name of Optimum GAT (Dupont, 2006). The acetyl transferase uses another bacterial enzyme which has been engineered in the laboratory to increase its effectiveness (Castle et al., 2004).

Transgenic plants with introduced cytochrome P450 genes have been shown to degrade various pesticides more efficiently than their unmodified counterparts (Inui and Ohkawa, 2005). This may be useful both for phytoremediation of sites with contamination and for enhancing resistance of plants to herbicides or insecticides. Some of the P450 enzymes (mixed function oxidases) have broad substrate range specificity; others are more restricted in their activity. "Stacking" resistance by introduction of multiple P450 genes has been reported by Kawahigashi et al. (2005), while one particular P450 (CYP2B6) alone gives resistance to 13 of 17 herbicides tested from several classes (Hirose et al., 2005). In these two instances, rice is the plant being modified and the potential for transfer of genes to weedy wild rice is a consideration for concern. Because the CYP2B6 is a human enzyme, selection of a comparable form directly from those naturally present in plants is not likely, although increased metabolism of herbicides could perhaps be selected directly.

Introduction of a somewhat more selective oxidase was used to produce cotton resistant to 2,4-D as described by Laurent et al. (2000). In this instance a dioxygenase from a bacterial species was able to oxidize 2,4-D to dichlorophenol, which in turn was rapidly conjugated by endogenous enzymes within the plant to yield more polar, less toxic products including a glucoside, malonyl glucoside and sulfatoglucoside. In this instance, transfer of the gene to weedy relatives of cotton is unlikely, because they are not indigenous to many cotton growing areas. Of course comparable engineering of oilseed rape would be more problematic, because weedy relatives are usually common in areas where it is grown (Lutman et al., 2005).

4.3 Introduction of natural pesticide

In the area of insect control, initially very broad spectrum, persistent insecticides such as DDT or arsenate of lead were often used. This resulted in a depauperate insect fauna in treated areas because they killed many nontarget and predator species. Introduction of specific insecticidal proteins into plants, rather than applying general insecticides, may well yield a richer fauna, even within a crop monoculture. The most prominent example of this approach is introduction of *Bacillus thuringensis* Cry protein genes into maize to prevent the infestation by European corn borer and corn earworm that is common in unprotected fields. (Product name commonly Yieldgard in U.S.). Specific stewardship agreements were developed in 2001, to assure that selective pressure does not elicit resistant forms of the targeted insects (Sloderbeck,

2006). Other lepidopterans including rootworms are also inhibited by related Cry proteins, and products have been introduced (e.g. Herculex XTRA by DuPont).

Similar benefits have been observed when Cry genes are introduced into cotton to prevent bollworm feeding. Monsanto Company markets Bollgard varieties of cotton, containing Cry genes, which are planted on a large fraction of all the cotton area in the U.S. In most of that area, the Roundup Ready trait is also present. India and China are rapidly adopting the same technology (James, 2006). Monsanto reported an estimated 8 million ha of Bollgard cotton in India and 2 million ha in China for 2006. China has other locally developed sources of Bt cotton (Huang et al., 2002), so two-thirds of all cotton grown there now is Bt cotton (China Daily, 2006). The advent of Bt cotton has resulted in increased concern for formerly minor insect pests in cotton in China, presumably because of decreased insecticide use (Jia, 2006).

4.4 Enhanced phytoremediation

Thus far a minor area of genetic engineering is the modification of plants to enhance phytoremediation. Plants have inherent capacity to deal with a wide range of chemical classes because they are exposed to numerous chemicals in their environment, whether from other plants or predators. Typically, for degradation, xenobiotics are hydroxylated by cytochrome P450 type oxidases and then conjugated to glutathione or sugars or organic acids to increase their polarity. Then they are transferred to the vacuole or out of the plant cell into the region of the cell wall. Many compounds end up incorporated into the insoluble wall material, often by lignification enzymes (Davis et al., 2002). If plants are deliberately used to speed up the degradation or sequestration of undesirable chemicals by these routes it is termed phytoremediation (Burken, 2003).

The actual mechanism of action for any particular instance of phytoremediation or herbicide detoxification has to be determined experimentally and only a few pathways are known in detail (Harms et al., 2003; Schwitzguebel and Vanek, 2003). Until the likely path is known, it is difficult to enhance it by engineering. In many cases of phytoremediation at the field scale, plants simply supply substrate for microbes that are the actual effectors of remediation (Davis et al., 2002).

The uptake and metabolism of heavy metals by plants has been extensively studied. Some species accumulate metals or metalloids to concentrations above those found in the soil and sequester the metal so that it does not produce toxicity to the plant. Specific examples are accumulation of cadmium with phytochelatins & metallothioneins, volatilization of metallic mercury or selenium, and accumulation of arsenate. Overexpression of the phytochelatins & metallothioneins may allow a plant to accumulate higher concentrations of the heavy metals (Cobbett and Goldsbrough, 2002; Eapen and D'Souza,

2005; Gratao et al., 2005). Expression of bacterial reductases, the Mer genes, has been shown to promote mercury volatilization, and by engineering the Mer genes into the chloroplast, uniparental inheritance is assured (Ruiz et al 2003), because chloroplast genes are generally not transmitted by pollen. Selenium volatilization has been enhanced too (Berken et al., 2002) though many plants are fairly effective in this process without any engineering.

Because in each instance the engineered plants used for phytoremediation are to be applied to very specific relatively small areas, they are unlikely to have anywhere near the environmental impact that would be had with a plant species engineered to be generally resistant to one or more herbicides. Nor does it appear that transfer of the genes in question to native populations is likely to happen, unless the engineered species is indigenous to the location being treated. Then the chloroplast transformation strategy of Ruiz et al. (2003) may be applied to prevent movement of engineered genes by pollen flow. A third concern would be if the introduced genes produced a significant advantage to the plants carrying them so that they might become more invasive. With the species used thus far, that is not likely to be an issue. Usually the engineered genes only provide an advantage to the plant under the conditions of contamination, if at all.

5. Ecosystem impacts of herbicides

Use of broad spectrum herbicides must of necessity have impact on more species than use of individual selective herbicides. However, when the objective is a clean crop, multiple applications of multiple fairly selective herbicides, necessary to control different classes of weeds, could lead to equivalent or more damage than use of a single broad-spectrum herbicide, depending on the residual action and the drift of the applied herbicides (Karthikeyan et al., 2004a,b). There is not a large amount of literature on herbicide effects on nontarget species, but there are some ecological studies, cited by Karthikeyan et al. (2004a,b) indicating that spray drift significantly alters species composition in areas not part of the intentionally managed area.

Large changes of the weed population occur within a herbicide treated crop area, as reviewed by Owen and Zelaya (2005). Use of herbicide may actually increase weed species diversity within treated areas, at least by some measures. Purecelli and Tuesca (2005) found that in Argentina, application of glyphosate did decrease the numbers and diversity of early-emerging weeds, but promoted the appearance of late-season broad-leaved weeds. In one instance the weed was known to be present at low levels prior to initiation of the study and became abundant, while another weed was not identified at the study location until after application of the herbicide. It has inherent glyphosate resistance, and the authors predict that with continued use of glyphosate in crop rotations the abundance of such weeds will increase. Similar trends are forecast for other herbicides in other areas (Owen and Zelaya, 2005).

There are suggestions that use of herbicide resistant crops will hasten changes already occurring under mechanized large-scale agriculture. For instance, modeling of production systems suggests that glyphosate resistant sugar beets may allow elimination of weeds that are essential food sources for birds in the U.K. (Watkinson et al., 2000). Although such sugar beets have been available since the late 1990s, no U.S. processor had accepted any transgenic beets through 2001 (Gianassi et al., 2002) and none are presently grown commercially in Europe, although they have been widely studied at the field scale (Teagasc, 2006).

As an example of how herbicide application may radically alter a weedy ecosystem consider the example of Tropical Spiderwort (*Commelina benghalensis* L.) in Florida (Ferrell et al., 2006). This weed was uncommon until about a decade ago and now is a major concern. It has an inherent high tolerance to glyphosate so that it is not well controlled in "Roundup Ready" crops, or in minimum tillage systems that depend on the herbicide to clear fields before planting. Because the plant rapidly reproduces from fragments, ordinary cultivation is relatively ineffective against it also. This makes it a serious difficult to manage weed, as an indirect effect of genetic engineering of crop species. It is not the GMO per se that alters the ecosystem, but rather its interaction with the herbicide management strategy that does so.

Reddy (2004) has examined the relative shift in weed populations under different herbicide regimes with cotton resistant to either bromoxynil (BR) or glyphosate (GR). Continuous BR led to increase of three main weeds ~15 to 350 fold over 3 seasons, compared to continuous GR. The most dramatic increase was with nutsedge (*Cyperus esculentus*) which had 373 plants per m2 in BR vs 1 in GR. This weed is difficult to control by cultivation also as noted above.

6. Crops as weeds from carryover of herbicide resistant crops

A non-trivial problem in some cropping sequences is the appearance of volunteer seedlings of the crop plant which may allow survival and multiplication of plant pests, including insects and viruses, even if the intended following crop in a rotation is not a host for the insect or virus. Herbicide resistance interacts with this in a subtle way. If we have "Roundup Ready" brassica or wheat, we can no longer control spontaneous seedlings with that herbicide which has often been used in minimum tillage systems to clear the field before planting. In particular with *B. rapa* or *B. napus*, herbicide tolerance of spontaneous seedlings can have significant impact on following crops. Lutman et al (2005) found that an

average of 3575 seeds per m2 was lost at harvest of *B. napus,* due to shattering of the seed pods. While 60% of these spilled seeds lost viability within a few months, the decline was slower over following years with 5 % persisting to 5 yr at several sites in the U.K. The volunteer seedlings that arise from these may be strongly competitive weeds for another crop grown in rotation, such as sugar beets. Their inherent herbicide resistance may increase the difficulty of control, unless the following crop is also herbicide resistant (for a different herbicide) as could be the case with sugar beets if the transgenic versions were accepted. Both plants genetically engineered for glyphosate or glufosinate resistance, and those conventionally selected for tolerance to imidazolinones had the same behavior in the study of Lutman et al. (2005). So it is not the genetic engineering per se that makes these "weeds" troublesome, but rather the dependence on a single herbicide.

Owen and Zelaya (2005) consider the problem of glyphosate resistant maize and soybeans in rotations of those crops. While maize may sometimes be a problem in a following crop of soybeans, soybeans have poor winter survival and are not competitive with a following crop of maize. Deployment of glyphosate resistant spring wheat has been delayed for economic and management reasons, including potential weediness which would have to be controlled by a comprehensive stewardship program (Dill, 2005).

7. Selecting natural herbicide resistance vs genetically engineered resistance

Many people express concern that genetically engineered crops may transfer genes to wild relatives resulting in either increased competition from the undesirable wild relatives, as for instance with shattering sorghum or rice, or that they may give new combinations of genes that yield unexpected results such as extreme competitiveness of a weed (e.g.Garcia and Altieri, 2005). In most instances it can be shown that for herbicide resistance at least, the resistant form is unlikely to out-compete the susceptible in the absence of strong selective pressure. With the exception of the deliberately introduced genes for resistance to, or degradation of, glyphosate and glufosinate, most forms of herbicide resistance arise spontaneously so far as we can tell. They only emerge with detectable frequency in a population under strong selective pressure. For instance in South-East Asia, there were four biotypes of various species resistant to paraquat and three resistant to 2,4-D after 20 years of intensive use of these herbicides (Itoh, 1994). It was also noted that a change to direct seeding of rice has resulted in a whole suite of different problem weeds, with a shift to grassy weeds, which have similar herbicide resistance as the crop plant.

As of June 2006, there are known to be three hundred biotypes of nearly 200 species with herbicide resistance that has resulted specifically from selective pressure of herbicide application (Heap, 2006). Just two dozen biotypes have an identified resistance to synthetic auxins, although the auxins have been in use for about 50 years. So far, 95 biotypes, representing 70 species, have developed resistance to acetolactate synthase inhibitors. There are five classes of such chemical inhibitors targeting one enzyme (Dupont, 2006). They are known as sulfonyl ureas, imidazolinones, triazolopyrimidines, pyrimidyl thiobenzoates and sulfonylamino- carbonyl-triazolinones. Some resistance includes multiple chemical classes, some does not. The high number of resistant types may relate to the popularity of these herbicides, or the single amino acid change required to produce resistance without detriment to the plant (Tranel and Wright. 2002). There are 65 biotypes of weeds resistant to triazine type herbicides that work at photosystem II. In this instance the altered plant is at a significant disadvantage in the absence of selective pressure from herbicide (Jordan, 1999). Twenty-three biotypes are resistant to photosystem I inhibitors like paraquat and 21 resist inhibition at photosystem II by chlortoluron or its relatives. The geographic distribution of resistant biotypes is related to the intensity of herbicide use with the U.S. having the greatest number of reports (112), followed by Australia with 47, Canada with 44, France with 30, Spain with 27, the U.K with 24 and Israel with 20.

Mutant forms of herbicide-binding proteins may arise (or be identified) repeatedly within one species, or at an equivalent site of the receptor or enzyme protein within different species. In some cases different amino acids at the herbicide binding site are altered in different biotypes. This is particularly clear and common for the ALS inhibitors (Tranel and Wright, 2002). Because the selected populations are not mutagenized, one must assume that the selected resistant biotypes are pre-existent within the population. They must be present at very low frequency or resistance would be observed more quickly than is typical for successfully introduced herbicides. Tranel and Wright (2002) discussed likely causes for the relatively high incidence of resistant ALS which include its dominant character, variety of active site modifications possible, and low fitness penalty to plants carrying mutant forms of the enzyme.

Some forms of resistance depend on changes in translocation or metabolism of the herbicide. For instance, the conjugation of atrazine to glutathione is enhanced in foxtail millet (*Setaria italica*) resistant to atrazine (Giminez-Espinosa et al., 1996). Cross-resistance of rigid ryegrass (*Lolium rigidum*) to quite different herbicides (diclofop-methyl and chlorsulfuron) depends on increased metabolism using a mixed function oxidase (Christopher et al., 1991). Wheat possesses sufficient activity of this enzyme that it is naturally resistant to chlorsulfuron at certain (field application rate) doses, although it is still sensitive to other inhibitors of acetolactate synthase beside chlorsulfuron. The newly

reported resistance of downy brome (*Bromus tectorum*) to chlortoluron, a photosystem electron transport inhibitor (Menendez et al., 2006) also depends on induction of oxidase(s). This kind of resistance is similar to that induced by use of safeners.

The nature of glyphosate resistance in two weed species has been characterized. For *Conyza canadensis* (horseweed), decreased translocation of 14C labeled glyphosate has been shown to be associated with the resistant phenotype (Koger and Reddy, 2005). Uptake is not altered in the source leaf. For *Lolium rigidum* (rigid rye) translocation is also affected (Lorraine-Colwill et al., 2003). The mutant rye was identified in a field after 15 years of repeated glyphosate use (Powles et al., 1998). In neither species is the detailed mechanism of resistance understood. There is no evidence for selective advantage of the weed in absence of the herbicide.

8. Producing herbicide resistance in crop plants without genetic engineering

In some countries, the major objections to GMOs are based on the construct rather than the consequence. Hence, advanced breeding strategies not using recombinant DNA have been applied. Sebastian et al. (1989) described successful selection of a soybean line resistant to sulfonyl urea herbicides, whose site of action is the acetolactate synthase (ALS) enzyme. They used chemical mutagenesis of 400,000 soybean seeds and obtained one line after selection with chlorsulfuron. This laborious strategy was used because at the time there was no reliable way to regenerate soybean plants from tissue culture, so neither engineering and transformation, nor selection in tissue culture were viable options to obtain resistance.

Recent patent applications claims that wheat resistant to glyphosate can be obtained by direct selection from hard red winter wheat cultivars (Davis, 2005b). Genes allowing the resistance are identified by name. In this instance a strong selective pressure was applied to identify uncommon genes already present in the population. If these genes are deployed, glyphosate can no longer be effectively used to control volunteer wheat seedlings (Lyon et al., 2002).

One large chemical company, BASF, already markets non-engineered (non-GM) crop plants with high resistance to a particular herbicide family that acts on the enzyme ALS. The Clearfield production system for wheat makes use of wheat that was selected for resistance to field application levels of Beyond herbicide (active ingredient imazamox, a member of the imidazolinone family). The resistance arises as a natural, selected mutation of the enzyme. Note that wheat is inherently resistant to another ALS inhibitor, chlorsulfuron, through oxidation and glycosylation (Christopher et al., 1991). The broad-spectrum herbicide Beyond is specifically intended to allow control of jointed goatgrass, (*Aegilops cylindrica*) a close relative of wheat, as well as other grasses and broad-leaf weeds. The entire stewardship program of the Clearfield system includes a number of restrictions on the grower, including purchase of certified seed each year, management with appropriate rotations to avoid selection of resistant weeds, and judicious use of other herbicides (Clearfield wheat , 2004).

The same general strategy, with slightly different herbicides and doses can be had for maize, canola and rice according to company literature, available on-line (Clearfield, 2006). Application of this production strategy promises great advantages in some specific regions. For instance, the Clearfield rice cultivars are grown on 30-40% of the entire acreage of rice in Arkansas and Louisiana after only a few years of availability. A weedy rice relative called red rice is a serious problem, because it freely interbreeds with commercial cultivars, and competes for space, greatly lowering yields and market quality of the desired cultivars wherever it is present. Thus far the red rice, which shatters, dropping seed prior to harvest, is susceptible to the herbicide, named Newpath, an imidazolinone. Under the stewardship agreement enforced by BASF, only certified seed free of red rice is to be grown with the Clearfield gene. Carelessness in using the herbicide, or reutilization of seed contaminated with red rice that has picked up the resistance, can lead to a rapid breakdown of the control strategy (Bennett, 2006). Outcrossing has happened within the first two years of growing the rice large-scale (Boyd, 2005). The weedy relative with resistance is highly competitive and produces a large yield of shattering seed. (See further discussion of gene flow below.)

Since 2004 a similar Clearfield production system has been available for sunflower (NSA 2005). A wild sunflower with resistance to herbicides of this class was found in Kansas in 1996. From this, a USDA breeder in Fargo, ND was able to breed by backcrossing and selection to produce oil-seed sunflowers with the desirable traits of the cultivated form. Use of Beyond herbicide in fields of resistant sunflower permits effective control of problem weeds like cocklebur and of course wild sunflower volunteers. However, as with rice, out-crossing to the weedy relative is a serious concern. Failure to use the herbicide at appropriate dose in a field with infestation by the wild relative will lead to development of a resistant population of the weedy form through natural gene flow by pollination. It may be that these advanced strategies as exemplified by rice and sunflower ALS inhibitor programs are only usable under highly mechanized, advanced agriculture with viable crop rotations and alternatives.

The Clearfield technology has captured a significant fraction of the Canadian market for canola (low erucic *Brassica*) with about a million hectares, and up to several million hectares of maize in the U.S. Equivalent mutations have been identified in other crops including sugarbeet, cotton, lettuce, tomato and tobacco indicating a potential for application to a further range of crops (Tan et a., l 2005). Whether their development is economically

justified in the view of herbicide manufacturers remains to be seen, because even non-engineered crops require major investment in regulatory compliance testing for use in some jurisdictions, such as Canada (Devine, 2005).

Very recently CIMMYT has developed imidazolinone-resistant (IR) maize for use in Africa (CIMMYT, 2006). Seed is now under offer for testing at research centers for control of *Striga,* a parasitic plant that is not, thus far, resistant to this class of herbicides. The same company that deployed Clearfield crops in the U.S. and Europe is offering to develop treated seed processes for Kenya, and in future presumably in other countries. As with the crops previously deployed, there are specific stewardship agreements to be signed so that the technology does not experience a quick breakdown through development of *Striga* resistance to the herbicide. The African Agricultural Technology Foundation, CIMMYT and BASF provide details for the StrigAwayR technology (AATF, 2006). Coating seed with imidazolinone herbicide allows normal growth of the maize crop while inhibiting any parasite that attaches to the treated seedling.

9. Risk analysis for modified crops

9.1. Gene flow from desired plant to others

One issue repeatedly raised as a concern is gene flow from engineered crops (Ellstrand and Hoffman, 1990; Eastham and Sweet, 2002). Engineering or selecting herbicide resistance is unlikely to make a cultivar into a troublesome weed, although volunteer wheat and brassica seedlings have raised some concerns. Much more likely is introgression of the trait into weedy relatives as mentioned above for red rice and sunflower. In those instances, where the weedy relative could be troublesome for following crops, a detailed stewardship program may be needed. For the Clearfield technologies, such a stewardship program has been designed by the company marketing the modified crop plant and all producers are supposed to sign and abide by the requirements of the agreement. Similarly, there are stewardship agreements for use of Bt maize (Sloderbeck 2006). Insect resistance may give weedy relatives an advantage (see discussion of Snow et al. 2003 below).

So far there is little evidence for differential gene flow. Rates of trait transfer are presumed to be independent of the nature of the gene being transferred (e.g. for virus, herbicide or insect resistance), but too few studies have been done to assure this. As discussed below (Halfhill et al., 2004), different constructs of the same resistance gene, with different chromosomal locations, do appear to migrate at different rates in the case of insect resistance (Bt toxin gene) during hybridization of different brassica species. Gene flow is one of the many topics considered extensively at meetings of the International Society for Biosafety Research. For the most recent published proceedings see ISBR (2004), available at their web site.

9.1.1. Virus resistance

One species containing virus resistance traits is in extensive production. That is the papaya resistant to ringspot virus, growing in Hawaii. There the need was extremely strong because the virus was rapidly devastating the standard cultivars being grown in intensive agricultural settings (Perry, 2005). Trees are traditionally grown from saved seed using repeated selection for desirable cultivars and landraces. Trees take only a few years to reach peak production. Little information was available on gene flow for that species, until a recent study undertaken by the organization GMO Free Hawaii, to detect contamination of traditional cultivars by the transgenes from pollen of the transgenic cultivars, which include a GUS gene amenable to rapid screening tests. (Bondera and Query, 2006). Results were different for various islands in the chain but indicated high levels of contamination. Later PCR tests for the 35-S promoter (driving expression of the virus coat protein transgene) ranged up to 50% of seeds in some bulked samples of organic and home garden fruits. While there is no evidence of human health hazard associated with the GMO, loss of markets such as Japan which did not accept GMO fruits, and loss of purity in traditional land races of papaya, have caused considerable dissatisfaction amongst parts of the local population.

As noted for herbicide resistance in brassicas below, contamination of certified seed stocks is an issue. It appears that between one and 10 seeds per 10,000 of the traditional non-GMO cultivars being distributed by the University of Hawaii are contaminated (transgenic). The widespread small-scale cultivation of papaya by individuals and the wide-spread, perhaps long-range, gene flow made evident by the studies of Bondera and Query (2006) indicate that this will be a very difficult system to control. Mechanisms and frequencies of gene flow could not be determined from the sampling design used. The selective advantage of resistant trees under virus infection pressure will encourage their spread, both in cultivation and as weedy feral trees. Views on the benefits and costs of transgenic papaya are highly divergent, with the Hawaii Papaya Industry Association very positive (Perry, 2005) and the GMO Free Hawaii group quite negative (Bondera and Query, 2006).

Another species complex with virus resistance genes introduced is the summer squash (*Cucurbita pepo*). As with the papaya, a viral coat protein is use to introduce viral resistance in the plants. Multiple genes, specific to several viruses were simultaneously introduced. Fuchs et al (2004a) monitored movement of the protein genes from commercial squash to a wild relative, *Curcurbita pepo* ssp *ovifera* var texana (*C. texana*). In field settings, gene transfer occurred only when the wild relative was not under severe virus selection. Once transferred, the genes were expressed, yielding resistance to the three viruses for which coat proteins had been introduced into the transgenic form. Under low disease pressure the wild *C*

texana out-performed all of the various hybrids and backcrosses (F1, BC1, BC2). Under high pressure, the transgenic backcrosses to *C texana* outperformed both parents, and only one back cross was needed to recover most traits of *C. texana*. So it appears that timing of the introgression event relative to viral disease pressure may be important to whether it spreads in the wild population or not.

The insect-pollinated cucurbits provide a complex pattern of natural gene flow with low frequencies at distances >1 km, but not exceeding 5% even when plants are close together (Kirkpatrick and Wilson 1988). Pollen of the pistillate parent received preference (Wilson and Payne 1994). Other species: such as rice and maize discussed below show different spatial effects on gene flow because they are wind pollinated. Implications of gene flow within the *C. pepo* complex between cultivated and feral types was discussed in some detail by Wilson (1993) during the time of approval and review for release of the transgenic cultivars. Issues raised at that time included extensive documentation that *C. texana* is a common noxious weed in many specific areas (soil types) over areas including those in which the transgenic squash would be grown. This raises a significant concern of increased weediness under viral disease pressure, where plants with the introgressed genes might well have a reproductive advantage (Fuchs et al., 2004b). The market acceptance (by producers) of transgenic squash is relatively low (<15 %) because it does not prevent all viral diseases and the seed is 78 % more costly (Sankula et al., 2005). Commercial production only partially overlaps the range of the feral gourd type. There is thus far no report of increased weediness of the wild *C. texana* due to introgression of viral resistance.

9.1.2. Herbicide resistance

In Canada, large quantities of oilseed brassicas are grown with herbicide resistance genes present. Beckie et al. (2003) examined gene flow between commercial fields of glyphosate and glufosinate resistant cultivars at distances up to 800 m. Eleven sites were studied in 1999, sampling seeds and testing for resistance. In the following year, volunteer seedlings that escaped herbicide control were tested for double resistance at three locations. Rates of gene flow at field edges were above 1%, but only 0.04% at 400 m in the 1999 sampling. However doubly-resistant volunteer plants were found to the maximum distance of 800 m. In two of the three locations sampled in 2000, it was concluded that the glyphosate-resistant seed used the previous year was adventitiously contaminated with glufosinate resistance. This provides an example of why a vigorous stewardship program is essential to maintain the integrity of herbicide resistant crops.

The central U.S. is an area with large production of sunflower for oil, and in addition the source of diversity of the crop species *Helianthus annuus*. An imidazolinone-resistant (IMI) biotype of wild sunflower was first identified in Kansas, and the group making that identification examined the facility of gene flow from improved, domesticated strains to wild relatives (Massinga et al., 2003). Both common sunflower (*H. annuus*) and prairie sunflower (*H. petiolaris*) were highly receptive to pollen from the IMI type in controlled crosses. In field studies, 11-22 % of seedling progeny were IMI resistant when wild sunflower was grown 2.5 m from a dense patch of IMI-resistant domestic sunflower while at 30 m 0.5 - 3% were resistant during one season. Somewhat lower levels of resistance gene transfer were seen in a second year of study. The maximum distance over which gene flow is likely was not determined in these studies, nor was the distance needed to reduce transfer below 0.1% which is significant for acceptance of a transgenic crop in the E.U., where contamination of food or feed grains is a major concern.

The likelihood of gene flow from modified crop plants to unmodified cultivars and weedy relatives has been examined in some considerable detail for both maize and oilseed rape (*Brassica* spp) by the U.K Department for Environment, Food and Rural Affairs (DEFRA). Extensive reports are available (Henry et al., 2003; Ramsay et al., 2003). For forage maize, some gene flow was detected at distances greater than 200 m, although the level of gene flow dropped rapidly within the first 20 m (Henry et al., 2003). With rapeseed there was some gene transfer detected at distances up to 26 km, although relatively short distances (tens of meters) were required to lower the level of contamination of surrounding crops to below 0.1% (Ramsay et al., 2003).The long distance transfer was attributed to a particular insect, the pollen beetle which travels over long distances compared to bees which forage over a few km. One important finding of this study was that for this crop, insect pollination is predominant over wind pollination.

DuPont Company, through its subsidiary Pioneer Seed, maintains a highly informative website which provides extensive reviews of several issues concerning genetically engineered plants, including concerns for gene flow. As discussed on that site, gene flow is a much smaller issue for a plant such as soybean which is almost exclusively self-pollinated prior to opening of the flower (DuPont, 2006). Presumably the same would be true of herbicide resistant lentils developed by BASF.

For rice, which is a major food crop for about half of the world, detailed studies of gene flow have been done (Zhang et al., 2003; Chen et al., 2004; Lu and Snow, 2005). Zhang et al. (2003) studied a glufosinate resistant (bar gene) cultivar of rice grown in Louisiana, and monitored several traits in spontaneous crosses with red (weedy) rice and a purple leafed rice grown in all combinations as 1:1 mixtures in large plots. All three types bloomed at the same time in the initial year. An out-crossing rate of 0.3 % was observed with the bar gene but the hybrid progeny showed decreased fitness, and

most bloomed too late to produce seed in a field. The purple marker trait was transferred at a frequency of <1%. Chen et al. (2004) studied rice in Korea and China. With weedy rice (called red rice in the U.S.) gene flow, as measured by appearance of marker genes, was relatively low at less than 1/1000 seeds, when the cultivated and wild rice were growing close together. For different types of weedy rice and different commercial cultivars the degree of introgression varied as a function of the timing of anthesis and height of plants. Wind pollination of plants with short pollen viability is expected to show such a pattern. With wild perennial rice (*Oryza rufipogon*), a higher frequency of >1/100 was observed. Chen et al. (2004) cite several earlier studies indicating that gene flow to weedy rice may also approach 1% or higher in some cropping systems. The observations on red rice in the U.S. discussed above under Clearfield technologies are consistent with this. If red rice is not fully controlled by herbicide application in the first year, it is likely that in the next year there will be resistant plants in herbicide-treated fields. If a plant produces 1000 seeds of which 10 are herbicide resistant, a major problem could rapidly ensue, unless an alternative herbicide is used. However, few of the hybrids may survive and establish if the observations on flowering time reported by Zhang et al. (2003) are of general application. They observed that hybrid plants bloomed too late to set seed.

Lu and Snow (2005) have provided a table showing that already by 2004 there were several dozen transgenic rice cultivars being tested for a wide range of applications beyond herbicide resistance. However, the gene flow properties of these transgenes are likely to be essentially the same as those for herbicide resistance. In some cases, only a strong selection would give an advantage to the form carrying the transgene. Lu and Snow suggest that a systematic examination of ecological risks is urgently needed, because few studies have examined changes of population fitness that might accrue from introduction of new traits into wild or weedy populations.

9.1.3. Insect resistance

Halfhill et al. (2002) tested the effect of Bt toxin gene movement from oilseed rape (*Brassica napus*, 2n=38) to its wild relative *B. rapa* (2n=20). Following repeated backcrosses to the *B. rapa*, progeny from half of the six transformant *B napus* lines that had been tested as donors had lost the Bt trait, along with the phenotypic characters of the donor parent. Ploidy level also declined to near that of the *B rapa* parent. Because *Agrobacterium* transformation was used to introduce the Bt genes into the *B napus*, it was expected that some lines (with different locations of DNA insertion) would give more effective gene transfer to *B rapa* than others. Also the levels of expression of the Bt toxin protein varied between lines. Generally, the surviving progeny expressed the protein at levels comparable to the *B napus* donor. This level was sufficient to deter feeding by corn earworm in controlled feeding tests. The general fitness of these transgenic *B. rapa* was not directly compared to that of its progenitors, so it is not possible to speculate on fitness in the absence of insect predation.

No studies were done under natural conditions by Halfhill et al (2002) to assess the extent to which expression of the Bt toxin protein might protect from insect predation in natural settings. However, field studies were done to show that *B rapa* growing in the midst of the oilseed *B napus* did produce hybrids with the resistance gene. The frequency varied from <1 % to nearly 17 %, depending on the *B napus* line used as donor. Further study of deliberately constructed hybrids and their backcrosses to the weedy *B rapa* showed that the Bt gene is stably expressed at levels comparable to those in the donor *B napus* (Zhu et al., 2004). Thus under insect pressure, the transgenic wild plants might have a significant fitness advantage. The implications for weediness in natural ecosystems are unclear. Because no brassicas containing the Bt toxin have been released commercially, no large scale studies have been done in agricultural settings.

In the U.S., part of the stewardship plan for use of Bt maize crops is to provide refuges for susceptible insects by planting non-Bt crops as a proportion of the entire acreage (Sloderbeck, 2006). In the northern U.S. the requirement is 20%, for cotton-growing regions it is 50% because maize is the alternate host of the cotton bollworm. Chilcutt and Tabashnik (2004) documented gene flow from transgenic maize carrying Bt genes. Pollen-mediated gene flow resulted in kernels containing the Bt gene at distances up to 31 m. At 3 m (three rows), over 15% of the kernels carried the trait, and at 8 m it was near 10%. This raises the concern that insects will be exposed to low levels of Bt within the refuge areas, giving a selective advantage to heterozygous insects which would be exposed to non-lethal doses of the Bt toxin. Refuge strips are permitted to be as narrow as four rows (4 m) (Sloderbeck, 2006) and must be within ½ mile of the Bt crop. The extent of gene flow reported here indicates that maize from refuge areas, or adjacent fields, not intended as refuge, and perhaps with a different owner, may well carry the Bt gene at significant levels, so that it ought not to enter the food chain as unengineered grain. In addition, many of the particular combinations of stacked insect resistance alone or stacked with herbicide resistance, do not qualify for food and feed in the E.U. Companies selling those products provide detailed stewardship plans and producer instructions (e.g. Monsanto, 2006b).

9.2. Potential insect population shifts in a Bt containing crop

When the Cry proteins (BT toxins) are expressed in all parts of the maize or cotton plant, there is some risk that other nontarget insects coming into contact with the protein could be affected. For instance the pollen of maize, which drifts a considerable distance from the plant, could potentially be toxic to lepidopteran larvae feeding on plants in the vicinity of maize. It was suggested that this could be a hazard for monarch butterflies consuming leaves of milkweed near fields of transgenic maize. However, the initial suggestion, though published in a prominent magazine, could not be reproduced, and there is likely to be only a small risk, as shown by a thorough risk analysis (Sears et al., 2001).

The work of Tabashnik and Carriere (2004) documents the important point that presence of Bt toxin protein in crops does not give a selective advantage to resistant insects. They reviewed studies that used natural populations selected in responses to sprayed Bt bacteria on crops. The authors cite two cases in which there was no difference between normal and resistant biotypes, and eight studies in which the resistant strains are disadvantaged when growing on Bt plants compared to unmodified plants. The most reasonable conclusion is that the resistance is not complete and transgenic Bt plants have on average much higher doses of Bt than applied by direct spraying of the bacteria.

Studies in the southeastern parts of the U.S. have indicated no adverse effect of cotton containing Cry genes on natural predators of the bollworm (Moar et al., 2002). Compared to fields sprayed with regular insecticides, predation on bollworm eggs was significantly increased. Populations of nontarget insects were not significantly affected in any of several test areas. More recent papers have confirmed and extended these observations (Sisterson et al., 2004; Hagerty et al., 2005). Sisterson et al (2004) scored all arthropods (except pink bollworm and nymphs of whitefly) on Bt and non-Bt cotton grown alone or as a mixture of 75 % Bt:25 % non-BT. The greatest abundance and diversity was observed in the mixed plot but there was not a significant difference between the Bt and non-Bt plots, although the Bt plots did have a lower abundance and diversity of arthropod families. Over 3300 individuals were found during 3 sampling dates (over 2 years at 2 sites) with a final total of 120 plants for each treatment. Thus about 10 arthropods per plant were found, even though some of the fields had been treated with insecticide for control of some insects other than pink bollworms.

Hagerty et al. (2005) considered the impact of transgenic cotton, with and without insecticide application, on arthropod abundance. Predators that feed on the bollworm were of particular interest. In both Bollgard (containing Cry1Ac) and Bollgard II (containing Cry1Ac + Cry2Ab) plots the populations of predators were as high as or higher than in the non-Bt cotton, when no insecticide was used early in the cropping season. This may have been because the non-Bt plants were severely damaged and could support only a smaller population of prey insects. Disruption of the predator population by broad spectrum insecticide treatment in one season resulted in the bollworm population reaching an economic threshold, even when Bollgard II was used, because the moths producing worms invade from maize, sometimes in great numbers. The authors concluded that their results were consistent with earlier cited studies in showing that the presence of Cry proteins does not reduce predator populations, and less use of insecticides leads to an increase in generalist predators (as well as prey).

Dutton et al (2003) describe a general approach to assessing the risk to nontarget insects from use of Bt transgenic plants. Their work is done in the European context, where there are relatively stringent assessments required for transgenic crops, with potentially a near infinite number of predators, competitors, symbionts and parasitoids to be considered. They considered in detail one entomophagous insect, the green lacewing (*Chrysoperla carnea*), on Bt-maize. Using a tiered approach they first tested the direct toxicity of Bt toxin protein, then the toxicity of prey insects fed with Bt toxin, and then the behavioral preferences of the predator. The preferred prey insects, aphids and spidermites, were not toxic to lacewings when feeding on Bt-maize. Nor was a nutrient solution containing the toxin. Lepidopterans, which are affected by the toxin, were not a preferred prey. Hence the risk to lacewings of Bt-maize is minimal. Field tests have confirmed that observation, as cited by the authors. In the case of the lacewing, testing could have stopped once it was shown that the toxin had no effect when fed directly at levels higher than likely to be encountered in the field. The authors propose that formal analysis can be done for other predators, by identifying those that feed on lepidopterans and hence would be exposed, and then determining their sensitivity to the toxin first in the laboratory, then if needed in plants in controlled environments and finally in field studies.

9.3. Demonstrable ecosystem impact of a transgene migration

Jenczewski et al. (2003) reviewed the literature on crop to wild plant gene flow and found that there were few examples providing unambiguous evidence on the relative fitness of specific genes. Examples were cited in which the hybrids showed increased vigor in the F1 and other instances of decreased fertility in the F2. Few studies have monitored a population beyond that time.

One example where movement of a transgene from cultivated plant to a wild relative has a demonstrable effect is a study by Snow et al. (2003) of wild sunflower carrying a Bt insect resistance gene. In this instance, insect predation was reduced and hence reproductive fitness was increased in a wild population. A similar scenario might be envisioned for a rice transgene enhancing insect resistance, if insect predation were a strong limiting factor in the weedy population. Snow et al. (2003) found that wild sunflower produced 55% more seeds at one site in Nebraska when carrying the *cry1Ac*

gene which reduced lepidopteran damage. Weevil and fly damage was unaltered. At a second site in Colorado the seed increase was 14% (but not significantly). T his appears to be the first experiment to show at a field scale that a transgene confers a clear fitness advantage in a natural (non-agricultural) setting. Whether seed production is a limiting factor in wild sunflower populations is not discussed.

9.4. Risk of novel traits vs risk of genetic engineering

All of the above engineered or custom-tailored crops, such as the imidazolinone resistant maize can be found described in more detail at the Canadian food inspection agency website under "Decision documents-determination of environmental and livestock feed safety" (Canada, 2006). In Canada, it is a government policy to examine novel traits whether they were produced by gene transfer to constitute GMOs or if they arose by selection of induced or spontaneous mutation. The rationale is that appearance of a novel trait is the key consideration, rather than if it was engineered by use of recombinant DNA techniques. Thus one can find somaclonal variants, induced and spontaneous mutants, and engineered modifications (typically via *Agrobacterium*-mediated transformation) all receiving the same review. At that site one finds that lentils (*Lens culinaris*) have been selected for resistance to imidazolinone, although none are indicated as commercially available yet. BASF has also reported to the Canadian authorities a maize line resistant to sethoxydim, a herbicide of a different family specific for grasses (acetyl-CoA carboxylase inhibitors). There are a number of crops with glufosinate resistance (bar gene, Basta or Liberty herbicide), sugar beet resistant to glyphosate, cotton resistant to bromoxynil, and various insect resistant crops. A few viruses are on the list also, as well as a few crops with altered lipid composition. The last of these do not appear to be marketed yet.

10. Perceived risks vs quantifiable effects

10.1. Altered composition

A fundamental difference in viewpoints between different parties to the process of development and deployment of modified crops may be seen in the following example. The organization Greenpeace responded to a notification for placing on the market glufosinate-tolerant rice (Monsanto LibertyLink LLRice62). They invoked the precautionary principle that because there was "no proof of no adverse effects on genome function", there should be no release of the rice. A quantitatively more substantial comment pertained to the risk of red rice acquiring herbicide resistance. Zhang et al. (2003) put this risk at about 0.3% per year. Also Greenpeace indicated that because the composition (protein, starch, lignin etc) of the rice was not reported, it could have an altered composition.

An altered composition of the plant not obviously related to the transgene in question had been posited for the impact of Bt toxin transgenes in maize. Jung and Schaeffer (2004) undertook an extensive study of lignification and digestibility of maize stover because there had been suggestions that the hybrids contain Cry1Ab might contain more lignin. Using four locations and 12 commercial hybrids (paired, half with and half without the Bt toxin) they examined yield, digestibility and lignin content (by three different assays). No consistent differences between Bt/ non-Bt pairs were observed in any of the measures, and for two pairs there was no difference in lignin at any location. There were environment, and hybrid x environment interactions as expected but not related to the presence or absence of the Bt trait. Differences in composition are anticipated for hybrids derived by conventional breeding and are often sought out deliberately.

10.2. Altered survival in natural conditions

Two early studies in the U.K. by Crawley et al.(1993, 2001) showed that transgenic crops had no increase in invasive potential compared to their unmodified counterparts, and none survived in natural conditions over several years. Included were sugar beet with Roundup Ready character, potato with the Bt insect resistance gene, maize and rapeseed with glufosinate tolerance (bar gene). This latter trait is for resistance to the herbicides Basta or Liberty. A dozen habitats in four regions of the U.K were studied up to 10 years, until extinction of the introduced plants. There are no reports that there is increased survival of transgenic plants in the absence of selective pressure for the trait in question.

For a virus-resistant transgenic cucurbit, summer squash, studies were done by the developer to show that it is not more likely to overwinter than non-transgenic cultivars of the same type (NRC, 2002). It is not considered a weed and so would not show increased weediness due to virus resistance.

10.3. Potential emergent traits

The European Union initially accepted a number of GM crops, including herbicide resistant tobacco, oilseed rape, soybean, chicory, carnations, and maize. However, a reaction set in and by 1999 there was a moratorium on approval of new crops. In 2002 a new directive on deliberate release of GM crops was in effect (Madsen and Sandoe, 2005). At this point there are specific regulations on labeling and traceability of GM in food and feed products. Although considerable quantities are imported, only a relatively small area of GM crops was grown by 2005 (~100,000 ha of maize in aggregate for five

countries) (James, 2006). Public perceptions and lack of trust in government have led to this situation (Madsen and Sandoe, 2005). With more than a dozen GM crops approved for growth, few are actually grown. It is suggested that both gene technology and herbicide use prompt a "dread" response, amongst the public in Europe. Concern for human health effects, such as allergies, and fears of invasiveness seem to be the major factors. There is thus little economic incentive to develop or introduce new crops even in those countries in which the activities are not expressly prohibited. The surveys upon which the attitudinal information is based (Eurobarometers) did not include the "novel traits" obtained without genetic engineering, so it is unclear how the general public in E.U. states might view Clearfield technology.

Quite recently, the E.U. has accepted Herculex I maize for import to use in animal and human food, although not to grow in Europe (Dupont, 2006). The Herculex trait is a Bt gene introduced with a bar gene so that the plants are resistant to European corn borer and the glufosinate herbicide. Not yet approved are maize lines with stacked resistance to glyphosate herbicide, or a Bt gene for root worm resistance alone (Herculex RW) or in combination with the previous Bt gene (Herculex XTRA). The most recent literature from Monsanto indicates that the majority of their modified maize lines are also awaiting approval (Monsanto, 2006b). It should be noted that small amounts of Bt maize (<100,000 ha) are already being grown in several member states of the E.U. (James, 2006).

11. Concluding comments

During the preparation of this review, many transgenic plants were noted in searches of websites and formal databases. Many agricultural and horticultural crop plants have been engineered for expression of genes that may enhance their resistance to insects or fungi, increase salt and drought tolerance, increase levels of essential nutrient and vitamin accumulation, amongst other traits. However, very few have been introduced to the commercial development stage. Almost all of those that have been are all mentioned above. As discussed by Devine (2004) with regard to herbicide resistance, high costs of regulatory clearance, and international trade issues are likely to delay introduction, perhaps indefinitely for many traits. As noted in a recent review of prospects for India (Bhat and Chopra, 2005), crops for which there is little or no external trade may be more amenable to engineering until such time as regulatory acceptance becomes more routine and less costly. Thus for Basmati rice which is extensively traded to Japan, transgenic forms might not be useful at this time because of market resistance, whereas for tomatoes there might be considerable benefit in using an already available technology to delay ripening (even though it was not a commercial success in the U.S. and was withdrawn from the European market in 1999 because of concerns over GMO foods).

Until such time as the general public in many countries is more willing to accept GMOs, further studies are needed at a field scale in the U.S. and elsewhere to provide additional documentation of the extent to which GMOs pose risks not strictly comparable to those of non-GMOs. Many GMOs have proven of great value in research, enhancing our understanding of metabolic pathways. In some cases, "traditional" breeding strategies may permit exploitation of that knowledge without use of GMOs per se.

12. Acknowledgments

Supported in part by the Kansas Agricultural Experiment Station. This is contribution number 7-11-j of the Kansas Agricultural Experiment Station.

13. References

AATF (2006). Deployment of IR-maize through the StrigAway R technology, available at http://www.africancrops.net/striga/Deployment-IR-Maize.pdf

Beckie HJ, Warwick SI, Nair H, Seguin-Swartz G (2003). Gene flow in commercial fields of herbicide-resistant canola (*Brassica napus*). Ecological Appl. 13:1276-1294.

Bennett D (2006). Louisiana Clearfield lawsuit points to stewardship requirements, Delta Farm Press, April 25, 2006, available at http://deltafarmpress.com/news/060425-clearfield-louisiana/

Berken A, Mulholland MM, LeDuc DL, Terry N (2002). Genetic engineering of plants to enhance selenium phytoremediation. Crit. Rev. Plant Sci. 21: 567-582.

Bhat SR, Chopra VL (2005). Transgenic crops: priorities and strategies for India, Curr. Sci. 88:886-889.

Bondera M, Query M (2006). Widespread contamination of papaya in Hawaii by gene-altered variety, Hawaii SEED 2006, available at http://www.organicproducers.org/2006/article_523.cfm for the executive summary or http://www.gmofreemaui.com/press_releases/Contamination_Report.pdf (full report)

Boyd V (2005). Red flag warning: follow stewardship plan to prolong red-rice fighting Clearfield system. Rice Farming, Feb. 2005, available at http://www.ricefarming.com/home/2005_FebOutcrossing.html

Buckingham WA (1983). Operation ranch hand: the Air Force and herbicides in southeast Asia. Air University Review 34:42-53, available at http://www.airpower.maxwell.af.mil/airchronicles/aureview/1983/Jul-Aug/buckingham.html

Burken JG (2003). Uptake and metabolism of organic compounds: green-liver model. In S.C. McCutcheon and J.L Schnoor (eds) Phytoremediation. Transformation and control of contaminants, Wiley-Interscience, Hoboken, NJ, p 59-84.

Byrne P, Ward S, Harrington J, Fuller L (2006). Transgenic crops: an introduction and resource guide, available at http://cls.casa.colostate.edu/TransgenicCrops/defunct.html

Canada (2006) Decision documents- determination of environmental and livestock feed safety. Canadian Food Inspection Agency- Plant Biosafety Office, available at http://www.inspection.gc.ca/english/plaveg/bio/dde.shtml

Castle LA, Siehl DL, Gorton R, Patten PA, Chen YH, Bertain S, Cho HJ, Duck N, Wong J, Liu D, Lassner MW (2004). Discovery and directed evolution of a glyphosate tolerance gene. Science 304:1151-1154.

Chen LJ, Lee DS, Song ZP, Suh HK, Lu BR (2004). Gene flow from cultivated rice (*Oryza sativa*) to its weedy and wild relatives. Ann. Bot. 93:67-73.

Chilcutt CF, Tabashnik BE (2004) Contamination of refuges by *Bacillus thuringensis* toxin genes from transgenic maize. Proc. Nat. Acad. Sci USA 101:7526-7529.

Christopher JT, Powles SB, Liljegren DR, Holtum JAM (1991). Cross-resistance to herbicides in annual ryegrass (*Lolium rigidum*) II. Chlorsulfuron resistance involves a wheat-like detoxification system. Plant Physiol. 95:1036-1043.

CIMMYT (2006). CIMMYT three-way IR-maize hybrids announcement, available at http://www.africancrops.net/striga/CIMMYT-IR-Maize-Hybrids.pdf

Clearfield (2004). Clearfield wheat stewardship guide, BASF, available at http://www.agproducts.basf.com/products/Beyond-Herbicide/Beyond-Herbicide.asp

Clearfield (2006) descriptions available at http://www.agproducts.basf.com/products/Beyond-Herbicide/Beyond-Herbicide.asp

Cobbett C, Goldsbrough P (2002). Phytochelatins and metallothioneins: roles in heavy metal detoxification and homeostasis. Annu. Rev. Plant Biol. 53:159-182.

Crawley MJ, Hails RS, Rees M, Kohn D, Burton J (1993). Ecology of transgenic oilseed rape in natural habitats. Nature 363:620-623.

Crawley MJ, Hails RS, Kohn DD, Rees M (2001). Transgenic crops in natural habitats. Nature 409:682-683.

CWMA (2006). Yellow nutsedge. Colorado Weed Management Association, available at http://www.cwma.org/nx_plants/yelnut.htm

Davis WH (2005a). Soybean seeds and plants exhibiting natural herbicide resistance. U.S. Pat. Appl. Publ. as cited in Chem Abstr 143:402779.

Davis WH (2005b). Natural glyphosate herbicide resistance in wheat comprising ngw1ngw1 gene pair and ngw2ngw2 gene pair. U.S. Pat. Appl. Publ. as cited in Chem Abstr 142:460274.

Davis LC, Castro-Diaz S, Zhang Q, Erickson LE (2002). Benefits of vegetation for soils with organic contaminants. Crit. Rev. Plant Sci. 21:457-491.

Devine, MD (2005). Why are there not more herbicide-tolerant crops? Pest Manag. Sci. 61:312-317.

Dill GM (2005). Glyphosate-resistant crops:history, status and future. Pest Manag. Sci. 61:219-224.

Duke, SO (2005). Taking stock of herbicide-resistant crops ten years after introduction. Pest Manag. Sci. 61:211-218.

Dupont (2006). available at http://www2.dupont.com/Biotechnology/en_us/index.html

Dutton A, Romeis J, Bigler F (2003). Assessing the risks of insect resistant transgenic plants on entomophagous arthropods: Bt-maize expressing Cry1Ab as a case study. BioControl 48:611-636.

Eapen S, D'Souza SF (2005). Prospects of genetic engineering of plants for phytoremdiation of toxic metals. Biotechnol. Adv. 23:97-114.

Eastham K, Sweet J (2002). Genetically modified organisms (GMOs): the significance of gene flow through pollen transfer. European Environment Agency, Environmental issue report # 28, available at http://reports.eea.eu.int/environmental_issue_report_2002_28/en

Ellstrand NC, Hoffman CA (1990) Hybridization as an avenue of escape for engineered genes: Strategies for risk reduction. BioScience 40:438-442.

Ferrell JA, Macdonald GE, Brecke, BJ (2006). Tropical spiderwort (*Commelina benghalensis* L.) identification and control. University of Florida Institute of Food and Agricultural Science SS-AGR-223 available at http://edis.ifas.ufl.edu/AG230

Fuchs M, Chirco EM, Gonsalves D (2004). Movement of coat protein genes from a commercial virus-resistant transgenic squash into a wild relative, Environ. Biosafety Res. 3: 5-16.

Fuchs M, Chirco EM, Mcferson JR, Gonsalves D (2004) Comparative fitness of a wild squash species and three generations of hybrids between wild x virus-resistant transgenic squash, Environ. Biosafety Res. 3:17-28.

Garcia MA, Altieri MA (2005) Transgenic crops: Implications for biodiversity and sustainable agriculture. Bull. Sci. Technol. Soc. 25:335-353.

Gianessi LP, Silvers CS, Sankula S, Carpenter JE (2002) Herbicide tolerant sugarbeet. Plant biotechnology: current and potential impact for improving pest management in U.S. agriculture An analysis of 40 case studies. National Center for Food and Agricultural Policy, Washington DC, available at http://www.ncfap.org/40CaseStudies/CaseStudies/SugarbeetHT.pdf

Gimenez-Espinosa R, Romera E, Tena M, DePrado R (1996) Fate of atrazine in treated and pristine accessions of three Setaria species. Pestic. Biochem. Physiol. 56:196-207.

Gratao LP, Prasad MNV, Cardoso PF, Lea PJ, Azevedo RA (2005). Phytoremediation: green technology for the clean-up of toxic metals in the environment. Braz. J. Plant Physiol. 17:53-64, available at http://www.scielo.br/

Hagerty AM, Kilpatrick AL, Turnipseed SG, Sullivan MJ, Bridges WC (2005) Predaceous arthropods and lepidopteran pests on conventional, Bollgard and Bollgard II cotton under untreated and disrupted conditions. Environ. Entomol. 34:105-114

Halfhill MD, Millwood RJ, Raymer PL, Stewart CN (2002). Bt-transgenic oilseed rape hybridization with its weedy relative *Brassica rapa*. Environ. Biosafety Res. 1:19-28, available at http://www.edpsciences.org/journal/index.cfm?edpsname=ebr

Harms H, Bokern M, Kolb M, Bock C (2003). Transformation of organic contaminants by different plant systems. In S.C. McCutcheon and J.L Schnoor, (eds) Phytoremediation. Transformation and control of contaminants, Wiley-Interscience, Hoboken, NJ, p 285-316.

Heap IM (2006). International survey of herbicide resistant weeds. Weed Science Society of America, available at httpP//www.weedscience.org/in.asp

Henry C, Morgan D, Weekes R, Daniels R, Boffey C (2003). Farm scale evaluations of GM crops: monitoring gene flow from GM crops to non-GM equivalent crops in the vicinity (contract reference EPG 1/5/138), Part 1: forage maize, available at http://www.defra.gov.uk/environment/gm/research/pdf/epg_1-5-138.pdf

Hirose S, Kawahigashi H, Ozawa K, Shiota N, Inui H, Ohkawa H, Ohkawa Y (2005). Transgenic rice containing human CYP2B6 detoxifies various classes of herbicides. J. Agric. Food Chem. 53:3461-3467.

Huang J, Hu R, Fan C, Pray CE, Rozelle S (2002). Bt cotton benefits, costs and impacts in China. AgBioForum 5(4): 153-166.

Inui H, Ohkawa H (2005). Herbicide resistance in transgenic plants with mammalian P450 monooxygenase genes, Pest Manag. Sci. 61: 288-291.

ISBR (2004) Proceedings of the 8th International Meeting on Biosafety of Genetically Modified Organisms (Sep. 26-30, 2004, Montpelier, Fr). International Society for Biosafety Research, available at http://www.isbr.info/document/proceedings_montpelier2004.pdf. p. 322.

Itoh K (1994). Weed ecology and its control in south-east tropical countries. Jpn. J. Tropic. Ag. 38:369-373.

James C (2006). Global status of commercialized biotech/GM crops:2005. International service for the Acquisition of Agri-Biotech Applications, available at http://www.isaaa.org/

Jenczewski E, Ronfort J, Chevre AM (2003), Crop to wild gene flow, introgression and possible fitness effects of transgenes. Environ. Biosafety Res. 2: 9-24.

Jia H (2006) China intends to push for GM crop studies, China Daily Feb 14, 2006 English edition available at http://www.chinadaily.com.cn/english/doc/2006-02/14/content_519769.htm

Jordan (1999). Fitness effects of the triazine resistance mutation in *Amaranthus hybridus*: relative fitness in maize and soyabean crops. Weed Res 39: 493-505.

Jung HG, Schaeffer CC (2004) Influence of Bt transgenes on cell wall lignification and digestibility of maize stover for silage. Crop Sci. 44: 1781-1789.

Karthikeyan R, Davis LC, Erickson LE, Al-Khatib K, Kulakow PA, Barnes PL, Hutchinson SL, Nurzhanova AA (2004a). Potential for plant-based remediation of pesticide-contaminated soil and water using nontarget plants such as trees shrubs and grasses, Crit. Rev. Plant Sci. 23: 91-101.

Karthikeyan R, Davis LC, Erickson LE, Al-Khatib K, Kulakow PA, Barnes PL, Hutchinson SL, Nurzhanova AA (2004b). Studies on responses of nontarget plants to pesticides: a review, available at http://www.engg.ksu.edu/HSRC/karthipesticide.pdf

Kawahigashi H, Hirose S, Inui H, Ohkawa H, Ohkawa Y. (2005). Enhanced herbicide cross-tolerance in transgenic rice plants co-expressing human CYPA1, CYP2B, and CYP2c19, Plant Science 168:773-781.

Kirkpatrick KL, Wilson HD (1988). Interspecific gene flow in Curcurbita: *C. texana* vs *C. pepo*, Am. J. Bot. 75: 519-527.

Koger CH, Reddy KN (2005). Role of absorption and translocation in the mechanism of glyphosate resistance in horseweed (*Conyza canadensis*). Weed Sci. 53: 84-89.

Konza (2002). available at http://www.mediarelation.k-state.edu/WEB/News/Webzine/konza/index.html

Laurent F, Debrauwer L, Rathahao E, Scalla R (2000). 2,4-dichlorophenoxyacetic acid metabolism in transgenic tolerant cotton (*Gossypium hirsutum*). J. Agric. Food Chem. 48: 5307-5311.

Lorraine-Colwill DF, Powles SB, Hawkes TR, Hollinshead PH, Warner SAJ, Preston C (2002). Investigations into the mechanism of glyphosate resistance in *Lolium rigidum*. Pestic. Biochem. Physiol. 74: 62-72.

Lu BR, Snow AA (2005). Gene flow from genetically modified rice and its environmental consequences. BioScience 55: 669-678.

Lutman PJW, Freeman SE, Pekrun C (2003) The long-term persistence of seeds of oilseed rape (*Brassica napus*) in arable fields. J.Agric. Sci. 141: 231-240.

Lyon DJ, Bussan AJ, Evans JO, Mallory-Smith CA, Peeper TF (2002). Pest management implications of glyphosate-resistant wheat (*Triticum aestivum*) in the western United States. Weed Technol. 16:680-690.

Madsen KH, Sandoe P (2005). Ethical reflections on herbicide-resistant plants. Pest Manag. Sci. 61: 318-325.

Massinga RA, Al-Khatib K, St Amand P, Miller JF (2003). Gene flow from imidazolinone-resistant domesticated sunflower to wild relatives. Weed Sci. 51: 854-862.

Menendez J, Bastida F, de Prado R (2006). Resistance to chlortoluron in a downy brome (*Bromus tectorum*) biotype. Weed Sci. 54: 237-245.

Moar WJ, Eubanks M, Freeman B, Tirnipseed S, Ruberson J, Head G (2002). Effects of Bt cotton on biological control agents in the southeastern United States. 1st International Symposium on Biological Control of Arthropods, available at http://www.bugwod.org/arthropod/day4/moar.pdf

Monsanto (2006a). Monsanto biotechnology trait acreage:fiscal years 1996-2006, available at http://www.monsanto.com/monsanto/content/investor/financial/reports/ 2006/Q32006Acreage.pdf

Monsanto (2006b). 2006 Technology Use Guide (39 pp), available at http://www.monsanto.com/monsanto/us_ag/content/stewardship/tug/2006TUGPDF.pdf

NRC (2002) Environmental Effects of Transgenic Plants: the Scope and Adequacy of Regulation, National Academy Press, Washington DC, p. 320.

NSA (2005). Manage wild sunflower in Clearfield sunflower. National Sunflower Asociation, available at http://www.sunflowernsa.com/magazine/details.asp?ID=406&printable=1

O'Neill G (2006). Fire: a burning issue. Society of American Foresters, available at http://forestry.about.com/library/saf/blsafire.htm.

Owen MDK, Zelaya IA (2005). Herbicide-resistant crops and weed resistance to herbicides. Pest Manag. Sci. 61:301-311.

Perry D (2005). Different applications for genetically modified crops: prepared remarks of Mr. Dolan Perry to U.S. House of Representatives, available at http://www.monsanto.co.uk/news/ukshowlib.phtml?uid=9151.

Powles SB, Lorraine-Colwill DF, Dellow JJ, Preston C (1998). Evolved resistance to glyphosate in rigid ryegrass (*Lolium rigidum*) in Australia. Weed Sci. 46:604-607.

Puricelli E, Tuesca D (2005). Weed density and diversity under glyphosate-resistant crop sequences. Crop Protection 24:533-542.

Ramsay G, Thompson C, Squire G (2003). Quantifying landscape-scale gene flow in oilseed rape. Final report of DEFRA project RG0216, available at http:www.defra.gov.uk/environment/gm/research/pdf/epg_rg0216.pdf

Reddy KN (2004). Weed control and species shift in bromoxynil- and glyphosate- resistant cotton (*Gossypium hirsutum*) rotation systems. Weed Technol. 18: 131-139.

Ruiz ON, Hussein HS, Terry N, Daniell H (2003). Phytoremediation of organomercurial compounds via chloroplast engineering. Plant Physiol. 132: 1344-1352.

Sankula S, Marmon G, Blumenthal E (2005). Biotechnology derived crops planted in 2004- impacts on U.S. agriculture. National Center for Food and Agriculture Policy, Washington DC, available at http://www.ncfap.org/whatwedo/pdf/2004biotechimpacts.pdf

Schwitzguebel JP, Vanek T (2003). Some fundamental advances for xenobiotic chemicals In S.C. McCutcheon and J.L Schnoor, (eds) Phytoremediation. Transformation and control of contaminants, Wiley-Interscience, Hoboken, NJ, p 123-157.

Sears MK, Hellmich RL, Stanley-Horn DE, Oberhauser KE, Pleasants JM, Mattila HR, Siegfried BS, Dively GP (2001). Impact of Bt corn pollen on monarch butterfly populations: A risk assessment. Proc. Nat. Acad. Sci. U.S. A. 98: 11937-11942.

Sebastian SA, Fader GM, Ulrich JF, Forney DR, Chaleff RS (1989). Semidominant soybean mutation for resistance to sulfonylureas. Crop Sci. 29: 1403-1408.

Sisterson MS, Biggs RW, Olson C, Carriere Y, Dennehy TJ, Tabashnik BE (2004) Arthropod abundance and diversity in Bt and non-BT cotton fields. Environ. Entomol. 33: 921-929.

Sloderbeck P (2006). Current status of Bt corn hybrids. Kansas Agricultural Experiment Station, available at http://www.oznet.ksu.edu/swao/Entomology/Bt_Folder/Bt%20Corns.html

Snow AA, Pilsen D, Rieseberg LH, Paulsen MJ, Pleskac N, Reagon MR, Wolf DE, Selbo SM (2003). A Bt transgene reduces herbivory and enhances fecundity in wild sunflowers. Ecological Appl. 13:279-286.

Tabashnik BE, Carriere Y (2004) Bt transgenic crops do not have favourable effects on resistant insects. J. Insect Sci. 4: 1-3.

Tan S, Evans RR, Dahmer ML, Singh BK, Shaner DL (2005). Imidazolinone-tolerant crops: history, current status and future. Pest Manag. Sci. 61: 246-257.

Teagasc (2006). Sugar beet. Information Centre for Genetically Modified Crops in Ireland, available at http://www.gmoinfo.ie/sugarbeet.php

Tranel PJ, Wright TR (2002). Resistance of weeds to ALS-inhibiting herbicides: what have we learned. Weed Sci. 50:700-712.

USDA (2006). Adoption of genetically engineered crops in the U.S.: Corn varieties. Economics Research Service, USDA, available at http://www.ers.usda.gov/Data/BiotechCrops/ExtentofAdoptonTable1.htm

Watkinson AR, Freckleton RP, Robinson RA, Sutherland WJ (2000) Predictions of biodiversity response to genetically modified herbicide-tolerant crops. Science 289: 1554-1557.

Wilson HD (1993). Free living *Cucurbita pepo* in the United States: viral resistance, gene flow and risk assessment [a review submitted to USDA]. Texas A & M University Biology, available at http://www.csdl.tamu.edu/FLORA/flcp/flcp1.htm

Wilson HD, Payne JS (1994). Crop/weed microgametophyte competition in *Cucurbita pepo* (Cucurbitaceae). Am. J. Bot. 81: 1531-1537.

Zhang J, Linscombe S, Oard J(2003). Outcrossing frequency and genetic analysis of hybrids between transgenic glufosinate resistant rice and the weed, red rice. Euphytica 130: 35-45.

Zhu B, Lawrence R, Warwick SI, Braun ML, Halfhill MD, Stewart CN (2004). Stable *Bacillus thuringiensis* (BT) toxin content in interspecific F1 and backcross populations of wild *Brassica rapa* after Bt gene transfer. Mol. Ecol. 13: 237-241.

Permissions

All chapters in this book were first published in BMBR, by Academic Journals; hereby published with permission under the Creative Commons Attribution License or equivalent. Every chapter published in this book has been scrutinized by our experts. Their significance has been extensively debated. The topics covered herein carry significant findings which will fuel the growth of the discipline. They may even be implemented as practical applications or may be referred to as a beginning point for another development.

The contributors of this book come from diverse backgrounds, making this book a truly international effort. This book will bring forth new frontiers with its revolutionizing research information and detailed analysis of the nascent developments around the world.

We would like to thank all the contributing authors for lending their expertise to make the book truly unique. They have played a crucial role in the development of this book. Without their invaluable contributions this book wouldn't have been possible. They have made vital efforts to compile up to date information on the varied aspects of this subject to make this book a valuable addition to the collection of many professionals and students.

This book was conceptualized with the vision of imparting up-to-date information and advanced data in this field. To ensure the same, a matchless editorial board was set up. Every individual on the board went through rigorous rounds of assessment to prove their worth. After which they invested a large part of their time researching and compiling the most relevant data for our readers.

The editorial board has been involved in producing this book since its inception. They have spent rigorous hours researching and exploring the diverse topics which have resulted in the successful publishing of this book. They have passed on their knowledge of decades through this book. To expedite this challenging task, the publisher supported the team at every step. A small team of assistant editors was also appointed to further simplify the editing procedure and attain best results for the readers.

Apart from the editorial board, the designing team has also invested a significant amount of their time in understanding the subject and creating the most relevant covers. They scrutinized every image to scout for the most suitable representation of the subject and create an appropriate cover for the book.

The publishing team has been an ardent support to the editorial, designing and production team. Their endless efforts to recruit the best for this project, has resulted in the accomplishment of this book. They are a veteran in the field of academics and their pool of knowledge is as vast as their experience in printing. Their expertise and guidance has proved useful at every step. Their uncompromising quality standards have made this book an exceptional effort. Their encouragement from time to time has been an inspiration for everyone.

The publisher and the editorial board hope that this book will prove to be a valuable piece of knowledge for researchers, students, practitioners and scholars across the globe.

List of Contributors

Olawole O. Obembe
Graduate School Experimental Plant Sciences, Laboratory of Plant Breeding, Wageningen University P O Box 386, 6700 AJ Wageningen, The Netherlands

Evert Jacobsen
Graduate School Experimental Plant Sciences, Laboratory of Plant Breeding, Wageningen University P O Box 386, 6700 AJ Wageningen, The Netherlands

Richard G.F. Visser
Graduate School Experimental Plant Sciences, Laboratory of Plant Breeding, Wageningen University P O Box 386, 6700 AJ Wageningen, The Netherlands

Jean-Paul Vincken
Graduate School Experimental Plant Sciences, Laboratory of Plant Breeding, Wageningen University P O Box 386, 6700 AJ Wageningen, The Netherlands

Raj Kumar Joshi
Centre of Biotechnology, Siksha O Anusandhan University, Bhubaneswar, India

Sanghamitra Nayak
Centre of Biotechnology, Siksha O Anusandhan University, Bhubaneswar, India

Hamed Kharrati Koopaei
Department of Animal Science, Shahid Bahonar University of Kerman, Iran

Ali Esmailizadeh Koshkoiyeh
Department of Animal Science, Shahid Bahonar University of Kerman, Iran

Valentina Tosato
Yeast Molecular Genetics Laboratory, International Centre for Genetic Engineering and Biotechnology, AREA Science Park – W, Padriciano 99, IT-34012 Trieste, Italy

Carlo V. Bruschi
Yeast Molecular Genetics Laboratory, International Centre for Genetic Engineering and Biotechnology, AREA Science Park – W, Padriciano 99, IT-34012 Trieste, Italy

S Preetha
Centre for Plant Breeding and Genetics (CPBG), Tamilnadu Agricultural University, Coimbatore-641001, Tamilnadu, India

T. S Raveendren
Centre for Plant Breeding and Genetics (CPBG), Tamilnadu Agricultural University, Coimbatore-641001, Tamilnadu, India

Jelili T. Opabode
Department of Plant Science, Obafemi Awolowo University, Ile-Ife, Nigeria

Murunwa Makwarela
School of Molecular and Cell Biology, University of the Witwatersrand, Private Bag 3, WITS 2050, Johannesburg, South Africa

Christine Rey
School of Molecular and Cell Biology, University of the Witwatersrand, Private Bag 3, WITS 2050, Johannesburg, South Africa

Maureen Fonji ATEMKENG
Leguminous Crop Program, Institute of Agricultural Research for Development (IRAD), P.O. Box 2067, Yaounde, Cameroon

Teboh Jasper MUKI
School of Plant, Environmental, and Soil Sciences, Lousiana State University – AgCenter, 238 Sturgis Hall, Baton Rouge, LA, 70803, USA

Jong-Won PARK
Texas AgriLife Research, Texas A&M University System, 2415 E Business 83, Weslaco, Texas 78596, USA

John JIFON
Texas AgriLife Research, Texas A&M University System, 2415 E Business 83, Weslaco, Texas 78596, USA

Nishawar Jan
Department of Biotechnology, the University of Kashmir-1900 06 (J&K), India

Mahboob-ul-Hussain
Department of Biotechnology, the University of Kashmir-1900 06 (J&K), India

Khurshid I. Andrabi
Department of Biotechnology, the University of Kashmir-1900 06 (J&K), India

Subbiah POOPATHI
Centre for Research in Medical Entomology (Indian Council of Medical Research), Ministry of Health and Family

Welfare, Govt of India, Chinna Chokkikulam, Madurai-625002, Tamil Nadu, India

Brij Kishore TYAGI
Centre for Research in Medical Entomology (Indian Council of Medical Research), Ministry of Health and Family
Welfare, Govt of India, Chinna Chokkikulam, Madurai-625002, Tamil Nadu, India

A. R. Varsale
Centre for Advanced Life Sciences, Deogiri College, Aurangabad, Maharashtra, India

A. S. Wadnerkar
Centre for Advanced Life Sciences, Deogiri College, Aurangabad, Maharashtra, India

R. H. Mandage
Centre for Advanced Life Sciences, Deogiri College, Aurangabad, Maharashtra, India

K. Muniswamy
Division of Veterinary Biotechnology, Indian Veterinary Research Institute, Izatnagar- 243 122, India

P. Thamodaran
Division of Veterinary Microbiology, TANUVAS, Chennai-600 007, India

Kazhila C. Chinsembu
University of Namibia, Faculty of Science, Department of Biological Sciences, Private Bag13301, Windhoek, Namibia

N. Manikanda Boopathi
Department of Plant Molecular Biology and Biotechnology, Coimbatore- 641003, India

R. Ravikesavan
Department of Cotton, Tamil Nadu Agricultural University, Coimbatore- 641003, India

Mohammad Sayyar Khan
Gene Research Center, University of Tsukuba, 1-1-1-Tennoudai, Ibaraki, Tsukuba, Japan, 305-8572
Institute of Biotechnology and Genetic Engineering (IBGE), Khyber Pukhtunkhwa Agricultural University Peshawar, Pakistan

Lawrence C. Davis
Department of Biochemistry,Kansas State University,141 Chalmers Hall,Manhattan, KS 66506

www.ingramcontent.com/pod-product-compliance
Lightning Source LLC
Chambersburg PA
CBHW080255230326

41458CB00097B/4998